Sociology of Families

READINGS

Cheryl M. Albers
Buffalo State College

with contributions by David Newman
DePauw University

Pine Forge Press
Thousand Oaks, California • London • New Delhi

For information, address:

Pine Forge Press
A Sage Publications Company
2455 Teller Road
Thousand Oaks, California 91320
(805) 499-4224
E-mail: sales@pfp.sagepub.com

SAGE Publications Ltd
1 Oliver's Yard
55 City Road
London EC1Y 1SP
United Kingdom

Sage Publications India Pvt. Ltd.
B-42, Panchsheel Enclave
Post Box 4109
New Delhi 110 017

Production Coordinator: Windy Just
Production Management: Scratchgravel Publishing Services
Copy Editor: Linda Purrington
Typesetter: Scratchgravel Publishing Services
Cover Designer: Ravi Balasuriya
Cover Image: "The Builders" by Jacob Lawrence. Courtesy of the artist and Francine Seders Gallery.

Printed in the United States of America
03 10 9 8 7 6 5 4 3

Library of Congress Cataloging-in-Publication Data

The sociology of families : readings / [edited by] Cheryl Albers.
p. cm.
Includes bibliographical references and index.
ISBN 0-7619-8610-3 (alk. paper)
1. Family. 2. Family—United States. I. Albers, Cheryl.
HQ734.S768 1999
306.85—dc21 99-6035
CIP

This book is printed on acid-free paper that meets Environmental Protection Agency standards for recycled paper.

ABOUT THE EDITOR

Cheryl M. Albers (Ph. D., State University of New York at Buffalo) is Assistant Professor of Sociology at Buffalo State College, where she teaches family, childhood and youth, gender roles, and social psychology. Professor Albers began her professional career as a secondary school teacher of family life and child development courses. She has previously taught at the University of Minnesota and the Queensland Institute of Technology in Australia. Current research interests focus on public policy regarding child care. She has written and lectured extensively on family studies in the United States and overseas. Her previous books include *Focus on You* and *Person to Person: Human Development and Relationships.*

ABOUT THE PUBLISHER

Pine Forge Press is a new educational publisher, dedicated to publishing innovative books and software throughout the social sciences. On this and any other of our publications, we welcome your comments. Please call or write us at:

Pine Forge Press

A Sage Publications Company

2455 Teller Road

Thousand Oaks, CA 91320

(805) 499-4224

E-mail: sales@pfp.sagepub.com

Visit our new World Wide Web site, your direct link to a multitude of online resources:

http://www.pineforge.com

In loving memory of a true family man—my father, Herman Albers

CONTENTS

PREFACE

Thirty years ago, I began my study of families as an undergraduate in a large midwestern public university. It was a time of cultural revolution, and my generation believed that, one way or another, we would tear down the social institutions. Along with the government, the military, and education, the family was targeted for reform. What I remember most clearly about this time in history was our perception of tension between the generations. We saw these social institutions as developed by our parents and grandparents to fulfill the needs of their generations. But we rejected their legacy. Through cohabitation, communal living, and blurred gender boundaries, we challenged the aspects of family life our elders held dear.

Today, as a member of the older generation, I sit on the other side of the desk, shepherding undergraduate students through a study of the sociology of families. I sense in these students a sentiment profoundly different from that I attach to my youth. Like the students of thirty years ago, today's students are eager to discuss change in the institution of family. However, their conversations are not filled with the convictions of agency that permeated such talks in the late 1960s and early 1970s. It is true that the reform rhetoric of my generation was often based in naive idealism regarding the restraints placed on family change by other social institutions such as church, government, law, and the economy. But in this idealism was also a strong belief that individuals shape society and that if enough of us acted on our beliefs we could implement social change.

By contrast, today's students have a much better understanding of the institutional restraints on promoting social change. However, this realization seems to have created a fatalistic attitude toward the future of the family. Students no longer speak optimistically of the changes in family life, and, what is more important, they do not feel empowered to direct these transformations. When students do make connections between human agency and family change, the conversations are pessimistic and *other*- oriented, gravitating to the roles of "deadbeat dads," never married mothers, and frequent divorce as contributing factors in the "breakdown" of family. These discussions inevitably turn to a lament for the "golden years" of family life in the 1950s. Ironically this was precisely the form of family that my generation wanted to end and that research has now revealed to be a short-lived anomaly experienced by few.

My primary goals for students who enroll in my sociology of families classes, and for those who read this book, are to develop an understanding of ways that families are shaped by and shape the larger social context and to nurture students' belief in the role of human agency in shaping society. I find these connections are most easily made when learning activities are focused on issues that students find relevant. Therefore, the framework of this book consists of nine issues that will engage students in thinking about families in a sociological context. This issue-based approach reflects Part I of the companion text, David Newman's *Sociology of Families*. Although the issues Newman selects differ from those in this reader, the concepts introduced and developed in the text will enrich students' understanding of the issues presented here.

The composition of this reader differs from that of other family readers, which normally follow one of two formats. Some present one or two readings on each of the topics commonly found in sociology of families texts. This approach provides breadth of coverage but does not facilitate exploring any one topic in depth. In the second format, readers take a point–counterpoint presentation, which provides more depth of coverage on a limited range of topical issues. The controversy stimulated by these issues in the popular media is incorporated into the readings. However, these readings are often selected for their timely and provocative treatment of issues rather than for their contribution to a sociological perspective.

This reader is structured to provide a third alternative. I have selected nine issues, faced by contemporary families, that have sociological relevance as well as high interest for students. Selected readings examine these issues from various levels of analysis. Authors introduce a variety of perspectives, including individual experience, the diversity of individual experience, the structural context, cross-cultural and historical comparisons, and policy implications. Note the absence of chapters addressing race, class, and gender. This is a calculated omission, based on the belief that these sociological constructs are embedded in all issues faced by families. Rather than being isolated in specific chapters, the concepts of race, class, and gender are woven into the selection of issues and readings.

I have adopted this structure as a means of addressing two recurring difficulties in teaching the sociology of families. First, instructors bring to their teaching a personal theoretical perspective of the study of families. My colleagues in teaching the sociology of families range from demographers to symbolic interactionists. As balanced as we may try to be in presenting our courses, we are inevitably more familiar with the literature we apply in our own research. In assembling these selections, I have included readings examining, through a variety of sociological lenses, issues that families face. In addition to the diverse perspectives evident within sociology, other disciplines also inform the study of families. When appropriate, I have included the insights of psychologists, historians, social commentators, and novelists on specific issues. Second, a common complaint about sociology of families courses, is that the time required to cover the breadth of relevant and critical material restricts the ability to provide an in-depth study of issues that interest students. Most instructors compensate for this problem through student choice in term paper topics. This solution gives students some depth but removes the instructor from the important role of monitoring the quality and diversity of the readings the student selects. This collection presents interrelated readings that provide an integral study of issues promoting higher-order thinking.

Two organizational tools have been used to guide students to this deeper understanding of family issues. Each chapter is introduced with an essay that provides a thought-provoking sociological frame for the issue considered. In addition to expanding students' thinking, these essays provide a rationale for the selection of readings and help students establish links among various perspectives on the issue. Each chapter concludes with a series of questions, which instructors can use in a variety of ways. I have used such questions to stimulate lively classroom discussion or debate. Although most students have strong opinions about the issues dealt with in these questions, not all students will voice their ideas in public. Therefore, instructors may prefer to use the questions as bases for writing assignments; for some students this approach is less threatening than discussion.

When class size permits, consider developing small discussion groups of six to eight students to share responses to these questions. Such groups can be formed in the second or third week of the term and remain together for the duration of the course. This allows students to develop a rapport with one another that facilitates discussion. Students can be given the option of reforming groups at midsemester. It is my experience that, given this option, students inevitably prefer to stay with "their" group—even when interpersonal differences are evident. Students often remark that the opportunity to get to know classmates in depth is rare in their studies and is a valuable aspect of this class.

Coming of age in the late 1960s and early 1970s has left me with an indelible imprint on my character, the belief that individual actions shape social life. Undoubtedly this belief played a significant role in my selection of a career in education. As a family sociologist, I have found that the best days in the classroom are those when students make the connections between personal issues and the social conditions in which the issues are experienced. In selecting these readings I have tried to provide a schema for students to understand their own experiences of family in relation to the wider historic and cultural contexts in which their experiences take place.

Acknowledgments

Many people have contributed their ideas and energy to the book you hold in your hands. I am indebted to each of them. First on my list is Steve Rutter. His insight, foresight, and hindsight all provided much needed direction. I am most grateful for his patience, his confidence in me, and his constant encouragement. Through cyberspace and telephone lines, he gave me the energy to see this project to completion.

Sometimes this manuscript felt like a heavy stone weighing me down. At such times I tried to visualize it as the stone used in the sport of curling. In curling, a big, shiny stone slides across the ice while a team member skates in front of the stone, "sweeping" the ice to give the stone speed and direction. Among my "sweepers" were Windy Just, Becky Smith, Mary Shields, David Newman, Paul O'Connell, and Jean Skeels. My thanks also to family sociologist Diane Lye for her invaluable comments and suggestions.

It is ironic that during the writing of this book about the importance of strong family relationships, my own family was often shortchanged. I am grateful to my husband, Bob Stevenson, for understanding the importance of this endeavor and making accommodations in our family life to help me complete this book. Our daughter, Anya Albers-Stevenson, remained her usual upbeat, understanding self as I stole time from her to give to writing.

Most important, I thank an uncountable number of students who have shared in my endeavors to better understand the complexities of family life. They have been good teachers.

Cheryl M. Albers

CREDITS

We gratefully acknowledge the following for permission to reprint the selections in this anthology:

1 From *Reviving Ophelia* by Mary Pipher, Ph.D., Copyright © 1994 by Mary Pipher, Ph.D. Used by permission of Putnam Berkley, a division of Penguin Putnam Inc.

2 Reprinted by permission of the publisher from *Dubious Conceptions: The Politics of Teenage Pregnancy* (pp. 81–94 and 106–108) by Kristin Luker, 1996, Cambridge, MA: Harvard University Press. Copyright © 1996 by the President and Fellows of Harvard College. Reprinted with permission.

3 From *Early Childbearing: Perspectives of Black Adolescents on Pregnancy, Abortion, and Contraception* (pp. 20–39) by Ellen W. Freeman and Karl Rickels, 1993, Thousand Oaks, CA: Sage Publications. Copyright © 1993 Sage Publications, Inc. Reprinted by permission of Sage Publications, Inc.

4 "Using Behavioral Theories to Design Abstinence Programs" by Kristin A. Moore and Barbara W. Sugland, 1997, *Children and Youth Services Review*, 19, pp. 485–500. Copyright © 1997. Reprinted with permission from Elsevier Science.

5 From *Teenage Pregnancy in Industrialized Countries* (pp. 228–240) by Elise F. Jones, Jacqueline Darroch Forrest, Noreen Goldman, Stanley Henshaw, Richard Lincoln, Jeannie I. Rosoff, Charles F. Westoff, and Deirdre Wulf, 1986, New Haven, CT: Yale University Press. Copyright © 1986 by Yale University Press. Reprinted with permission.

6 "Values, Attitudes, and the State of American Marriage," by Norval Glenn, from *Promises to Keep: Decline and Renewal of Marriage in America* (pp. 15–33), David Popenoe, Jean Bethke Elshtain, and David Blankenthorn (eds.), 1996, Lanham, MD: Rowman and Littlefield. Copyright © 1996. Reprinted with permission.

7 "Marital Status, Gender, and Perceptions of Well-Being" by Harsha Mookherje, 1997, *The Journal of Social Psychology*, 137, pp. 95–106. Published by Heldref Publications, 1319 Eighteenth St. N.W., Washington, DC 20036-1802. © 1997 Heldref Publications. Reprinted with permission.

8 "A Comparison of Marriages and Cohabitation Relationships" by Steven L. Nock, 1995, *Journal of Family Issues*, 16, pp. 53–76. Copyright © 1995 Sage Publications, Inc. Reprinted by permission of Sage Publications, Inc.

9 From *The Decline in Marriage Among African-Americans* (pp. 145–171) by M. Belinda Tucker and Claudia Mitchell-Kernan, 1995, New York: Russell Sage Foundation. Copyright © 1995 Russell Sage Foundation, New York, NY. Reprinted with permission.

10 From *Pricing the Priceless Child: The Changing Social Value of Children* (pp. 189–205) by Viviana A. Zelizer. Copyright © 1985 by BasicBooks, Inc. Reprinted by permission of BasicBooks, a subsidiary of Perseus Books Group, LLC.

11 Reprinted by permission of Transaction Publishers. "Stolen Children and International Adoptions" by Mary Ellen Fieweger, 1991, *Child Welfare*, 70, pp. 285–291. Copyright © 1991 by Transaction Publishers; all rights reserved.

12 From *Barren in the Promised Land: Childless Americans and the Pursuit of Happiness* (pp. 211–259) by Elaine Tyler May. Copyright © 1995 by Elaine Tyler May. Reprinted by permission of BasicBooks, a subsidiary of Perseus Books Group, LLC.

13 From *Lesbian Mothers: Accounts of Gender in American Culture* (pp. 47–74) by Ellen Lewin, 1993, Ithaca, NY: Cornell University Press. Copyright © 1993 Cornell University Press. Used by permission of Cornell University Press.

14 "The Politics of Reproductive Benefits: U.S. Issuance Coverage of Contraceptive and Infertility Treatments" by Leslie King and Madonna Harrington Meyer, 1997, *Gender and Society*, 11, pp. 8–30. Copyright © 1997 Sage Publications, Inc. Reprinted by permission of Sage Publications, Inc.

15 From *Thinking Out Loud* by Anna Quindlen, Copyright © 1993 by Anna Quindlen. Reprinted by permission of Random House, Inc.

16 "Women, Work and Family in America" by Suzanne M. Bianchi and Daphne Spain, 1996, *Population Bulletin*, Dec., 51, pp. 2–47. Copyright © 1996 Population Reference Bureau. Reprinted with permission.

17 From *Family Man: Fatherhood, Housework and Gender Equity* (pp. 110–115) by Scott Coltrane, 1996, NY: Oxford University Press. Copyright © 1996 by Oxford University Press, Inc. Used by permission of Oxford University Press, Inc.

18 Reprinted by permission of Transaction Publishers. "Parental Androgeny" by David Popenoe, 1993, *Society*, 30, pp. 5–11. Copyright © 1993 by Transaction Publishers; all rights reserved.

19 "Understanding the Future of Fatherhood: The 'Daddy Hierarchy' and Beyond" by Arlie Russell Hochschild, in Mirjam van Dongen, Gerard Frinking, and Menno Jacobs (eds.), *Changing Fatherhood: A Multidisciplinary Perspective* (pp. 219–230), 1995, Amsterdam: Thesis Publishers. Copyright © 1995. Reprinted with permission.

20 Reprinted by permission of *Harvard Business Review* from "Business and the Facts of Family Life" by Fran Sussner Rodgers and Charles Rodgers, 1989, Nov.–Dec., pp. 121–129. Copyright © 1989 by the President and Fellows of Harvard College; all rights reserved.

21 "Vanishing Dreams of America's Young Families" by Marian Wright Edelman, 1992, *Challenge*, May–June, 35, pp. 13–19. Copyright © 1992 M. E. Sharp, Inc. Reprinted with permission from M. E. Sharp, Inc., Armonk, NY 10504.

22 From F*amilies on the Fault Line* (pp. 103–125) by Lillian B. Rubin, 1994, New York: HarperCollins Publishing. Copyright © 1994 HarperCollins Publishing. Reprinted with permission.

23 From *Making Ends Meet: How Single Mothers Survive Welfare on Low-Wage Work* (pp. 218–235) by Kathryn Edin and Laura Lein, 1997, New York: Russell Sage Foundation. Copyright © 1997 Russell Sage Foundation, New York, NY. Reprinted with permission.

24 "Mothers with Children and Mothers Alone: A Comparison of Homeless Families" by Jennifer E. Glick, 1996, *Journal of Sociology and Social Welfare*, 23, pp. 85–94. Copyright 1996. Reprinted with permission.

25 "A New Look at Poverty in America" by William P. O'Hare, 1996, *Population Bulletin*, Sept., 51, pp. 2, 11, 31–43. Copyright © 1996 Population Reference Bureau. Reprinted with permission.

26 From *Race: How Blacks and Whites Think and Feel About the American Obsession* (pp. 245–248) by Studs Terkel, 1992, New York: The New Press. Copyright © 1992 The New Press. Reprinted with permission.

27 From *Paddy Clark Ha Ha Ha* by Roddy Doyle. Copyright © 1993 by Roddy Doyle. Used by permission of Viking Penguin, a division of Penguin Putnam Inc.

28 "Violent Men or Violent Women? Whose Definition Counts?" by Dawn H. Currie, from *Issues in Intimate Violence*, Raquel Kennedy Bergen (ed.), 1998, pp. 97–111. Thousand Oaks, CA: Sage Publications. Copyright © 1998 Sage Publications, Inc. Reprinted by permission of Sage Publications, Inc.

29 "Behind Closed Doors: Domestic Violence" by Steve Friess, 1997, *The Advocate*, Dec. 9, #748, pp. 48–52. Copyright © 1997 by Liberation Publications, Inc. Reprinted with permission.

30 "Violence Against Women and Girls: The Intolerable Status Quo" © 1997 Charlotte Bunch. UNICEF, Progress of Nations, 1997. Reprinted with permission.

31 "Through a Sociological Lens: Social Structure and Family Violence" by Richard Gilles, from *Current Controversies on Family Violence*, Richard Gilles and Donileen Loseke (eds.), 1993. Thousand Oaks, CA: Sage Publications. Copyright © 1993 Sage Publications, Inc. Reprinted by permission of Sage Publications, Inc.

32 From *Heroes of Their Own Lives* by Linda Gordon. Copyright © 1988 by Linda Gordon. Used by permission of Viking Penguin, a division of Penguin Putnam Inc.

33 From *Women of Belize: Gender and Change in Central America*, by Irma McClaurin, 1996, pp. 80–89, 120–121. New Brunswick, NJ: Rutgers University Press. Copyright © 1996 by Rutgers University Press. Reprinted with permission.

34 "Old Problems and New Directions in the Study of Violence Against Women" by Demie Kurz, from *Issues in Intimate Violence*, Raquel Kennedy Bergen (ed.), 1998, pp. 197–208. Thousand Oaks, CA: Sage Publications. Copyright © 1998 Sage Publications, Inc. Reprinted by permission of Sage Publications, Inc.

35 From *Uncoupling: Turning Points in Intimate Relationships* (pp. 79–102) by Diane Vaughan, 1986, New York: Oxford University Press. Copyright © 1986. Reprinted with permission.

36 Reprinted by permission of the publisher from *Divided Families: What Happens to Children When Parents Part* (pp. 45–61) by Frank F. Furstenberg, Jr., and

Andrew J. Cherlin, 1991, Cambridge, MA: Harvard University Press. Copyright © 1991 by the President and Fellows of Harvard College. Reprinted with permission.

37 "After Divorce: Investigations into Father Absence" by Terry Arendell, 1992, *Gender and Society*, 6, pp. 562–596. Copyright © 1992 Sage Publications, Inc. Reprinted by permission of Sage Publications, Inc.

38 "Divorcing Reality" by Stephanie Coontz, 1997, *The Nation*, 265, Nov. 17, pp. 21–25. Copyright © 1997 The Nation Company. Reprinted with permission.

39 "Stepfamilies Over the Life Course: Social Support" by Lynn White in Alan Booth and Judy Dunn (eds.), *Stepfamilies: Who Benefits, Who Does Not?* (pp. 109–138), 1994, Hillsdale, NJ: Lawrence Erlbaum Associates. Copyright © 1994. Reprinted with permission.

40 "The Stepfamily Requires a New Public Policy" by Mary Ann Mason and Jane Mauldon, 1996, *Journal of Social Issues*, 52, pp. 11–27. Copyright © 1996 Plenum Publishing Corp. Reprinted with permission.

41 "Some Economic Complexities of Child Care Provided by Grandmothers" by Harriet B. Presser, 1989, *Journal of Marriage and Family*, 51, pp. 581–591. Copyright © 1989 by the National Council on Family Relations, 3989 Central Ave. N.E., Suite 550, Minneapolis, MN 55421. Reprinted by permission.

42 "Legacy, Aging, and Succession in Farm Families" by Norah C. Keating, 1996, *Generations*, 20, pp. 61–65. Copyright © 1996 The American Society on Aging, San Francisco, California. Reprinted with permission.

43 "Social Security and Medicare Policy: A Personal, Intergenerations Story" by Mal Schechter, 1997, *Generations*, 21, pp. 51–59. Copyright © 1997 The American Society on Aging, San Francisco, California. Reprinted with permission.

44 "Aging and the Family: Present and Future Demographic Issues" by Kevin Kinsella in Rosemary Blieszner and Victoria Hilkevitch Bedford (eds.) *Aging and the Family: Theory and Research* (pp. 32–56), 1996, Westport, CT: Praeger. Copyright © 1996 Greenwood Publishing Group, Inc. Reprinted with permission.

45 "Programming for Family Care of Elderly Dependents: Mandates, Incentives and Service Rationing" by Amanda Smith Barusch, 1995, *Social Work*, 40, pp. 315–322. Copyright © 1995 National Association of Social Workers. Reprinted with permission.

46 From *In the Name of the Family: Rethinking Values in the Postmodern Age* (pp. 83–104) by Judith Stacey. Copyright © 1996 by Judith Stacey. Reprinted by permission of Beacon Press, Boston.

47 "Family Privacy: Issues and Concepts" by Felix M. Berardo, 1998, *Journal of Family Issues*, 19, pp. 4–19. Copyright © 1998 Sage Publications, Inc. Reprinted with permission.

48 "Wishful Thinking and Harmful Tinkering?" by Carol Smart, 1997, *Journal of Social Policy*, 26, pp. 301–321. Copyright © 1997 Cambridge University Press. Reprinted with the permission of Cambridge University Press.

Sociology of Families

READINGS

1

Adolescent Sexuality

What factors contribute to and hinder teenage pregnancy?

Teen pregnancy may seem an unusual topic with which to launch a book of readings on the sociology of families. A little background on competing theories or methodological issues or an overview of changes in family over time might seem like a more logical place to start. However, this book is not like other family readers, with the goal of touching on all the basic concepts. Rather, my goal is to build an in-depth sociological understanding of a select number of contemporary issues faced by families. I have sequenced these issues to reflect the order of events in a family's lifetime, and according to my definition of families, many are begun when an adolescent gives birth to a baby.

I also chose to begin with the topic of teen pregnancy because it is an excellent example of what I believe to be the most important aspect of any topic you study in sociology: the interplay of individual choice and powerful social forces. I became a sociologist after having a number of other careers, one of them as a counselor. One reason I left counseling was that I was uncomfortable with its assumption that an individual's problems are within his or her control. My counselor training taught me that individuals make choices that shape their lives. But my training as a sociologist taught me that individuals' lives, including their problems, are also shaped by the larger community in which they live. You probably have no trouble seeing how your family and friends affect the choices you make in your life. But understanding the impact of the government, church, or media on your decisions is not always so simple. You may have even more trouble seeing the power of your actions to shape the society in which you live. But this two-directional interaction is at the heart of understanding issues from a sociological perspective, and so I begin by considering both how society is related to adolescent pregnancy and how adolescent pregnancy is changing society.

As a group, these articles provide you with information to consider a social situation from several perspectives, and to inform your opinions on some very important sociological concerns regarding teen pregnancy. The first issue, understanding the impact of social norms on behavior, is considered by Mary Pipher in "Then and Now." Pipher, a counselor who works with adolescent girls, contrasts the community norms and values of her youth with those surrounding girls today. The second issue deals with why society has labeled teen pregnancy as a problem. The United States has always had a high rate of adolescent pregnancy. However, in the past most of these pregnancies resulted in marriage. Today they result in abortion or single-parent families. What about this change has caused social concern? Is this concern justified? Kristin Luker's work, "Constructing an Epidemic," helps us understand the history and politics of defining teenage pregnancy in the United States as a problem. A third sociological concern is to understand factors that contribute to teen pregnancy, and identify prevention programs that can successfully address these factors. Of particular interest in considering cause and prevention is the fact that teenagers in the United States are far more likely than teens in other Western countries to become pregnant, yet this seems more attributable to their nonuse of contraceptives than to their level of sexual activity. In "Risking Pregnancy: Avoidance, Ignorance, and Delay of Contraceptive Use," Freeman and Rickels look at the very personal issue of why some sexually active teens use contraception to prevent pregnancy, whereas others don't.

Now let me share my stance on these issues and challenge you to agree or disagree based on what you read. I believe three main factors work together to produce a high teen pregnancy rate in the United States. The first is that many young people have little meaningful purpose in their lives. When your school, your community, and your family cannot provide hope and direction for your future, you can take charge by having a child and pinning your hopes for success on being a good parent. In the bargain, you might produce a child who will love you with that unending love you may not have from anyone else in your life. Adolescents surrounded by poverty and violence are not the only people who might feel this way, although surely this explanation applies to them. Teens in well-to-do suburban families might feel similarly directionless; many rural teens who expected to spend their adult years working the family farm have found their hopes dashed through foreclosure. The constantly escalating rate of suicide among teens is a powerful testament to the lack of meaning or purpose in the lives of teens of all races and backgrounds. Unfortunately, most teenage parents find out too late that being a parent is more challenging and less rewarding than they imagined.

A second major factor that contributes to teenage pregnancy is the exploitation of sex by media such as film, magazines, music, and television. Now, how could someone like me, who was a college student during the sexual revolution, when "free love" and "make love, not war" were more than just slogans, sound like such a prude? I would never want to return to the unhealthy time when people, even married people, seldom discussed sex; the ignorance and guilt produced by not understanding human sexuality is deplorable. However, the way society treats sexuality today provides an equally distorted picture, and in many ways it is a more harmful one. Teens 35 or 45 years ago had few reliable resources for sexual information, but today people of all ages are bombarded with more sexual information than most of us can process. So, in my lifetime, we have moved from a culture where adolescents felt abnormal when they experienced and acted on sexual drives, to one where they feel abnormal when they *don't* act on their sexual drives.

People who produce lyrics, scripts, and advertisements certainly have the freedom to use images and language that serve their purpose. However, should any child old enough to sit in front of a television, look at pictures in a magazine, or listen to the car radio be subjected to the same sophisticated and sometimes explicit sexual material produced for adults? Although very young children are probably oblivious to the true purpose of advertisements for sex talk lines, children's natural curiosity eventually draws their attention to such ads. Parents who have trained their toddlers to observe the world around them and encouraged them to develop verbal skills have always encountered questions regarding sexuality from preschool children. But the nature of those questions changes when children see and hear information more appropriate for an older audience.

Being exposed to sexual information at a much earlier age than in the past may not in itself be negative. However, what is seen and heard shapes a young person's developing attitudes and beliefs about sex. The decisions a teenager makes about becoming sexually active, using a contraceptive, and carrying a pregnancy to term are influenced by what society presents as the norm in these areas. Have today's teens heard from song lyrics that sex will bring them happiness?

Have they watched movies implying that multiple partners and unprotected sex have few repercussions? Have they grown up with television shows where the only problems single parents faced were resolved successfully, with humor, within half an hour? Of course, young people are often exposed to different messages from parents, teachers, and the spiritual leaders, but the images of sexuality presented in our media are pervasive and powerful.

Some of you may argue that the media are reflecting, not shaping, the sexual practices of Americans. I challenge you to think about that argument as you read the selections in this chapter. I am concerned about the ways in which society contributes to young people becoming parents before they have had a chance to live their own lives and to find out who they are and where they want to go. Accepting the responsibilities of parenthood during adolescence cuts short the time people need to make sense of their own lives. And I think we need to support social changes that give that important time back to young Americans, which leads into what I see as the third factor contributing to teen pregnancy.

The third major factor contributing to adolescent pregnancy is our cultural bias toward individual privacy. At the same time that we have allowed the public media to saturate our public space with sexual images, we have continued to view sexual behavior as a private matter for most individuals. The exception is the sexual behavior of those who rely on government assistance. But we regard the sexual practices of financially self-sufficient individuals as very private, unless they pose a danger to others. In "Using Behavioral Theories to Design Abstinence Programs," Kristin Moore and Barbara Sungland provide examples of prevention programs grounded in this approach. This approach contrasts with the public support provided for health and education programs in many countries, as described by Elise Jones and her coauthors in "Teenage Pregnancy in Industrialized Countries: Policy Implications for the United States." As you read these two articles, ask yourself if America might have a lower teen pregnancy rate if we accepted more public responsibility for addressing social factors that contribute to teen pregnancy. Why do Americans tend to simplify the issue by placing all the responsibility for change on the individual?

Before you start reading what others have to say about teen pregnancy, let me challenge you to be aware of criteria used to determine the statistics presented in anything you read or hear about adolescent pregnancy. For example, is a 19-year-old considered a teen in the statistics cited? Is a 19-year-old married mother or only an unmarried one included in the data? Some researchers talk about pregnancy, which may be terminated through abortion and miscarriage as opposed to babies born to teen moms. The numbers used to support any argument on teen pregnancy can be misleading, so ask questions and read between the lines. Most of all, don't be convinced by evidence that doesn't really support the position.

Then and Now

Mary Pipher

Cassie reminded me of myself as a girl. She even looked like me, with long brown hair, blue eyes and a gawky, flat-chested body. Like me, she loved to walk in the woods and cried when she read poetry. She wanted to visit the Holocaust Museum and join the Peace Corps. She preferred books to clothes and didn't care a fig for money. Like me, she was the oldest daughter of a doctor. She loved both her parents even though they were now divorcing and had little energy to care for her. At school she was shy and studious. Kids with problems could talk to her.

But Cassie also wasn't like me. I was fifteen in 1963, she was fifteen in 1993. When I was fifteen, I'd never been kissed. She was in therapy because she'd been sexually assaulted. Her hands folded in her lap, she whispered the story.

She'd been invited to a party by a girl in her algebra class whose parents were out of town. The girl was supposed to stay with a friend, but she had worked out a way to be home. The kids could use her parents' hot tub and stereo system.

Cassie didn't get invited to many parties, so she accepted the invitation. She planned to leave if things got out of control. She told her mother the truth about her plans, except she didn't mention that the parents were gone. Because her mother had been to her lawyer's that day, she was preoccupied by the divorce proceedings and didn't ask for more details.

The party was okay at first—lots of loud music and sick jokes but Cassie was glad to be at a party. A guy from her lunch period asked her to dance. A cheerleader she barely knew asked her to go to the movies that weekend. But by eleven she wanted to go home. The house was packed with crashers and everyone was drinking. Some kids were throwing up, others were having sex or getting rowdy. One boy had knocked a lamp off a desk and another had kicked a hole through a wall.

Cassie slipped away to the upstairs bedroom for her coat. She didn't notice that a guy followed her into the room. He knew her name and asked for a kiss. She shook her head no and searched for her coat in the pile on the bed. He crept up behind her and put his hands under her shirt. She told him to quit and tried to push him away. Then things happened very fast. He grabbed her and called her a bitch. She struggled to break free, but he pinned her down and covered her mouth. She tried to fight but was not strong or aggressive enough. He was muscular and too drunk to feel pain when she flailed at him. Nobody downstairs heard anything over the music. In ten minutes it was over.

Cassie called her mother and asked her to come get her. She shivered outside until her mother arrived. Cassie told her what had happened and they cried together. They called her father and the police, then drove to a nearby hospital. Cassie was examined and she met with a crisis counselor.

Two weeks later Cassie was in my office, in part because of the rape and in part because of the flak she'd taken at school. The guy who sexually assaulted her had been suspended from the track team pending his trial. His friends were furious at her for getting him in trouble. Other kids thought she led him on, that she had asked for it by being at that party.

Cassie awakened me to an essential truth: In 1993, girls' experiences are different from those of myself and my friends in the 1960s. When I tried to understand them based on my own experience, I failed. There was some common ground, enough to delude me that it was all common ground, but there was much new, uncharted territory. . . .

During my adolescence, I lived in a town of 400 people where my mother practiced medicine and my father sold seed corn and raised hogs. I spent my days riding my bike, swimming, reading, playing piano and drinking limeades at the drugstore with my friends. . . .

. . . Everyone "doctored" with my mother and bought corn from my father. All the children played at the same places—the swimming pool, the school yard, the swing across Beaver Creek and the fairgrounds. Everyone knew who was related to whom. When people met, the first thing they did was establish a connection. People on the street said hello to someone with whom they had a rich and complicated lifelong relationship. My pottery teacher, Mrs. Van Cleave, was the grandmother of my good friend Patti and the mother of our next-door neighbor. She was my mom's patient, and her husband went fishing with my dad. Her son was the football coach and his children were in my Methodist youth group.

I had eleven aunts and uncles and thirty cousins who showed up for long visits. The women cooked and watched babies, the men played horseshoes and fished. We all played cards in the evening. My grandfather recited limericks and demonstrated card tricks. Conversation was the main entertainment. We cousins would compare stories about our towns and families. The older cousins would impress the younger ones with their worldly wisdom. Children sat and listened as grown-ups told stories and talked politics. My fondest memory is of falling asleep to laughter and talk in the next room.

The word "media" was not in our language. I saw television for the first time when I was six, and I hid behind the couch because the cowboys' guns scared me. I was eight before we had a black-and-white television on which we watched one grainy station that showed a test pattern much of the day.

As a young teenager I watched "The Mickey Mouse Club," "American Bandstand" and "The Ed Sullivan Show." I wasn't allowed to watch "Perry Mason" or "Gunsmoke" because my parents thought these shows were too violent. We had one movie theater with a new movie every other week. The owner of the theater was a family man who selected our town's movies carefully. His wife sold us salty popcorn, Tootsie Rolls and Cokes. Kids went to the movies on Saturday afternoons and spent most of their time spying on other kids or giggling with their friends.

I loved *Tammy, Seven Brides for Seven Brothers, The Chartroose Caboose* and *South Pacific*. I scanned these movies for information about sex. Rock Hudson, Doris Day, Debbie Reynolds and Frank Sinatra fought and flirted until the end of the movie, when they kissed against backdrops of sunsets to the sounds of swelling violins. This was the era of biblical epics. In *The Story of Ruth*, a demure young Ruth lies down on Boaz' pallet for the night and the camera zooms to the stars. I asked myself, What were they doing on that pallet?

Forty-five RPM records were big in the late fifties. I listened to mushy songs by the Everly Brothers, Roy Orbison and Elvis. My favorite song was Elvis' "Surrender," a song whose lyrics gave me goose bumps and filled me with longing for something I couldn't name. My parents forbade me to listen to Bobby Darin's hit "Multiplication" because it was too suggestive. I learned to twist, a dance that was considered daring.

As Garrison Keillor said, "Nobody gets rich in a small town because everybody's watching." Money and conspicuous consumption were downplayed in my community. Some people were wealthier than others, but it was bad taste to flaunt a high income. We all shopped at the Theobald's grocery and the Rexall and ordered our clothes from Sears and JC Penney catalogs. The banker ordered a new Oldsmobile every year, and my family drove to Mexico at Christmas. A rancher's widow with asthma had the only home air-conditioning unit. The only places to spend money foolishly were the Dairy King and the pool hall.

Particularly children were outside the money economy. Most of our pleasures were free. Most of us had the same toys—Schwinn bikes, Hula-Hoops, basketballs, Monopoly games and dolls or toy soldiers. We could buy Sugar Babies or licorice at the pool, and makeup, comics and *Mad* magazines at the drugstore.

After school I worked for my mother at her clinic. I sterilized syringes and rubber gloves and counted pills. The money I earned went into a college account. By junior high, gifts went into my hope chest—good china, luggage, a dictionary and tatted pillowcases.

Elsewhere mass marketing had begun. Women were encouraged to fix up their homes and dress themselves and their children smartly. Via commercials and advertisements, they were fed a distorted image of themselves and their place in society. This image was less focused on their sexuality and more on their femininity. But because of our distance from a city, mass marketing barely touched our town.

Our town was a dry town and our state had "blue laws," which kept liquor from being advertised, sold on Sundays or served in restaurants. Even our pool hall served nothing stronger than root beer. My father brought tequila back from Mexico and would open a bottle and share it with other men on a Saturday night. Teenage boys had a difficult time finding alcohol. Once my cousin Roy drove fifty miles, convinced a stranger to buy him a six-pack, returned home and hid the six-pack in a culvert.

The Surgeon General had yet to issue his report on smoking, and cigarettes were everywhere, but marijuana and other drugs were unheard of in my town. My father told me that during World War Two a soldier had offered him a marijuana cigarette. He said, "I turned him down and it's a good thing. If I'd said yes, I probably wouldn't be alive today."

At Methodist Youth Fellowship we saw films about the deterioration of people who drank or used marijuana. Women in particular were portrayed as degraded and destroyed by contact with chemicals. After these films we signed pledges that we would never drink or smoke. I didn't break mine until I was in college.

As Tolstoy knew so well, in all times and places there have been happy and unhappy families. In the fifties, the unhappiness was mostly private. Divorce was uncommon and regarded as shameful. I had no friends whose parents were divorced. All kinds of pain were kept secret. Physical and sexual abuse occurred but were not reported. Children and women who lived in abusive families suffered silently. For those whose lives were going badly, there was nowhere to turn. My friend Sue's father hanged himself in his basement. She missed a week of school, and when she returned we treated her as if nothing had happened. The first time Sue and I spoke of her father's death was at our twenty-fifth-year class reunion.

There was cruelty. The town drunk was shamed rather than helped. Retarded and handicapped people were teased. The Green River Ordinance, which kept undesirables—meaning strangers—out of town, was enforced.

I was a sheltered child in a sheltered community. Most of the mothers were homemakers who served brownies and milk to their children after school. Many of them may have been miserable and unfulfilled with their lives of service to men, children and community. But, as a child, I didn't notice.

Most of the fathers owned stores downtown and walked home for lunch. Baby-sitters were a rarity. Everyone went to the same chili feeds and county fairs. Adults were around to keep an eye on things. Once I picked some lilacs from an old lady's bush. She called my parents before I could make it home with my bouquet.

Teenagers fought less with their parents, mostly because there was less to fight about—designer clothes and R-rated movies didn't exist. There was consensus about proper behavior. Grown-ups agreed about rules and enforced them. Teenagers weren't exposed to an alternative value system and they rebelled in milder ways—with ducktails, tight skirts and rock and roll. Adults joked about how much trouble teenagers were, but most parents felt proud of their children. They didn't have the strained faces and the anxious conversations that parents of teenagers have in the 1990s. . . .

Sexuality was seen as a powerful force regulated by God Himself. There were rules and euphemisms for everything. "Don't touch your privates except to wash." "Don't kiss a boy on your first date." "Never let a guy go all the way or he won't respect you in the morning."

Sex was probably my most confusing problem. I read Pat Boone's *Twixt Twelve and Twenty*, which didn't clarify anything. I wasn't sure how many orifices women had. I knew that something girls did with boys led to babies, but I was unable to picture just what that was. I misunderstood dirty jokes and had no idea that songs were filled with sexual innuendo. Well into junior high, I thought that the word "adultery" meant trying to act like an adult.

One of my girlfriends had an older cousin who hid romance magazines under her bed. One day when she was away at a twirling competition, we sneaked up to her room to read them. Beautiful young women were overwhelmed by lust and overpowered by handsome heroes. The details were vague. The couple fell into bed and the woman's blouse was unbuttoned. Her heart would flutter and she would turn pale. The author described a storm outside or petals falling from flowers in a nearby vase. We left the house still uneducated about what really happened. Years later, when I finally heard what the sex act entailed, I was alarmed.

I was easily embarrassed. Tony, the town hoodlum, was my particular curse. Tony wore tight jeans and a black leather jacket and oozed sexual evil. In study hall he sketched a naked woman, scribbled my name on her and passed her around the room. Another time he told me to hold out my hand, and when I did, he dropped a screw into it and shouted, "You owe me a screw."

There was a scary side of sexuality. One friend's dad told her, "Don't get pregnant, but if you do, come to me and I'll load up my gun." A second cousin had to marry because she was pregnant. She whispered to me that her boyfriend had blackmailed her into having sex. She was a homecoming queen candidate and he said he'd go to homecoming with her only if she gave in. He claimed that he was suffering from "blue balls," a painful and unhealthy condition that only sex would remedy.

Lois and Carol taught me my most important lessons. Lois was a pudgy, self-effacing fourteen-year-old whose greatest accomplishment was eight years of perfect attendance at our Sunday school. One Sunday morning she wasn't there, and when I remarked on that fact, the teacher changed the subject. For a time no one would tell me what had happened to Lois. Eventually, however, I was so anxious that Mother told me the story. Lois was pregnant from having sex with a middle-aged man who worked at her father's grocery store. They had married and were living in a trailer south of town. She was expelled from school and would not be coming to church anymore, at least not until after the baby was born. I never saw her again.

Carol was a wiry, freckled farm girl from a big family. She boarded with our neighbors to attend high school in town. In the evenings, after she had the chores done, Carol came over to play with me. One night we were standing in our front yard when a carload of boys came by and asked her to go for a ride. She hesitated, then agreed. A month later Carol was sent back pregnant to her farm. I worried about her because she'd told me her father used belts and coat hangers on the children. My father told me to learn from Carol's mistake and avoid riding with boys. I took him literally and it was years before I felt comfortable riding in cars with any boys except my cousins.

In my town the rules for boys were clear. They were supposed to like sex and go for it whenever they could. They could expect sex with loose girls, but not with good girls, at least not until they'd dated them a long time. The biggest problem for boys was getting the experience they needed to prove they were men.

The rules for girls were more complicated. We were told that sex would ruin our lives and our reputations. We were encouraged to be sexy, but not sexual. Great scorn was reserved for "cockteasers" and "cold fish." It was tough to find the right balance between seductive and prim.

The rules for both sexes pitted them against their Saturday-night dates. Guys tried to get what they could and girls tried to stop them. That made for a lot of sweaty wrestling matches and ruined prom nights. The biggest danger from rule breaking was pregnancy. This was before birth control pills and legal abortion. Syphilis and gonorrhea were the most common sexually transmitted diseases, and both were treatable with the new miracle drug, penicillin.

Sexual openness and tolerance were not community values. Pregnant teachers had to leave school as soon as they "showed." I had no girlfriends who admitted being sexually active. There was community-wide denial about incest and rape, which undoubtedly occurred in my small town as they did all over America. The official story was kept G-rated.

There was a great deal of hypocrisy. A wealthy man in my town was known for being a pincher. We girls called him "the lobster" among ourselves and knew to avoid him. But because his family was prominent, no one ever told him to stop his behavior.

I didn't know that pornography existed until I was a senior in high school. My parents took me to Kansas City and we stayed near the Time to Read bookstore. It was two bookstores in one: on the left, classics, best-sellers and newspapers from all over the world, and on the right, an eye-popping display of pornography....

As I recall my childhood, I'm cautioned by Mark Twain's line, "The older I get, the more clearly I remember things that never happened." Remembering is more like taking a Rorschach test than calling up a computer file. It's highly selective and revealing of one's deep character. Of course, others had different experiences, but I recall small-town life as slower, safer and less sexualized. Everyone did know everyone. Sometimes that made the world seem safe and secure and sometimes that made the world seem small and oppressive.

Cassie attends a high school with 2,300 students. She doesn't know her teachers' children or her neighbors' cousins. When she meets people she doesn't try to establish their place in a complicated kinship network. When she shops for jeans, she doesn't expect the clerk to ask after her family.

Cassie sees her extended family infrequently, particularly since her parents' divorce. They are scattered all over the map. Most of the adults in her neighborhood work. In the evening people no longer sit on their front porches. Instead they prefer the privacy of backyard patios, which keep their doings invisible. Air-conditioning contributes to each family's isolation. On hot summer days and nights people go inside to stay cool. Cassie knows the Cosby family and the people from "Northern Exposure" better than she knows anyone on her block.

Cassie fights with her parents in a more aggressive way than the teens of my youth. She yells, swears, accuses and threatens to run away. Her parents tolerate this open anger much more readily than earlier generations would have....

Cassie is much more politically aware of the world than I was. By the time she was ten she'd been in a protest march in Washington, D.C. She's demonstrated against the death penalty and the Rodney King trial. She writes letters to her congressmen and to the newspapers. She writes letters for Amnesty International to stop torture all over the world. She is part of a larger world than I was and takes her role as an active participant seriously.

Cassie and her friends all tried smoking cigarettes in junior high. Like most teenagers today, Cassie was offered drugs in junior high. She can name more kinds of illegal drugs than the average junkie from the fifties. She knows about local drug-related killings and crack rings. Marijuana, which my father saw once in his lifetime, wafts through the air at her rock concerts and midnight movies.

Alcohol is omnipresent—in bowling alleys, gas stations, grocery stores, skating rinks and Laundromats. Alcohol advertising is rampant, and drinking is associated with wealth, travel, romance and fun. At sixteen, Cassie has friends who have been through treatment for drugs or alcohol. The schools attempt alcohol and drug education, but they are no match for the peer pressure to consume. Cassie knows some Just Say No leaders who get drunk every weekend. By eighth grade, kids who aren't drinking are labeled geeks and left out of the popular scene.

Spending money is a pastime. Cassie wants expensive items—a computer, a racing bike and trips to Costa Rica with her Spanish class and to the ski slopes of Colorado. She takes violin and voice lessons from university professors and attends special camps for musicians.

Cassie's been surrounded by media since birth. Her family owns a VCR, a stereo system, two color televisions and six radios. Cassie wakes to a radio, plays the car stereo on the way to school, sees videos at school and returns home to a choice of stereo, radio, television or videocassettes. She can choose between forty channels twenty-four hours a day. She plays music while she studies and communicates via computer modem with hackers all over the country in her spare time.

Cassie and her friends have been inundated with advertising since birth and are sophisticated about brand names and commercials. While most of her friends can't identify our state flower, the goldenrod, in a ditch along the highway, they can shout out the brand of a can of soda from a hundred yards away. They can sing commercial jingles endlessly.

Cassie's been exposed to years of sophisticated advertising in which she's heard that happiness comes from consuming the right products. She can catch the small lies and knows that adults tell lies to make money. We do not consider that a sin—we call it marketing. But I'm not sure that she catches the big lie, which is that consumer goods are essential to happiness....

There are more magazines for girls now, but they are relatively unchanged in the thirty years since I bought my copies of *Teen*. The content for girls is makeup, acne products, fashion, thinness and attracting boys. Some of the headlines could be the same: TRUE COLORS QUIZ, GET THE LOOK THAT GETS BOYS, TEN COMMANDMENTS OF HAIR, THE BEST PLACES TO MEET AVAILABLE MEN and TEN WAYS TO TRIM DOWN. Some headlines are updated to pay lip service to the themes of the 1990s: TWO MODELS CHILL OUT AT OXFORD UNIVERSITY IN SEASON'S GREATEST GRAY CLOTHES or ECO-INSPIRED LOOKS FOR FALL. A few reflect the greater stress that the 1990s offer the young: REV UP YOUR LOOKS WHEN STRESS HAS YOU DOWN, THE STD OF THE MONTH, GENITAL WARTS and SHOULD I GET TESTED FOR AIDS? Some would never have appeared in the 1950s: WHEN YOU'RE HIGHLY SEXED, IS ONE PARTNER ENOUGH? and ADVICE ON ORGASMS.

Cassie listens to music by The Dead Milkmen, 10,000 Maniacs, Nirvana and They Might Be Giants. She dances to Madonna's song "Erotica," with its sadomasochistic lyrics. The rock-and-roll lyrics by 2 Live Crew that make Tipper Gore cringe don't upset her. Sexist lyrics and the marketing of products with young women's naked bodies are part of the wallpaper of her life.

Cassie's favorite movies are *The Crying Game, Harold and Maude* and *My Own Private Idaho*. None of these movies would have made it past the theater owner of my hometown.

Our culture has changed from one in which it was hard to get information about sexuality to one in which it's impossible to escape information about sexuality. Inhibition has quit the scene. In the 1950s a married couple on TV had to be shown sleeping in twin beds because a double bed was too suggestive. Now anything—incest, menstruation, crotch itch or vaginal odors—can be discussed on TV. Television shows invite couples to sell their most private moments for a dishwasher.

The plot for romance movies is different. In the fifties people met, argued, fell in love, then kissed. By the seventies, people met, argued, fell in love and then had sex. In the nineties people meet, have sex, argue and then, maybe, fall in love. Hollywood lovers don't discuss birth control, past sexual encounters or how a sexual experience will affect the involved parties; they just do it. The Hollywood model of sexual behavior couldn't be more harmful and misleading if it were trying to be.

Cassie has seen *Playboys* and *Penthouses* on the racks at local drugstores and Quick Stops. Our city has adult XXX-rated movie theaters and adult bookstores. She's watched the adult channels in hotel rooms while bouncing on "magic fingers" beds. Advertisements that disturb me with their sexual content don't bother her. When I told her that I first heard the word "orgasm" when I was twenty, she looked at me with disbelief.

Cassie's world is more tolerant and open about sex. Her friends produced a campy play entitled *Vampire Lesbians of Sodom*. For a joke she displays Kiss of Mint condoms in her room. She's a member of her school's branch of Flag—Friends of Lesbians and Gays—which she joined after one of her male friends "came out" to her. She's nonjudgmental about sexual orientation and outspoken in her defense of gay rights. Her world is a kinder, gentler place for girls who have babies. One-fifth of all babies today are born to single mothers. Some of her schoolmates bring their babies to school.

In some ways Cassie is more informed about sex than I was. She's read books on puberty and sexuality and watched films at school. She's seen explicit movies and listened to hours of explicit music. But Cassie still hasn't heard answers to the questions she's most interested in. She hasn't had much help sorting out when to have sex, how to say no or what a good sexual experience would entail.

Cassie is as tongue-tied with boys she likes as I was, and she is even more confused about proper behavior. The values she learned at home and at church are at odds with the values broadcast by the media. She's been raised to love and value herself in a society where an enormous pornography industry reduces women to body parts. She's been taught by movies and television that sophisticated people are sexually free and spontaneous, and at the same time she's been warned that casual sex can kill. And she's been raped.

Cassie knows girls who had sex with boys they hardly knew. She knows a girl whose reason for having sex was "to get it over with." Another classmate had sex because her two best girlfriends had had sex and she didn't want to feel left out. More touching and sexual harassment happens in the halls of her school than did in the halls of mine. Girls are referred to as bitches, whores and sluts.

Cassie has been desensitized to violence. She's watched television specials on incest and sexual assaults and seen thousands of murders on the screen. She's seen *Fatal Attraction* and *Halloween II*. Since Jeffrey Dahmer, she knows what necrophilia is. She wasn't traumatized by *The Diary of Anne Frank*.

Cassie can't walk alone after dark. Her family locks doors and bicycles. She carries Mace in her purse and a whistle on her car keys. She doesn't speak to men she doesn't know. When she is late, her parents are immediately alarmed. Of course there were girls who were traumatized in the fifties, and there are girls who lead protected lives in the 1990s, but the proportions have changed significantly. We feel it in our bones.

I am not claiming that our childhoods are representative of the childhoods of all other females in America. In some ways Cassie and I have both had unusual childhoods. I grew up in a rural, isolated area with much less exposure to television than the average child of the times. My mother was a doctor instead of a homemaker. Compared to other girls, Cassie lives in a city that is safer than most and has a family with more money. Even with the rape, Cassie's situation is by no means a worst-case scenario. She lives in a middle-class environment, not an inner city. Her parents aren't psychotic, abusive or drug-addicted.

Also, I am not claiming that I lived in the good old days and that Cassie lives in the wicked present. I don't want to glorify or to "Donna Reedify" the fifties, which were not a golden age. They were the years of Joe McCarthy and Jim Crow. How things looked was more important than how things really were. There was a great deal of sexual, religious and racial intolerance. Many families had shameful secrets, and if revealed, they led to public disgrace rather than community help.

I left my town as soon as I could, and as an adult, I have been much happier in a larger, less structured environment. Many of my friends come from small towns, and particularly the smart women among them have horror stories of not fitting in.

What I am claiming is that our stories have something to say about the way the world has stayed the same and the way it has changed for adolescent girls. We had in common that our bodies changed and those changes caused us anxiety. With puberty, we both struggled to relate to girls and boys in new ways. We struggled to be attractive and to understand our own sexual urges. We were awkward around boys and hurt by girls. As we struggled to grow up and define ourselves as adults, we both distanced ourselves from our parents and felt some loneliness as a result. As we searched for our identities, we grew confused and sad. Both of us had times when we were moody, secretive, inarticulate and introspective.

But while some of our experiences are similar, many are radically different. Cassie's community is a global one, mine was a small town. Her parents were divorcing, mine stayed together. She lives in a society more stratified by money and more driven by addictions. She's been exposed to more television, movies and music. She lives in a more sexualized world.

Things that shocked us in the 1950s make us yawn now. The world has changed from one in which people blushed at the term "chicken breast" to one in which a movie such as *Pretty Woman* is not embarrassing. We've gone from a world with no locks on the doors to one of bolt locks and handguns. The issues that I struggled with as a college student—when I should have sex, should I drink, smoke or hang out with bad company—now must be considered in early adolescence.

Neither the 1950s nor the 1990s offered us environments that totally met our needs. My childhood was structured and safe, but the costs of that security were limited tolerance of diversity, rigid rules about proper behavior and lack of privacy. As one man from a small town said, "I don't need to worry about running my own business because there are so many other people who are minding it for me." Although my community provided many surrogate parents and clear rules about right and wrong, this structure was often used to enforce rigid social and class codes and to keep people in their place.

Cassie lives in a town that's less rigid about roles and more supportive of autonomy, but she has little protected space. Cassie is freer in some ways than I was. She has more options. But ironically, in some ways, she's less free. She cannot move freely in the halls of her school because of security precautions. Everyone she meets is not part of a community of connected people. She can't walk alone looking at the Milky Way on a summer night.

The ideal community would somehow be able to combine the sense of belonging that small towns offer with the freedom to be oneself that small towns sometimes inhibit. Utopia for teenage girls would be a place in which they are safe and free, able to grow and develop in an atmosphere of tolerance and diversity and protected by adults who have their best interest at heart.

Constructing an Epidemic

Kristin Luker

By the early 1980s Americans had come to believe that teenagers were becoming pregnant in epidemic numbers, and the issue occupied a prominent place on the national agenda. "Teenager pregnancy," along with crack-addicted mothers, drive-by shootings, and the failing educational system, was beginning to be used as a form of shorthand for the country's social ills.[1] Everyone now agreed that it was a serious problem, and solutions were proposed across the ideological spectrum. Conservatives (members of the New Right, in particular) wanted to give parents more control over their daughters, including the right to determine whether they should have access to sex education and contraception.[2] Liberals, doubting that a "just say no" strategy would do much to curtail sexual activity among teenagers, continued to urge that young men and women be granted the same legal access to abortion and contraception that their elders had. Scholars debated the exact costs of early pregnancy to the individuals involved and to society, foundations targeted it for funding and investigation, government at all levels instituted programs to reduce it, and the media gave it a great deal of scrutiny.[3] In the early 1970s the phrase "teenage pregnancy" was just not part of the public lexicon. By 1978, however, a dozen articles per year were being published on the topic; by the mid-1980s the number had increased to two dozen; and by 1990 there were more than two hundred, including cover stories in both *Time* and *Newsweek*.[4]

Ironically (in view of all this media attention), births to teenagers actually *declined* in the 1970s and 1980s. During the baby boom years (1946–1964), teenagers, like older women, increased their childbearing dramatically: their birthrates almost doubled, reaching a peak in 1957. Subsequently, the rates drifted back to their earlier levels, where they have pretty much stayed since 1975.[5] The real "epidemic" occurred when Dwight Eisenhower was in the White House and poodle skirts were the height of fashion.[6] But although birthrates among teenagers were declining, other aspects of their behavior were changing in ways that many people saw as disturbing. From the vantage point of the 1970s, the relevant statistics could have been used to tell any one of a number of stories. For example, when abortion was legalized in 1973, experts began to refer to a new demographic measure, the "pregnancy rate," which combined the rate of abortion and the rate of live births. In the case of teenagers an increasing abortion rate meant that, despite a declining birthrate, the pregnancy rate was going up, and dramatically so.[7]

Since the rise in the pregnancy rate among teenagers (and among older women as well) was entirely due to the increase in abortions, it is curious that professionals and the public identified pregnancy, rather than abortion, as the problem. It is likewise curious that although the abortion rate increased for all women, most observers limited their attention to teenagers, who have always accounted for fewer than a third of the abortions performed. Teenagers *are* proportionately overrepresented in the ranks of women having abortions. But to pay attention almost exclusively to them, while neglecting the other groups that account for 70 percent of all abortions, does not make sense.

A similar misdirection characterized the issue of illegitimacy. In the 1970s teenagers were having fewer babies overall than in previous decades, but they—like older women—were having more babies out of wedlock. Compared to other women, teenagers have relatively few babies, and a very high proportion of these are born to unmarried parents (about 30 percent in 1970, 50 percent in 1980, and 70 percent in 1995). But although most babies born to teenagers are born out of wedlock, most babies born out of wedlock are not born to teens. In 1975 teens accounted for just under a half of all babies born out of wedlock; in 1980 they accounted for 40 percent; and in 1990 they accounted for fewer than a third.[8] Obviously, teens should hardly be the only population of interest.

Thus, in the 1970s and early 1980s the data revealed a number of disquieting trends, and teenagers became the focus of the public's worry about these trends. More single women were having sex, more women were having abortions, more women were having babies out of wedlock, and—contrary to prevailing stereotypes—older women and white women were slowly replacing African Americans and teens as the largest groups within the population of unwed mothers. These trends bespeak a number of social changes worth looking at closely. Sex and pregnancy had been decoupled by the contraception revolution of the 1960s; pregnancy and birth had been decoupled by the legalization of abortion in the 1970s; and more and more children were growing up in "postmodern" families—that is, without their biological mother and father—in part because divorce rates were rising and in part because more children were being born out of wedlock. But these broad demographic changes, which impinged on women and men of all ages, were seen as problems that primarily concerned *teenagers.* The teenage mother—in particular, the black teenage mother—came to personify the social, economic, and sexual trends that in one way or another affected almost everyone in America.

A number of different responses might have been devised to meet the challenge of these new trends. It would have been logical, for example, to focus on the problem of abortion, since more than a million abortions were performed each year despite the fact that people presumably had access to effective contraception. Or the problem might have been defined as the increase in out-of-wedlock births, since more and more couples were starting families without being married.[9] Or policymakers could have responded to the way in which sexual activity and childbearing were, to an ever greater extent, taking place outside marriage (in 1975 about three-fourths of all abortions were performed on single women).[10] Yet American society has never framed the problem in any of these broader terms. The widest perspective was perhaps that of the antiabortion activists, who saw the problem as abortion in general. A careful reading of the specialist and nonspecialist media suggests that, with a few exceptions, professionals and the general public paid scant attention to abortion and out-of-wedlock childbearing among older women, while agreeing that abortion and illegitimate births among teenagers constituted a major social and public-health problem. Why did Americans narrow their vision to such an extent? How did professionals, Congress, and the public come to agree that there was an "epidemic" of pregnancy among teenagers and that teenagers were the main (if not the only) population worth worrying about?[11]

A Story That Fits the Data

...Advocates for young people had used Congress and the media to publicize an account of teenagers and their circumstances that seemed to make sense of the emerging demographic data and that was extremely persuasive. In essence, they claimed that teenagers, like older women, were increasingly likely to have sex and that their sexual activity was increasingly likely to take place outside marriage. Teens, however, like poor women of earlier generations, had been left out of the contraceptive revolution that had so changed the lives of other American women. They were having babies they did not want and could not support. Many of them were too inexperienced to know how to avoid conception, to appreciate the difficulties of childrearing, or to obtain an abortion (besides, abortion was expensive). And most gave birth without the support of the partner who had impregnated them. Unless they were granted access to affordable contraception and abortion, they would continue to have babies out of wedlock and would be mired in a life of poverty. Advocates noted that most babies born to teenagers were born out of wedlock, and that babies who lived with one parent were obviously less well off than those who lived with two. Moreover, black teenagers, who have always been disadvantaged in American society, had much higher rates of childbearing and illegitimacy than whites, although the reproductive behavior of white teenagers was beginning to resemble that of blacks. And in this account, teens who gave birth were much more likely to drop out of school than those who did not, so that as adults they were less well educated and hence poorer than women who postponed their childbearing.

Taken together, the data added up to a story that made sense to many people. It convinced Americans that young mothers . . . who gave birth while still in high school and who were not married—were a serious social problem that brought a host of other problems in its wake. It explained why babies . . . were born prematurely, why infant mortality rates in the United States were so high compared to those in other countries, why so many American students were dropping out of high school, and why AFDC costs were skyrocketing. Some people even believed that if teenagers in the United States maintained their high birthrates, the nation would not be able to compete internationally in the coming century. Others argued that distressing racial inequalities in education, income, and social standing were in large part due to the marked difference in the birthrates of white and black teenagers.

Yet this story, which fed both on itself and on diffuse social anxiety, was incomplete; the data it was based on were true, but only partial. Evidence that did not fit the argument was left out, or mentioned only in passing. Largely ignored, for example, was the fact that a substantial and growing proportion of all unmarried mothers were not teenagers. And on those rare occasions when older unwed mothers were discussed, they were not seen as a cause for concern.[12] Likewise, although the substantially higher rates of out-of-wedlock childbearing among African Americans were often remarked upon, few observers pointed out that illegitimacy rates among blacks were falling or stable while rates among whites were increasing. Few noted that most of the teenagers giving birth were eighteen- and nineteen-year-olds, or that teens under fifteen had been having babies throughout much of the century.[13]

This story, as it emerged in the media and in policy circles in the 1970s and 1980s, fulfilled the public's need to identify the cause of a spreading social malaise. It led Americans to think that teenagers were the only ones being buffeted by social changes, whereas these changes were in fact pervasive; it led them to think that heedless, promiscuous teenagers were responsible for a great many disturbing social trends; and it led them to think that teenagers were doing these things unwittingly and despite themselves. When people spoke of "children having children" or of "babies having babies," their very choice of words revealed their belief that teenage mothers, because of their youth, should not be held morally responsible for their actions. "Babies" who had babies were themselves victims; they needed protection from their own ungovernable impulses.

In another sense, limiting the issue to teenagers gave it a deceptive air of universality; after all, everyone has been or will be a teenager. Yet the large-scale changes that were taking place in American life did not affect all teenagers equally. The types of behavior that led teenagers to get pregnant and become unwed mothers (engaging in premarital sex, and bearing and keeping illegitimate children) were traditionally much more common among African Americans than among whites, and more common among the poor than among the privileged.

For average Americans in the 1970s, life had undergone profound changes in just a few short years. Unmarried couples were engaging more readily in sex, and doing so much more openly. Many of them were even living together, instead of settling for . . . sex in the back seats of cars. When an unmarried woman got pregnant, she no longer made a sudden marriage or a hasty visit to a distant aunt; now she either terminated her pregnancy or openly—even proudly—had her baby. Often she chose to live as a single parent or to set up housekeeping with her partner, rather than allowing her child to be adopted by a proper, married middle-class couple. In the 1970s people of all ages began to follow this way of life, but the . . . fears of the public coalesced in large part exclusively around teenagers. The new patterns of sexual behavior and new family structures were simply more visible among younger people, who had not committed themselves to the older set of choices. At the same time, teenagers, especially those who had children, were defined as people who were embarking on a lifetime of poverty. The debate, in centering on teenagers in general, thus combined two contrasting features of American society: it permitted people to talk about African Americans and poor women (categories that often overlapped) without mentioning race or class; but it also reflected the fact that the sexual behavior and reproductive patterns of white teenagers were beginning to resemble those of African Americans and poor women—that is, more and more whites were postponing marriage and having babies out of wedlock.

The myriad congressional hearings, newspaper stories, and technical reports on the "epidemic" of pregnancy among teenagers could not have convinced the public to subscribe to this view if other factors in American life had not made the story plausible. The social sciences abound with theories suggesting that the public is subject to "moral panics" which are in large part irrational, but in this case people were responding to a particular account because it helped them make sense of some very real and rapidly changing conditions in their world.[14] It appeared to explain a number of dismaying social phenomena, such as spreading signs of poverty, persistent racial inequalities, illegitimacy, freer sexual mores, and new family structures.[15] It was and continues to be a resonant issue because of the profound changes that have taken place in the meanings and practices associated with sexuality and reproduction, in the relations among sex, marriage, and childbearing, and in the national and global economies. Through the story of "teenage pregnancy," these revolutionary changes acquired a logic and a human face.[16]

The Sexual Revolution

In the 1950s and 1960s (as those who long for the good old days are fond of telling us) sex was a very private matter.[17] Like childbearing, it was sanctioned only within marriage. Respectable women were careful lest their behavior earn them a reputation for being "loose," which would limit their ability to marry a "nice" man. True, in 1958 about four out of ten unmarried women were sexually active before their twentieth birthday, but in those days premarital sex was in a strict sense *premarital*, for the most part occurring within a committed relationship that soon led to marriage.[18] Though the data collected by Alfred Kinsey and his colleagues in the 1940s and 1950s are not nationally representative, they do show that for earlier generations of American women, most premarital sex was in large part "engagement" sex—sex with the man the woman was planning to marry, and then for only a relatively short period before the wedding. In the Kinsey report, almost half of the married women who had engaged in premarital sex had done so only with their fiancés, and for less than two years prior to their marriage.[19] More recent and more representative data suggest that this pattern continued for some time: in the 1960s half of all women who engaged in premarital sex did so with their fiancés. By the mid-1980s, this proportion had fallen to less than a fourth.[20]

Many people recall the transformation in sexual behavior that took place in the 1970s, but they may well have forgotten the rapidity of that change. In 1969 the overwhelming majority of Americans—almost 70 percent—agreed that having sex before marriage was wrong; three out of four agreed that magazine photos of nudes were objectionable; and more than four out of five agreed that nudity in a Broadway show (for example, "Hair" or "Oh! Calcutta") was unacceptable.[21] A mere four years later, only traces of these values remained: the percentages of Americans who objected to premarital sex and to nudes in magazines had both dropped an astonishing twenty points, and the percentage of those who objected to nudity on the stage had dropped eighteen points.[22] Similarly, a Roper poll conducted in 1969 found that only 20 percent of the public approved of premarital sex; four years later, the respondents were equally divided on the issue. The General Social Survey conducted by researchers at the University of Chicago asked the question in a slightly different way: in 1972 it found that only 26 percent of the public thought premarital sex was "not wrong at all"; but a decade later this figure had jumped to 40 percent, while the percentage of those who said it was "almost always wrong" had correspondingly declined.[23]

Not surprisingly, as more and more people engaged in premarital sex or extramarital sex (after being divorced, separated, or widowed), it became increasingly difficult to claim that sexual activity should be limited to adults. By what logic could sex be declared taboo for the young? And how young was too young? This created a genuine dilemma. In 1969 the rules about sex were clear, even if they were often ignored in practice. Sex was for married people, and if society sometimes turned a blind eye to sex between unmarried partners, it did so only for those who had attained or were close to attaining legal adulthood. Minors, unless they were deemed mature or emancipated, could not obtain contraception, and in most states "minors" included everyone under twenty-one. Moreover, under the age-of-consent laws that were in force in many states, young women could not legally consent to have sex.[24] In challenging these rules in the courts and in Congress, advocates had been successful in claiming that teenagers had a right to contraception, and therefore a right to have sex. But the new concept of rights for teenagers created a "bright-line" problem. Once adults accepted that unmarried people could have sex and that teenagers had a right to contraception, by what logic was an unmarried thirteen-year-old too young to have sex? What bright line separated the too young from the old enough? The category "teenagers" or "adolescents" included people who were barely out of childhood as well as people who were legal adults. And if teenagers had rights, why not even younger people?

Ever since the late nineteenth century, Americans have assumed that individuals in this amorphous category are not emotionally or physically mature enough to have sex. Furthermore, nowadays few teenagers have the financial resources or the educational preparation necessary to raise a family. The public thus evinces [perceived] worries about the *emotional* capacities of teenagers, combined with realistic concern about their *social* capacities—capacities to deal with marriage and children, both of which often follow sexual activity. Many people take it for granted that teenagers are capable only of infatuation or puppy love. And in most cases teenage couples today, unlike their counterparts in the 1940s and 1950s, are not engaged and ready to settle down in a year or two. In short, adults tend to think that teenagers are unprepared for the serious business of building a family and are capable only of careless premarital sex that is rooted in pure pleasure. And the very fact that the category "teenagers" is so broad, ranging from seventh-graders to legal adults, exacerbates the problem.

The few data available tend to confirm the commonsense notion that American adults consider premarital sex acceptable for themselves but not for teenagers. In 1977, for example, 63 percent of respondents to a public-opinion poll said that they believed sex between unmarried teenagers was morally wrong; in 1994, in answer to a slightly different question, 50 percent said it was wrong. Polls conducted between 1986 and 1994 found that a consistent 85 percent of respondents considered premarital sex unacceptable for people aged fourteen to sixteen: nearly 70 percent thought it was "always wrong," and nearly 20 percent thought it was "almost always wrong."[25]

But in the late 1960s and early 1970s, whatever adults may have preferred, patterns of sexual behavior were changing for everyone, and that included teenagers. Although it is debatable whether these changes constituted a "revolution" among adults, they unquestionably did among teenagers. Current statistics show that in a typical group of forty teenagers (with an equal number of each sex), five of the women and ten of the men will have had intercourse by the time they enter the tenth grade; twelve of the women and fourteen of the men will have done so by the time they are seniors; and fully fifteen of the women and seventeen of the men will have done so before they are twenty.[26] Virtually all studies confirm that the young people of today are more likely to have premarital sex than were those of earlier generations. In 1970 it was estimated that slightly more than a fourth of all unmarried women aged fifteen to nineteen were sexually active; by 1984 the proportion had risen to just under a half.[27] In 1982 the National Survey of Family Growth found that whereas in the 1950s 40 percent of teenagers had reported engaging in premarital sex, among women who had turned twenty between 1979 and

1981 the figure had jumped to 70 percent. There was a substantial increase in sexual activity among younger women as well: in the mid-1950s only three out of a hundred engaged in premarital sex before the age of fifteen; in the mid-1970s one in ten did so. Among more recent groups of teens, the increase in activity is even more striking.[28]

These changes in the statistics, dramatic though they are, do not begin to capture the extent of the transformation that has actually taken place in teenagers' sexual behavior. For example, we tend to speak of their involvement in "premarital sex," and this is technically correct: today 96 percent of American teenagers have sex before they get married. But this is not the "engagement sex" that young women allowed themselves in the 1950s. Now teenagers are sexually active whether or not they have immediate plans to marry. And for reasons that no one fully understands, Americans of all ages are retreating from marriage. As a result, many of the teenagers who are engaging in sex and having babies are doing exactly what teenagers did in the 1950s, but the nontrivial change is that they are doing so without the benefit of wedlock.

On the one hand, the median age at first marriage has been rising throughout the century and today is virtually the highest it has ever been since the United States began keeping accurate records: twenty-four for women and twenty-six for men. On the other hand, age at first menstruation has slowly been going down, probably because of better nutrition. Thus, young people are becoming fertile a bit earlier than they used to but getting married much later, and hence face very long periods of time during which they are physically capable of sex and childbearing but unable or unwilling to marry. Furthermore, public attitudes toward out-of-wedlock sex have become increasingly liberal. As a result of all these factors, young white Americans in general face about a decade in which they are sexually active but not married. For blacks, who have very low marriage rates, the period is even longer: twelve years for women and nineteen for men.[29] So today's teenagers, in contrast to those of earlier generations, face a whole new set of issues and dilemmas not easily resolved by slogans such as "just say no" to sex or childbearing. It probably isn't realistic to ask today's teens to abstain from sex for a decade or two, but we have no clear guidelines on when young people are "ready" for sex. At the same time, longer periods of sexual activity outside marriage surely change the nature and meaning of sex for the people involved. To take just one example, teenage women face real dilemmas negotiating sex, intimacy, and plans for babies over a number of partners. In 1971 approximately six out of ten sexually active teenage women in metropolitan areas had had only a single partner; by 1979 this figure had declined to only one in two. In the early 1990s more than 70 percent of such women had had more than one partner, and one in five had had six or more partners. Counter to the stereotype, however, most of these teens, like most adults, engaged in serial monogamy.[30]

These broad demographic data hint at the dismay that many Americans feel in response to such radical social change, especially when teenagers are involved. Sexual behavior, like any other behavior, is situated within a complex web of social ties based on factors such as race, class, gender, and ethnicity. When women in the 1950s engaged in premarital sex, it was thought of simply as marital sex beginning a little before the wedding. Today, however, teenagers become sexually active not when they move out of the parental home and into a family of their own making but while they are still defined primarily as children. Since they are under the control (however nominal) of their parents, their sexual activity raises troubling questions about the purposes and meaning of sex, particularly in the case of young women. Many Americans object to the idea of "casual" sex, meaning sex that is not closely linked to the process by which people form couples and settle down. Yet teenagers, especially young teenagers, are almost universally regarded as too young to "get serious" and contemplate marriage. The kinds of sex that are appropriate for them (short-term relationships for the purpose of pleasure, not procreation) run counter to the basic values espoused by many adults. This double bind, according to which serious commitments are premature but casual sex is immoral, makes sexual activity among teens inherently troubling for many adults and . . . can make it very difficult for teens to manage their sexual lives.

In addition, sexual behavior in the United States has long been governed by notions of propriety that depend on gender and racial distinctions, and teenagers' sexual activity has altered some of these traditional notions. For example, the statistics show that young women are almost as likely as young men to be sexually active before marriage. And the patterns of sexual activity among blacks and whites are converging to a remarkable extent. In 1988 the rate of premarital sexual activity among young white women was only three points lower than the rate for young black women. Never before had the margin been so narrow—an indication that such activity was increasing among whites, and perhaps decreasing among blacks.[31] Furthermore, scattered data suggest that as sexual activity among teenagers becomes more widespread, differences in the sexual experiences of people from various classes are diminishing. Many studies, beginning with the Kinsey report, have demonstrated that the likelihood of engaging in premarital sex and the age at first occurrence are both linked to class: the higher an individual's class background, the older he or she will be when first engaging in premarital sex, and the more likely it is that he or she will never engage in such sex at all. Since more and more people are postponing marriage and since teens of all classes are becoming sexually active, the relationship between sexual activity and class has greatly diminished among the young.[32]

To the extent that choices about sexual behavior (like choices about any other form of social behavior) both reflect and constitute social roles, these developments are bound to be troubling. Sociologists from Max Weber to Pierre Bourdieu have noted how groups of individuals use behavior patterns as "markers" to distinguish themselves from others. Behavior that people may have no objection to when they observe it in a young working-class man (particularly a young working-class man of color) will affect them very differently when they observe it in a young white upper-middle-class woman.[33] Whereas sex outside marriage was once appropriate only for "unruly" types (adult men, and young people from working-class or minority backgrounds) or for "nice" girls in love with their fiancés, the boundaries are no longer clear. Although adults now tend to think that premarital sex is acceptable when it is accompanied by emotional commitment, society seems to want young people—young middle-class white women in particular—to sustain this kind of sex throughout almost a decade without "getting too serious" or "being cheap." A difficult task, indeed.[34]

Public-opinion polls, when read carefully, suggest that adults have complex preferences about the best way to deal with sexual activity among teenagers. Most adults don't want teens to be sexually active, but for a surprisingly long time they have agreed that teenagers who *are* sexually active should have access to birth control information and contraceptives. Most have also long favored providing sex education in the schools, but they are remarkably skeptical about its ability to curtail sexual activity or pregnancy among teenagers. They disapprove of unmarried teenage mothers and consider them a source of social problems, but a majority are strongly in favor of laws that require parental approval before a teenager can have an abortion. In fact, about 40 percent of Americans think that a young woman should not be permitted to have an abortion even if pregnancy would cause her to drop out of school.[35] In short, most adults seem to have a clear first choice—namely, that teens should not have sex. At the same time, a large majority of them support contraceptive and sex education programs for teens, a fact that suggests they doubt they will get their first choice. In general, adults want teens to have access to services and programs that will reduce the problems associated with sexual activity, but they do not wish this access to be unrestricted. Thirty to 40 percent believe that contraceptives should be made available to teenagers only if they have their parents' approval, and 70 to 80 percent believe that such approval is needed if a teen wants an abortion.[36]

In general, when it comes to teenagers Americans are much more liberal than they used to be on issues such as the availability of birth control information, the provision of contraceptive services, access to abortion, and openness on matters of sexuality and reproduction. A small, fairly constant number of people still long for the old sexual order; most Americans do not. What parents *do* seem to want is some measure of control over how their children behave in the world of sexual freedom that opened in the 1970s. Many Americans, however, think that their authority over their children is precarious at best. A Harris poll undertaken for Planned Parenthood in 1985 found that almost half of the parents surveyed felt they didn't have much control over their teenagers' sexuality, and an additional 18 percent felt they had no control at all. The data also suggest that *daughters'* decisions are of particular concern to parents. In 1986 a poll of fathers found that 26 percent were "very worried" about their teenage daughters' sexual activity, and an additional 32 percent were "somewhat worried." The figures for teenage sons were, respectively, 10 percent and 36 percent.[37] In addition, people are alarmed about the prevalence of sexually transmitted diseases, including AIDS. To put the matter simply: the rules of the game have changed, and they have done so in ways that are particularly troubling for parents. In the 1970s, as the current generation of parents was coming of age, the old order ended. Premarital sex became more common, and public opinion on the issue shifted dramatically.

Today's parents want to protect their children from the myriad dangers—seen and unseen, life-threatening and emotionally bruising—that sex entails these days. And they want to set their own timetable, so that they themselves can decide when their children are old enough to have sex. Often parents find it difficult to allow a child to be sexually active while he or she is still living as a dependent under their roof. Yet the point at which many parents consider a child old enough to be sexually active—whether they define it as when the child marries, or moves out of the parental home, or becomes self-supporting—is occurring ever later in American life, due to societal and economic changes over which individuals have minimal control. Except for the relatively few people who think that sex outside marriage is always wrong (and whose problem is chiefly one of finding a way to promulgate their values in an unsympathetic society), Americans have numerous questions relating to teenagers' sexual behavior. Should teens be sexually active? At what age? With whom? How are parents to encourage the use of contraception without seeming to push a teenager into having sex before he or she is ready? How can individuals reconcile their antipathy toward abortion with their desire to see fewer children born out of wedlock? In short, the contradictions inherent in teenagers' sexual activity make it hard for adults to give a clear, precise, and unambiguous message to today's young people.

Teenagers, however, are simply the most visible aspect of a far larger problem. Nowhere has public or private life caught up with the sexual revolution of the 1970s, and most Americans do not yet fully appreciate how far-reaching the changes really were. Now that sex seems to have been permanently disconnected from marriage (or as permanently as anything ever is in social life), private citizens as well as policymakers must grapple with a host of legal, ethical, medical, and social issues. Teenagers are a focus of anxiety because so many of them are participating in the new world of sexual freedom and because most adults are (often rightly) doubtful about the skills and resources these young people possess. The challenge facing parents is to find a way to protect their children and their children's children without making unrealistic or impractical demands, yet still maintain some authority over them. As a consequence, public attitudes toward teenagers' sexual activity are an awkward amalgam of attempts to come to terms with vague fears and a sense that young people are out of control. The American public supports sex education because it has long thought that providing knowledge and skills can modify behavior. At the same time, people are skeptical about the ability of education alone to change patterns of sexual behavior, which are so strongly motivated. They believe that early pregnancy can best be prevented if parents readily communicate with their offspring, but they are confused and ambivalent about what sort of information parents should provide and how they should go about conveying it, as evidenced by the relatively small number of parents who actually talk to their children about sex.[38] Adults disapprove of sexual activity and childbearing among unmarried teenagers, but are generally resigned to the fact that such activity takes place. Many adults prefer the lesser of two evils—contraception as opposed to pregnancy—even to the point of allowing schools to dispense contraceptives. But a substantial minority strive to maintain some degree of control in this area and continue to favor parental-consent laws for teenagers who wish to obtain contraception.[39] . . .

In short, pregnant teenagers made a convenient lightning rod for the anxieties and tensions in Americans' lives. Economic fortunes were unstable, a post-industrial economic order was evolving, sexual and reproductive patterns were mutating. Representing such teenagers as the epitome of society's ills seemed one quick way of making sense of these enormous changes. This was particularly true as poverty was becoming ever more visible and being poor appeared to be the direct result of immoral or unwise behavior. Pregnant teenagers seemed to embody the very essence of such behavior. Indeed, the phrase "teenage pregnancy" continues to be a powerful shorthand way of referring to the problem of poverty.

The rhetoric of the 1970s, generated in good faith by advocates who wanted to ensure that young women had access to contraception, created a comforting but unrealistic fantasy to explain the fact that some people were getting poorer in an uncertain economy. By noting that young mothers were poor mothers, advocates persuaded the public that young mothers are poor *because* they had untimely pregnancies and births. This in turn led to the conclusion that if young poor women simply did what young affluent women do, then they, too, would be affluent. It is not surprising that when affluent people dramatically change their attitudes and behavior toward marriage and childbearing but poor people do not, the well-to-do would try to explain the existence of poverty by saying the poor have failed to adapt. In recent years, both liberals and conservatives have tended to ascribe poverty to the sexual and reproductive decisions that poor women make. What gives this argument resonance is the fact that the affluent are postponing their childbearing and early motherhood is increasingly the province of the "left behind"—poor women who realistically know that postponing their first birth is unlikely to lead to a partnership in a good law firm. But the deep cultural belief that it *might* continues to attract people of every ideological persuasion. Commentators as diverse as Charles Murray and David Ellwood, one a conservative bent on undoing the welfare system and the other a liberal bent on saving it, agree on the foolishness of early pregnancy.[40]

There is no arguing the case that teenagers who bring a child into the world put a strain on public patience, values, and funds. The public assumes that teenagers are unable to support a child financially, and in the overwhelming majority of cases this is true. Moreover, poor mothers tend to have children who will themselves grow up to be poor. Not surprisingly, teenagers and their babies have come to be perceived (to use the words of a *Time* essayist) as "the very hub of the U.S. poverty cycle," often creating up to three generations of poor people who will depend on the public purse. Congress, the media, reports by the National Academy of Sciences, and statements by private voluntary groups all associate poverty with childbearing among teenagers. But this linkage depends on an assumption that reducing pregnancy among teenagers, specifically among unmarried teenagers, can reduce poverty.[41]

In the opinion of many well-meaning middle-class people, the trouble with poor and pregnant teenagers is that they do not do what middle-class people do: invest in an education, establish themselves in a job, marry a sensible and hardworking person, and only then begin to think about having a baby. Many poor people do these things, of course, and so do many poor teenagers. But the deck is stacked against people at the lower levels of a world in which the job distribution has been hollowed out. People who lack an education are less well off than ever before, and thus find it ever harder to maintain a marriage and support a family. Even if they work at one or more of the "lousy jobs" at the bottom of the wage structure, full-time year-round employment is insufficient to keep a family out of poverty. . . .[42] The idea that young people would be better off if they worked harder, were more patient, and postponed their childbearing is simply not true—and is unlikely to become true in the foreseeable future—for a great many people at the bottom of the income scale. Even when poor people obtain more education, for example, they only displace other people at the end of the queue, and the problem of poverty and childbearing among young people continues.

A compelling body of scholarship now shows that although people who become parents as teenagers will eventually be poorer than those who do not, a very large proportion of that difference is explained by preexisting factors. Well over half of all women who give birth as teenagers come from profoundly poor families, and more than one-fourth come from families who are slightly better off but still struggling economically. Taken together, more than 80 percent of teenage mothers were living in poverty or near-poverty long before they became pregnant.[43] Teenage parents are not middle-class people who have become poor simply because they have had a baby; rather, they have become teenage parents because they were poor to begin with. More than two decades of research, summarized in the National Academy of Sciences' report *Risking the Future* (1987), make clear a point not highlighted in the report itself: at every step of the process that leads to early childbearing, social and economic disadvantage plays a powerful role. Poor kids, not rich ones, have babies as teenagers, and their poverty long predates their pregnancy. By the same token, poor kids, not rich ones, have babies without being married. In part this is also a product of the hollowing out of the income structure. Low-wage jobs rarely pay enough to support a family if only the father works; and if both parents work, they are likely to face daunting childcare problems—problems exacerbated by the fact that such jobs are often episodic, with unpredictable hours and swing shifts.

In addition, conservatives are right: AFDC as it is structured in most states exacts a subtle, and in some cases not so subtle, marriage penalty. Only about half the states permit AFDC funds to go to families in which the father lives in the home, and these programs (known as AFDC-UP) are open only to unemployed men who have an employment record—a difficult criterion to satisfy in communities where unemployment can run to more than 50 percent and where many of the low-wage jobs do not meet the eligibility criteria for AFDC-UP. As a result, two-parent families make up only 7 percent of AFDC cases. Still, states that permit poor fathers married to poor mothers to obtain AFDC seem to have higher rates of marriage than do states without AFDC-UP, suggesting that AFDC itself penalizes people for getting married.[44]

But if teenage mothers are poor before they ever become mothers, if in many cases they would be poor and in need of welfare at whatever age they had their first child, and if marriage brings its own set of problems to poor people, much of the easy equation that identifies early pregnancy as a cause of poverty breaks down. If the real problem is poverty, not the age or marital status of young women when they give birth, then it is not surprising that poor women tend to have children and even grandchildren who grow up to be poor. Preventing teenagers from getting pregnant and persuading them to delay their childbearing would merely postpone the problem of poor women and their dependence on welfare. Childbearing among teenagers has relatively little effect on the levels of poverty in the United States. But income disparities have become a pervasive fact of American life, and it is scarcely surprising that when experts in the 1970s labeled "teenage pregnancy" a fundamental cause of poverty, Americans were willing to listen.

NOTES

1. For example, see William Bennett, *The Index of Leading Cultural Indicators: Facts and Figures on the State of American Society* (New York: Simon & Schuster, 1994).
2. Although much of the rhetoric on the Right is about "children," conservatives and even many liberals think of pregnancy among teenagers as something fundamentally affecting "girls" or young women. The issue is usually framed in such a way that half of the people involved—namely, young men—are excluded, and this selectivity is an enormous handicap in the effort to find a solution. As we will see, thinking about the problem in terms of two sexes rather than one opens up a number of new possible solutions.
3. For an overview, see U.S. House of Representatives, 99th Congress, Select Committee on Children, Youth and Families, "Teen Pregnancy: What Is Being Done? A State-by-State Look" (Washington, D.C.: Government Printing Office, 1986); Charles Stewart Mott Foundation, *A State-by-State Look at Teenage Childbearing in the United States* (Flint, Mich.: Charles Stewart Mott Foundation, 1991); Gloria Magat, ed., *Adolescent Pregnancy: Still News in 1989* (New York: Grantmakers Concerned with Adolescent Pregnancy, Women and Foundations/Corporate Philanthropy, 1989); Junior League, *Teenage Pregnancy: Developing Life Options* (New York: Association of Junior Leagues, 1988). For the National Urban League's program with Kappa Alpha Psi, see Cheryl Hayes, ed., *Risking the Future: Adolescent Sexuality, Pregnancy, and Childbearing* (Washington, D.C.: National Academy Press, 1987), vol. 1, p. 178.
4. Prior to the mid 1970s, pregnant teenagers were treated by the media as a subset of "school-age mothers" or of the larger set of "unwed mothers." See *Reader's Guide to Periodic Literature,* 1968–1994. A tabulation of these stories by title and content has been compiled by Kristin Luker.

5. In 1955, out of every thousand adolescent women of all races, 90 gave birth. By 1975 the rate had fallen until it was approximately equal to that of 1915: 60 per thousand. And by 1985 it had declined even further, to only 50 per thousand. Interestingly, the fertility of teenagers has always been remarkably similar to that of older women; the birthrates for both groups rise and fall in tandem. (The similarities are most marked, of course, between the rates for teens and the rates for women who are just a little older—twenty to twenty-four.) Clearly, the fertility of American women tends to respond to large, society-wide forces. See National Center for Health Statistics, *Advance Report of Final Natality Statistics* (Hyattsville, Md.: Public Health Service, various years).
6. Robert L. Heuser, *Fertility Tables for Birth Cohorts by Color: United States, 1917–1973,* DHEW Publication no. (HRA) 76-11182 (Rockville, Md.: National Center for Health Statistics, 1976); National Center for Health Statistics, *Advance Report of Final Natality Statistics,* 1987 (Rockville, Md.: National Center for Health Statistics, 1989), vol. 38, no. 3. Even the post-1988 upturn in birthrates among teenagers is still within the range of historical fluctuation, although whether this will continue to be so is uncertain.
7. In 1973, among teenage women of all races, 60 out of every thousand gave birth and 21 per thousand had abortions; thus, a total of 81 out of every thousand were becoming pregnant. In 1980, in contrast, the rate of live births was 52 per thousand and the abortion rate had more than doubled, to 44 per thousand; the pregnancy rate had thus increased to 96 per thousand.
8. For an overview, see *Statistical Abstract of the United States* (Washington, D.C.: Government Printing Office, 1993), Table 101, "Births to Unmarried Women, by Race of Child and Age of Mother, 1970–1990"; U.S. Center for Health Statistics, *Vital Statistics of the United States,* various years; idem, *Monthly Vital Statistics,* various years.
9. This has led to a set of new social practices unanticipated by Emily Post. People now speak of "my baby's father" or "my baby's mother." One proud father even placed a notice in his local paper announcing that his fiancée had just given birth to their baby (I am indebted to Sheldon Messinger for this information). In the late 1980s commentators did begin to take note of the rising rate of out-of-wedlock births in general; but even within this broader context, experts and the media still focused on teenage mothers.
10. Larry Bumpass and James A. Sweet, "Children's Experience in Single-Parent Families: Implications of Cohabitation and Marital Transition," *Family Planning Perspectives* 21 (November–December): 256–260.
11. In 1986 polls revealed that more than 84 percent of Americans considered pregnancy among teenagers a "major" problem facing the country. Harris poll for PPFA, 1985. See also Roper Report 86-3, 1986 R37XE.
12. Some people, among them demographers such as Phillips Cutwright and polemicists such as Charles Murray, argue that the proportion or ratio of out-of-wedlock births is much more important than the rate. In demographic terms, a "rate" is an event that is standardized over a specified population for a particular period of time. Thus, the birthrate is defined as the number of births (the numerator) per thousand women aged fifteen to forty-four (the denominator) in a year. But many commentators speak of the "illegitimacy rate" or the "abortion rate" when what they really have in mind is a proportion or ratio, a figure that compares two sets of *events* rather than an event to a population. What many people call the "illegitimacy rate" is really a measure that compares the number of out-of-wedlock births (the numerator) to the total number of births (the denominator). The problem here is that there can be wide fluctuations in *both* of the events being charted, and these fluctuations can lead to dramatic changes in the measure. (Populations fluctuate, too, of course, but much less sharply.) The illegitimacy *rate* (the number of out-of-wedlock births per thousand unmarried women aged fifteen to forty-four) went from 25.4 in 1970 to 43.8 in 1990, an increase of about 70 percent, while the illegitimacy *ratio* (the proportion of out-of-wedlock births to legitimate births) went from 11 percent to 28 percent of all births during that same period, an increase of more than 250 percent. The dramatically larger increase in the ratio, compared to the increase in the rate, was due to an increase in the propensity of American women to bear children out of wedlock, and, simultaneously, a declining propensity to bear children in wedlock. Among African Americans, virtually all of the increase in the illegitimacy ratio was due to declining marital fertility (the denominator), and in fact illegitimacy rates for African American women declined for most of the 1970–1990 period. As Cutwright says, what the majority of a cohort is doing matters. Still, commentators tended to emphasize troubling statistics (changes in the proportion of babies born out of wedlock) over more comforting ones

(such as the decreases in the incidence of pregnancy per sexually experienced woman and in the rate of out-of-wedlock births among African Americans).

13. One could make the case that this *was* the real story: the fact that birthrates among very young women had not changed much. Since the period of childhood had gradually lengthened in the course of the nineteenth and twentieth centuries, one would have expected a reduction in births to very young women. Birthrates among fourteen-year-olds for the calendar years 1925–1990 were as follows:

1925	3.9 per thousand
1930	3.8
1935	3.7
1940	3.8
1945	3.9
1950	5.8
1955	6.1
1960	6.0
1965	5.2
1970	6.6
1975	7.1
1980	6.5
1985	6.2
1990	7.8

Source for 1925–1970: Heuser, *Fertility Tables for Birth Cohorts by Color,* "Central Birth Rates for All Women during Each Year 1917–73 by Age and Live-Birth Order for Each Cohort from 1888 to 1959," p. 37, Table 4a. Source for 1975–1990: *Vital Statistics of the United States: Natality,* "Central Birth Rates by Live-Birth Order, Current Age of Mother, and Color for Women in Each Cohort," p. 1-32, Table 1-16 (1975); p. 1-42, Table 1-18 (1980); p. 1-36, Table 1-18 (1985); p. 1-45, Table 1-19 (1990).

14. The classic example is Stanley Cohen, *Folk Devils and Moral Panics* (Oxford: Basil Blackwell, 1987). For another view, one that is more in line with the position presented here, see John Kingdom, *Agendas, Alternatives, and Public Policies* (Boston: Little, Brown, 1984).

15. This does not imply that stories told by advocates are necessarily right. Indeed, as in this case, advocates typically confront contradictory data and must strive to make sense of them long before the whole pattern of the phenomenon is clear. On the issue of pregnancy among teenagers, advocates and policymakers were wrong in several important respects, and their errors had profound implications for social policy.

16. Rosalind Petchesky has made the astute point that social scientists often speak of "revolutions" when only white and middle-class behavior has changed. See Petchesky, *Abortion and Women's Choice: The State, Sexuality and Reproductive Freedom* (Boston: Northeastern University Press, 1990).

17. John D'Emilio and Estelle Freedman, *Intimate Matters: A History of Sexuality in America* (New York: Harper & Row, 1988).

18. Calculated from Sandra L. Hofferth, Joan R. Kahn, and Wendy Baldwin, "Premarital Sexual Activity among U.S. Teenage Women over the Past Three Decades," *Family Planning Perspectives* 19, no. 2 (1987): 46–53.

19. Alfred Kinsey et al., *Sexual Behavior in the Human Female* (Philadelphia: W. B. Saunders, 1953), p. 336, Table 78.

20. Melvin Zelnik, "Sexual Activity among Adolescents: Perspectives of a Decade," in E. R. McAnarney. ed., *Premature Adolescent Pregnancy and Parenthood* (New York: Game and Stratton, 1983); Melvin Zelnik and F. K. Shah, "First Intercourse among Young Americans," *Family Planning Perspectives* 15 (1983): 64–70.

21. American Institute of Public Opinion (AIPO), Gallup Poll, July 1969 (sex before marriage); idem, May 1969 (nudes in magazines); idem, May 1969 (Broadway shows). See Mayer, *Changing American Mind,* p. 385.

22. William G. Mayer, *The Changing American Mind: How and Why American Public Opinion Changed between 1960 and 1988* (Ann Arbor: University of Michigan Press, 1992), p. 385.

23. Floris W. Wood, *An American Profile: Opinions and Behavior, 1972–1989* (New York: Gale Research, 1990), p. 597.

24. These laws were the modern-day remnants of the ones that women reformers had campaigned for in the nineteenth century. When they were challenged in 1981 as a form of reverse discrimination (because a boy under the age of consent was charged with criminal sanctions for having sex with a girl his own age), they were legitimated by the Supreme Court—as a remedy for pregnancy among teenagers! The Court's reasoning was that statutory-rape penalties would discourage teenage men from having sex, just as the risk of pregnancy would discourage young women. Overall, a most curious case. It shows how profoundly the concept of an "epidemic" of teenage pregnancy had permeated judicial thinking at the highest levels. See *Michael M. v. Superior Court of Sonoma County,* 101 S. Ct. 1200 (1981).

25. Yankelovich, Skelly, and White, poll reported in *Public Opinion Online*, LEXIS, Market Library, R-Poll File, Accession no. 0132089 (1977). National Opinion Research Center, poll reported in *Public Opinion Online*, LEXIS, Market Library, R-Poll File, Accession no. 0092411 (1994).
26. Estimated from James Trussell and Barbara Vaughan, "Selected Results Concerning Sexual Behavior and Contraceptive Use from the 1988 National Survey of Family Growth and the 1988 National Survey of Adolescent Males," Office of Population Research Working Paper Series, Princeton University, working paper no. 91-12 (September 1991).
27. See Melvin Zelnik and John Kantner, "Sexual Activity, Contraceptive Use and Pregnancy among Metropolitan Area Teenagers, 1971–1979," *Family Planning Perspectives* 12 (1980): 230; Kathleen Ford, *Sex and Pregnancy in Adolescence* (Beverly Hills, Calif.: Sage, 1981).
28. Hofferth et al., "Premarital Sexual Activity," p. 49; Jacqueline D. Forrest and Susheela Singh, "The Sexual and Reproductive Behavior of American Women, 1982–1988," *Family Planning Perspectives* 22 (1990): 206–214; Trussell and Vaughan, "Selected Results Concerning Sexual Behavior."
29. Robert Hatcher and his colleagues estimate that the average interval between menarche (first menstruation) and marriage is 13.5 years for all teens. See Robert A. Hatcher et al., *Contraceptive Technology* (New York: Irvington, 1994), p. 131. On menarche, see Phyllis B. Eveleth, "Timing of Menarche: Secular Trend and Population Differences," in Jane Lancaster and Beatrix Hamburg, eds., *School-Age Pregnancy and Parenthood: Biosocial Dimensions* (New York: Aldine, 1986), pp. 39–52. See also J. D. Forrest, "Timing of Reproductive Life Stages," *Obstetrics and Gynecology* 82 (1993): 105–111; and A. F. Saluter, "Marital Status and Living Arrangements," *Current Population Reports*, Series P-20, no. 461 (March 1991). Saluter gives estimates for the length of time between first *intercourse* and marriage. Data on *male* sexual maturity are scattered and contradictory.
30. Melvin Zelnik and John Kantner, "Sexual and Contraceptive Experience of Young Unmarried Women in the United States, 1971 and 1976," *Family Planning Perspectives* 9 (1977): 55–73. See also K. Kost and J. D. Forrest, "American Women's Sexual Behavior and Exposure to Sexually Transmitted Diseases," *Family Planning Perspectives* 24 (1992): 244–254; and Trussell and Vaughan, "Selected Results."
31. J. D. Forrest and S. Singh, "The Sexual and Reproductive Behavior of American Women, 1982–1988," *Family Planning Perspectives* 22 (1990): 206–214; Trussell and Vaughan, "Selected Results Concerning Sexual Behavior and Contraceptive Use."
32. According to the Kinsey data, which are not nationally representative and in which middle-class people are overrepresented, in the 1940s and 1950s 18 percent of women who obtained only a grade school education (education being a proxy for class) had been sexually active before marriage and prior to the age of fifteen, compared to only 1 percent of the women who went on to college or graduate school. See Kinsey et al., *Sexual Behavior in the Human Female*, p. 295.
33. For an overview of the social construction of symbolic boundaries, see Michele Lamont and Marcel Fournier, *Cultivating Differences: Symbolic Boundaries and the Making of Inequality* (Chicago: University of Chicago Press, 1992), pp. 1–17.
34. Ira Reiss, *Premarital Sexual Standards in America* (New York: Free Press, 1960).
35. Harris Poll for PPFA, 1985; Yankelovich/Time/CNN, 1990; General Social Survey, 1983 (access to birth control information); Contemporary American Family, September 18, 1981, p. 9F (access to birth control devices). In the language of rational-choice theory, adults' preferences on how to deal with sexual activity among teens are "nontransitive."
36. ABC–Washington Post Poll, 1990; Yankelovich/Time/CNN, 1990; Gallup Poll, *Newsweek*, February 1987.
37. Harris Poll for PPFA, 1985; "Speaking of Kids: A National Survey of Children and Parents," September 17, 1990; Gordon S. Black, *U.S.A. Today*, June 1986.
38. Yankelovich, 1978, Health Survey.
39. Louis Harris for PPFA, 1985.
40. Charles Murray, *Losing Ground: Social Policy, 1950–1980* (New York: Basic Books, 1984); David Ellwood, *Poor Support: Poverty in the American Family* (New York: Basic Books, 1988).
41. This framing of the problem admittedly begs a second question—namely, why teenagers don't wait until they are married to have children, so that a husband rather than welfare could support them and their babies. The answer to this is complex and draws on all three of the revolutions cited here: the new sexual mores, the increase in childbearing outside marriage, and the dramatic changes in the world economy. . . .

42. See Gary Burtless, ed., *A Future of Lousy Jobs? The Changing Structure of U.S. Wages* (Washington, D.C.: Brookings Institute, 1990).
43. Alan Guttmacher Institute, *Sex and America's Teenagers* (New York: Alan Guttmacher Institute, 1994), p. 58.
44. Robert Moffitt, "Incentive Effects of the U.S. Welfare System: A Review," *Journal of Economic Literature* 30 (1992): 1–61.

Risking Pregnancy: Avoidance, Ignorance, and Delay of Contraceptive Use

Ellen W. Freeman and Karl Rickels

Introduction

Each year in the United States, about 4.5 million female teenagers obtain family planning services (Horn & Mosher, 1984). Although some young women seek these services before or within the same month as initial intercourse, about three-fourths delay an average of 23 months after initial intercourse (Mosher & Horn, 1988). For too many teenagers, pregnancy occurs during this interval of delay: more than one-third of a teenage sample made their first family planning visit because they thought they were pregnant (Zabin & Clark, 1981). More than one-fifth of premarital pregnancies occurred in the first month after initial intercourse, and half occurred within the first 6 months (Zabin, Kantner, & Zelnik, 1979). These facts prompt two questions:

1. Why do teenagers delay obtaining contraceptive services, particularly when they do not want pregnancy?
2. Are there differences in information and attitudes about contraception between teenagers who obtain services before or after pregnancy occurs?

Why female teenagers wait so long to use family planning services is a continuing puzzle. Public family planning programs are widely available (Horn & Mosher, 1984). The length of delay is nearly identical in white and black women, and background variables fail to explain why contraceptive services are not obtained more immediately (Mosher & Horn, 1988). Teenagers say that they do not think they will get pregnant, but even this faulty perception was not among the leading reasons for delaying the first family planning visit. The predominant responses given by teenagers when asked why they did not obtain contraceptive services were "just didn't get around to it," "afraid my family would find out," and "waiting for a closer relationship with my partner" (Zabin & Clark, 1981). These responses suggested simple procrastination, but they also implied psychological issues of ambivalence, guilt, and romantic notions of sex that deny the need to prepare for sex.

The documented delay in obtaining contraception has been difficult to change, particularly among teenagers, for reasons that are not well understood. Although psychological factors might help explain sexual behavior and contraceptive use, they are not readily amenable to programmatic change. Relatively few studies have examined psychological factors, and none has provided strong predictors of contraceptive use or teenage pregnancy.

Cobliner (1974) studied a group of teenagers who had terminated unwanted pregnancy and identified reasons that their intentions to avoid pregnancy had failed. The largest single group (43%) were "risk-takers," whose reasons for not using contraception were "I didn't think I could get pregnant" or "I didn't expect to have sex." The researchers viewed these responses as a probabilistic appraisal of risks, "a protective psychological mechanism . . . that is commonly practiced to make life more bearable" (p. 24). The researchers noted that the explanation was too frequent to be ascribed to emotional difficulty and that it was not correlated with educational level.

Luker (1975) described risk-taking behavior in relation to sex and contraception by weighing the costs of contraception against the benefits of pregnancy. The cost–benefit analysis illustrated that women who took a successful risk then faced the next contraceptive decision with the fact that they had not become pregnant. This weighted subsequent decisions about contraceptive use further toward risk-taking, because they had risked successfully and obtained little new information.

Other psychological studies assessed teenagers' lack of ability to plan ahead or anticipate future events (Keller, Sims, Henry, & Crawford, 1970; MacDonald, 1970; Mindick, Oskamp, & Berger, 1977; Rader, Bekker, Brown, & Richard, 1978; Rovinsky, 1972). A common theme of these studies is that preventing pregnancy is an abstraction that requires a more mature level of cognitive development. Concrete thought and the recognition and processing of directly experienced events characterize young people. Less than 75% of 15-year-olds exhibited the ability to understand abstractions in a study of U.S. teenagers (Dale, 1970). A concrete thinker does not consider an abstract future and does not link future consequences to present events. A concrete thinker is unlikely to weigh the possibilities of alternative behaviors ("If I do this rather than that, then this rather than that will happen"). Researchers generally concluded that teenagers risked pregnancy not because of "any form of pathology, moral or otherwise" but because of normal factors of adolescent development (Cvetkovich & Grote, 1980). They also indicated that pregnant teenagers did not lack intelligence, but lacked reasoning ability to analyze abstract issues realistically (Kreipe, Roghmann, & McAnarney, 1981).

Little evidence suggests that emotional distress or psychopathology are significant in teenage sexual behavior. Intercourse experience even in early adolescence is no longer exceptional. Studies using the Minnesota Multiphasic Personality Inventory (MMPI), a measure to identify neurotic personality concepts, failed to identify differences between pregnant and nonpregnant teens (Brandt, Kane, & Moan, 1978; Gispert & Falk, 1976; Kane, Moan, & Bolling, 1974).

However, the effects of common emotional factors, such as anxiety and depression, and related psychological concepts, such as self-esteem, remain unclear in relation to teenage sexual behavior and pregnancy. Gabrielson, Klerman, Currie, Tyler, and Jekel (1970) identified an increased rate of suicide among pregnant teenagers. Others suggested that depression in teenagers commonly is unrecognized (Inamdar, Siomopoulos, Osborn, & Bianchi, 1979) and consequently its frequency and effects are not evaluated....

The Penn Study

We assumed that our teenage sample, which had experienced intercourse before study enrollment, was no different from the many other reported samples that showed limited information about contraceptive use and delay in family planning enrollment. This clinical sample, which compared never-pregnant contraceptive users and pregnant teenagers who chose abortion or delivery of a first pregnancy, would address the following question ...:

... Did the three study groups differ in the levels of contraceptive information, sexual attitudes, and the length of time after first intercourse before obtaining family planning services? ...

Sexual Behavior

The teenagers in the Penn Study had all started dating: 30% at age 12 or younger, 49% at ages 13 to 14, and 21% at ages 15 to 16. About 40% had been dating their current boyfriend for more than 1 year, nearly 50% had been with their current boyfriend for less than 1 year, and 10% said they had no boyfriend when they entered the study.

The mean age at first sexual intercourse was 14.2 years, nearly identical in all three study groups. About 10% first had sex at age 12 or younger, 74% at ages 13 to 15, and 16% at ages 16 to 17. At study enrollment, nearly one-fourth of all the teenagers said they had had sex only a few times, but 72% reported having sex more than five times, an indication that it was more than a "one-time" experience. Nevertheless, sporadic rather than regular sexual activity was characteristic: 42% said they had no sex in the past month, 44% had sex one to five times, and only 14% had sex more than five times in the past month.

Contraceptive Use At the initial interview, 9% of the abortion group and 63% of the delivery group said they had never used contraception. In contrast, 81% of the never-pregnant teenagers (all of whom were enrolled in family planning) reported using contraception the last time they had had sex. Almost all (89%) of the teenagers who became pregnant (both the abortion and delivery groups) said they were not using contraception when pregnancy occurred. Of those who said they were using contraception when they became pregnant, 2% said they were using oral contraceptives but missed taking some pills, 6% said their method failed, and 3% said they didn't know why pregnancy occurred. Although these latter numbers describe contraceptive failure, most of the teenagers simply said they were not using any method when they became pregnant.

Delay After Initial Intercourse One-fourth of the teenagers said that some contraceptive method was used at the first sexual intercourse, but this was nearly three times more likely in the never-pregnant group (44%) than in the abortion and delivery groups, where only 16% in each group said they used contraception initially ($p < 0.001$). About 20% began using contraception up to 3 months after initial intercourse, 6% began using it 4 to 6 months later, 13% used it 7 to 12 months later, and 8% used it 13 to 48 months later. About 28% had never used any contraceptive method before pregnancy occurred.

Contraceptive Information

Many teenagers have only limited and often incorrect information about reproduction and contraception (Freeman, Rickels, Huggins, Mudd, et al., 1980; Zelnik & Kantner, 1977). This puts them at high risk for pregnancy because they believe they know how to prevent pregnancy when in fact they do not. The teenagers in this study were no exception. At study enrollment the teenagers were asked to identify the most fertile time in the menstrual cycle, a question that is particularly important because many teenagers believe they can avoid pregnancy by having intercourse during their "safe" (nonfertile) time. Only 28% knew the most fertile time in the cycle at their first study interview. More than half (56%) answered incorrectly, and 16% said they did not know.

Two-thirds of the teenagers said pregnancy was extremely likely if they had intercourse without using contraception. However, the other one-third—including many who already had become pregnant—said pregnancy was *not* a very likely outcome of unprotected intercourse. Perhaps realistically, the teenagers did not express much confidence in their own use of contraception. Only 38% said that if they used contraception, it was "extremely likely" to prevent pregnancy. The rest said it was "somewhat likely" (24%), "not at all likely" (9%), or "didn't know" (29%). Even never-pregnant teenagers who received family planning services were uncertain: Only 61% said that by using contraception they were "extremely likely" to prevent pregnancy, compared to 45% in the abortion group and only 16% in the delivery group ($p < 0.0001$). Although the responses in the pregnant groups may have reflected their experiences of having become pregnant, many of these teenagers also said they had never used contraception at all, and nearly all said they were not using contraception when pregnancy occurred. Rather, many teenagers, including those in family planning services, appeared to lack sufficient information to use contraception effectively.

Contraception to these teenagers almost invariably meant oral contraceptives, or "the pill." Taken daily, the pill's theoretical effectiveness exceeds 99% (Garcia & Rosenfeld, 1977). Nonetheless it failed to prevent pregnancy for many teenage users. Teenage sexual activity often is sporadic, defined by shifting relationships and weekend social activity. Occasional or unpredictable intercourse may make daily pill-taking seem unnecessary and difficult to remember. Missed pills—even just a few in a monthly cycle—increase the risk of pregnancy. At study enrollment, one-fourth of the teenagers in the delivery and abortion groups did not know that pregnancy was possible if just a few pills were missed during the cycle. Also, starting the pill after intercourse (a common practice among teenagers) cannot prevent a pregnancy that has already occurred.

The teenagers' knowledge of contraceptive methods other than oral contraceptives was meager. Only 25% had used condoms, 36% knew about them but had never used them, and 39% had no information about them. Only 15% reported ever using spermicidal foam, 33% said they knew about it but had not used it, and 52% had no information about it. Most disturbing of all, fully 29% of these sexually active teenagers—many of them pregnant and expecting to carry to term—said they had never used *any* contraceptive method, including limiting intercourse to their "safe times" and withdrawal, and appeared to have little or no knowledge about contraception.

Access to Contraception The teenagers' knowledge of where to obtain contraceptives was greater than their reported use. Overall, 88% identified a place that they could obtain oral contraceptives, and 66% knew where to obtain an abortion or counseling for an unwanted pregnancy. (Omitting the abortion group, 63% of the never-pregnant teenagers and 47% of the delivery group knew where they could obtain an abortion—a significant difference [$p < 0.05$].) In addition, a high percentage knew where to obtain condoms (82%), intrauterine devices (78%), and spermicidal foam (79%). These data do not tell us what the teenagers knew before they became pregnant, but even at this time, the never-pregnant and abortion group teenagers had more information about where they could obtain contraceptives, abortions, or counseling for unwanted pregnancy than did those in the delivery group.

Sources of Information We asked the teenagers who had contraceptive information (71% of the total) where they had first obtained it. The most frequent response was that contraceptive information had been obtained from their mothers (22%) or another family member (23%). The never-pregnant and abortion group teens were more than twice as likely as the delivery group teenagers to have obtained contraceptive information from their mothers ($p < 0.05$). Although mothers were the primary source of contraceptive information, the numbers were small. Less than half the teenagers with any contraceptive information (and less than one-third of all the teenagers in the study) had obtained contraceptive information from family members, even though the family is widely viewed as the appropriate source of sex education. Other sources combined—friends (18%); school classes (18%); a clinic or doctor (13%); and books, television, and other media (6%)—were more likely to be the initial source of contraceptive information for these teenagers than were family members.

Because mothers were the single leading source of contraceptive information for these teenagers, we further explored the range of sexual topics that mothers and daughters discussed. At study enrollment, 92% of the teenagers said they had talked about sex with their mothers. (The daughter's pregnancy may have spurred such discussion *after* the pregnancy occurred.) However, the specific topics discussed varied greatly. Menstruation was the most common topic (discussed by 79% of the teenagers and their mothers), followed by contraception (discussed by 64%), pregnancy (discussed by 62%), and body changes (discussed by 59%). The delivery-group teenagers were the *least* likely to have ever talked with their mothers about pregnancy compared to the abortion- and never-pregnant-group teenagers (55%, 63%, and 73%, respectively; $p < 0.05$).

More behaviorally specific topics were discussed even less frequently. Abortion was discussed by about half the sample—predominantly by those who had recently had an abortion (74%), but *least* by the delivery-group teenagers (40%)—again a significant difference ($p < 0.001$). Where to obtain contraceptives (50%), how to use contraceptives (40%), and how to avoid sexually transmitted diseases (37%) were the topics least frequently discussed by the mothers and daughters. Contraceptive use had been discussed more in the never-pregnant and abortion groups than in the delivery group (47%, 45%, and 31%, respectively; $p < 0.05$). The delivery-group teenagers clearly were less likely to have talked with their mothers beyond the most general aspects of these sexual behavior issues.

Although nearly all of the teenagers reported talking with their mothers about sex, the data show the considerable range in the specific topics discussed. Talking about menstruation and body changes clearly is important for a young girl entering puberty, but this information does not provide guidance for preventing pregnancy. Talking about birth control may provide information, but does not necessarily mean that a young teenager adequately understands where to obtain and how to use contraceptive methods. Although these were all sexually active teenagers, only a minority reported any communication with their mothers about contraceptive use and sexually transmitted diseases. The teenagers who were planning to deliver were the least likely to have talked about pregnancy, contraception, and abortion with their mothers.

Sexual Information and Attitudes

In addition to answering interview questions, the teenagers completed a 14-item self-report Sexual Information and Attitudes Questionnaire (SIAQ) that we used in high school classes and in other family planning samples (Freeman, Rickels, Huggins, Mudd, et al., 1980)....

Sexual Information At study enrollment the SIAQ results indicated that the never-pregnant and abortion-group teenagers had more contraceptive information than the delivery-group teenagers ($p < 0.05$). However, only three items on this questionnaire discriminated among the three study groups with significantly different responses ($p < 0.05$). More never-pregnant teenagers (89%) than abortion (79%) or delivery (75%) teenagers knew that contraception was needed even with infrequent sex. More never-pregnant teenagers (88%) than abortion (78%) or delivery (73%) teenagers knew that missed pills during the cycle could result in pregnancy. And more never-pregnant (71%) and abortion (78%) compared to delivery teenagers (63%) agreed that it was *not* "showing more love" to risk unprotected sex. Expectedly, the never-pregnant teenagers, who were enrolled in family planning, were more likely to have this information, and the two groups of teenagers who became pregnant had less information.

Although the remaining SIAQ items did not elicit differential responses from the three study groups, some items are noteworthy for responses that described the limited basis these teenagers had for preventing pregnancy. The teenagers' responses overall are shown in Table 1.

TABLE 1 *Sexual Information and Attitudes of Female Teenagers at Study Enrollment,* n = *326*

	% Agree
1. Pregnancy can occur after unprotected sex even if it did not occur the first time.	87
2. If sex is less than once a week, pregnancy could occur.	80
3. Birth control needs to be used even if sex is infrequent.	80*
4. A girl can get pregnant the first time she has sex.	80
5. Birth control methods (except sterilization) can be stopped when pregnancy is wanted.	79
6. If the pill is missed a few days, a girl can get pregnant.	79*
7. Teenagers do not need parental consent to obtain birth control.	76
8. Sex without birth control does not show more love.	71*
9. Pregnancy can occur without orgasm.	65
10. It is difficult to know when the "safe" time is.	60
11. It is not too much trouble to prevent pregnancy.	56
12. The most fertile time is about 2 weeks before the period begins.	55
13. Boys respect girls who use birth control.	49
14. A girl will not feel "used" if her boyfriend knows she uses birth control.	41

Note: *$p < 0.05$ in chi-square tests of never-pregnant, abortion, and pregnant groups. All items are stated here to present "agree" as the response supporting contraceptive use and are listed in order of percent agreement. Items were stated on the questionnaire such that both "agree" and "disagree" responses supported contraceptive use.

Only about half could correctly identify the most fertile time in the cycle—a major problem when they think it is a "safe" time to have sex. Their responses to the attitudinal items suggested their ambivalence about contraceptive use. More than half agreed they felt "used" if their boyfriends knew they used contraception. Nearly half thought it was "too much trouble" to prevent pregnancy, and half indicated that boyfriends did not respect girls who used contraception. These ambivalent attitudes about contraceptive use, combined with limited knowledge of the methods, would seem to logically and not surprisingly lead to "user failures" in preventing pregnancy.

The sex information scores as measured by the SIAQ increased significantly over time in all study groups ($p < 0.0001$). At the 1-year follow-up assessment, the delivery-group teenagers still had lower scores than the never-pregnant and abortion-group teenagers ($p < 0.05$). By the 2-year follow-up, however, the three study groups had similar scores with no significant differences. The abortion- and delivery-group teenagers enrolled in family planning after their pregnancies, and all three study groups gradually reached about the same level of contraceptive information.

Sexual Attitudes and Feelings The teenagers in the three study groups uniformly endorsed a set of attitudes about sex, contraception, and pregnancy, as assessed by the 12-item Sexual Attitudes Questionnaire (SAQ) of Cvetkovich and Grote (1980). The items were not hypothetical to these teenagers, but were directly relevant to their experience, because they all had had sex. The teenagers typically responded in ways consistent with their experience. They agreed with the acceptability of premarital sex with "someone they loved or had good feelings about," but not with someone they didn't "know well." They agreed that "men lie" and women "manipulate" to get sex. Most teenagers agreed that it was "a good idea to experiment sexually before marriage" and did not want to "marry a virgin." They endorsed the male partner's lack of responsibility should pregnancy occur and did not think that marriage was the answer. None of these responses differed among the three study groups.

The above responses can be viewed as expected on the basis that they are consistent with the teenagers' behavior. It was more intriguing to observe that despite their sexual activity, only 37% of the teenagers indicated that it was easy to become sexually excited, and that a majority (63%) indicated that it was difficult to understand their sexual feelings. This suggests that at the emotional level, they were less involved in sexual activity than evidenced by their behavior.

Only two SAQ items had statistically significant differences in responses among the three study groups. First, the abortion-group teenagers (45%) were much more likely than the never-pregnant and delivery-group teenagers (24%) to agree that "having an abortion is a good way to cope with a premarital pregnancy" ($p < 0.001$). Second, the delivery-group teenagers (71%) were the most likely to indicate that it was difficult to understand feelings about sex, compared to never-pregnant (60%) and abortion-group teenagers (56%) ($p < 0.05$).

These sexual attitudes as assessed by the SAQ remained notably stable over the 2-year study. At the 1- and 2-year follow-ups, there were no significant changes in SAQ scores in any study group.

Additional questions were asked at the enrollment interview about what was most important to these teenagers about having sex. About two-thirds reported relatively positive reasons: 34% said they engaged in intercourse mainly because they "wanted to experiment," they enjoyed the "excitement," or they "felt sexy"; 27% said they had sex because they were in love or wanted to please their boyfriend. On the negative side, 26% "didn't know" why they had sex, and 13% blamed alcohol, drugs, or loneliness for their sexual activity. Responses did not significantly differ among the three study groups.

Few of these teenagers viewed their initial sexual experience positively. Only 5% said it was pleasurable, 11% said it made them feel closer to their boyfriend, and 12% said it made them feel more grown up. Fully 72% reported no overall satisfaction or pleasure associated with their initial sexual experience: 38% said they felt nothing, 22% said they didn't know what they felt, and 12% said they felt used or cheapened by the experience. When asked to describe their feelings about their most recent sexual experience, more than half (58%) said it was pleasurable, but 33% were still indifferent, 5% disliked it, and 5% had no answer. Again, these responses did not differ significantly among the study groups.

Only one-third of these teenagers said they had worried about becoming pregnant, and, interestingly, the never-pregnant group using contraception worried least (26% compared to 35% of the abortion group and 50% of the delivery group, $p < 0.001$). Worry appeared related to unprotected intercourse. When asked what they would do if they became pregnant, 35% said they never thought about it, 24% said they would have an abortion, and 41% said they would have the baby. Forty percent of the delivery group compared to 36% of the abortion group and 28% of the never-pregnant group said they had never thought about what they would do if pregnancy occurred ($p < 0.001$).

As another means of evaluating the teenagers' sexual maturity, after each interview the interviewer rated each teenager's overall degree of comfort in discussing sexual feelings and behaviors. At the initial interview, more than half of the teenagers (58%) had ratings below the midpoint of the rating scale, indicating low levels of comfort discussing the sexual topics or difficulties in talking about abstract feelings or decisions. Typical of these responses is that many of the teenagers stated that they "didn't know" what they thought about having sex or what they would do if they became pregnant.

The ratings became increasingly more positive during the 2 years of the study ($p < 0.001$ for the time effect). A significant interaction between group and time ($p < 0.05$) signified that the delivery-group teenagers, who had the lowest ratings initially, had the greatest improvement during the study. At endpoint there were no differences among the three study groups in these ratings that reflected feelings about discussing sexual issues and behavior.

It is noteworthy that although the mean ratings increased throughout the study, they remained below the midpoint in all study groups. This was consistent with the young ages and mid-adolescent development of these teenagers. Despite their sexual activity, most were not comfortable talking about sex and could not formulate future plans and goals. These global ratings do not detail the complexity of psychological growth and development, but they are descriptive of teenagers who were involved in but not comfortable with sexual activity. Although the teenagers had sex, many indicated little understanding of their feelings or behavior....

Future Orientation Teenagers commonly have difficulty making decisions about abstract situations. To most of them, pregnancy and parenthood are mere abstractions. Even those who become pregnant may have no specific plans related to childbearing issues. They may be intelligent but lack the reasoning ability to analyze and process abstract, future-oriented issues realistically.

From this perspective, the interviewer rated how well each teenager appeared to connect her sexual behavior and future goals and to formulate plans for pregnancy, child care, and related issues if pregnancy should occur. These ratings of future orientation significantly differed ($p < 0.0001$) among the three study groups at each assessment. Throughout the study, the never-pregnant and abortion groups had similar ratings, and the delivery group had the lowest ratings. The ratings significantly increased during the 2-year study in all groups ($p < 0.0001$ for the time effect), an expected result of developing with age and experience. However, having a child did not differentially affect this development. All study groups had higher ratings at the end of the 2 years, but the delivery group did not reach the same level of planning for future goals as the teenagers who avoided childbearing.

Summary

The data from these teenagers thus far describe few differences among the three study groups in terms of sexual behavior, information, [and] attitudes, ... although the differences that do occur show a consistent pattern that describes the delivery-group teenagers as the least prepared for sexual activity and childbearing.

Did sexual behavior predict pregnancy risk? At study enrollment, the three groups appeared extremely similar in terms of their sexual behavior. The age at first sex (14.2 years), the reported frequency of sex (which typically was sporadic), and the length of the relationship with the current boyfriend (which was less than 1 year) were comparable. However, the never-pregnant teenagers were nearly three times more likely to have used contraception at the first intercourse experience. By sampling definition, the entire never-pregnant group had used contraception, but 63% of the delivery group had never used any contraceptive method. Overall the teenagers shared a similar risk of pregnancy in terms of their sexual activity, but for many pregnancy occurred before they obtained contraceptive services.

Did information and attitudes about contraception differ between the teenagers who did and did not have pregnancies? Lack of information is often cited as a cause of unintended pregnancies, but there is little evidence of relationship between contraceptive knowledge and contraceptive behavior (Chilman, 1979) or between sex education and premarital teenage pregnancy (Marsiglio & Mott, 1986). Among these teenagers, there were initial differences in contraceptive information, but they started from different baselines. The never-pregnant group, which was selected from teenagers enrolled in family planning services, had more correct information than the abortion or delivery groups, many of whom had not used contraception prior to pregnancy. But nearly all pregnancies occurred because the teenagers did not *use* contraception, although many knew generally that contraceptives were available. As would be expected, experience using contraception tended to increase information. At the end of the 2-year study, when all the teenagers had used contraception, there were no differences in contraceptive information among the three original study groups.

Noteworthy differences did appear in communication with the mother. The never-pregnant teenagers were nearly three times more likely than delivery-group teenagers to report that they had learned about contraceptives from their mothers. Furthermore, the never-pregnant teenagers were the most likely to say that they had discussed pregnancy, contraceptive use, and abortion with their mothers. Among these teenagers from similar backgrounds, it appeared that those who had more discussions with their mothers about specific sexual issues and behavior were more successful in preventing pregnancy. This is consistent with a recent analysis of national survey data, which found that the family is effective in encouraging sexually active girls to use contraceptives (Casper, 1990).

What of the teenagers' sexual attitudes? Although excitement or pleasure derived from sexual activity, avoidance of consequences, and lack of concern about pregnancy appeared typical of adolescents, these attitudes did not differentiate the three study groups. Only the global ratings, which evaluated the teenagers' overall ability to answer and be comfortable with questions about sex and pregnancy, showed significant differences among the groups, with the delivery group having the lowest ratings at the outset. However, by the 2-year follow-up, the differences disappeared, and the three groups appeared more similar than different....

The delivery group also exhibited the least future orientation as rated by the interviewers at every assessment point, whereas the never-pregnant and abortion groups were more likely to recognize that postponing motherhood was important for achieving other educational and career goals.

REFERENCES

Brandt, C. L., Kane, F. J., & Moan, C. A. (1978). Pregnant adolescents: Some psychosocial factors. *Psychosomatics, 19* (12), 790–793.

Casper, L. M. (1990). Does family interaction prevent adolescent pregnancy? *Family Planning Perspectives, 22*(3), 109–114.

Chilman, C. (1979). Teenage pregnancy: A research review. *Social Work, 24*(6), 492–498.

Cobliner, W. G. (1974). Pregnancy in the single adolescent girl: The role of cognitive functions. *Journal of Youth and Adolescence,* 3(1), 17–29.

Cvetkovich, G., & Grote, B. (1980). Psychological development and the social problem of teenage illegitimacy. In C. Chilman (Ed.), *Adolescent pregnancy and child-*

bearing: Findings from research (pp. 15–41) (NIH Publication No. 81-2077). Washington, DC: Department of Health and Human Services.

Dale, L. G. (1970). The growth of systematic thinking: Replication and analysis of Piaget's first chemical experiment. *Australian Journal of Psychology, 22*(3), 277–286.

Freeman, E. W., Rickels, K., Huggins, G., Mudd, E. H., Garcia, C.-R., & Dickens, H. O. (1980). Adolescent contraceptive use: Comparisons of male and female attitudes and information. *American Journal of Public Health, 70*(8), 790–797.

Gabrielson, I. W., Klerman, L. V., Currie, J. B., Tyler, N. C., & Jekel, J. F. (1970). Suicide attempts in a population pregnant as teenagers. *American Journal of Public Health, 60*(12), 2289–2301.

Garcia, C.-R., & Rosenfeld, D. L. (1977). *Human fertility: The regulation of reproduction*. Philadelphia: F.A. Davis.

Gispert, M., & Falk, R. (1976). Sexual experimentation and pregnancy in young black adolescents. *American Journal of Obstetrics and Gynecology, 126*(4), 459–466.

Horn, M. C., & Mosher, M. D. (1984). Use of services for family planning and infertility, United States, 1982. *Advance Data From Vital and Health Statistics*, No. 102 (DHHS Publication No. [PHS] 85-1250). Hyattsville, MD: National Center for Health Statistics.

Inamdar, S. C., Siomopoulos, G., Osborn, M., & Bianchi, E. C. (1979). Phenomenology associated with depressed moods in adolescents. *American Journal of Psychiatry, 136*(2), 156–159.

Kane, F. J., Moan, C. A., & Bolling, B. (1974). Motivational factors in pregnant adolescents. *Diseases of the Nervous System, 35*(3), 131–134.

Keller, R., Sims, J., Henry, W. K., & Crawford, T. J. (1970). Psychological sources of resistance to family planning. *Merrill-Palmer Quarterly, 16*(3), 285–302.

Kreipe, R. E., Roghmann, K. J., & McAnarney, E. R. (1981). Early adolescent childbearing: A changing morbidity. *Journal of Adolescent Health Care, 2*(2), 127–131.

Luker, K. (1975). *Taking chances: Abortion and decision not to contracept*. Berkeley: University of California Press.

MacDonald, A. P. (1970). Internal-external locus of control and the practice of birth control. *Psychological Reports, 27*(1), 206.

Marsiglio, W. K., & Mott, F. L. (1986). The impact of sex education on sexual activity, contraceptive use, and premarital pregnancy among American teenagers. *Family Planning Perspectives, 18*(4), 151–162.

Mindick, B., Oskamp, S., & Berger, D. E. (1977). Prediction of success or failure in birth planning: An approach to prevention of individual and family stress. *American Journal of Community Psychology, 5*(4), 447–459.

Mosher, W. D., & Horn, M. C. (1988). First family planning visits by young women. *Family Planning Perspectives, 20*(1), 33–40.

Rader, G. E., Bekker, D., Brown, L., & Richard, T. C. (1978). Psychological correlates of unwanted pregnancy. *Journal of Abnormal Psychology, 87*(3), 373–376.

Rovinsky, J. J. (1972). Abortion recidivism. *Obstetrical Gynecology, 39*(5), 649–659.

Zabin, L. S., & Clark, S. D. (1981). Why they delay: A study of teenage family planning clinic patients. *Family Planning Perspectives, 13*(5), 205–217.

Zabin, L., Kantner, J., & Zelnik, M. (1979). The risk of adolescent pregnancy in the first months after intercourse. *Family Planning Perspectives, 11*(4), 215–222.

Zelnik, M., & Kantner, J. (1977). Sexual and contraceptive experience of young unmarried women in the United States, 1976 and 1971. *Family Planning Perspectives*, 9(2), 55–71.

Using Behavioral Theories to Design Abstinence Programs

Kristin A. Moore and Barbara W. Sugland

If adolescents abstain from sex, or stop having sex, the risk of pregnancy is essentially zero and the odds of getting a sexually transmitted disease (STD) are vastly reduced. Regardless of the public health or moral reasons one may have for preferring abstinence as an approach, simple mathematics support the notion of abstinence. Unfortunately, the political and public debate has focused more on whether abstinence is *the* desirable goal and less on figuring out the most effective ways for securing abstinence among adolescents....

To address this issue, we need to answer several initial questions on the basis of theory and previous research. Specifically, we need to consider: (1) what theoretical frameworks are appropriate for influencing sexual behavior among adolescents; (2) which adolescents should be the focus of the intervention; (3) ... what specific program activities or services should be provided....

The Theoretical Framework

... [A] theoretical framework lays out the underlying assumptions regarding the factors that influence behavior (that is, the influence of self-esteem on sexual activity) and identifies how such factors affect that behavior (that is, how diminished self-esteem reduces a teen's ability to withstand peer pressure to have sex).

Researchers have used several theories to explain adolescent sexual and fertility behavior and, to a lesser extent, to develop pregnancy prevention programs. Some theories are quite narrow and presume that a small set of individual or personal characteristics are key to human behavior. For example, the *social and cognitive skills model* that Gilchrist and Schinke (1983) developed and tested posits that for behavior to change, individuals need specific cognitive and social skills to resist pressures and to negotiate interpersonal interactions successfully. They do not address personal values or attitudes toward the behavior or whether other factors may influence behavior change.

Other theories provide a somewhat broader framework for how people learn varied behaviors. For example, the social learning theory (Bandura, 1977, 1986) assumes that whether an individual will engage in or avoid a behavior is determined by a sequence of factors. First, the individual must understand the association of a behavior with an outcome, for example, that unprotected sex carries a high risk of pregnancy. Second, the person must believe that he or she is capable of either engaging in or avoiding the behavior and that the specific strategy chosen can be implemented effectively. For instance, individuals must believe that they have the capacity to abstain from sex and that they can effectively employ a strategy to avoid sex. Finally, people must believe that avoiding the outcome is beneficial, for example, that delaying sex will make their lives better in ways that matter to them. Individuals develop their specific attitudes and feelings about behaviors for themselves by observing the behaviors of others, by observing the rewards and punishments the behavior (and the avoidance of the behavior) elicits, and then by developing the necessary skills through practice that enable them to behave in accordance with the beliefs they develop.

A number of other *value-expectancy* models also take account of the costs and benefits associated with engaging in or avoiding a specific type of behavior. According to the *health belief model*, for example, the probability that persons will engage in a particular preventive behavior, such as abstinence, is based on several personal perceptions (Janz & Becker, 1984; Rosenstock, Strecher, & Becker, 1988). These include (1) their perception of the probability of an outcome as a result of the behavior (for example, pregnancy as a result of unprotected sex); (2) the perceived seriousness of experiencing the outcome (for example, not being able to complete school); and (3) the perceived benefits minus the perceived costs of avoiding the outcome (that is, completing school outweighs the difficulty of saying no). The health belief model proposes that a person consider each of these criteria before engaging in a protective or preventive behavior. Thus, protective behavior is most likely to occur if the adolescent perceives himself or herself as vulnerable to an outcome, perceives the outcomes as negative, and perceives the benefits of protection to outweigh the costs of protection.

Other theories such as the *theory of reasoned action* emphasize individual perceptions (Fishbein & Ajzen, 1980, 1975). This theory emphasizes the importance of an intention to engage in a behavior and attempts to explain the factors that determine that intention. Factors presumed to influence such intentions consist of (1) one's belief regarding the outcome of the behavior in question; (2) one's assessment that the outcome of the behavior is good or desirable; (3) one's assessment that the outcome is desired by significant others; and (4) the individual's motivation to comply with the preferences of these significant others. According to this model, an adolescent would have to believe that avoiding sex will prevent pregnancy and sexually transmitted diseases, that avoiding pregnancy and STDs is desirable, that the significant persons in their lives want them to avoid pregnancy and STDs, and that they want to comply with the views of the significant persons in their lives.

The *opportunity-cost perspective* also takes a cost–benefit accounting approach and puts specific emphasis on whether an adolescent feels a particular behavior will have negative consequences for him (Moore, Simms, & Betsey, 1986). This theory emphasizes the notion that adolescents in different segments of the socioeconomic distribution face very different costs to pregnancy if it occurs. Thus, pregnancy represents a much more substantial cost to a college-bound adolescent than to an adolescent whose future does not realistically include a good education, a good job, a good income, or a good marriage. The motivation to prevent parenthood is therefore substantially lower for adolescents from disadvantaged families and communities.

The *culture of poverty* perspective (Lewis, 1959, 1961, 1966) also focuses on the role that poverty and socioeconomic disadvantage play and argues that early sex and childbearing among impoverished persons represent "both an adaptation and a reaction of the poor to their marginal position in society" (Lewis, 1968, 168). The distinction of this theory, however, is the argument that such behavior becomes normative and is passed on from generation to generation.

Utility maximization perspectives, such as the opportunity-cost perspective (Moore, Simms, & Betsey, 1986), tend to focus less on long-term norms and more on the varied individual costs and benefits associated with sex, contraception use, and fertility (Philliber & Namerow, 1990). Studies based on such frameworks have examined the utility derived from sex among adolescent males as well as females and have explored the role of a wide array of benefits, not just economic ones (Hingson, Strunin, Berlin, & Heeren, 1990). The authors find that social utilities, such as popularity with peers, also affect adolescent decision making. Thus, the notion of relationships emerges, even among the more traditional utility maximization paradigms.

In general, cost–benefit approaches to teen sexual behavior are fairly persuasive, theoretically. Various interventions employing these approaches, however, capture *costs* and *benefits* in very different ways. Some interventions have attempted to alter adolescents' perceptions of costs and benefits (for example, bringing in a teen mother to talk about the difficulties of adolescent parenthood); some interventions have worked to enhance the real gains to an individual from engaging in one type of behavior and avoiding another (for example, providing funds for postsecondary education for nonparents). More rarely, interventions have attempted to generally and broadly alter the actual rewards or gains associated with behavior in a community, changing the actual employment prospects of adolescents (for example, the Youth Incentive Entitlement Project). At this time, it is not known which approach, if any, shows more promise than another or for whom a given approach is more effective. Changing the perceived, the individual, and the broader social opportunities in the same intervention represents a very challenging approach. It is a strategy that, nonetheless, warrants serious consideration.

On Whom Should an Intervention Focus?

. . . Studies employing a life course perspective . . . draw our attention to the notion that development occurs across the stages of the life span and suggest the possibility that interventions may and perhaps ought to focus on life-cycle stages other than adolescence. For instance, while most pregnancy prevention interventions focus on adolescents (for example, junior and senior high school students), studies indicate that many of the factors that predispose adolescents to early sex (for instance, early problem and school behaviors) begin before adolescence (Moore et al., 1995a). In addition, studies consistently indicate that adolescents, both male and female, who are positively engaged in school and who eschew problem behaviors, such as acting out in class, are at a lower risk of early parenthood (Moore, Manlove, Glei, & Morrison, 1997; Zabin, 1994). Thus, it might make considerable sense to focus on children of elementary school age, and even preschoolers, to reduce risk factors such as early school failure and early behavior problems that so frequently precede early sexual activity among youth. Programs that focus on young children might enhance their preschool or early educational experiences, improve the chances for educational success and school engagement during preadolescence and adolescence, and thus reduce the risk of sexual activity. There are, obviously, many other reasons to encourage stronger preschool and early childhood educational programs. Delaying sex and pregnancy may be an added benefit to such early and sustained investments in children.

Scientific research also indicates that multiple factors influence the transition to first sex, and the manner in which such factors do so is particularly complex. Determinants of sexual behavior include families, peer groups, schools, and communities, as well as individual factors. Thus, targeting solely the adolescent ignores the full range of factors that influence adolescent sexual behavior, and diminishes the potential for success.

Other population subgroups may be an important focus of abstinence interventions. One such subgroup appropriate for involvement in efforts to prevent pregnancy and encourage abstinence is young adults, particularly young adult males. Many prevention programs acknowledge that the partners of sexually active female adolescents are often somewhat older than the adolescent, often in their late teens or early twenties. Whether these are coercive or exploitative relationships or just relationships that undermine the prospects for an adolescent to avoid sex, an important focus may be not just the young adolescent but the somewhat older partner. Few studies have examined the factors that may contribute to sexual relationships among somewhat older young adult males with adolescent females, particularly those relationships that are not overtly abusive or exploitative. Several pregnancy prevention efforts, however, have targeted at-risk males specifically to address the importance of sexual and contraceptive responsibility, to help young men redefine and understand the meaning of manhood and fatherhood, and to offer opportunities for education, training, and employment.

Another important subgroup to target is older siblings. Studies show that younger siblings are more likely to have initiated sex at any given age than their older siblings when they were at the same age (Rodgers & Rowe, 1988). These differences are larger for same-sex than for opposite-sex sibling pairs and stronger for whites than for blacks. The reasons for the earlier age at first sex among younger siblings are unclear, however. According to one argument, siblings tend to be powerful role models and confidantes as they occupy a relatively similar status in the family power structure. Older siblings (who may themselves be sexually active or have started to express interest in members of the opposite sex) serve as models for younger siblings. Younger siblings may strive to model the more "mature" or sophisticated behavior of their older brothers or sisters. Programs might involve older siblings to educate them about the ways their behavior can influence the behavior of their younger brothers and sisters and as a way to develop strategies to help siblings encourage their younger brothers and sisters to delay sex.

In addition to older partners, siblings, and the individual teen, adults, especially parents, play an important role in the adolescents' predisposition toward early sex. Parents with children, particularly preadolescent or adolescent children, need to recognize their potential for helping adolescents delay sex. Indeed, studies indicate that healthy and positive family and parent–child relationships are important for delaying the transition to first sex (Moore et al., 1995a). Specifically, parental attitudes toward sex, family rules and monitoring, and parent–child communications are some of the factors that influence adolescent sexual behavior. In particular, when problems within the family or between parent and child increase, the influence of others, such as peers, also increases (Benda & DiBlasio, 1991). If such influences are negative or if those influences act to predispose the adolescent toward sex, the likelihood of sexual activity increases.

Other adults such as teachers, youth workers, religious leaders, and coaches, who work with adolescents, may be important intervention targets. Social workers or youth service providers—particularly those dealing with adolescent runaways or teens with alcohol or drug problems or those working with children in foster care—often face multiple-problem adolescents at high risk for early sexual activity and pregnancy. In some cases, such people may be important and appropriate targets of intervention efforts.

A related issue is whether to focus on an individual (whether the teen or adult) or on a group. Because children become increasingly focused on peers as they enter adolescence, it may be valuable to focus on peers as a target group. For example, interventions might work with peer groups to change the values and activity patterns of individuals as well as their friends, to increase the support for abstinence received by a youth, and to provide alternatives to sexual activity.

Interventions might focus on families. Although parent–child relationships are important during adolescence, those relationships often exist within a broader context. More important, for some youth, particularly those at an increased risk for early sex, the family unit may consist of single parents as well as extended family members who also play a role in the adolescent's daily life. Parents and other kin can be enlisted to help families develop more positive patterns of interaction, communication, discipline, and activity that can help adolescents delay sex.

What Should Be the Focus of an Intervention?

Depending on the underlying assumptions of an intervention, programs may employ a variety of approaches or strategies to foster abstinent behavior. For instance, interventions could include an education or information component, if one presumes that knowledge and information about sexuality or sexual and reproductive health and the risks of sex are sufficient or helpful to adolescents to avoid engaging in intercourse. In fact, many abstinence-focused programs (as well as teen pregnancy prevention programs in general) include information-based instruction. Evaluation studies, however, clearly document that didactic approaches alone are not effective in changing behavior, particularly avoiding sex (Kirby, 1997). Rather, programs that combine information with skill-building activities demonstrate somewhat stronger and more sustained impacts. Thus, providing information can be an important component in an intervention, particularly when combined with other strategies.

Modules can also address attitudes or values supportive of abstinence or the development of skills to help teens avoid peer pressure and negotiate difficult interpersonal relationships. Studies suggest, however, that *motivation* to abstain is key and may thus be more important for avoiding sexual activity than simply knowledge or even skills. That is, armed with information and skills to avoid sex, some teens may still put themselves at risk if the underlying desire to use the information and skills is insufficient or nonexistent. We know relatively little, though, about what truly motivates a teen to postpone having sex. From an ecological perspective, the factors that predispose one to engage in sexual activity range from the individual adolescent and his or her partner, to the nuclear family, to the extended family, peer group, neighborhood, religious organization, and school, and to the larger culture including the media, the economy, and laws and social policies.

Studies show that proximal factors (for example, individual or family) have a stronger effect on sexual behavior than more distal factors (such as policies), though factors in the broad social context have been found to play a role in teen pregnancy as well (Moore et al., 1995a). In addition, studies have also documented dramatic changes in attitudes and values regarding sex, marriage, and gender roles over the past several decades (Thornton, 1995), suggesting that larger social changes, whether socioeconomic or attitudinal, can influence individual sexual behavior.

Recognizing that sexual behavior is affected by multiple influences points up the need for interventions to focus on more than one level to get adolescents to avoid sex. Thus, interventions should provide knowledge, address attitudes and values relating to the avoidance of sex, and provide skills to help adolescents delay sexual intercourse and maintain their abstinent behavior over time.

What Activities or Services Should Be Provided?

The question of what type of activities or services should be provided to help adolescents abstain from sex has not been systematically addressed. That is, while studies are fairly clear about the limited benefits of an education-only approach relative to an education and skill-building combination, for example, we know little about what types of activities best transmit the information, skills, and desired attitudes to adolescents, who should lead those activities, and to whom they should be delivered. For instance, does a lecture-style education session combined with role-playing help adolescents retain the information and skills better than a 30-minute information video followed by group discussion or better than a computer simulation or interactive game? In addition, to what extent do "intangibles" or interpersonal dynamics of the program influence the degree of success of a particular intervention? That is, what role does the charismatic program administrator or dynamic young mentor play in the transmission of the program activities? To what extent are program effects a result of the intervention strategy or the result of the relationships that develop in the course of engaging in those activities?

Moreover, our understanding of which activities or services are most appropriate for various population subgroups is virtually nil. Given the diversity of the adolescent population and the communities in which they live, we suspect that a variety of strategies and combination of strategies should be explored and evaluated.

One strategy that has been found to be helpful in the implementation of pregnancy prevention programs is the use of peer educators or youth slightly older than the target population, as in *Postponing Sexual Involvement* (Moore, Sugland, Blumenthal, Glei, & Snyder, 1995b).

Experience, scientific research, and conventional wisdom all suggest that comprehensive programs are needed for disadvantaged and at-risk youth. Hard evidence in support of a comprehensive approach, however, is lacking. While numerous factors influence sexual behavior and evidence shows that some factors play a more significant role than others, programs rarely target the multiple factors deemed by scientific evidence, or even conventional wisdom, to be important.

In general, we posit that the intensity of activities should vary with the degree of need of a youth. While some teens from advantaged families suffer from an overload of activities, teens from disadvantaged families and those from communities that lack recreational, artistic, and academic opportunities may require a substantial array of services and activities. The nature of the specific activities is probably less important or may depend on the individual adolescent, so long as teens are positively engaged in one or more activities....

Conclusion

There is no shortage of opinions about what will reduce adolescent pregnancy, nor is there a shortage of program models. What is in short supply, however, is objective empirical evidence identifying specific programs or policies that will reduce teen pregnancy, either through delaying sexual intercourse or improving contraceptive use among sexually active adolescents. Furthermore, not only has no one found a single silver-bullet program, but attention to previous research and theory suggests that a single silver-bullet solution is unlikely. Program planners, however, should take time to consider several factors before implementing a pregnancy prevention initiative, irrespective of the desired behavioral outcome. First, it is important to define clearly what behavior is desired (for example, no sex until marriage; no sex until midtwenties), the program's underlying assumptions about the behavior desired and the factors that influence the behavior, and which key factors the program will address.

Second, one should decide whom the intervention should target. Will the intervention focus on adolescents, preadolescents, or children of elementary school age? Will the intervention also include other individuals who may be important to the teen's behavior, such as peers, the teen's family, or the teen's potential sexual partners? Will the intervention address the larger community or neighborhood context in which the adolescent lives, either by collaborating with local institutions such as youth service organizations or local churches or by addressing socioeconomic or other opportunities that may influence adolescent behavior?

Third, which strategies and activities are most appropriate for securing behavior change given the desired behavior outcome and the target populations? Which components are most appropriate or most likely to be supported by the teens and their local community? What type of individuals should be involved in program implementation to secure a reasonably high participation over time?...

While considering such questions, providers need to remember that the U.S. population is highly heterogeneous. Different programs may appropriately emphasize different issues and approaches, especially as they focus on different populations and age groups. Policy makers and program planners need to acknowledge that the origins of adolescent sexual behavior accumulate over the course of life and reflect the force of numerous influences that pose costs and benefits to the adolescent in the short term and the long term. We should develop programs that recognize this complex reality.

REFERENCES

Bandura, A. (1986). *Social foundations of thought and action.* Englewood Cliffs, NJ: Prentice-Hall.

Bandura, A. (1977). *Social learning theory.* Englewood Cliffs, NJ: Prentice-Hall.

Benda, B. B., & DiBlasio, F. A. (1991). Comparison of four theories of adolescent sexual exploration. *Deviant Behavior, 12,* 235–257.

Berger, K. S. (1983). *The developing person through the life span.* New York: Worth Publishers, Inc.

Fishbein, M., & Ajzen, I. (1980). *Understanding attitudes and predicting social behavior.* Englewood Cliffs, NJ: Prentice-Hall.

Fishbein, M. & Ajzen, I. (1975). *Belief, attitude, intention & behavior: An introduction to theory and research.* Reading, MA: Addison-Wesley.

Gilchrist, L. D., & Schinke, S. P. (1983). Coping with contraception: Cognitive and behavioral methods with adolescents. *Cognitive Therapy and Research, 7,* 379–388.

Hingson, R. W., Strunin, L., Berlin, B. M., & Heeren, T. (1990). Beliefs about AIDS, use of alcohol, drugs and unprotected sex among Massachusetts adolescents. *American Journal of Public Health, 80,* 295–299.

Janz, N., & Becker, M. (1984). The health belief model: A decade later. *Health Education Quarterly,* 11, 1–47.

Kirby, D. (1997). *No easy answers.* Washington, DC: The National Campaign to Prevent Teen Pregnancy.

Lewis, O. (1968). The culture of poverty. In Moynihan, D. (Ed.) *On understanding poverty: perspectives from the social sciences* (pp. 187–200). New York: Basic Books.

Lewis, O. (1966). *La Vida: A Puerto Rican family in the culture of poverty—San Juan and New York.* New York: Random House.

Lewis, O. (1961). *The children of Sanchez.* New York: Random House.

Lewis, O. (1959). *Five families: Mexican case studies in the culture of poverty.* New York: Basic Books.

Maynard, R. A. (1996). *Kids having kids: Economic costs and social consequences of teen pregnancy.* Washington, D.C.: The Urban Institute Press.

Moore, K. A., Manlove, J., Glei, D. A., & Morrison, D. R. (1997). *Nonmarital school-age motherhood: family, individual, and school influences.* Washington, D.C.: Child Trends, Inc.

Moore, K. A., Miller, B. C., Glei, D., & Morrison, D. R. (1995a). *Adolescent sex, contraception, and childbearing: A review of recent research.* Washington, D.C.: Child Trends, Inc.

Moore, K. A., Simms, M. C., & Betsey, C. L. (1986). *Choice and circumstances: Racial differences in adolescent sexuality and fertility.* New Brunswick, NJ: Transaction Books.

Moore, K. A., Sugland, B. W., Blumenthal, C., Glei, D. A., & Snyder, N.O. (1995b). *Adolescent pregnancy prevention programs: Interventions and evaluations.* Washington, D.C.: Child Trends, Inc.

Philliber, S., & Namerow, P. B. (1990). Using the Luker model to explain contraceptive use among adolescents. *Advances in Adolescent Mental Health, 4,* Jessica Kingsley Publishers, Inc.

Rodgers, J. L., & Rowe, D. C. (1988). Influence of siblings on adolescent sexual behavior. *Developmental Psychology,* 24, 722–728.

Rosenstock, I., Strecher, V., & Becker, M. (1988). Social learning theory and the health belief model. *Health Education Quarterly, 15,* 175–183.

Thornton, A. (1995). Attitudes, values, and norms related to nonmarital fertility. In U.S. Department of Health and Human Services, *Report to Congress on out-of-wedlock childbearing* (DHHS Pub. No. 95-1257, pp. 201–216). Washington, D.C.: U.S. Department of Health and Human Services.

Zabin, L. S. (1994). Addressing adolescent sexual behavior and childbearing: self-esteem or social change? *Women's Health Issues, 4,* 92–97.

Teenage Pregnancy in Industrialized Countries: Policy Implications for the United States

Elise F. Jones, Jacqueline Darroch Forrest, Noreen Goldman, Stanley Henshaw, Richard Lincoln, Jeannie I. Rosoff, Charles F. Westoff, and Deirdre Wulf

[This reading reports the results of a] large study of 37 countries . . . and case studies of individual countries . . . undertaken to better understand the reasons why birthrates and pregnancy rates among teenagers in the United States are so much higher than they are in other developed countries, and to determine whether it is possible to learn from the experience of countries with low teenage pregnancy rates how to reduce those rates in the United States. . . .

One of the most important policy implications to come from this study is the discrediting of certain beliefs, some widely held by social conservatives and others held by liberal reformers. Both sets of beliefs—each in its own way—have tended to paralyze efforts aimed at reducing the relatively high rates of pregnancy experienced by American adolescents. What are those beliefs?

- Adolescent pregnancy rates are higher in the United States than in Europe mainly because of the high pregnancy rate among U.S. blacks.
- American adolescents begin sex much earlier and more of them are sexually active than their counterparts in other developed countries.
- Teenagers are too immature to use contraceptives effectively.
- Unwed adolescents want to have babies in order to obtain welfare assistance.
- Making abortions and contraceptives available and providing sex education only encourages promiscuity and, therefore, increases teenage pregnancy.
- As long as there is no clear path for unemployed teenagers to improve their economic condition, they will continue to have high pregnancy rates because having babies is one of the few sources of satisfaction and accomplishment available to them.

None of these beliefs explains the differences between teenage pregnancy rates in the United States and other developed countries.

Lessons from the 37-Country Study

The study of 37 countries was restricted to an examination of the factors affecting teenage birthrates rather than pregnancy rates. . . . Of the 13 countries for which both abortion and birth data are available, however, those with high teenage birthrates also have high abortion rates, and vice versa. Thus it is reasonable to infer that factors affecting fertility differentials found to be significant in the 37-country study also apply to differences in pregnancy. It is notable that in those countries where abortion *and* fertility data are available, the gap between the United States and the

FIGURE 1 *Percentage Distribution of Pregnancies and Pregnancy Rates, by Outcome, for Women Aged 15–19, 1980–1981*

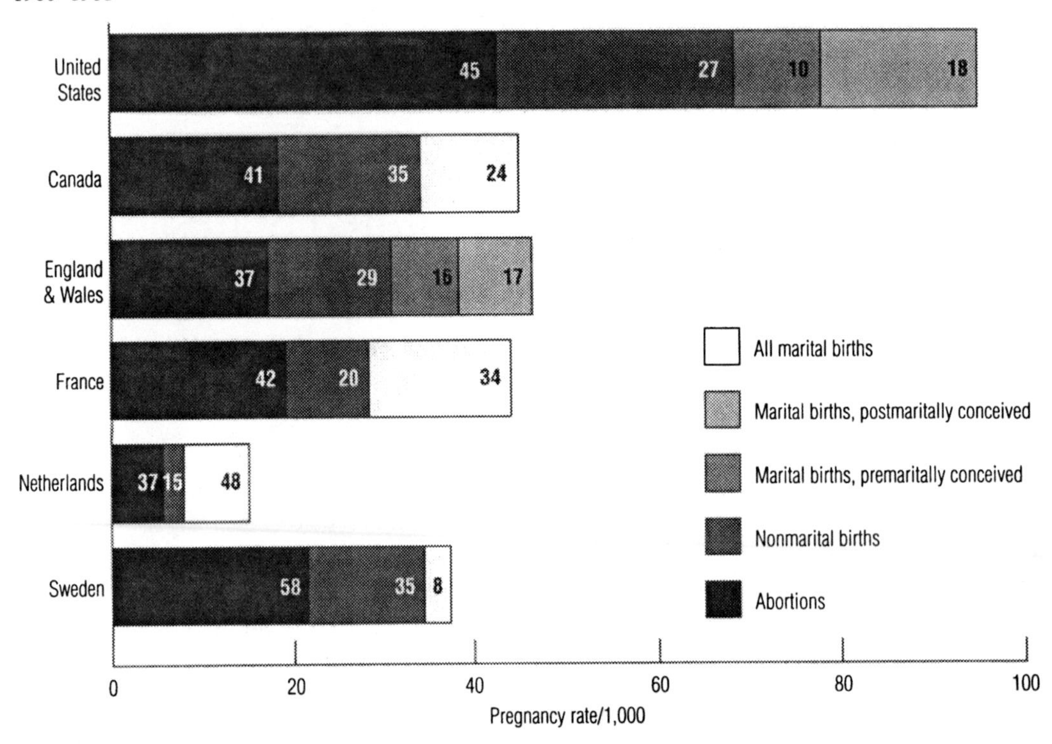

Sources: Canada: Abortions—Statistics Canada, 1984; *Therapeutic Abortions, 1981,* 71. Births—Statistics Canada, 1983; *Vital Statistics,* vol. 1, "Births and Deaths," table 8. France: Abortions—Tietze, 1984; unpublished data from DAGSPR (deaging program) output (age at term) for *Induced Aboriton: A World Review, 1983,* The Population Council, New York, 1994. Births—Les Collections d'INSEE [D94], table 18. England and Wales: Abortions and Births—OCPS, 1984; *Birth Statistics 1981, England and Wales,* Series FMI, No. 8. HMSO, London. Netherlands: Abortions—Tietze, 1984. Births—Central Bureau of Statistics, 1984, unpublished statistics, The Hague. Sweden: Abortions and total births—Tietze, 1984. Legitimate births—United Nations, 1982. *United Nations Demographic Yearbook 1982,* table 33. United States: Table 3.1.
Note: The numbers inside the bars represent the percentage distributions.

other countries with regard to their pregnancy rates is at least the same as, or even greater than, the differentials for birthrates alone.

Six factors were found to be most important in their effect on teenage fertility.... Those associated with low teenage fertility rates are:

- High levels of socioeconomic modernization
- Openness about sex
- A relatively large proportion of household income distributed to the low-income population (important mainly for younger teenagers)
- A high minimum legal age at marriage (important for older teenagers only)

Factors associated with high teenage fertility are:

- Generous maternity leaves and benefits
- Overall pronatalist policies designed to raise fertility

The United States differs from most of the countries with comparably high adolescent fertility with regard to four of these factors. The position of the United States is anomalous, however, with regard to socioeconomic development, one of the most important factors associated with low teenage fertility. Although it is one of the most highly developed countries examined, the United States has a teenage fertility rate much higher than the rates observed in countries that are comparably modernized, and considerably higher than the rates found in a number of much less developed countries. The inconsonance applies particularly to fertility among younger teenagers, where the U.S. rate falls between that of Romania and Hungary.... The relatively high adolescent birthrate in the United States would suggest that it has a pronatalist fertility policy (which it does not), high levels of maternity leaves and benefits, and a low minimum age at marriage. Maternity benefit policies on average are less liberal than those in most European countries (Kamerman, Kahn, and Kingston, 1983), and in most states women can marry on their own consent by age 18, an age similar to that found in most of the countries studied.

The United States fits the pattern for high teenage fertility in that it is far less open about sexual matters than most countries with low teenage birthrates, and a smaller proportion of its income is distributed to families on the bottom rungs of the economic ladder. All of these findings suggest that two factors are key to the location of the United States with regard to high teenage fertility: an ambivalent, sometimes puritanical attitude about sex, and the existence of a large, economically deprived underclass.

With better or more complete information, it is likely that at least some of the factors found significantly correlated with adolescent fertility in the bivariate analysis would have survived into the multivariate analysis.... These include the relationship between high levels of adolescent fertility and restrictions placed on teenagers' access to contraception and with high levels of religiosity in the country, and the association of low adolescent fertility with teaching about contraceptives in the schools. It is notable that government subsidy of abortions is not associated with teenage fertility.

In the 37-country study, the United States does not appear to be more restrictive than low-fertility countries in the provision of contraceptive services to teenagers; however, comparable data could not be obtained on the provision of contraceptives free of charge or at very low cost—a factor that appears to be very important in terms of accessibility in the case studies. In this respect, the United States is more restrictive than the other 5 countries studied in detail—all of which have much lower adolescent fertility and pregnancy rates than the United States. The very high levels of religiosity reported for the United States (the highest of any of the 13 countries for which there are data) is probably one, but certainly not the only, factor underlying the low rating of the United States on openness about sex. It is also noteworthy that the United States scores relatively low among the countries studied on the measures of availability of contraceptive education in the schools.

The results from the comparative study also indicate that intercountry differences in teenage fertility and in teenage pregnancy are *not* mainly due to the fact that these rates are much higher among U.S. black than white teenagers. The birthrates of white American teenagers are higher than those of teenagers in any Western European country (Westoff et al., 1983). (It is notable that many of these countries have substantial numbers of nonwhite inhabitants—who tend to be poorer and to have higher fertility than the general population—e.g., West Indians and Asians in England, Algerians in France, and Surinamese in the Netherlands.) Birthrates for adolescents under age 18 are higher than those reported in any developed country except Bulgaria, Cuba, and Hungary.... Similarly, U.S. pregnancy rates for those young teenagers are higher than those in all developed countries for which there are data.[1]...

The Six Country Case Studies

The individual country studies confirm the findings of the 37-country study, help explain some of the ambiguities in those study findings, and cast light on factors for which data were not available for the larger study. Two of the commonly held beliefs about teenage pregnancy that were noted earlier in this chapter are refuted in the case studies: teenagers are *not* precluded by their presumed immaturity from using contraceptives consistently and effectively; and teenagers in the United States are *not* more sexually

[1]Canada, Czechoslovakia, Denmark, England and Wales, Finland, France, Hungary, the Netherlands, New Zealand, Norway, Scotland, Sweden.

experienced than adolescents in other countries with lower pregnancy rates.

With the possible exception of younger teenagers in Canada, national differences in the percentage of teenagers who have ever had intercourse appear to contribute little or nothing to the variations in teenage pregnancy rates. Disparities in age at initiation of sexual relations appear to be too small to play a large part in determining the pregnancy differentials between the United States and the other countries.

In the five countries other than the United States, contraceptive use, and particularly use of the pill, appears to be greater than it is in the United States—among younger as well as older teenagers. Physicians and clinics in these countries actively encourage pill use among teenagers; and in some of the countries, there have been active campaigns addressed to young men to encourage the use of condoms. There is not the same level of concern that pill use may involve serious health complications. In fact, the health risks to teenagers have been found to be exceedingly low and outweighed by the health benefits (Ory et al., 1983). IUD use is not especially high among teenagers in the five countries, but neither is it automatically considered to be contraindicated for teenagers—largely because pelvic inflammatory disease is less prevalent and tends to be treated more promptly than it is in the United States.

Does the availability of contraceptives, sex education, and abortion services in the United States encourage sexual promiscuity and thereby account for the higher teenage pregnancy rates in the United States? The findings from the case-study countries suggest that this cannot be the case, since availability is generally greater in countries with lower teenage pregnancy rates.

Teenagers who obtain contraceptives from *any* source are assured of confidentiality of services in Sweden, the Netherlands, and France. It is notable that in England the government went to the House of Lords asking that a previous court decision be reversed which would have required those under age 16 to have parental consent to obtain contraceptive services. The House of Lords law panel, Britain's highest court, agreed with the government that physicians may provide contraceptive services to minors under 16 on their own consent. Conversely, in the United States, the U.S. Justice Department went to court, unsuccessfully, to defend a Department of Health and Human Services regulation that would have *required* that the parents of all teenagers under 18 be notified if their children obtained a prescription contraceptive. (Another branch of government, the U.S. Congress, had rejected a proposal that parental consent or notification be mandated.)

Abortions are available free of charge in England and Wales and in France. The cost to the woman is low in Canada and Sweden and is small enough in the Netherlands not to be prohibitive. In more than two-thirds of U.S. states, the full price of an abortion must be paid by the woman, whether or not she can afford it; and, because abortion services tend to be concentrated in populous metropolitan areas, costs can include not only the fee for the abortion itself, but also travel and hotel expenses. Access to abortion in some of the countries studied, however, is more restrictive than in the United States with regard to parental consent, gestational age, required waiting period, legally required overnight hospital stays, and required approval by an additional doctor or doctors or a medical committee.

It is noteworthy that liberalization of abortion has often been accompanied by facilitating the access of teenagers to contraceptive services to minimize the need for abortions among women in this young age group.

Sex and contraceptive education in the schools differs widely among the case-study countries. Sweden has by far the most extensive program. In the Netherlands, although there is little formal school sex education beyond the facts of reproduction, widespread public education via all the media is superior to that of any of the countries studied except, possibly, Sweden.

In short, teenagers living in countries where contraceptive services, sex education in and out of the schools, and abortion services are widely available have lower rates of adolescent pregnancy and do not have appreciably higher levels of sexual experience than do teenagers in the United States.

The findings from the 37-country study suggest that generous maternity leaves and benefits—mostly in pronatalist countries of Eastern Europe—are associated with high fertility rates among older teenagers. The United States was not included in those comparisons because policies differ among the states and because such benefits are largely provided through private rather than government programs. Overall, however, the United States does not appear to have generous policies compared to other countries, even when private programs, as well as public subsidy, are taken into account (Kamerman, Kahn, and Kingston, 1983). Nevertheless, it has been suggested that U.S. Aid to Families with Dependent Children—a benefit that is largely limited to single mothers, a very high proportion of whom are teenagers—encourages out-of-wedlock fertility in the United States.

All of the countries in the case studies have more generous health and welfare provisions for the general population and for mothers than does the United States; and several of them, like the United States, provide special financial assistance for single mothers. Yet these countries all have lower teenage fertility rates than the United States. Differences in welfare assistance to mothers, or to single mothers, do not appear to explain the differentials in teenage birthrates between the United States and the other countries studied, and certainly they do not explain the differences in abortion rates. Although this study does not address the question of whether differential welfare payments can encourage adolescent fertility within individual countries, a number of American studies have failed to confirm this hypothesis (Cutright, 1970 and 1973; Moore, 1978; Moore and Caldwell, 1977; Placek and Hendershot, 1975; Hendershot and Placek, 1974; Presser and Salsberg, 1975; Cutright et al., 1974). Yet it continues to be put forward (Fuchs, 1983; Murray, 1984; Sklar and Berkov, 1974). One recent study finds that although welfare does not encourage out-of-wedlock fertility, it does influence the decision of single parents to move out of the parents' home and, to a lesser extent, may encourage young married women to divorce or separate (Ellwood and Bane, 1984).

Teenagers' prospects for economic improvement do not appear to be appreciably greater in the five case-study countries than in the United States; nor is the educational achievement of young people greater. However, more extensive health, welfare, and unemployment benefits in the other countries keep poverty from being as deep or as widespread as it is in the United States. The findings from the 37-country study suggest that more equitable distribution of household income is associated with lower fertility among younger teenagers. The inequality of income distribution appears to be a contributing factor to the differences between teenage birthrates in the United States and those in other developed countries. Unfortunately, data are not available to compare teenage pregnancy rates by income between countries. However, it does not seem likely, given the fact that the U.S. white adolescent pregnancy rate is so much higher than those in the other countries, and that poverty, while not so extensive or so deep as in the United States, also exists in the other countries studied, that differentials in socioeconomic status can explain the differences in adolescent pregnancy between the United States and the other five countries. Differences in economic status among teenagers *within* individual countries, however, may very well contribute substantially to pregnancy and especially to fertility differentials.

To summarize, the 37-country study and the individual country studies provide convincing evidence that many widely held beliefs about teenage pregnancy cannot explain the large differences in adolescent pregnancy between the United States and other developed countries: teenagers in these other countries apparently are *not* too immature to use contraceptives consistently and effectively; the availability of welfare services does *not* seem correlated with higher adolescent fertility; teenage pregnancy rates are *lower* in countries where there is *greater* availability of contraceptive and abortion services and of sex education; adolescent sexual activity in the United States is not very different from what it is in countries that have much *lower* teenage pregnancy rates; although the pregnancy rate of American blacks is much higher than that of whites, the white rate is still much higher than the overall teenage pregnancy rates in the other case-study countries; teenage unemployment appears to be at least as serious a problem in all the countries studied as it is in the United States, and American teenagers have more or at least as much schooling as those in most of the countries studied that have lower pregnancy rates. Because the other case-study countries have more extensive public health and welfare benefit systems, however, they do not have so extensive an economically deprived underclass as does the United States. However, the differences in teenage pregnancy rates would probably not be eliminated if socioeconomic status could be controlled.

Clearly, then, it *is* possible to achieve a lower teenage pregnancy rate than that experienced in the United States, and a number of countries with comparable levels of adolescent sexual activity have done so. Although no single factor has been found responsible for the differences in adolescent pregnancy rates between the United States and the other five countries, is there anything to be learned from these countries' experiences to improve the situation in the United States?

A number of factors affecting teenage pregnancy rates, of course, are not easily transferable or are not exportable at all to the United States. Each of the five case-study countries has a considerably smaller population, and all but Canada are geographically more compact than the United States—making rapid dissemination of innovations easier; their populations are less heterogeneous ethnically (though not so homogeneous as is commonly assumed, since most have substantial minority nonwhite populations, usually with higher-than-average fertility); religion and the influence of conservative religious bodies are less pervasive than they are in the United States, even in countries like Sweden and England that have officially established churches; governments tend to be more centralized; the provision of wide-ranging social and welfare benefits is firmly established, whether the country is led by parties labeled conservative or liberal; income distribution is more equal; there appears to be less of a tradition of political confrontation and, possibly, a more widespread respect for authority and for public order; and constituencies that oppose contraception, sex education, and legal abortion are not so powerful and well funded as they are in the United States.

Some other factors are, at least theoretically, transferable—and here, it is important to note that some of the factors associated with low pregnancy rates may differ between countries. For example, school sex education appears to be a much more important factor in Sweden than it is in the other countries; high levels of media exposure to contraceptive information and sex-related topics is more prominent in the Netherlands; condoms are more widely available in England, the Netherlands, and Sweden. Use of the pill by teenagers is most extensive in the Netherlands.

By and large, Sweden has been the most active of the countries studied in developing programs and policies to reduce teenage pregnancy. It is notable that Sweden has *lower* teenage pregnancy rates than have all of the countries examined, except for the Netherlands, although teenagers begin intercourse at earlier ages in Sweden. It is also notable that Sweden is the only one of the countries observed to show a rapid decline in teenage abortion rates in recent years, even after its abortion law was liberalized.

It is also noteworthy that none of the five case-study countries has developed government-sponsored programs designed to discourage teenagers from engaging in sexual relations—even at young ages—a program intervention officially advocated in the United States and rewarded through government subsidies. Although in Sweden committed relationships and responsible sexual behavior are advocated in the school sex education program, most of the countries have preferred to leave such matters to parents and churches.

Theoretically, universal health insurance, which would include contraceptive (and abortion) services, could be made available in the United States. Practically, there is little likelihood that legislation of this kind will be passed in the near future. Most persons, however, who are employed or who are students in higher educational institutions are covered by health insurance and, although most insurance policies do not now cover contraception, there is an increasing trend toward the inclusion of preventive health services, including contraception, in insurance contracts. It is possible that the American answer to national health insurance, at least for the foreseeable future, is to supplement private insurance coverage for those who are unemployed or otherwise uninsured and to create or continue publicly supported programs to meet special needs and to reach specific population groups. (Coverage for abortion services in any publicly funded health program would require the reversal, at the federal and state levels, of prohibitions now in effect.) Even if this trend were to accelerate more rapidly than seems likely, the problem of teenagers' access to effective contraceptive services would not be solved. In addition to the problem of cost, many teenagers in the United States, as well as in the other countries studied, do not want their parents informed that they are obtaining contraceptive services. For young people who are not themselves employed or enrolled in an institution of higher learning, coverage under their parents' health insurance policy could violate their privacy and constitute a barrier to effective services.

Possible Approaches

Several U.S. communities have instituted school-based health clinics that provide contraceptive services—usually in partnership with health, youth-serving, or other nonprofit agencies. In many cases, parental consent is required to enroll in the health clinics (Dryfoos, 1985). Contraception, however, is only one of the many health services offered, so that the parent is not specifically informed when contraceptive services or advice are being obtained. The school, which has a continuing relationship with the young person, is in a position to monitor both continuation and any possible medical complications; the student is not lost to followup because she has dropped out of the clinic.[2]

A complementary approach would be to enhance the current family planning clinic system, by increasing government subsidy, to provide free or low-cost contraceptive services to all teenagers who want them, not just to those from poor families. This is already permissible under federal law and, to some extent, the process has already begun. In 1982, 6 out of 10 contraceptive visits by U.S. teenagers were visits to family planning clinics; and 44 percent of teenagers using the pill, IUD, or diaphragm (and 54 percent of those aged 15–17) first obtained their method at a family planning clinic (Pratt et al., 1984, tables 9 and 10). In point of fact, however, although the high unmet need for family planning services among teenagers in the United States is well documented, federal subsidies in real dollars have declined. Moreover, in many communities family planning clinics—especially those operated by hospitals and health departments—tend to be stigmatized by teenagers who have not actually been to such clinics; they are likely to avoid them in the belief that they are restricted to very poor patients, that services are not confidential, that the surroundings are shabby and unclean, the services poor, and the treatment of patients disrespectful (Kisker et al., 1985). Advertising that portrays the clinic services as inviting, professional, confidential, and available to all segments of the community can do much to counteract this negative image. In Planned Parenthood and some neighborhood health clinics, however, there is a trend toward charging a flat fee to all patients; such a fee policy is thought to be likely to discourage teenage enrollment.

The growing reliance on health maintenance organizations (HMOs) to increase health coverage while reducing health costs provides another opportunity to extend family planning services to teenagers. There is no reason why HMOs cannot establish special adolescent clinics on the Swedish model to provide contraceptive services confidentially as part of a general health-care service. Youth-serving agencies, the Society for Adolescent Medicine, and physician groups might well consider developing models for the provision of adolescent health care in the United States that are not linked to poverty programs.

Unlike the governments of most of the countries in the case studies, the U.S. federal government does not mandate or even encourage the inclusion of sex education courses in public schools. Only two states and the District of Columbia require sex education, and hardly any encourage such courses. For the most part, state and even local governments leave the question of whether sex education should be offered at all to the nearly 16,000 local school districts. The extent to which contraception should be discussed—and at

---

[2]It might also be feasible to develop demonstration sex education programs in local school districts that are closely integrated with teenage birth control clinics as has been done in Sweden. A pilot program of this nature was developed by Johns Hopkins University in Baltimore (Zabin et al., 1984).

what grade—is entirely a matter for the local school district to decide. Although numerous public opinion polls show that American parents overwhelmingly approve of sex education in the schools, including education about contraception . . . local school districts have tended to be timid about establishing courses because of fear of minority, but highly vocal, opposition. Although not all the countries studied put much emphasis on school sex education, the evidence from Sweden at least suggests that comprehensive sex education programs can help to reduce teenage pregnancy.

Local school districts in the United States can be encouraged to institute programs and experiment with new approaches. Open discussions of sexuality may also encourage more rational contraceptive decision making. Although sex education is usually left to the school districts, both the federal and state governments are in a position to influence the development and establishment of school sex education courses. Simply by asserting that sex education is desirable, they could help to legitimate the inclusion of sex education courses in the curricula. Congress, by providing subsidies for the development of curricula, for teacher-training programs, and perhaps for some demonstration programs, could further encourage such instruction. State governments similarly can promote sex education efforts, as a few have by taking a clear position, offering selected subsidies, and providing practical help in curriculum development.

Openness about sex is increasing in the United States. (The days when even married couples in motion pictures could not be shown occupying the same bed are long past.) Cohabitation among unmarried couples is rising rapidly. Some restrictive laws relating to sexual information have been struck down by the Supreme Court. Sex, nevertheless, is treated far less openly and is surrounded by more ambivalence than it is in most of the countries in the case studies. In virtually all of the countries examined, for example, information about contraception and sexuality is far more available through the media than it is in the United States; condoms are more widely distributed; and advertisements for contraceptives are far more ubiquitous. Shops providing sex-related materials in other countries, such as the Netherlands, are not as sleazy as they are in the United States.

The self-imposed restrictions on contraceptive advertising in the media—especially on television—are incongruous in an era when virtually every other product, including vaginal douches, sanitary napkins, and hemorrhoid preparations, is advertised everywhere and without protest. At least one cable television network in the United States has begun to carry advertisements for spermicides; and sunbathers at New York area beaches during the summer of 1985 could look up and see a popular condom brand advertised via streamers from an airplane. It seems likely that if the restrictions on advertising were lifted, some aggressive manufacturers would develop and promulgate effective advertising campaigns. A recent study sponsored by the U.S. Food and Drug Administration suggests that such advertising may be feasible (Morris, 1984). Of course, governmental restrictions on advertising prescription drugs except in medical journals also preclude advertising the most widely used reversible contraceptive method—the pill.

There is also a need to disseminate more realistic information among the general public and health professionals about the health risks of the pill (which are minimal for teenagers) and about its extensive benefits (Ory et al., 1983). Most Americans are badly misinformed on this subject (Gallup, 1985). Although teenagers in other countries have experienced much lower pregnancy rates than U.S. adolescents while using currently available methods, it is probable that the development of new methods more appropriate for teenagers who have episodic sex—such as a once-a-month pill—could greatly reduce teenage pregnancies in the United States, and further reduce them in other countries, too. Yet funds for contraceptive development have declined in real terms in recent years in the United States (the major funder of contraceptive research); and research into a monthly pill is further hampered by governmental restrictions on abortion-related expenditures.

In general, American teenagers seem to have inherited the worst of all possible worlds insofar as their exposure to messages about sex is concerned: movies, music, radio, and television tell them that nonmarital sex is romantic, exciting, and titillating; premarital sex and cohabitation are visible ways of life among the adults they see and hear about; their own parents or their parents' friends are likely to be divorced or separated but involved in sexual relationships. Yet, at the same time, young people get the message (now subsidized by the federal government) that good girls should say no. Little that teenagers see or hear about sex informs them about contraception or the consequences of sexual activity. (They are much more likely to hear about abortions than contraception on the daily television soap opera.) Increased exposure to messages about sex has not meant more realistic exposure or exposure to messages about responsible sex. (Nonmarital sex, though it may be irresistible, is branded irresponsible.) Such mixed messages lead to the kind of ambivalence about sex that stifles communication between partners and exposes young people to increased risk of pregnancy, out-of-wedlock births, and abortions. Increasing the legitimacy and availability of contraception and of sex education in its broadest sense is likely to result in declining pregnancy rates, without raising teenage sexual activity rates to any great extent. That has been the experience of most countries of Western Europe, and there is no reason to think it would not also occur in the United States.

Application of any of the program and policy measures that appear to have been effective in other countries is admittedly more difficult in the United States where governmental authority is far more dispersed, but it may, in fact, be as easy or easier in some states and communities. Efforts need to be directed not just to the federal executive branch of government, but to Congress, the courts, state houses, state legislatures, local authorities, and school superintendents and principals—as well as to such private-sector and charitable enterprises as insurance companies, broadcast and publishing executives, church groups, and youth-serving agencies. Because of its complexity, the task may require considerable effort and ingenuity, but clearly it can be accomplished, and there is a broad consensus that the need to reduce teenage pregnancy in the United States is high on the social agenda.

REFERENCES

Cutright, Phillips. 1970. "AFDC Family Allowances and Illegitimacy." *Family Planning Perspectives* 2, No. 4:4–9.

———. 1973. "Illegitimacy and Income Supplements." In *Studies in Public Welfare*, eds. R. Lerman and A. Townsend. Paper No. 12, Pt. I, 90. Joint Economic Committee of the Congress. Washington, D.C.: U.S. Government Printing Office.

Cutright, Phillips, Frank F. Furstenberg, Jr., June Sklar, and Beth Berkov. 1974. "Teenage Illegitimacy: An Exchange." *Family Planning Perspectives* 6:132.

Ellwood, David T. and Mary Jo Bane. 1984. "The impact of AFDC on Family Structure and Living Arrangements." Report to the U.S. Department of Health and Human Services. Cambridge, Mass.: Harvard University.

Fuchs, Victor. 1983. *How We Live: Economic Perspectives on Americans from Birth to Death.* Cambridge, Mass: Harvard University Press.

Gallup Poll. 1985b. "Premarital Sex." *The Gallup Report.* No. 237 (June). Princeton, N.J.

Hendershot, Gerry E., and Paul J. Placek. 1974. "Use of Contraceptive Services in Periods of Receipt and Nonreceipt of AFDC." *Public Health Reports* 89:533.

Kamerman, Sheila B., Alfred J. Kahn, and P. Kingston. 1983. *Maternity Policies and Working Women.* New York: Columbia University Press.

Kisker, Ellen, Stanley K. Henshaw, Aida Torres, Margaret Terry Orr, and Jacqueline Darroch Forrest. 1985. "Teenagers Talk about Sex, Pregnancy and Contraception." *Family Planning Perspectives* 17:83.

Moore, Kristin A. 1978. "Teenage Childbirth and Welfare Dependency." *Family Planning Perspectives* 10:233.

Moore, Kristin A., and Steven B. Caldwell. 1977. "The Effect of Government Policies on Out-of-Wedlock Sex and Pregnancy." *Family Planning Perspectives* 9:164.

Morris, L. A. 1984. "Prescription Drug Advertising to Consumers: Brief Summary Format for Television and Magazine Advertising." Food and Drug Administration, U.S. Department of Health and Human Services.

Murray, Charles. 1984. *Losing Ground.* New York: Basic Books.

Ory, Howard W., Jacqueline Darroch Forrest, and Richard Lincoln. 1983. *Making Choices: Evaluating the Health Risks and Benefits of Birth Control Methods.* New York: Alan Guttmacher Institute.

Placek, Paul J. and Gerry E. Hendershot. 1975. "Public Welfare and Family Planning: An Empirical Study of the 'Brood Sow' Myth." *Social Problems* 23:226.

Pratt, William F., William D. Mosher, Christine A. Bachrach, and Marjorie C. Horn. 1984. "Understanding U.S. Fertility: Findings from the National Survey of Family Growth, Cycle III." *Population Bulletin*, Vol. 39, No. 5.

Presser, Harriet B., and L. S. Salsberg. 1975. "Public Assistance and Early Family Formation: Is There a Pronatalist Effect?" *Social Problems* 23:226.

Sklar, Jane, and Beth Berkov. 1974. "Teenage Family Formation in Postwar America." *Family Planning Perspectives* 6:80.

Westoff, Charles F., Gérard Calot, and Andrew D. Foster. 1983. "Teenage Fertility in Developed Nations: 1971–1980." *Family Planning Perspectives* 15:105.

CHAPTER 1 Questions for Writing, Reflection, and Debate

READING 1 Then and Now • *Mary Pipher*

1. How did attitudes toward dating, gender roles, and sexuality differ between the two times and places described in "Now and Then"? How did the communities described contribute to these differences? What impact did these environments have on the author in her youth and on her young client?
2. Why do different individuals react differently to the social contexts in which they grow up? What makes some adolescents today resist peer pressure, whereas others conform?

READING 2 Constructing an Epidemic • *Kristin Luker*

3. What is the difference between social concern over teenage pregnancy and teenage childbearing? How do the statistical trends for pregnancies and births for adolescents differ? What class and race differences exist in these trends?
4. What are the politics behind focusing on teen childbirth as opposed to teen pregnancy? What are the politics behind focusing on adolescent pregnancy as opposed to pregnancy to unmarried women?

READING 3 Risking Pregnancy: Avoidance, Ignorance, and Delay of Contraceptive Use • *Ellen W. Freeman and Karl Rickels*

5. What does this study reveal regarding the differences in the level of sexual activity and information between teenagers who become pregnant and those who don't?
6. Why is it difficult to draw conclusions regarding connections between sexual information and sexual behavior from this study? What do other studies show? What conclusions can be drawn from this study regarding the connections between the family as a source of information and risk of pregnancy?
7. What prevents young people from using contraceptives?

READING 4 Using Behavioral Theories to Design Abstinence Programs • *Kristin A. Moore and Barbara W. Sugland*

8. What is offered, in this article and other places, as support for the policy of promoting abstinence among unmarried teens?
9. How does this article view the role of parents and older siblings in determining adolescent sexual behavior? What factors, other than family, contribute to the potential success or failure of abstinence programs?

READING 5 Teenage Pregnancy in Industrialized Countries: Policy Implications for the United States • *Elise F. Jones et al.*

10. In determining how to address the problem of adolescent pregnancy in the United States, some people express the fear that the availability of

contraceptive services, sex education, and access to abortion will lead to more sexual activity and higher adolescent pregnancy rates. How does this study address these concerns?

11. What conclusions does this study draw about the availability of health and welfare provisions and adolescent pregnancy rates?
12. What factors might contribute to adolescent pregnancy that have not been considered in this study?
13. What are the difficulties in applying conditions that suppress teen pregnancy in one country to other countries?

Commitment: Cohabitation and Marriage

Is marriage a dying institution?

Formal, social norms in the United States have historically dictated that the route to establishing a union between two individuals is to become legally married. However, many Americans have moved away from this socially sanctioned means of establishing family. As you read in Chapter 1, increasingly families are started through pregnancy rather than marriage. In addition, many couples bypass marriage and move in together as a public declaration of their commitment; they too may consider themselves a family.

Some Americans are disturbed by this trend. They interpret the fact that couples of all ages are living together in committed relationships without the legal sanction of marriage as proof that the institution of marriage is falling out of favor in this country. Do you agree? Does the fact that two people have set up a household and made a public commitment to each other without marriage mean that marriage may soon no longer be the normative way to start a family in America?

When sociologists speak of marriage as an institution, they mean many things. The most straightforward explanation of "institution" is that each society creates established practices for the way individuals in that society are expected to deal with common human situations. Cultures establish rewards or benefits, referred to as positive sanctions, to encourage people to comply with these norms. In addition, negative sanctions are often applied to behaviors that do not comply with the accepted practices. Not too many years ago, unmarried couples bore the stigma of illegitimacy, and only married couples were positively sanctioned. But somewhere along the line, the benefits of cohabitation began to outweigh society's disapproval. What tipped the balance? Today, what factors motivate people to choose either marriage or cohabitation? And how might the actions of individuals be shaping the larger society?

One answer to this question is provided in the first reading in this chapter by Norval Glenn, "Values, Attitudes, and the State of American Marriage." His work indicates that the value placed by Americans on involvement in a good marriage has remained strong, but we have at the same time lost the ideal of marriage permanence. Glenn's article also supports the argument that "flight from marriage" is partially caused by a rise in the priority placed on self-interest. People who choose not to be legally married are concerned with their own desire to share their life with someone, but they don't want to risk involvement in a legal contract. Or maybe they are motivated by a belief that the larger society should not be involved in sanctioning of people's commitments to each other. Both explanations are based on individualism.

However, selfishness could explain what keeps drawing people to marriage. Most people want to marry, and they do. Even those who become disillusioned and get a divorce may try a second, third, or fourth marriage. What keeps drawing people back to marriage? For most people the answer, quite simply stated, is "I want to be happy. If I'm not happy with the first person I marry, I'll try another." Americans define a successful marriage in terms of happiness and self-fulfillment. Harsha Mookherjee's piece "Marital Status, Gender, and Perception of Well-Being" supports the idea that married people express more satisfaction, with life in general, than do unmarried people.

Although today we expect personal happiness in marriage, happiness was not always the primary focus of marriage. In the past, people viewed marriage as based on responsibilities that accompanied tightly scripted roles. In American colonial times, marriage was a partnership for producing children and the things the family needed to stay alive. The payoff of living in a family, where every person knew and accepted his or her responsibilities, was survival. With prosperity, self-sacrifice for the good of the family became less important. However, society still pressured people to conform to roles of husband and wife, mother and father. As compensation for accepting these responsibilities, society provided social sanction for a sexual union. The government also made certain tax benefits and insurance options available only to married couples.

If we accept that people in committed relationships decide whether to marry or cohabitate in cost–benefit terms, then the increase in cohabitation since the middle of this century reflects a change in the power of the social sanctions attached to marriage. In "A Comparison of Marriages and Cohabiting Relationships," Steve Nock finds this lack of social sanction of cohabitation still has some impact. He cites it as contributing to difficulties in relationships between cohabitating couples and their parents. He also attributes problems to the lower level of relationship commitment that is evident in cohabitating partners. This ranking of personal fulfillment over obligation is viewed as a positive development by some Americans. No longer are the social pressures to live up to the responsibilities and expectations of marriage so strong that people will commit to destructive or unsatisfying relationships.

The societal concern that has accompanied this change is how to ensure that responsibility and commitment to family are not sacrificed for self-fulfillment. However, self-fulfillment is not incompatible with family life; indeed, some people argue that you can only be a good family member when you are self-fulfilled. Perhaps, then, the institution of marriage is not dying but is merely undergoing a transformation that will allow people to balance freedom and mutual responsibility in their relationships. The hope that within a family we can find a remarkable blend of self-fulfillment and belonging is, in fact, what motivates people to continue to form families.

The final reading in the chapter, "Marital Behavior and Expectations: Ethnic Comparisons of Attitudinal and Structural Correlates" by M. Belinda Tucker and Claudia Mitchell-Kernan, is included to remind us that more than personal motivations and cultural norms are at work in stimulating marriage and cohabitation trends. As these authors do, we must all consider the impact on individual behavior of the conditions in the wider society. Specifically, this study highlights the importance, in influencing choice, of perceived availability of mates and economic conditions.

Values, Attitudes, and the State of American Marriage

Norval D. Glenn

This [reading] deals with a paradox, namely, that marriage remains very important to adult Americans—probably as important as it has ever been—while the proportion of Americans married has declined and the proportion successfully married has declined even more. Most people say that having a good marriage is one of their most important goals in life, and no other variable is more predictive of the health, happiness, and general well-being of adults than whether or not they are in satisfactory marriages. The importance high school seniors say they place on marriage has increased in recent years, even though journalistic and social scientific observers of adults continue to see signs of a "retreat from marriage."[1]

This paradox can be resolved by assuming that the decline in the probability of marital success has resulted from forces external to values, attitudes, and feelings concerning marriage. For instance, if economic and demographic changes have erected new barriers to marital success, a continued high motivation to achieve that success is unlikely to be sufficient to prevent a decline in achievement. Indeed, most authorities on American marriage rely partly on such trends as the declining earnings of young men and the increasing financial independence of women to explain the decrease in the proportion of adults who are married.

Most of these same authorities also believe, however, that changes in values, attitudes, and norms have affected American marriage. Rarely do discussions of contemporary marriage fail to mention, for instance, that spouses now expect more from marriage than they once did and that the roles of husband and wife have been redefined. A few authors refer to a decline in commitment to marriage as an institution and similar cultural and psychological changes that tend to weaken the institution and lower the probability that individual marriages will succeed.

Critics of the latter view cite national survey data on the importance of marriage to Americans as evidence that the alleged cultural undermining of marriage has not occurred. However, having a good marriage could remain a salient goal while the values and norms conducive to attainment of that goal become weaker. People could want and expect more from marriage while they become less willing to make the sacrifices and "investments" needed for marital success.

My purpose here is to consider whether or not such cultural changes have recently occurred in American society—whether or not the resolution of the paradox mentioned above is substantially within the realm of values and attitudes. For evidence I turn to data from recent national surveys of adults and adolescents, first to review the trends in American marriage that need to be explained and then to assess the attitudes that may help to explain them.

The State of American Marriage

The initial reaction of American family social scientists to the "divorce boom" that began in the mid-1960s and continued through the 1970s was generally positive. Most discussions of this trend emphasized that it did not indicate a corresponding increase in the tendency for marriages to go bad, since it reflected primarily, if not entirely, a decreased willingness of spouses to endure unsatisfactory marriages. And the latter change, according to the prevailing view, indicated that people were coming to place more, not less, importance on marriage.

If this view (which remained virtually unchallenged among family social scientists until I began reporting evidence inconsistent with it early in the 1990s) had been correct, the average quality of intact marriages would have increased steadily and rather sharply as the divorce rate climbed and as persons in the older and less divorce-prone cohorts became a smaller percentage of the married population—a trend that continued after the divorce rate leveled off in the early 1980s. However, the predicted increase in marital quality did not occur, as the 1973–93 data for currently married persons show. [2]

Rather, the proportion of married persons who reported that their marriages were "very happy" declined slightly—an indication that the probability of marital failure increased substantially. Furthermore, a downward trend in the probability of marital success is clearly indicated by the declines from 1973 to 1993 in the proportions of ever-married persons, and of all persons age thirty and older, who were in marriages they reported to be "very happy." [3]

One might think that the lowering of legal, moral, and social barriers to divorce would at least have diminished the proportion of adults in poor marriages, but the 1973–93 data in Figure 1 on the percent of all persons age eighteen and older who were in marriages they reported to be less than "very happy" (labeled "unhappily married") show virtual stability. The percent in "very happy" marriages (labeled "happily married") declined substantially while the percent unmarried (never-married, divorced, separated, or widowed) increased proportionately.

A major reason for concern about the decline in marital success is its effects on children, and the trends suggest that those effects have been more than trivial. Virtually everyone agrees that the best situation for children, all else being equal, is for them to live with biological (or adoptive) parents who have a good marriage. There is no agreement on the relative

FIGURE 1 *Percent of Persons Age Eighteen and Older in Each of Three Marital Situations, by Year*

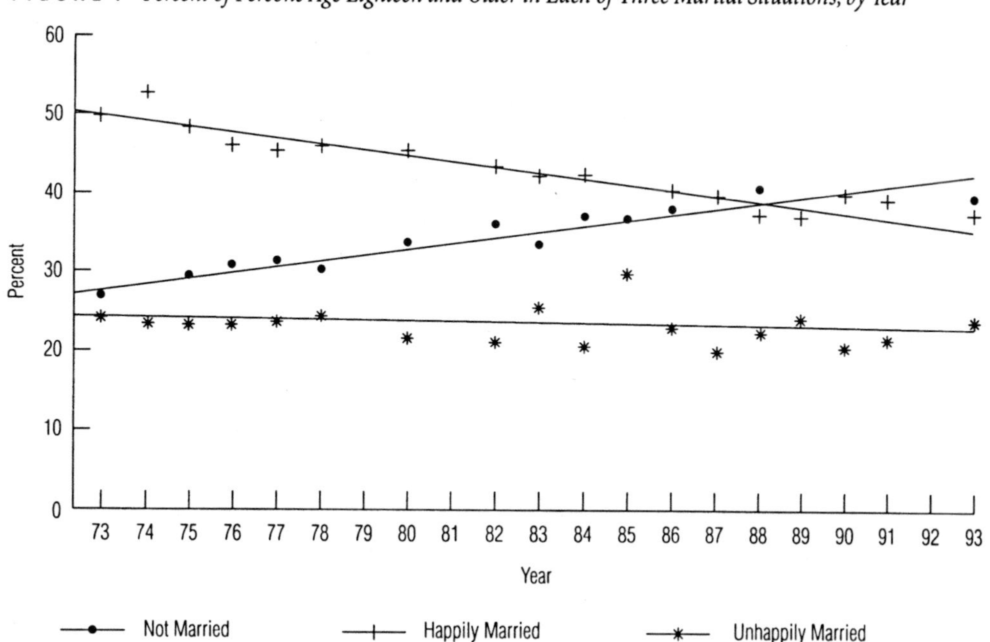

badness of other situations—single-parent, stepfamily, and unhappily married parent situations all being considered less than ideal. The percent of persons under age eighteen living with a less-than-happily married parent remained about the same, the percent living with a happily married parent declined, and the percent living with a single parent increased. The negative changes were even greater than the data indicate, since some of the pre-adults living with happily married parents were in stepfamilies, which is less than ideal, and the percent in such situations is known to have increased.

The percent of persons who were in successful (that is, intact and happy) first marriages at any given number of years after they first married has declined substantially in recent years,[4] the proportion after ten years now being about one-third. When the tendency for some survey respondents to overreport the quality of their marriages is taken into account, along with the fact that some persons who report their marriages to be "very happy" have spouses who disagree, the estimate of the proportion of first marriages that are successful after ten years almost certainly should not exceed about 25 percent.

This evidence, though inconsistent with the sanguine views that have prevailed among family social scientists, is congruent with what most laypersons think has happened to American marriage. For instance, a majority of the respondents to the Virginia Slims American Women's Opinion Polls in 1974, 1979, 1985, and 1989 said they thought the institution of marriage was weaker than it was ten years earlier, although the percent was lower in 1989 (61) than in 1974 or 1979 (70).[5]

The Importance of Marriage

An observer exposed only to the data in the preceding section might be inclined to suspect that the "retreat from marriage" has been to a large degree psychological, that Americans are marrying less and succeeding less often at marriage because alternatives have become more attractive, relative to marriage, than they once were. Such a change could have occurred because persons perceive that marriage has become less effective in meeting their needs and desires, and/or because they perceive that alternatives to marriage have become more effective. One cannot be certain that no such psychological retreat from marriage has occurred, but survey data on attitudes toward marriage gathered over the past quarter of a century provide scant evidence for it.

Unfortunately, there are no strictly comparable data gathered at regular intervals over a period of years concerning the importance that American adults place on marriage. However, all of the relevant data from the past thirty or so years show that adults of all ages say that having a "happy marriage" is one of their most important life goals. Some of the most sophisticated evidence on this topic is from the Quality of American Life Study conducted in 1971 by researchers in the Institute for Social Research at the University of Michigan. Respondents were asked to rate twelve "life domains" (ranging from "an interesting job" and "a large bank account" to "having good friends" and "a happy marriage") on a five-point scale ranging from "extremely important" to "not at all important." The highest percentage of "extremely important" ratings (74) were given to "having a happy marriage," followed by "being in good health and in good physical condition" (70) and "having a good family life" (67). When the respondents were asked to pick the two most important of the twelve domains, "a happy marriage" was selected most frequently (by 55 percent of the respondents), followed by "a good family life" (36) and "being in good health and in good physical condition" (35).[6]

More recent studies have yielded similar findings. For instance, the Massachusetts Mutual American Family Values Study in 1989 asked 1,200 respondents, who were interviewed by phone, to rate twenty-nine "values" on a five-point scale ("one of the most important," "very important," "somewhat important," "not too important," and "not at all important"). Among the "values" that could reasonably be considered life goals, "having a happy marriage" ranked first, being indicated as "one of the most important" by 39 percent of the respondents and "one of the most important" or "very important" by 93 percent. In contrast, the percent giving the "one of the most important" rating to each of the individualistic and materialistic goals was much smaller, being 18 for "earning a good living," 16 for "being financially secure," 8 for "having nice things," and 6 for "being free from obligations so I can do whatever I want to."[7]

Each year since 1976, the Monitoring the Future Survey conducted by the Institute for Social Research at the University of Michigan has asked a sample of high school seniors to rate fourteen life goals on a four-point scale ranging from "extremely important" to "not important." In Table 1, I report the percent of the 1992 respondents who rated each of the goals "extremely important." "Having a good marriage and family life" ranked first, being given the highest rating by almost four-fifths of the students, although "being able to find steady work" was a close second.[8]

The trend data from the annual Monitoring the Future Survey are generally inconsistent with the hypothesis that there has been a psychological retreat from marriage among young persons on the threshold of adulthood. For instance, the trend in the "extremely important" ratings given to "having a good marriage and family life" was slightly upward for both males and females from 1976 to 1992. There were also slight upward trends in the percent who said they definitely would prefer to have a mate most of their lives and in the percent who said they most likely would choose to marry or who were already married. Of course, the period covered by these data began after most of the increase in divorce that started in the mid-1960s had already occurred, and there could have been attitudinal changes prior to 1976 opposite in direction from those shown in the figures. However, it seems unlikely that any such trends that were substantial in the 1960s and early 1970s would have completely ceased by the late 1970s.

TABLE 1 *Percent of High School Seniors Who Said Certain Life Goals Were "Extremely Important," 1992*

Goal	%	Rank
Having a good marriage and family life	78	1
Being able to find steady work	77	2
Being successful in my line of work	66	3 (tie)
Being able to give my children better opportunities than I've had	66	3 (tie)
Having strong friendships	62	5
Finding purpose and meaning in my life	58	6
Having plenty of time for recreation and hobbies	30	7
Having lots of money	29	8
Making a contribution to society	22	9
Discovering new ways to experience things	21	10
Working to correct social and economic inequalities	15	11
Getting away from this area of the country	13	12 (tie)
Being a leader in my community	13	12 (tie)
Living close to parents and relatives	12	14

Source: Monitoring the Future Project, Institute for Social Research, University of Michigan. Data are from approximately 2,700 respondents.

The importance that people say they place on marriage does not necessarily mean, of course, that marriage continues to have important effects on their lives, but there is ample evidence that it does. As a whole, persons in satisfactory marriages are happier, healthier, more productive, and less inclined to engage in socially disruptive behavior than other adults, and at least among persons beyond the earliest stages of adulthood, there is no evidence of appreciable recent decline in these differences.[9] For instance, the data in Figure 2 on the reported personal happiness of persons age thirty-five or older who were not married, happily married, and unhappily married show virtual stability from 1973 through 1993.[10] The relationships indicated by these and similar data reflect to some degree the selection of happy, well-adjusted, and healthy persons into successful marriages, but most researchers who have studied them believe that they also result to a large extent from effects of marital situation on well-being and behavior.

Although the importance Americans say they place on marriage is consistent with the strength of the effects that marital situation seems to have on them, what survey respondents say about their attitudes and values concerning marriage should not necessarily be taken at face value. Many people may tend to give socially desirable responses or to respond in terms of what they think their values and attitudes should be rather than in terms of what they are—a topic to which I return below.

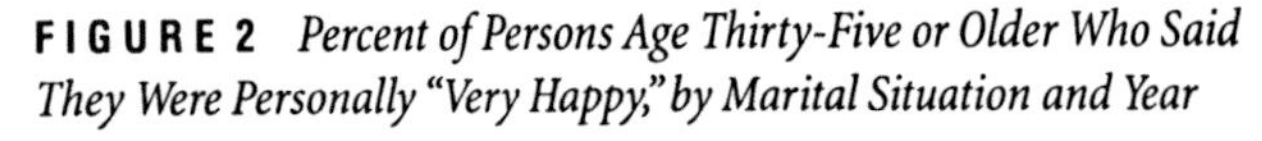

FIGURE 2 *Percent of Persons Age Thirty-Five or Older Who Said They Were Personally "Very Happy," by Marital Situation and Year*

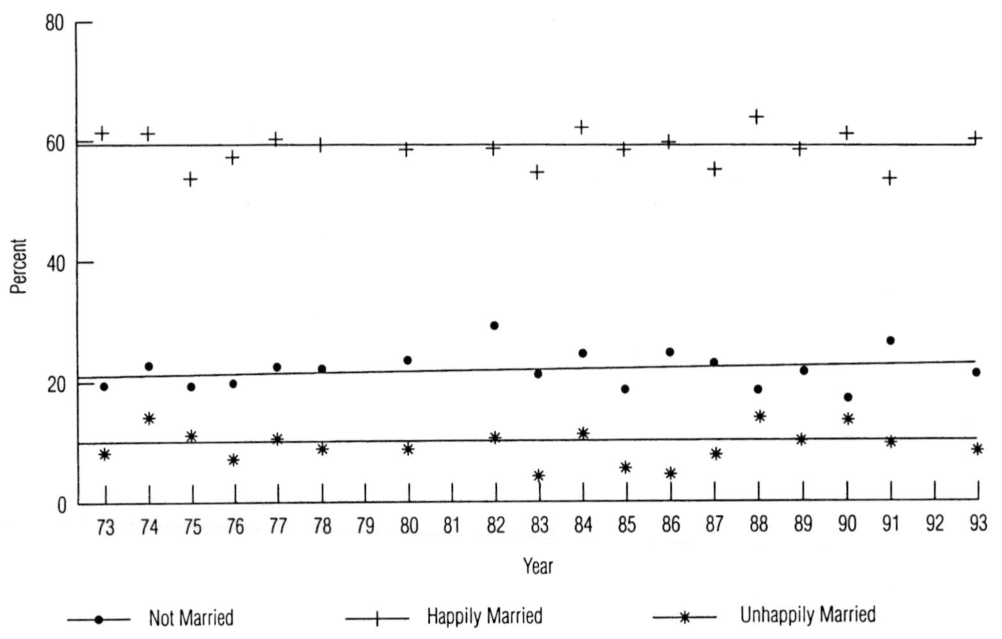

Some Evidence of Anti-Marriage Influence

The evidence I have presented so far would seem to support the view that values and attitudes supportive of marriage are strong and that any "retreat from marriage" must have resulted largely from situational influences, such as changes in job opportunities and economic pressures. However, there is also evidence of attitudinal and value changes that are likely to have lessened the probability of marital success.

The trends in the attitudes of high school seniors shown in Figures 3 and 4 can be considered anti-marriage, although the indicated changes are not large and their precise meaning is unclear. For instance, in view of the generally positive views of marriage expressed by the students in responses to other questions, it is not clear what one should make of the fact that a substantial and increasing proportion said they were inclined to question marriage as a way of life (Figure 3). The downward trend in the expectation of marital permanence . . . is more understandable and is not inconsistent with other trends, but the meaning of the decline in pro-marriage responses from females shown in Figure 4 is uncertain. It could reflect primarily a greater acceptance of being single for others, but not for oneself, or changes in views of nonmarital cohabitation rather than of solitary living.

The limited trend data on adults' attitudes toward marriage suggest changes likely to have weakened the institution. For instance, the Americans View Their Mental Health Surveys, conducted in 1957 and 1976, asked respondents how a person's life is changed by being married and classified the responses into positive, neutral, or negative. The positive responses declined from 43 to 30 percent from 1957 to 1976 and the negative ones increased from 23 to 28 percent.[11]

Probably the most important change in attitudes toward marriage has been a weakening of the ideal of marital permanence—a change that virtually all observers of American marriage agree has occurred even though there apparently are no national trend data on the topic. The best "hard" evidence on the topic is from the Study of American Families, a panel

FIGURE 3 *Percent of High School Seniors Who Said They Agreed or Mostly Agreed That "One Sees So Few Good Marriages That One Questions It as a Way of Life," by Sex and Year*

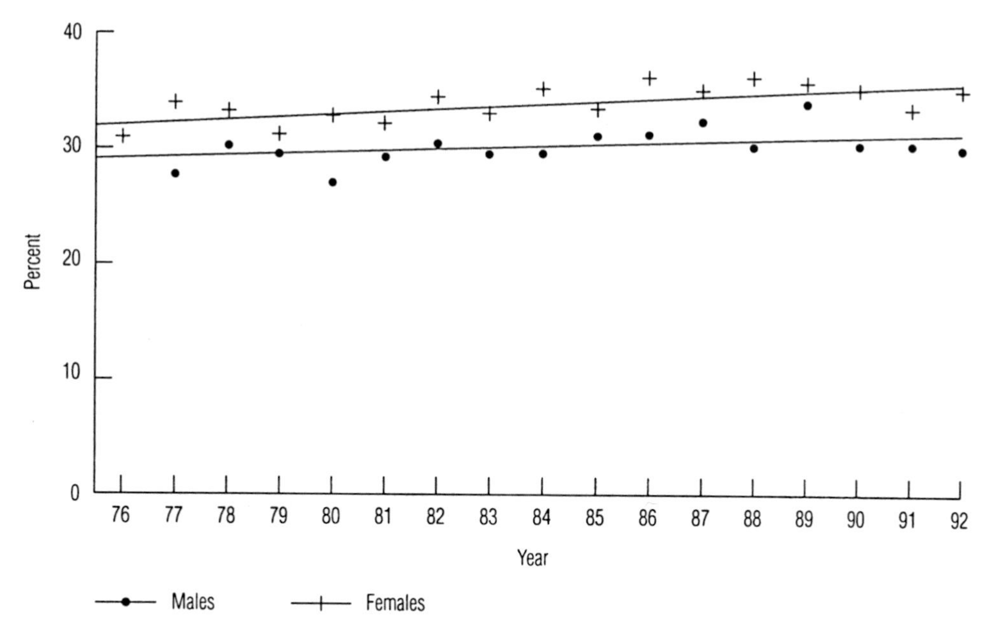

FIGURE 4 *Percent of High School Seniors Who Said They Agreed or Mostly Agreed That Most People Will Have Fuller and Happier Lives If They Choose Legal Marriage Rather Than Staying Single, or Just Living with Someone, by Sex and Year*

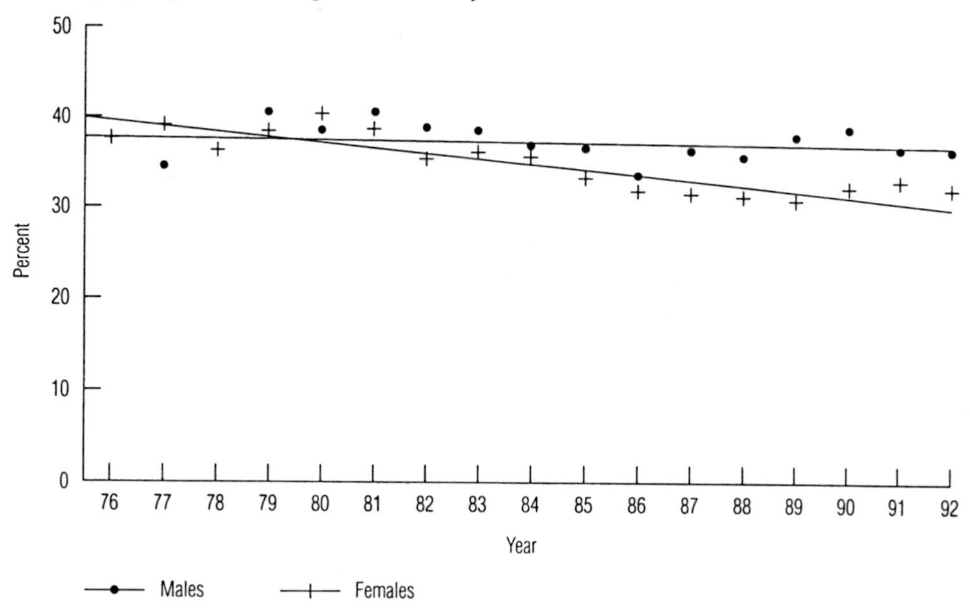

study that interviewed the same sample of mothers at four different dates. In 1962, 51 percent of the respondents said parents who do not get along should *not* stay together because there are children in the family, compared with 80 percent in 1977 and 82 percent in 1980 and 1985.[12] Since there is no reason to think that the respondents' growing older would cause such a change, it almost certainly reflects a similar but probably larger change in the entire adult population.

Almost as important as changes in values and attitudes about marriage are changes in the strength of other values and goals, such as materialistic and achievement ones, that may detract from the pursuit of marital success. The best evidence on trends in such values is, again, for late adolescents and very young adults. Among the high school seniors who responded to the Monitoring the Future Surveys, the percent who said "having lots of money" was extremely important went from 15 in 1976 to 29 in 1992, and the percent rating "being successful in my line of work" extremely important went from 53 to 66. The Cooperative Institutional Research Program of the American Council on Education and UCLA found that only 40 percent of college freshmen in the early 1970s, compared with 70 percent in 1985, said "to be very well off financially" was a very important or essential life goal.[13] It is not known whether these changes were part of a longer-term trend or were merely a return to the values that existed prior to an atypically antimaterialistic period in the late 1960s and early 1970s.

There are no comparable trend data for American adults, but there is evidence that many of these persons let values they say are less important interfere with their pursuit of marital success and other family values. The respondents to the 1989 Massachusetts Mutual American Family Values Study, who as a whole rated "having a happy marriage" their most important life goal and such goals as "having nice things" and "being financially secure" much lower, were asked to imagine that they were thirty-eight years old and were offered a new job in a field they liked, that the job would require more work hours and take them far away from their families more often, and that it would be more highly rewarded in certain ways than their present job. Just how the new job would be more highly rewarded differed among three versions of the question, each of which was asked of one-third of the sample (400 respondents). For two of the subsamples, the increment in reward included higher pay, but for the third subsample it was only greater prestige. The respondents were then asked how likely it would be that they would take the new job—very likely, somewhat likely, somewhat unlikely, or very unlikely. The responses are shown in Table 2.

TABLE 2 *Responses (in Percent) of Three Adult National Subsamples to Questions About the Likelihood of Their Taking a More Highly Rewarded Job That Would Take Away Family Time, 1989*

	Increment in Reward		
	More Prestigious Only	Plus 15% More Pay	Plus 35% More Pay
Very likely	38	34	32
Somewhat likely	27	33	37
Somewhat unlikely	36	34	31
Very unlikely	0	0	0
Total	101*	101*	100
(n)	(400)	(400)	(400)

*Percentages do not add to 100 because of rounding.
Source: The Massachusetts Mutual American Family Values Study conducted in June of 1989 by Mellman and Lazarus, Inc.

In each of the three subsamples, around one-third or more of the respondents said it was very likely they would take the new job, and almost another third or more said it was somewhat likely they would do so. Not a single respondent said it was very unlikely that he or she would take the job. Belief that the spouse and other family members would benefit might help account for the willingness of the respondents to accept a job with higher pay, but prestige is primarily a personal reward rather than one readily shared with family members.

These data take on added meaning in light of the fact that the question about the new job came in the middle of the interview after many questions about family values and almost directly (with only one intervening question) after the following ones:

1. Do you think most people today put a higher value on family, or do most people put a higher value on material things?
2. Over the course of an average week, please tell me what percentage of your waking time you spend being with your family.
3. Would you say that you spend too much, about the right amount, or not enough time with your family?

It is well established that responses to survey questions can be influenced by the content of preceding questions, and in this case the expectation is that having responded to the earlier questions should have lowered the respondents' tendency to say "very likely" or "somewhat likely." This is especially the case in view of the fact that 85 percent said most people today put a higher value on material things, and 46 percent admitted they did not spend enough time with their families. Indeed, a "question order effect" is the only reasonable explanation for the "very likely" responses not being higher when greater pay was part of the increment in rewards associated with the new job than when the increment was only greater prestige. The earlier question about "material things" seems to have predisposed respondents not to choose greater pay over family time but apparently had less effect on their tendency to choose prestige.

These findings suggest that survey respondents are inclined to exaggerate the importance they place on marriage and the family and that many people will risk sacrificing marital success in pursuit of goals they say are less important.

Reasons for the Decline of Marriage

The survey data reviewed above provide only a very sketchy picture of what has happened to American marriage. Many of the questions that could have provided insight into changes in marriage-related values and attitudes have not been asked,[14] and structured survey questions are inherently limited in their ability to deal with the subtleties and complexities of cultural phenomena. Furthermore, since it is impossible to demonstrate conclusively just how the trends in marital success and in attitudes toward marriage are causally interrelated, the data are amenable to differing interpretations.

To me, however, the data, when considered in light of theory and along with other kinds of evidence, suggest one major conclusion: The very importance that people place on marriage as a source of gratification has contributed to the decline of marriage as an institution. Explanation is in order.

A conjugal family system, of which the United States has an extreme form, is centered around the marriage relationship, in contrast to a consanguine family system, in which "blood" relationships are the crucial ones. In a conjugal system, spouses choose one another instead of their marriages being arranged by others, and providing for the needs and desires of the spouses is considered a primary purpose, if not *the* main purpose, of marriage. Of course, marriage in any family system performs societal functions, such as providing much of the early care and socialization of children, but in conjugal systems marital success tends to be defined in hedonistic terms and from the perspective of the spouses. The successful marriage is one that provides happiness, satisfaction, and other positive feelings to the husband and wife.

The United States has always had a conjugal family system, and by the time social scientists began writing about American marriage early in this century, hedonistic and individualistic criteria of marital success already prevailed. However, at that time persons were encouraged to pursue marital success by choosing spouses wisely and working to maintain the marriage—not, except in extraordinary circumstances, by moving from one marriage to another. Furthermore, the happiness and satisfaction pursued and attained through marriage were to a large degree through such practical benefits as economic security, social standing, and the receipt of domestic services, and these benefits were more obviously enhanced by marital stability than the less tangible ones resulting from companionship and the pleasures of associating with the spouse. Only in the past few decades has the single-minded pursuit of marital happiness through the attraction and retention of an intrinsically desirable spouse received strong and virtually unqualified social encouragement.

The greater emphasis on having an intrinsically good marriage has been accompanied by a decline in the ideal of marital permanence. To many progressive thinkers, it has seemed reasonable that lowering the legal, moral, and social barriers to divorce would enhance the average quality of intact marriages and facilitate movement from poor marriages to better ones. Although the increase in divorce that began in the mid-1960s may have come about largely for other reasons, such as the decline in economic dependence of wives on husbands, professionals interested in improving marital quality, such as social scientists and therapists, provided a strong rationale, based on the goal of enhancing personal happiness, for a greater social acceptance and even encouragement of divorce.

According to Bernard Farber, the decline in the ideal of marital permanence has taken us substantially toward a condition he calls "permanent availability." By this he means that all adults, regardless of marital status, tend to remain on the marriage market. That is, married persons as well as unmarried ones tend to assess the marital desirability of members of the opposite sex they know and meet, whether those persons are married or not, and will consider moving from their current marriage to one they anticipate will be more satisfactory. If permanent availability were to become universal, few persons would remain married for life to their first spouse.[15]

The progressives who believed that lowering the barriers to divorce and moving toward permanent availability on the marriage market would necessarily enhance the quality of marriages and contribute to personal happiness ignored what the most astute social philosophers have always known, namely, that a completely unfettered pursuit of self-interest by individuals does not lead to the maximization of the well-being of the population as a whole.[16] They ignored the fact that the freedom of one spouse to leave the marriage at will is the other spouse's insecurity, and that without a reasonable degree of security, it is unlikely that a spouse will commit fully to the marriage and make the sacrifices and investments needed to make it succeed.[17]

Furthermore, marital discontent will almost certainly result when a person constantly compares his or her marriage to real or imagined alternatives to it. Persons are hardly aware of needs well served by their marriages but are acutely aware of those not very well served, and there always are some. Therefore, the grass will always tend to look greener on the other side of the marital fence; people will always tend to imagine they would be happier married to someone else or not married. It is also relevant that persons not intimately known often appear to be more desirable as prospective spouses than they really are. The person who has revealed all or most of his or her faults and weaknesses to a spouse is always at a disadvantage when competing with the well-cultivated public images of other men or women.

Although the weakening of the ideal of marital permanence is likely to be a crucial reason for the decline in marital success, I do not believe it is the only one. Increased expectations of marriage and a breakdown in consensus on the content of marital roles are almost certainly involved, and other cultural trends may have had an effect. For instance, if, as some social critics maintain, there has been a general increase in American society of a sense of entitlement—in what people believe they should receive from others—while there has been a decline of a sense of duty—of what people believe they should give to others—all institutions, including marriage, must have suffered.[18] A related change may have been an increased tendency for people to feel they can, and deserve, to have it all—including success at marriage, parenthood, and work—and a decreased recognition that a relentless pursuit of career goals and financial success is likely to interfere with attainment of marital and parental goals.

Situational influences, such as the decline in the earnings of young men, are also undeniably responsible for some of the changes in marriage, but it seems to me that the resolution of the paradox addressed by this chapter is largely in the realm of culture. As the purpose of marriage is coming to be defined more exclusively as the gratification of the married persons, and as marriage is becoming more nearly just a personal relationship, the nature of which is determined largely by the private negotiations of each married couple, the traditional institutional functions of marriage are being less well performed. The consequences of this change for children are now widely recognized, and those for adults, while less severe, also seem to be distinctly negative. An increasingly hedonistic form of marriage seems to be decreasingly able to facilitate the hedonistic strivings of those who participate in it.

NOTES

1. E.g., Bryce Christensen, ed., *The Retreat from Marriage: Causes and Consequences* (Lanham, MD: University Press of America, 1990).
2. These and all other data from the General Social Surveys are weighted on the number of persons in the household age 13 and older because the sampling is representative of households and not adult individuals. Unweighted data do not accurately indicate the trends of interest here.
3. A false appearance of a decline in the probability of marital success could have been created by persons' becoming more willing to admit their marriages were not of the highest quality, but apparently no such change occurred. If the reports of marital happiness had become more nearly accurate, their correlation with other variables, such as reports of personal happiness, should have increased, and that did not happen.

   The alternatives presented to the respondents by the marital happiness question are "very happy," "pretty happy," and "not too happy." There is evidence that most respondents who fail to give their marriages the highest rating have rather serious marital problems.
4. Norval D. Glenn, "The Recent Trend in Marital Success in the United States," *Journal of Marriage and the Family* 53 (1991): 261–70.
5. These data were provided by the Roper Public Opinion Research Center at the University of Connecticut.
6. Angus Campbell, Philip E. Converse, and Willard L. Rodger, *The Quality of American Life: Perceptions, Evaluations, and Satisfactions* (New York: Russell Sage Foundation, 1976), ch. 3.
7. The data were provided by the Roper Public Opinion Research Center at the University of Connecticut.
8. All data from the Monitoring the Future Surveys reported here are from an 18-volume set authored/compiled by Jerald G. Bachman, Lloyd D. Johnston, and Patrick M. O'Malley (order of names varies by year) entitled *Monitoring the Future: Questionnaire Responses from the Nation's High School Seniors [year]* (Ann Arbor, MI: Institute for Social Research, 1975 through 1992). The data for 1975 are not reported here, since changes in question order after 1975 make the 1975 data incomparable with the later data.
9. Walter Gove, "The Relationship between Sex Roles, Marital Status, and Mental Illness," *Social Forces* 51 (1972): 34–44; James S. House, Karl R. Landis, and Debra Umberson, "Social Relationships and Health," *Science* 241 (1988): 540–45.
10. The differences did decline noticeably for very young adults due to an increase in the reported happiness of young never-married persons.
11. Joseph Veroff, Elizabeth Douvan, and Richard A. Kulka, *The Inner American: A Self-Portrait from 1957 to 1976* (New York: Basic Books, 1981), ch. 4.
12. Arland Thornton, "Changing Attitudes toward Family Issues in the United States," *Journal of Marriage and the Family* 51 (1989): 873–93.
13. Alexander W. Astin, Kenneth C. Green, and William S. Korn, *The American Freshman: Twenty Years of Trends* (Los Angeles: Higher Education Institute, 1987), p. 23.
14. My review here of the survey data is by no means comprehensive, but adding the missing evidence would not make the picture appreciably less sketchy.
15. Bernard Farber, "The Future of the American Family: A Dialectical Account," *Journal of Family Issues* 8 (1987): 431–33.
16. It is ironic that many persons who are adamant in their rejection of the notion of an "invisible hand" in the economic marketplace (that produces social benefits from the self-interested strivings of economic actors) believe that something akin to an invisible hand can be depended on to enhance general well-being in the realm of marriage and other family relations.
17. I do not claim that the lowering of barriers to divorce was not beneficial to a point. People need the freedom to leave abusive and extremely unsatisfactory marriages, but the barriers can be, and I believe in American society now are, too weak.
18. These critics include, but are not limited to, many of the leaders of the communitarian movement. See various issues of the journal *The Responsive Community.*

READING 7

Marital Status, Gender, and Perception of Well-Being

Harsha N. Mookherjee

Bernard (1972) claimed that marriage in the United States is beneficial to men but not beneficial to women; subsequent findings have not corroborated Bernard's claim. Although most research findings support a positive relationship between marital status and various measures of physical and psychological well-being (Campbell, Converse, & Rodgers, 1976; Glenn, 1975; Glenn & Weaver, 1979; Gove, 1972, 1973, 1979b; Gove, Hughes, & Style, 1983; Gove, Style, & Hughes, 1990; Lowe & Smith, 1987; Pearlin & Johnson, 1977; Ross, Mirowsky, & Goldsteen, 1990; White, 1992; Williams, 1988), gender differences in well-being among married and unmarried people in the United States are well documented (Aneshensel, Frerichs, & Clark, 1981; Bernard, 1972; Glenn, 1975; Glenn & Weaver, 1988; Gove, 1979a; Gove, Style, & Hughes, 1990; Gove & Tudor, 1973; Radloff, 1975). Williams (1988) did not support the contention that "being married is of considerably greater importance for the well-being of men in comparison with women" (p. 465); rather, she found the quality of marital interaction to be more important for individual well-being than the fact of being married. In addition, Williams found marital quality to be positively associated with well-being in men and even more strongly associated with well-being in women. Similarly, Glenn and Weaver (1988) found that married women were happier than were married men.

The author thanks the National Opinion Research Center for providing the data used in this study and Donna K. Darden for valuable comments on an earlier version of this article.

Several explanations have been offered for the findings of gender differences in well-being among married and unmarried individuals. Some relate to the fundamental question of whether there exists a causal relationship between well-being and marital status. Glenn (1975) and White (1992) did not consider marriage a causal factor in relation to well-being, but Gove et al. (1990) concluded that marriage had a strong causal relationship to well-being.

The proponents of the selection explanation argue that healthy, happy people are more likely to be married than unhealthy, unhappy people are. In addition, age at marriage could be another factor in well-being (Gove et al., 1990). Proponents of social roles explanations suggest that men derive greater benefits from marriage than women do because men's roles are less stressful and more gratifying than those of women (Bernard, 1972; Gove, 1972, 1979b; Gove & Tudor, 1973). Study results support the view that social role may be associated with a variety of gender differences in psychological outcomes (Gore & Mangione, 1983; Gove & Geerken, 1977; Kessler & McRae, 1982; Pearlin, 1975; Ross, Mirowsky, & Huber, 1983; Vanfossen, 1981). Proponents of the social roles explanation suggest that "men's well-being is thought to be enhanced within marriage because men derive instrumental, role related benefits from the marital state that are not available to women" (Williams, 1988, p. 454).

The literature supports the explanation of marriage as an interpersonal relationship contributing to physical, mental, and social well-being (Gove et al., 1983; House & Kahn, 1984). Studies also support the contention that marital happiness significantly affects mental health, global happiness, and overall home and life satisfaction for women (Brown & Harris, 1978; Gove et al., 1983; Vanfossen, 1981). Affective quality of marriage may be more strongly related to women's perception of well-being than to that of men because women may be more likely than men to rely on their marital roles for a sense of personal value and self-worth (Frieze, Parsons, Johnson, Reble, & Zellman, 1978; Gove et al., 1983; Vanfossen, 1981).

Gender differences in the impact of marital quality on well-being may be due to the differential orientations and expectations that men and women bring to the marital relationship. Results of earlier studies suggest that because women tend to relate to others on a more intimate basis (Williams, 1988), they are more sensitive to emotional nuance in interpersonal relationships (Rossi, 1985) and have more intimate involvements with others than men do (Williams, 1988); hence, their expectations for intimacy and emotional support within marriage are likely to be high. On the other hand, men who depend primarily on their spouses for emotional support are less affected by the quality of marital interaction because they have fewer bases for comparison than do women (Komarovsky, 1962; Levinson, 1978; Williams, 1988). Therefore, women who experience a sense of relative emotional deprivation in their marital relationships may experience a greater loss in well-being than similarly affected men would.

My primary purpose in the present study was to reexamine gender differences in perception of well-being among the married and unmarried population in the United States. Because the results of earlier studies (Andrews & Withey, 1976; Campbell et al., 1976; Glenn & Weaver, 1981; Herzog, Rodgers, & Woodworth, 1982; Mookherjee, 1988, 1992; Olson, 1980) have indicated that race and financial status are significant variables in influencing an individual's perception of well-being, I used these variables as control variables in the present study.

Method

I used data from the combined 10-year (1982–1991) "General Social Surveys" conducted by the National Opinion Research Center (NORC, 1991). Because these surveys were independently conducted and indicated no differences in the perception of well-being among the sampled populations, I pooled the data for the analysis, which was thus based on information collected from 12,168 adults between 18 and 89 years of age from the United States.

Gender and marital status were the main independent variables; race and financial status were the control variables. Gender and race were dichotomized into male and female and White and non-White, respectively.

I defined financial status not as the family's actual annual income but rather as the respondent's perceived comparative financial standing in his or her community at the time of the survey. I measured this perception with the question "Compared with American families in general, would you say your family income is far below average, below average, average, above average, or far above average?" I divided the responses into three categories: below average, average, and above average.

I divided marital status into five categories: (a) married, (b) widowed, (c) divorced, (d) separated, and (e) never married.

I based measurement of the dependent variable perception of well-being on the premises that (a) measures that reflect cognitive or evaluative dimensions of well-being are more sociologically interesting and meaningful than measures that do not (Ellison, Gay, & Glass, 1989; George, 1981), (b) a combination of domain-specific indicators of life satisfaction is the best measure of subjective well-being (Campbell et al., 1976; Ellison et al., 1989; George, 1981), and (c) a simple summary index comprising the dimensions of social life would be as good a measure of perceived well-being as any other.

I defined perception of well-being as the cumulative score derived from a respondent's answers to six attitudinal questions in the "General Social Surveys" asking how satisfied the respondent was with regard to (a) then-current place of residence, (b) nonworking activities (hobbies), (c) family life, (d) friendship, (e) health and physical condition, and (f) financial situation. I divided each response into three categories: (a) not satisfied at all, (b) somewhat satisfied, and (c) well satisfied. These life satisfaction responses ranged from least satisfied (6) to most satisfied (18) and were considered as the scores for perception of well-being, $M = 14.48, SD = 2.50$.

Results

Table 1 contains the mean scores of perception of well-being by gender and marital status. The results of analysis of variance (ANOVA) indicated a significant difference in the mean scores of perception of well-being between gender and marital status, overall $F(5, 12158) = 109.69, p < .0001$. The effect of marital status on perception of well-being was statistically significant, $F(4, 12158) = 136.60, p < .0001$, but that of gender was not, $F(1, 12158) = 9.90, p < .002$, with .001 considered to be the level of significance.

The differences in mean scores of perception of well-being between married and widowed, divorced, separated, or never married respondents were also statistically significant. A comparison between married persons and widowed persons, 14.91 vs. 14.32, $t(\sim) = 7.49, p < .001$, and between married persons and separated persons, 14.91 vs. 13.18, $t(\sim) = 13.39$, $p < .001$, indicated that partner loss or marital dissolution had a substantial negative effect on perception of well-being. I found similar statistically significant differences between the mean scores of perception of well-being for men and women separately. These findings reveal the difference in perception of well-being between married and previously married respondents. Results of further comparisons between widowed persons and divorced persons, 14.32 vs. 13.73, $t(\sim) = 5.84, p < .001$, and widowed persons and separated persons, 14.32 vs. 13.18, $t(\sim) = 7.81$, $p < .001$, support the contention that marital stress affects perception of well-being. These results corroborate earlier findings (Bloom, Asher, & White, 1978; Booth & Amato, 1991; Bowling, 1987).

When the married respondents were compared with those who never married, the mean difference score on perception of well-being was statistically significant, 14.91 vs. 14.06, $t(\sim) = 14.41, p < .001$. Similarly, a comparison of mean difference scores on perception of well-being for married and never married men, 14.89 vs. 13.97, $t(\sim) = 14.29, p < .001$, and married and never married women, 14.92 vs. 14.15, $t(\sim) = 9.06, p < .001$, was statistically significant. A comparison of total married respondents with the remaining respondents, who were not married at the time of surveys, revealed a statistically significant difference of mean scores on perception of well-being, 14.91 vs. 13.97, $t(\sim) = 21.03, p < .001$.

However, the differences of mean scores on perception of well-being between men and women and between married men and women were not statistically significant, 14.45 vs. 14.51, $t = -1.31$, and 14.89 vs. 14.92, $t = -0.53$, respectively. In both cases, the negative signs indicate that the women were more satisfied than the men. On the other hand, a comparison

TABLE 1 *Mean Scores on Perception of Well-Being, by Marital Status and Gender*

	Male			Female			Total		
Marital status	*n*	*M*	*SD*	*n*	*M*	*SD*	*n*	*M*	*SD*
Married	3122	14.89	2.30	3539	14.92	2.32	6661	14.91	2.31
Widowed	213	13.87	2.96	1138	14.41	2.65	1351	14.32	2.71
Divorced	462	13.44	2.81	908	13.88	2.47	1370	13.73	2.60
Separated	164	13.04	2.85	291	13.25	2.60	455	13.18	2.69
Never married	1209	13.97	2.52	1122	14.15	2.55	2331	14.06	2.53
Total	5170	14.45	2.52	6998	14.51	2.49	12168	14.48	2.50

between married and unmarried men and between married and unmarried women indicated that in both cases married individuals were more satisfied than unmarried individuals, 14.89 vs. 13.77, $t(\sim) = 14.55, p < .001$, and 14.92 vs. 14.09, $t(\sim) = 13.17, p < .001$, respectively. These findings indicate that married people reported they were more satisfied in life than unmarried people were, irrespective of gender. In other words, marriage was beneficial for both men and women.

Both race and financial status were significant intervening variables in the relationship among marital status, gender, and perception of well-being (see Table 2). Overall, White respondents were more satisfied with life than non-Whites were, 14.69 vs. 13.64, $t(\sim) = 17.74, p < .001$. Similarly, the differences of mean scores on perception of well-being between married White and non-White men, 15.00 vs. 14.20, $t(\sim) = 6.05, p < .001$, and married White and non-White women, 15.08 vs. 14.00, $t(\sim) = 8.85, p < .001$, were

TABLE 2 *Mean Scores on Perception of Well-Being, by Marital Status, Gender, Race, and Financial Status*

	Married			Not Married			
Race/ financial status	***n***	***M***	***SD***	***n***	***M***	***SD***	***t***
				Men			
White							
Below average	600	14.09	2.51	454	12.80	2.65	8.002*
Average	1324	15.05	2.12	727	14.21	2.44	7.80*
Above average	780	15.61	1.99	368	14.76	2.43	5.85*
Total	2704	15.00	2.24	1549	13.93	2.61	13.90*
Non-white							
Below average	145	13.44	2.48	219	12.40	2.84	3.70*
Average	220	14.51	2.50	226	13.96	2.53	23.11*
Above average	53	15.00	2.50	53	13.91	2.90	2.07
Total	418	14.20	2.55	499	13.27	2.81	5.20*
				Women			
White							
Below average	649	13.91	2.38	979	13.51	2.52	3.24
Average	1712	15.16	2.08	1167	14.80	2.34	4.24*
Above average	665	16.01	1.92	341	15.27	2.27	5.16*
Total	3026	15.08	2.23	2486	14.36	2.50	11.25*
Non-white							
Below average	186	13.12	2.69	523	12.66	2.65	2.01
Average	281	14.45	2.39	387	14.20	2.41	1.33
Above average	46	14.87	2.68	63	14.56	2.59	0.61
Total	513	14.00	2.61	973	13.40	2.67	1.34
Overall total	6661	14.91	2.31	5507	13.97	2.62	21.03*

*$p < .001$.

statistically significant. The same was true between married and unmarried White men, 15.00 vs. 13.93, $t(\sim) = 13.90, p < .001$, White women, 15.08 vs. 14.36, $t(\sim) = 11.25, p < .001$, and non-White men, 14.20 vs. 13.27, $t(\sim) = 5.20, p < .001$. But, among the non-White women, although the mean perception of well-being score for those who were married was higher than that of the unmarried respondents, the difference of mean scores was not statistically significant.

Financial status had significant effects on perception of well-being scores, $F(2, 12167) = 647.44, p < .00001$; respondents who perceived their financial status as average or above average expressed more satisfaction in life. The mean differences on perception of well-being scores between men and women were also significant for financially average and above average respondents, overall $F(4, 12156) = 404.27, p < .00001$; for marital status $F(1, 12156) = 265.05, p < .00001$; for gender $F(1, 12156) = 37.10, p < .00001$; and for financial status $F(2, 12156) = 559.38, p < .00001$. All two-way and three-way interactions were nonsignificant, except for the two-way interaction between marital status and gender, $F(1, 12156) = 28.45, p < .0001$, which confirmed the earlier findings that marital status affected perception of well-being of both men and women.

Table 3 contains the results of the ANOVA for perception of well-being scores by marital status, gender, race, and financial status. The mean difference scores on perception of well-being were significantly affected

TABLE 3 *Results of ANOVA for Perception of Well-Being Scores, by Marital Status, Gender, Race, and Financial Status*

Source	*F*	*df*
Main effects	355.44***	5
Marital status	214.91***	1
Gender	39.85***	1
Race	141.21***	1
Financial status	499.25***	2
2-way interactions	5.79***	9
Marital status x gender	32.51***	1
Marital status x financial status	3.14*	2
Marital status x race	1.31	1
Gender x financial status	2.01	2
Gender x race	6.11**	1
Race x financial status	1.87	2
3-way interactions	1.02	7
Marital status x gender x financial status	3.43*	2
Marital status x race x financial status	0.13	2
Marital status x gender x race	0.32	1
Gender x race x financial status	0.07	2
4-way interactions		
Marital status x gender x race x financial status	0.83	2

*$p < .05$.
**$p < .01$.
***$p < .0001$.

by all the variables considered. Because none of the interaction effects—the four-way, three-way, and two-way interactions among the variables—were statistically significant, except for the interaction of marital status and gender, the variables must have had individual independent effects on perception of well-being. In addition to their independent effects on perception of well-being, marital status and gender had a combined dual effect on perception of well-being. We can conclude that because perception of well-being was affected by marital status, irrespective of gender, the race and financial status of the respondents must have had additional effects on it.

Discussion

Because the research findings on gender differences in perception of well-being among married and unmarried people are not consistent, I reexamined differences in perception of well-being among currently married and unmarried adults in the United States. I considered race and financial status as additional control variables in this study. The results indicated a significant difference in the mean scores of perception of well-being between married and unmarried men and women. However, a comparison between men and women, or between married men and women, revealed no significant differences in the mean scores on perception of well-being. On the other hand, a comparison between unmarried men and women revealed a significant difference in mean scores on perception of well-being, 13.77 vs. 14.09, $t(\sim) = 4.38, p < .001$. The results further indicated that in all cases mean scores of perception of well-being for women were higher than those for men, suggesting that women reported being more satisfied with life than men did, whether they were married or unmarried. Because the present results indicated that women expressed themselves as more satisfied with life than men did, irrespective of their marital status, it would be difficult to conclude that marriage was beneficial or detrimental to women's perception of well-being relative to that of men. But, the results indicate that marriage enhanced perception of well-being for both men and women, and married women scored higher than married men on measures of perception of well-being.

When race was introduced as a control variable, except for the non-White women, the differences on mean scores for perception of well-being between married and unmarried men and women were more prominent, in addition to the significant difference in perception of well-being scores between White and non-White respondents. White respondents scored higher on the perception of well-being scale than the non-White respondents did, and their mean difference was statistically significant. The same was true when financial status was considered as an additional control variable. The results confirmed that perception of well-being was significantly affected by financial status for both married and unmarried men and women, irrespective of racial background. Although there were apparent differences in mean scores on perception of well-being because of financial status among the married non-White women, those were not statistically significant. We can conclude, therefore, that marital status did not make any significant difference in the perception of well-being among the non-White women, but financial status did have a significant effect on their perception of well-being, especially for those who perceived their financial status as average or above average. In addition, the results indicated that perception of well-being was significantly affected by marital status, gender, race, and financial status. Furthermore, because perception of well-being was affected by the marital status of the respondent, irrespective of gender, respondent's race and financial status must have had additional effects on it.

The results of this study suggest several important points. First, the effective multifactor measurement of perception of well-being/life satisfaction/global happiness (Campbell et al., 1976; Ellison et al., 1989; George, 1981) indicates that marriage seems to be beneficial for both men and women and that women are more satisfied in life than men are. This finding contradicts those of Bernard (1972) and other researchers (Gove, 1972, 1979a, 1979b; Gove et al., 1990; Gove & Tudor, 1973), but it supports the findings of Glenn and his associates (Glenn, 1975; Glenn & Weaver, 1979, 1981, 1988) and others (Lowe & Smith, 1983, 1987). Second, although I did not attempt to test any hypothesis, the results partially support the stress/crisis model and life strain model in explaining the relationship between marital status and subjective well-being, wherein it is argued that partner loss or marital dissolution is a stressful event that may have negative effects on perception of well-being (Bloom, Asher, & White, 1978; Booth & Amato, 1991; Bowling, 1987). Last, although the 10-year trend of gender differences in marital status and perception of well-being among the U.S. population was not appraised, the results show that marital status has a strong influence on perception of well-being for both men and women, and, in fact, marriage is beneficial to both men and women. In addition, race and financial status further contribute to the perception of life satisfaction.

REFERENCES

Andrews, F. M., & Withey, S. B. (1976). *Social indicators of well-being: Americans' perceptions of life quality.* New York: Plenum.

Aneshensel, C. S., Frerichs, R. R., & Clark, V. A. (1981). Family roles and sex differences in depression. *Journal of Health and Social Behavior, 22,* 379–393.

Bernard, J. (1972). *The future of marriage.* New York: Bantam.

Bloom, B. L., Asher, S. J., & White, S. W. (1978). Marital disruption as a stressor: A review and analysis. *Psychological Bulletin, 85,* 867–894.

Booth, A., & Amato, P. (1991). Divorce and psychological stress. *Journal of Health and Social Behavior, 32,* 396–407.

Bowling, A. (1987). Mortality after bereavement: A review of the literature on survival periods and factors affecting survival. *Social Science and Medicine, 24,* 117–124.

Brown, G. W., & Harris, T. (1978). *Social origins of depression: A study of psychiatric disorder in women.* New York: Free Press.

Campbell, A., Converse, P., & Rodgers, W. (1976). *The quality of life: Perceptions, evaluations and satisfactions.* New York: Russell Sage.

Ellison, C. G., Gay, D. A., & Glass, T. A. (1989). Does religious commitment contribute to individual life satisfaction? *Social Forces, 68,* 100–123.

Frieze, I. H., Parsons, J. E., Johnson, P. B., Reble, D. N., & Zellman, G. L. (1978). *Women and sex roles: A social psychological perspective.* New York: Norton.

George, L. K. (1981). Subjective well-being: Conceptual and methodological issues. *Annual Review of Gerontology and Geriatrics, 2,* 345–382.

Glenn, N. D. (1975). The contribution of marriage to psychological well-being of males and females. *Journal of Marriage and the Family, 37,* 594–601.

Glenn, N. D., & Weaver, C. N. (1979). A note on family situation and global happiness. *Social Forces, 57,* 960–967.

Glenn, N. D., & Weaver, C. N. (1981). Education's effects on psychological well-being. *Public Opinion Quarterly, 45,* 22–39.

Glenn, N. D., & Weaver, C. N. (1988). The changing relationships of marital status and happiness. *Journal of Marriage and the Family, 50,* 317–324.

Gore, S., & Mangione, T. W. (1983). Social roles, sex roles and psychological distress: Additive and interactive models of sex differences. *Journal of Health and Social Behavior, 24,* 300–312.

Gove, W. R. (1972). The relationship between sex roles, mental illness and marital status. *Social Forces, 51,* 34–44.

Gove, W. R. (1973). Sex, marital status and mortality. *American Journal of Sociology, 70,* 45–67.

Gove, W. R. (1979a). Sex differences in the epidemiology of mental disorder: Evidence and explanations. In E. S. Gomberg & V. Franks (Eds.), *Gender and disordered behavior: Sex differences in psychopathology* (pp. 23–68). New York: Brunner/Mazel.

Gove, W. R. (1979b). Sex, marital status and psychiatric treatment: A research note. *Social Forces, 58,* 89–93.

Gove, W. R., & Geerken, M. R. (1977). The effect of children and employment on the mental health of married men and women. *Social Forces, 56,* 66–76.

Gove, W. R., Hughes, M., & Style, C. B. (1983). Does marriage have positive effects on the well-being of the individual? *Journal of Health and Social Behavior, 24,* 122–131.

Gove, W. R., Style, C. B., & Hughes, M. (1990). The effect of marriage on the well-being of adults: A theoretical analysis. *Journal of Family Issues, 11,* 4–35.

Gove, W. R., & Tudor, J. F. (1973). Sex differences in mental illness. *American Journal of Sociology, 78,* 812–835.

Herzog, A. R., Rodgers, W. L., & Woodworth, J. (1982). *Sub-*

jective well-being among different are groups. Ann Arbor, MI: University of Michigan.

House, J. S., & Kahn, R. L. (1984). Measures and concepts of social support. In S. Cohen & L. Syme (Eds.), *Social support and health* (pp. 83–108). San Diego: Academic Press.

Kessler, R. C., & McRae, J. A. (1982). The effects of wives' employment on the mental health of married men and women. *American Sociological Review, 47,* 217–227.

Komarovsky, M. (1962). *Blue-collar marriage.* New York: Random House.

Levinson, D. (1978). *The seasons of a man's life.* New York: Alfred A. Knopf.

Lowe, G. D., & Smith, R. R. (1983). Gender, marital status, and mental well-being: A look at "his and her" marriages. Paper presented at the annual meeting of the Mid-South Sociological Association, Birmingham, AL.

Lowe, G. D., & Smith, R. R. (1987). Gender, marital status, and mental well-being: A retest of Bernard's his and her marriages. *Sociological Spectrum, 7,* 301–307.

Mookherjee, H. N. (1988). Assessment of life satisfaction in the United States. *Man and Life, 14,* 1–15.

Mookherjee, H. N. (1992). Perceptions of well-being by metropolitan and nonmetropolitan populations in the United States. *Journal of Social Psychology, 132,* 513–524.

National Opinion Research Center. (1991). General social surveys, 1972–1991. Chicago: University of Chicago.

Olson, J. K. (1980). The effect of change in activity in voluntary associations on life satisfaction among people 60 and over who have been active through time. Unpublished doctoral dissertation, University of Maryland, College Park.

Pearlin, L. I. (1975). Sex roles and depression. In N. Datan & L. Ginsberg (Eds.), *Life span developmental psychology: Normative and life crises* (pp. 191–197). San Diego: Academic Press.

Pearlin, L. I., & Johnson, J. S. (1977). Marital status, life strains and depression. *American Sociological Review, 42,* 704–715.

Radloff, L. (1975). Sex differences in depression: The effects of occupational and marital status. *Sex Roles, 1,* 249–265.

Ross, C. E., Mirowsky, J., & Goldsteen, K. (1990). The impact of the family on health: The decade in review. *Journal of Marriage and the Family, 52,* 1059–1078.

Ross, C. E., Mirowsky, J., & Huber, J. (1983). Dividing work, sharing work, and in-between: Marriage patterns and depression. *American Sociological Review, 48,* 890–923.

Rossi, A. S. (1985). Gender and parenthood. In A. Rossi (Ed.), *Gender and the life course* (pp. 161–192). New York: Aldine.

Vanfossen, B. E. (1981). Sex difference in the mental health effects of spouse support and equity. *Journal of Health and Social Behavior, 22,* 130–143.

White, J. M. (1992). Marital status and well-being in Canada: An analysis of age group variations. *Journal of Family Issues, 13,* 390–409.

Williams, D. G. (1988). Gender, marriage, and psychosocial well-being. *Journal of Family Issues, 9,* 452–468.

A Comparison of Marriages and Cohabiting Relationships

Steven L. Nock

There were over 3 million unmarried cohabiting couples in America in 1991, half a million more than in 1988 (U.S. Bureau of the Census, 1989, 1992). The increase in nonmarital cohabitation has been rapid and sustained for two decades, prompting social scientists to speculate about the implications of this trend for the institution of marriage.

Although we know more about cohabiting unions now than we did 10 years ago, one central question remains largely unexplored. Specifically, we know little about how individuals and couples differ in terms of the nature and quality of their relationships. This article addresses that issue by examining various dimensions of relationships of married and cohabiting individuals predicted to differ because of basic sociological processes.

Cohabitation is an increasingly common prelude to marriage. By their early 30s, almost one-half of the U.S. population has cohabited at some time. The majority of marriages begun since 1985 began as cohabitation (Bumpass & Sweet, 1989). As an element of "mate-selection," cohabitation is probably at least as commonplace today as going steady was 30 years ago.

Cohabitation is also gaining popularity as an alternative to marriage—especially for those who have already been married and divorced. Of all cohabiting unions in 1991, only one-third (30%) involved two never-married adults whereas almost all others (69%) included at least one divorced person (U.S. Bureau of the Census, 1992, Table 8). Most divorced individuals will eventually remarry. However, time spent in a cohabiting relationship delays remarriage just as it does first marriage; that is, it is an alternative to marriage or remarriage even if temporary (see Sweet & Bumpass, 1987, on first marriage).

There are good reasons to suspect that the relationships of married and cohabiting couples differ in important ways. We know, for example, that people who cohabit prior to marriage have considerably higher divorce rates regardless of whether they marry their cohabiting partner or someone else. This finding has been replicated so many times that it has taken on the "status of an empirical generalization" (DeMaris & Rao, 1992, p. 189). Those who cohabit are believed to be more accepting of divorce and less committed to marriage *to begin with* (i.e., regardless of any effect of cohabitation) than those who marry without ever having cohabited. This is described as a "selection" effect. Booth and Johnson (1988) concluded that the most likely explanation for the poorer *marital* quality of those who cohabited prior to marriage is their "deviant lifestyle or a disregard for the traditional norms of society," (p. 270) not something about cohabitation per se. DeMaris and MacDonald (1993), on the other hand, found evidence to suggest that cohabitation appeals to men who are *more conventional* regarding parental obligations and to women who tend to value a *traditional* lifestyle. Whether such differences existed prior to cohabitation or were a result of it could not be determined. Schoen and Weinick (1993) showed that partner-choice criteria differ between cohabiting relationships and marriages, suggesting that "a different kind of relationship calls for a different kind of partner" (p. 413).

There is evidence that the experience of cohabitation fosters values that make divorce more acceptable as a solution to problems. In their analysis of a 23-year, 7-wave panel, Axinn and Thornton (1992) found that "the same attitudes which increase the rate of cohabitation decrease the rate of marriage" (p. 367)—supporting the "selection" explanation. However, these same authors found that "cohabitation has a causal influence on susceptibility to divorce" (p. 371). In sum, cohabitation has been shown to attract a different type of couple than marriage, and to foster attitudes that contribute to divorce.

Findings such as these suggest that cohabitation and marriage are qualitatively different types of relationships (i.e., cohabitation attracts a different type of person than marriage does, or cohabitation fosters attitudes and beliefs inconsistent with marriage).

In sum, reliable evidence on the factors that contribute to cohabitation (i.e., the *causes*) and on some of the important consequences now exists. Between the causes and the consequences of cohabitation, however, is a relationship of two persons. Are the relationships in the two types of unions similar or different? This work addresses that question.

More specifically, this research tests the basic proposition that marriage and cohabitation are qualitatively different forms of relationships as a consequence of sociological processes related to the difference in the two types of unions.

Cohabitation and Marriage: Qualitatively Different Relationships

That people who cohabit are more divorce-prone than those who do not suggests that the nature of cohabiting and married relationships is, in some ways, incompatible. Cohabitation may attract the divorce prone and may produce attitudes and values that lead to divorce. Both possibilities have received empirical support (see introduction). Knowing this, however, still leaves us unclear about how and why the two types of relationships differ. Drawing from the limited empirical research on this subject, as well as from established sociological theory, the following are offered as dimensions that are predicted to differ qualitatively between cohabitation and marriage. Some dimensions are specific to the relationship whereas others are characteristics of the individual with potential relevance for the relationship.

These dimensions are not presumed to be exhaustive. Rather, each is suggested by prior research or theory on marriage or cohabitation. Broadly, the attempt here is to identify conspicuous features of relationships that are predicted to differ because of the *institutionalization* of the partnership (e.g., legal *vs.* extralegal; normatively approved *vs.* emerging and novel). The following section describes each dimension and provides theoretical justification for its inclusion.

Commitment The absence of any formally recognized status makes cohabitation quite different from marriage legally. Legal unions are more stable than nonlegal unions. That is, regardless of how a union began (by marriage or cohabitation), the current legal status is a strong predictor of its stability (Teachman, Thomas, & Paasch, 1991). Marriage differs so much from cohabitation legally because of the durability of the commitments involved. Even after a divorce, one may be held legally obligated to an ex-spouse (and to children). There are occasional instances in which courts have awarded limited support to unmarried estranged partners. However, such awards are extremely rare compared to the comparable situation in divorces. Moreover, the legal events of marriage (e.g., formalization of the union requiring significant effort to terminate it or legal assumptions about joint property) serve as what George Levinger (1976) called *barriers* that hold the relationship together. Barriers are things that hold two persons in their relationship in addition to, or even in the absence of, their interpersonal attraction. Property interests embodied in matrimonial law are a significant example of such barriers. Therefore, we expect lower levels of commitment between cohabiting rather than married partners.

Intergenerational Relationships Beyond the absence of formal legal recognition of cohabitation, there is also the absence of clearly defined normative social patterning. Andrew Cherlin (1978) once referred to remarriage as an "incomplete institution" because there were so few socially agreed upon standards governing it. Surely, cohabitation is also an incomplete institution suffused with ambiguity about even such simple issues as what one calls one's cohabiting partner, or the nature of the relationship between a child and the parent's cohabiting partner. American society may lack complete consensus on what it means to be a husband or a wife, but there are clearly traditional standards of propriety and decorum associated with one's relationships with married individuals. Newly married couples quickly realize they are treated differently once they are married than they were before, especially by parents. By comparison, we lack consensus over what it means to be a cohabiting partner. Cherlin argued that the lack of social norms governing remarriage contributed to high dissolution rates in such relationships. The same may be true of cohabiting relationships. Those in a relationship that is less socially recognized or governed by clear normative standards are less likely to be tightly integrated into networks of others who are in more traditional relationships. This is particularly problematic to the extent that it involves relations between parents and adult children. Cohabitation may, in fact, be a barrier to close relationships across generations.

Of the many forms of possible close social connections, "the parent/(adult) child relationship is a particularly strong and unique source of social integration for parents and adult children" (Umberson, 1992, p. 665). As Umberson notes, the permanence and involuntary origin of the relationship make it particularly significant. Her research revealed significant differences in the extensiveness and nature of such relationships, depending on adult children's marital status. Divorced children (and therefore those in a less socially patterned relationship) received less support from both parents and experienced more strained relationships with mothers. Never-married children were also found to express strained relationships with their mothers (Umberson, 1992, p. 668). Similar results were found by Cooney and Uhlenberg (1992) in their analysis of the National Survey of Families and Households (NSFH). They found that married children are more likely than unmarried children to cite parents as a source of potential help.

Relationships with parents are potentially important for the overall quality of the husband–wife or cohabiting couple partnership. Those who have poor relationships with their parents lack a basic emotional (and possibly economic) resource. Should the parent–child relationship suffer as a result of a marriage or cohabiting relationship (as a result of parental disapproval of a partner or other aspects of a union) the quality of the partner's affectionate bonds may suffer.

Cohabiting individuals are predicted to report poorer quality intergenerational relationships than married individuals.

Relationship Quality Research already cited has revealed that prior cohabitors have poorer quality marriages than those who did not cohabit prior to marriage. This has been attributed to the probable *selectivity* of cohabitation. That is, those who cohabit eschew tradition and are less committed to a traditional lifestyle. An equally plausible and complementary explanation, however, focuses on the *enforced intimacy* of marriage. There is greater commitment required in marriage. Further, there are stronger social sanctions associated with deviations from tradition in marriage than in cohabitation (e.g., there is stronger social disapproval of "deviant" marital behaviors such as adultery than there is of comparable behavior among unmarried individuals). Together, this means that the relationship between a married couple is harder to dissolve and is supported by stronger social norms than is true for cohabiting individuals. As such, married individuals are more likely to resolve their problems, or at least arrive at acceptable compromises, than cohabiting individuals whose relationships are less *enforced* by social and legal constraints. Indeed, one of the arguments offered for the higher divorce rates of prior cohabitors is their belief that escape is a solution to relationship problems (Axinn & Thornton, 1992). Thus the quality of the relationships is predicted to be higher among married than cohabiting individuals.

Ideal Fertility Marriage has traditionally been viewed as the acceptable arrangement for the bearing and rearing of children. Even in the presence of higher rates of out-of-wedlock childbearing, nonmarital fertility is still viewed as less desirable than marital fertility. It has been argued that having a child, or desiring one, is a fundamental difference between cohabiting and married couples (Bachrach, 1987; Rindfuss & VandenHeuvel, 1990). Cohabiting individuals may desire several children, but not with the current partner. Still, in the aggregate we would expect that those who intend to have children will be more likely to marry than those who do not, and those who intend to remain childless will be more likely than those who desire children to enter a cohabiting relationship. Indeed, in light of the traditional association of marriage and childbearing, those who do not intend to have children may have considerably less incentive to marry at all. Research by Rindfuss and VandenHeuvel (1990) found that the percentage of cohabitors intending to have a child in the next 2 years was almost 40% lower than the percentage of married individuals. Indeed, they suggest that on this and other important issues, cohabitors are more similar to single people than married people. Therefore, married individuals are predicted to express higher fertility intentions than cohabiting individuals.

A Note on Racial and Gender Differences

Although there are many racial differences in patterns of marriage and divorce, virtually no research on cohabitation makes racial comparisons. However, in their analysis of the NSFH, Bumpass, Sweet, and Cherlin (1991) showed that cohabitation compensated more for the overall drop in marriage rates among Blacks than among Whites (i.e., increases in cohabitation offset more of the declines in marriage among Blacks than among Whites). Among Blacks, cohabitation compensated for 83% of the overall decline in marriage rates by age 25. Among Whites, it compensated for only 61% (p. 916). Estimates from the NSFH show one-quarter (25.4%) of Blacks, 20.2% of Hispanics, and 13.6% of Whites currently cohabiting.

In her analysis of the NSFH, Manning (1993) found that race and age interact to influence the likelihood of legitimating a nonmarital pregnancy. Cohabitation (vs. not cohabiting) is not associated with differential probabilities of marrying to legitimate a pregnancy for teenage girls (regardless of race). However, older pregnant White women who are cohabiting are more likely to marry than comparable women who are not cohabiting (p. 847). Such patterns suggest that research on cohabitation should consider possible moderating influences of race on the distinction between marriage and cohabitation.

Because cohabitation substitutes more for marriage among Blacks than Whites (Bumpass et al., 1991), and because higher rates of cohabitation among Blacks have been interpreted to reflect the fact that cohabitation "has come to be acceptable and almost respectable" in the Black community (Billingsley, 1992, p. 38), smaller cohabitation/marriage differences were predicted among non-Whites than among Whites.

A moderating influence of gender was expected in light of research suggesting that for men, cohabitation is associated with lower levels of commitment to and responsibility for their partner (i.e., unconventional attitudes). The same research found that among women, cohabitation is associated with more conventional family lifestyle attitudes (e.g., disapproval of divorce, individual freedom in marriage, approval of mutual aid among family members). "Although somewhat counterintuitive, this may simply reflect that cohabitation is attractive to women primarily because it is felt to ensure the stability of subsequent marriage" (DeMaris & MacDonald, 1993, p. 404). In light of these findings, smaller cohabitation/marriage differences were expected among females than among males.

Data and Measures

Data for this research were taken from the National Survey of Families and Households (NSFH), a national sample of 13,017 individuals interviewed between March 1987 and May 1988. Cohabiting couples were oversampled to augment the number of such cases. In each selected household, a randomly selected adult was interviewed and (when appropriate) a self-administered questionnaire was completed by the spouse or cohabiting partner. In all, 6,881 married couples and 682 cohabiting couples were included. Of those, spouse questionnaires were completed by 5,684 married spouses and 519 cohabiting partners.

Cohabiting relationships do not endure as long as marriages. For that reason, the analysis is restricted to relationships of no more than 10 years duration. This restriction excludes a large number of marriages, but very few cohabiting relationships. If this is not done, however, a comparison of marriages and cohabitation will be confounded by the fact that the average marriage is much "older" than the average cohabitation. By imposing this restriction, two samples are produced that are closer in terms of duration than would be obtained in a sample of all cohabiting and married couples (average duration of marriages in the analysis sample = 5.4 years; of cohabiting relationships = 2.9 years). The samples may be regarded as representative of individuals in marriages and cohabiting relationships of no more than 10 years duration in the United States. The analysis sample consists of 2,493 married pairs and 499 cohabiting pairs of individuals.

Measures

Commitment

Answers to five questions were aggregated to form an additive index (alpha = .797) of the perceived *costs and benefits* of separation. Such *exit* costs are indicative of the commitment to the relationship. Those who perceive great costs as a consequence of separation are presumably more committed to maintaining the relationship.

The question to elicit these responses was "Even though it may be very unlikely, think for a moment about how various areas of your life might be different if you separated. For each of the following areas, how do you think things would change?" Answers ranged from 1 = *much worse* to 5 = *much better:*

1. Your standard of living
2. Your social life
3. Your career opportunities
4. Your overall happiness
5. Your sex life

The range of values for the index is 5–25. Higher scores on this index indicate a belief that separation would lead to improvements (or smaller declines) in quality of life.

Intergenerational Relationships

Two questions are considered as measures of the quality of relationships with parents. These two questions, differing only by reference to father or mother, are "How would you describe your relationship with your [father] [mother]?" Answers ranged from 1 = *very poor* to 7 = *excellent.*

Relationship Quality

Three measures are used to assess the perceived quality of the relationship between the two individuals. These measures are drawn from research that indicates that relationship quality as typically measured (most often *marital quality*) involves two primary dimensions: a *positive* dimension that includes satisfaction and happiness; and a *negative* dimension that includes disagreements (Johnson, White, Edwards, & Booth, 1986). In addition, research has shown that feelings of equity are significant in the overall quality of a relationship, especially equity in child care and household tasks (Yogev & Brett, 1985).

First is an index of disagreements on various issues, because frequent disagreements typically indicate a poorer quality relationship. Answers to six questions were aggregated to form an additive index (alpha = .786) indicating the *frequency of disagreement.* The items and the questions that elicited them are "The following is a list of subjects on which couples often have disagreements. How often, if at all, in the last year have you had open disagreements about each of the following?" Answers ranged from 1 = *never* to 6 = *almost every day:*

1. Household tasks
2. Money
3. Spending time together
4. Sex
5. Having a(nother) child
6. In-laws

The index ranges from 6 [to] 36. Higher scores indicate more frequent disagreements on these items.

Second, direct responses to a question about the overall *happiness with the relationship* were examined. This question asks "Taking all things together, how would you describe your marriage (or 'relationship' for cohabiting individuals)?" Answers ranged from 1 = *very unhappy* to 7 = *very happy.*

Third, the *perceived fairness* of the relationship was investigated. Each partner in a relationship was asked to evaluate the "fairness" in their relationship in four areas (household chores, working for pay, spending money, and child care). Answers to these questions do not correlate strongly (or produce a clear factor solution) and there was no way to combine them to form an index or scale (maximum alpha reliability was only .51). Therefore, the two areas that are most central to issues of fairness in relationships are investigated: the performance of household chores and caring for children (Yogev & Brett, 1985). The question used for this follows:

"How do you feel about the fairness in your relationship in each of the following areas?" Answers ranged from 1 = *very unfair to me* to 5 = *very unfair to spouse/partner:*

1. Performance of household chores
2. Caring for the children

Ideal Fertility

The question about ideal fertility is "If you could have just the number of children that you would like to have, how many would that be?" Answers were recorded as the actual number given. (Less than 1% of the respondents gave answers over 7 and they were recorded to "7" to eliminate the biasing effect of such few extreme values.) Only those who have had no children with their current partner are analyzed in an attempt to minimize the influence of completed fertility on responses to this question....

Results

Table 1 presents the descriptive statistics for all variables used in the analysis.

The minimal restriction placed on the sample (i.e., selecting only those in relationships of 10 or fewer years) minimized the difference between married and cohabiting individuals on the length of their relationships, but there still remains a 2-year difference in average duration. Differences in age, income, and education are consistent with other research on this topic. Cohabitors of both sexes report fewer years of completed schooling and are younger; cohabiting males report lower incomes than married men (see Sweet & Bumpass, 1987).

Table 2 includes results of the OLS regressions of each dependent variable on the distinction between cohabiting and the two types of married relationships (with all covariates entered simultaneously). Rather than present the entire equations, only the adjusted averages are shown for partners in cohabiting and married relationships. The difference among adjusted averages reported in Table 2 (i.e., the differences between adjusted averages for married individuals of either type and cohabitors) are the same as the regression coefficients for dummy variables distinguishing between cohabitors and the two groups of married individuals in an equation.[1]

Commitment The results reported in Table 2 show several consistent differences between married and cohabiting individuals. Because cohabitation is constrained by fewer social and legal rules than marriage, it was anticipated that the *exit costs* (a measure of commitment) of leaving a cohabiting relationship would be less than those associated with ending a marriage. Cohabiting males and females report that ending their relationship would have more positive (and/or fewer negative) consequences than do either group of married individuals. Commitment, in short, is lower in cohabitation than in marriage.

Intergenerational Relationships Cohabiting individuals report poorer relationships with both mothers and fathers than married individuals. The one exception to this general finding is that males who cohabited with their current spouse do not differ significantly from cohabiting males in their description of the relationships they have with their fathers. Still, bearing this one exception in mind, there is clear and

TABLE 1 *Descriptive Statistics for Analysis Sample (mean with standard deviation in parentheses)*

Variable N (male/female)	Married Did Not Cohabit (n = 1,514/1,524)	Married Did Cohabit (n = 948/969)	Cohabiting (n = 499/499)
Dependent variables			
Exit costs: (5 = *high*, 25 = *low*)			
Male partner	12.47 (3.47)	12.73 (3.26)	13.79 (3.34)
Female partner	11.50 (3.31)	11.83 (3.25)	13.06 (3.35)
Relations with mother (1 = *very poor*, 7 = *excellent*)			
Male partner	5.94 (1.32)	5.80 (1.43)	5.80 (1.54)
Female partner	6.04 (1.28)	5.78 (1.54)	5.59 (1.69)
Relations with father			
Male partner	5.58 (1.64)	5.33 (1.70)	5.06 (1.96)
Female partner	5.55 (1.62)	5.06 (1.80)	4.79 (1.99)
Disagreements (6 = *never*, 36 = *every day*)			
Male partner	11.21 (4.34)	11.74 (4.61)	11.46 (4.66)
Female partner	10.95 (4.06)	11.44 (4.35)	11.27 (4.45)
Relationship happiness (1 = *very unhappy*, 7 = *very happy*)			
Male partner	6.15 (1.15)	6.07 (1.16)	5.93 (1.16)
Female partner	6.14 (1.22)	6.04 (1.21)	5.87 (1.35)
Fairness—household tasks (1 = *very unfair to me*, 5 = *very unfair to partner*)			
Male partner	3.23 (0.59)	3.29 (0.61)	3.16 (0.58)
Female partner	2.71 (0.64)	2.64 (0.68)	2.68 (0.69)
Fairness—child care (1 = *very unfair to me*, 5 = *very unfair to partner*)			
Male partner	3.21 (0.55)	3.24 (0.56)	3.11 (0.49)
Female partner	2.78 (0.60)	2.74 (0.62)	2.81 (0.56)
Ideal fertility (number)			
Male partner	2.70 (1.74)	2.69 (1.92)	2.54 (1.91)
Female partner	2.72 (1.35)	2.58 (2.60)	2.37 (1.51)
Control variables			
Education (years)			
Male partner	13.25 (2.68)	13.11 (2.69)	12.45 (2.65)
Female partner	13.18 (2.43)	13.07 (2.48)	12.21 (2.68)
Total earnings ($1,000s)			
Male partner	25.86 (23.55)	28.30 (37.54)	20.22 (25.03)
Female partner	11.87 (28.15)	12.79 (13.79)	12.56 (21.63)
Age (years)			
Male partner	33.88 (11.46)	33.76 (8.75)	31.35 (9.84)
Female partner	31.29 (10.36)	30.92 (7.70)	28.40 (8.82)
Relationship years	5.69 (2.83)	5.07 (2.83)	2.91 (2.3)
Youngest child <4	53%	50%	24%
Youngest child 5–14	11%	15%	13%
Youngest child 15+	2%	2%	2%

(continued on next page)

TABLE 1 *Descriptive Statistics for Analysis Sample (continued)*

Variable N (male/female)	Married Did Not Cohabit (n = 1,514/1,524)	Married Did Cohabit (n = 948/969)	Cohabiting (n = 499/499)
Times married			
Male partner	1.34 (0.62)	1.49 (0.71)	0.52 (0.75)
Female partner	1.32 (0.62)	1.42 (0.64)	0.56 (0.72)
White			
Male partner	82%	83%	73%
Female partner	82%	84%	76%
Black			
Male partner	8%	7%	13%
Female partner	7%	7%	14%
Hispanic			
Male partner	6%	6%	10%
Female partner	7%	6%	7%

Note: Sample sizes vary across questions, depending on eligibility and nonresponse. Reported values are based on weighted sample. Sample sizes are unweighted.

persuasive evidence of poorer intergenerational relations among cohabiting than among married individuals.

Partner Relationship Quality The first measure of relationship quality refers to the frequency of disagreements. For both males and females, there does not appear to be any significant difference between married and cohabiting individuals. The second measure of relationship quality is the response to the direct question about overall happiness with the relationship. Relationship happiness is found to differ significantly, depending on whether one is married or cohabiting. Cohabitors report significantly lower levels of happiness than married individuals. This is particularly noteworthy because very few people in relationships describe themselves as *un*happy. As Table 1 revealed, mean levels of happiness on a 1 to 7 scale cluster near the extreme positive end (ranging from 5.87 to 6.14) for all groups. Given the extremely small amount of variation in responses to this question, a significant difference is particularly noteworthy.

The third measure of relationship quality pertains to fairness in the relationship. There are no differences between married and cohabiting males or females in their perceptions of fairness as indicated by these two questions.

Fertility Intentions For males, ideal fertility is the same for married and cohabiting individuals. However, cohabiting women report lower levels of ideal fertility than their married counterparts who did not cohabit prior to marriage.

To summarize, on issues of commitment, intergeneration relations, and expressed happiness with the relationship, there are broad differences between most married and cohabiting individuals in the direction hypothesized. Consistent differences were not discovered on measures of relationship fairness, frequency of disagreements, and intended fertility.

If we compare responses to questions about the mundane aspects of the present relationship with those having to do with the future or parents, we see a consistent pattern. When asked to speculate about how a separation would affect their lives, cohabiting individuals reported significantly fewer negative consequences than married persons. Similarly, when asked about their relationships with their parents, cohabiting

TABLE 2 *Adjusted Averages of All Dependent Variables*

Variable *N* (male/female)	Married Did Not Cohabit (*n* = 1,514/1,524)		Married Did Cohabit (*n* = 948/969)	Cohabiting (n = 499/499)	Equation *N*
Commitment					
Exit costs					
Male partner	12.47*		12.70*	13.80	2,346
Female partner	11.44*		11.75*	13.13	2,398
(5 = *high*, 25 = *low*)					
Intergenerational relationships					
Relations with mother					
Male partner	6.00*		5.87*	5.58	2,223
Female partner	6.16*	+	5.90*	5.28	2,288
(1 = *very poor*, 7 = *excellent*)					
Relations with father					
Male partner	5.55*	+	5.31	5.07	1,806
Female partner	5.65*	+	5.12*	4.56	1,954
(1 = *very poor*, 7 = *excellent*)					
Relationship quality					
Disagreement frequency					
Male partner	11.02	+	11.79	11.62	2,372
Female partner	10.90	+	11.37	11.32	2,359
(6 = *never*, 36 = *every day*)					
Relationship happiness					
Male partner	6.20*	+	6.07*	5.86	2,534
Female partner	6.22*		6.12*	5.72	2,510
(1 = *very unhappy*, 7 = *very happy*)					
Fairness household chores					
Male partner	3.20	+	3.27	3.22	2,557
Female partner	2.72	+	2.63	2.68	2,536
(1 = *very unfair to me*, 5 = *very unfair to partner*)					
Fairness child care					
Male partner	3.19		3.23	3.15	1,612
Female partner	2.80		2.75	2.80	1,547
(1 = *very unfair to me*, 5 = *very unfair to partner*)					
Ideal fertility (actual number desired —currently childless persons)					
Male partner	2.27		2.08	2.38	962
Female partner	2.45*	+	2.17	2.12	972

*Adjusted difference from "Cohabiting" is statistically significant ($p < .05$).
+Adjusted difference between "Married Did Not Cohabit" and "Married Did Cohabit" is statistically significant ($p < .05$).

individuals reported them to be of poorer quality than most of their married counterparts (there was but one exception). By way of contrast, issues of relationship fairness (as studied here) pertain to the two partners in their everyday circumstances. The same is true of disagreements. On these measures, there are no differences between married and cohabiting individuals. To the extent that these measures tap the ordinary life of partners, it may be that marriages and cohabiting relationships do not differ very much in their routine and mundane aspects.

The absence of consistent differences in intended fertility is open to several interpretations. One is that fertility intentions are independent of marital status. More likely, however, is that individuals who desire children and are presently cohabiting expect to marry (either their current partner or another) before beginning their childbearing.

Interactions involving sex, race, and marital/cohabiting status were investigated.[2] No significant three-way interactions were found. Three interaction effects involving race and marital status were discovered. In all three cases, the effects of cohabitation were different for Hispanics than for non-Hispanic Whites or Blacks (determined by using "non-Hispanic White" and "Black," in turn, as the reference category with associated interaction terms). In all three cases, cohabitation was associated with poorer outcomes for Hispanics than for others. With respect to commitment and relationship happiness, the difference between cohabiting and married Hispanics (both those who did and did not cohabit before marriage) was significantly greater than it was for Whites or Blacks. Cohabitation is also associated with significantly poorer quality relations with fathers (compared to marriage following cohabitation) for Hispanics than for others.

Only one interaction involving sex was found. Female Hispanics reported significantly greater levels of commitment (i.e., lower scores) than White or Black females. No sex by marital/cohabiting status interactions were significant.

The distinction between the present partner relationship and concerns about future or more distant issues may be part of a critical underlying difference between marriage and cohabitation. When focused on the immediate issues of daily life, partners in the two types of relationships may be similar. When confronted with concerns about the future, or relations with parents, there may be consistent differences.

Whether these differences are viewed as cause, consequence, or correlate of the type of relationship, the indisputable difference in reported happiness with the relationship is a central finding. Cohabitors report lower levels of happiness with their relationship than married people. Whether overall happiness is a reflection of the current state of affairs, a reflection of long-term concerns, or both, is not really known. Yet if this global assessment is taken to indicate concerns about both, then cohabitors' lower commitments to their relationships and poorer familial relationships are consistent with a poorer assessment of the present relationship.

To investigate this possibility (i.e., reported happiness is a function of the other dimensions considered in the research), a final equation was estimated in which relationship happiness was regressed on all variables described earlier (in Table 2) and the measures of commitment, intergenerational relations, and intended fertility. For each of these three, there was at least some evidence of differences associated with marital/cohabiting status. This model takes relationship happiness as a consequence of commitment, intergenerational relations, and intended fertility though it could be argued that all such factors are reciprocally linked. Without denying such a possibility, the logic of this research views happiness as subsequent to the other factors because of the effect of the degree of institutionalization of the relationship itself (i.e., marriage *creates* enduring commitments, through law, though it does not create happiness). Table 3 presents the results of this analysis.

The first thing to note is that the simple distinctions between cohabitation and marriage (with or without prior cohabitation) are not significant as they were in Table 2, which did not include the measures of commitment, intergenerational relations, and fertility intentions. When the effects of these three factors are considered, significant effects for commitment and intergenerational relations are found, but not for intended fertility. Moreover, for both males and females, the measure of commitment has the largest standardized effect in the equations (betas = –.33 for men, –.42 for women). Finally, for men, better relations with their mothers are of much greater importance for relationship happiness than those with their fathers. For women, there is no appreciable difference (i.e., better relations with either parent are associated with comparably better evaluations of the present relationship).

Together, these results suggest that much, if not most, of the difference in relationship quality found to

TABLE 3 *Regression of Relationship Happiness (1 =* very unhappy, *7 =* very happy*) on Commitment, Intergenerational Relationships, Ideal Fertility, and Marital/Cohabitation Status*

Independent Variable	Males		Females	
	b	**Beta**	**b**	**Beta**
Commitment				
Exit costs+ (5 = *high*, 25 = *low*)	–.117*	–.333	–.162*	–.415
Intergenerational relationships				
Relations with mother+	.126*	.139	.092*	.100
(1 = *very poor*, 7 = *excellent*)				
Relations with father+	.041*	.049	.109*	.134
Ideal fertility				
Number of children desired+	–.012	–.018	–.007	–.010
Marital/cohabiting status				
Married, did not cohabit (1 = *yes*, 0 = *no*)	.116	.050	.057	.022
Married, did cohabit (1 = *yes*, 0 = *no*)	.043	.017	.074	.028
Currently cohabiting (deleted)	—		—	
Controls				
Times married	.088	.056	.001	.001
Education, male partner	–.010	–.023	–.001	–.002
Education, female partner	.005	.010	.017	.034
Relationship duration (years)	–.035*	–.087	–.043*	–.099
Income, male partner ($1,000s)	.001	.021	–.001	–.025
Income, female partner ($1,000s)	–.001	–.021	–.001	–.009
Age, male partner	–.007	–.064	.003	.029
Age, female partner	.003	.027	–.008	–.058
Preschoolers in household?				
(1 = *yes*, 0 = *no*)	–.171*	–.074	–.212*	–.085
Middle-school children in household?	–.090	–.025	–.121	–.031
High-school age children in household?	.051	.006	–.319	–.036
Any nonbiological offspring?	-.050	–.017	.051	.015
Respondent is Black? (1 = *yes*, 0 = *no*)	.158	.036	.033	.007
Respondent is Hispanic? (1 = *yes*, 0 = *no*)	.148	.032	.188	.036
Respondent is White (deleted)				
Constant	6.92	7.10		
R^{2*}	.167	.235		
N	2,163		2,149	

+Subgroup means (by sex and marital/cohabiting status) were substituted for missing values when this variable was undefined (e.g., respondents whose fathers or mothers were not living).

*Indicates that coefficient is statistically significant ($p < .05$).

be associated with cohabitation, as opposed to marriage, is actually due to different levels of commitment and differences in the quality of relationships with parents. Cohabitation, that is, appears to take its toll on relationship quality because cohabitors have poorer relations with their parents and have lower levels of commitment—and these foster poorer assessments of the relationship. (The other measures of relationship quality studied in this research were similarly related to commitment and intergenerational relations [not shown], lending further support to the significance of these factors for relationship quality.)

Conclusion

This research takes us one step closer toward understanding the nature of the difference between cohabitation and marriage. Married and cohabiting individuals describe their relationships differently. Specifically, cohabitors report lower levels of happiness with their partnerships, express lower degrees of commitment to their relationships, and have poorer quality relationships with their parents. These differences are consistent with the sociological processes hypothesized to produce them: the lack of formal legal or normative structure for cohabitation and the enforced intimacy of marriage. One interpretation of such findings is that cohabitation and marriage do not differ so much in terms of the ordinary, everyday partnerships as they do with respect to long-term concerns and relationships with people beyond the immediate dyad.

The unanswered question raised repeatedly in this and other research is whether cohabitation attracts a different type of person initially or whether the experience itself should be credited with producing observed differences between cohabitation and marriage. In fact, both processes are relevant. Some clues to the relative importance of the two may be found in results comparing those who married after cohabiting with their partner and those who married someone with whom they had never cohabited. In all but one case, when one category of married persons differed from cohabitors, so did the other. Moreover, tests for differences between the two types of *married individuals* (indicated by a + in Table 2) supported this general conclusion. Although there were many cases in which the two groups of married individuals differed significantly, the magnitude of those differences was generally less than one-half of that found between cohabitors and either married group. The one notable exception is for disagreement frequency. Married individuals differ significantly on this measure, depending on whether they did or did not cohabit with their current spouse before marriage, even when neither group differs significantly from cohabitors. Within the limits of the analysis conducted, those who married after cohabiting appear more similar to those who married without cohabiting than to those who are currently cohabiting.[3] This suggests that the structural and institutional aspects of marriage discussed at the outset of this article define much of the differences between marriage and cohabitation. The results also lend further support to prior research (see Thompson & Colella, 1992) indicating that selection effects are relevant (though perhaps less so than the nature of the relationship) in understanding differences between married and cohabiting individuals.

In considering the differences between marriages and cohabiting relationships, it is important to recognize the durability of commitments embodied in matrimonial law and the enforced intimacy in marriage (i.e., the greater effort required to terminate it). Moreover, although mate selection is certainly not strongly controlled by parents, mothers and fathers are, nonetheless, conspicuous parties to most marriages. Relations with parents and in-laws in marriage may be difficult at times, but both sets of relatives are recognized to have legitimate interests in their offspring's marriages. In contrast, cohabitation is an incomplete institution. No matter how widespread the practice, nonmarital unions are not yet governed by strong consensual norms or formal laws. What is the legitimate role of a parent in her daughter's cohabiting union? What is the nonmarital equivalent of an in-law? Answers to such questions will emerge if cohabitation persists as a popular form of intimate relationship. For the time being, however, the absence of such institutional norms is a plausible explanation for much of the poorer quality of cohabiting relationships.

NOTES

1. The equations used to produce these adjusted averages included all variables discussed as controls in the text. Overall R^2s in statistically significant equations ranged from .02 to .20, with most between .05 and .10.

   When multiple comparisons are made, the cumulative probability of a Type I error increases. And although all comparisons in this section of the analysis are orthogonal (within equations) and independent (across equations), the number of equations increases the likelihood of falsely rejecting a null hypothesis. Procedures to adjust for "family-wise" cumulative Type I error probabilities within an equation (e.g., the

Bonferroni inequality; the Least Significant Difference) produced identical results within equations. Inflating standard errors by 1.20 had no effect on the pattern of statistically significant results across equations.

2. For tests of interaction effects, the samples were restructured. Rather than analyze males and females separately, a sample of primary respondents (males and females) and one of partners/spouses was assembled. Interaction effects were tested on both samples with, essentially, similar results (i.e., those effects significant in one sample were also significant in the other, although the magnitude of effects differed somewhat).
3. The tests that compare the two married groups with the cohabiting group are not independent of those comparing the two married groups, but see Note 1 above.

REFERENCES

Axinn, W. G., & Thornton, A. (1992). The relationship between cohabitation and divorce: Selectivity or causal influence? *Demography, 29*, 357–374.

Bachrach, C. A. (1987). Cohabitation and reproductive behavior in the U.S. *Demography, 24*, 623–637.

Billingsley, A. (1992). *Climbing Jacob's ladder.* New York: Simon & Schuster.

Booth, A., & Johnson, D. (1988). Premarital cohabitation and marital success. *Journal of Family Issues, 9*, 255–272.

Booth, A., & White, L. (1980). Thinking about divorce. *Journal of Marriage and the Family, 42*, 605–36.

Bumpass, L. L., & Sweet, J. A. (1989). National estimates of cohabitation. *Demography, 26*, 615–625.

Bumpass, L. L., Sweet, J. A., & Cherlin, A. J. (1991). The role of cohabitation in declining rates of marriage. *Journal of Marriage and the Family, 53*, 913–927.

Cherlin, A. J. (1978). Remarriage as an incomplete institution. *American Journal of Sociology, 84*, 634–50.

Cherlin, A. J. (1992). *Marriage, divorce, and remarriage* (rev. ed.). Cambridge, MA: Harvard University Press.

Cooney, T. M., & Uhlenberg, P. (1992). Support from parents over the life course: The adult child's perspective. *Social Forces, 71*, 63–84.

DeMaris, A., & MacDonald, W. (1993). Premarital cohabitation and marital instability: A test of the unconventionality hypothesis. *Journal of Marriage and the Family, 55*, 399–407.

DeMaris, A., & Rao, V. (1992). Premarital cohabitation and subsequent marital stability in the United States: A reassessment. *Journal of Marriage and the Family, 54*, 178–190.

Johnson, D. R., White, L. K., Edwards, J. N., & Booth, A. (1986). Dimensions of marital quality. *Journal of Family Issues, 7*, 31–49.

Langman, L. (1987). Social stratification. In M. B. Sussman & S. K. Steinmetz (Eds.), *Handbook of marriage and the family* (pp. 211–249). New York: Plenum.

Levinger, G. (1976). A social psychological perspective on marital dissolution. *Journal of Social Issues, 35*, 50–78.

Manning, W. D. (1993). Marriage and cohabitation following premarital conception. *Journal of Marriage and the Family, 55*, 839–850.

Nock, S. L. (1979). The family life cycle: Empirical or conceptual tool? *Journal of Marriage and the Family, 40*, 15–26.

Rindfuss, R. R., & VandenHeuvel, A. (1990). Cohabitation: A precursor to marriage or an alternative to being single? *Population and Development Review, 40*, 703–726.

Rossi, A. S., & Rossi, P. H. (1990). *Of human bonding.* New York: Aldine de Gruyter.

Schoen, R., & Weinick, R. M. (1993). Partner choice in marriages and cohabitation. *Journal of Marriage and the Family, 54*, 408–414.

Schumm, W. R., & Bugaighis, M. A. (1986). Marital quality over the marital career: Alternative explanations. *Journal of Marriage and the Family, 48*, 165–168.

Suitor, J. J. (1991). Marital quality and satisfaction with the division of household labor across the life cycle. *Journal of Marriage and the Family, 53*, 221–230.

Sweet, J. A., & Bumpass, L. L. (1987). *American families and households.* New York: Russell Sage.

Teachman, J. D., Thomas, J. J., & Paasch, K. (1991). Legal status and the stability of coresidential unions. *Demography, 28*, 571–586.

Teachman, J. D., Polonko, K. A., & Scanzoni, J. J. (1987). Demography of the family. In M. B. Sussman & S. K. Steinmetz (Eds.), *Handbook of marriage and the family* (pp. 3–36). New York: Plenum.

Thompson, E., & Colella, U. (1992). Cohabitation and marital stability: Quality or commitment? *Journal of Marriage and the Family, 54*, 259–267.

Umberson, D. (1992). Relations between adult children and parents: Psychological consequences for both generations. *Journal of Marriage and the Family, 54,* 664–674.

U.S. Bureau of the Census. (1989). *Marital status and living arrangements: 1989* (Series P-20, No. 433). Washington, DC: U.S. Government Printing Office.

U.S. Bureau of the Census. (1992). *Marital status and living arrangements: 1992* (Series P-20, No. 468). Washington, DC: U.S. Government Printing Office.

Yogev, S., & Brett, J. (1985). Perceptions of the division of housework and child care and marital satisfaction. *Journal of Marriage and the Family, 47,* 609–18.

Marital Behavior and Expectations: Ethnic Comparisons of Attitudinal and Structural Correlates

M. Belinda Tucker and Claudia Mitchell-Kernan

Recent dramatic changes in marital patterns in the United States generally, and among African Americans most particularly, have given rise to a renewed emphasis on the determinants of marital behavior and family formation more generally. As certain of the changes first became evident in the black population in the early 1960s, earlier discussions focused on the seemingly distinctive nature of African American family organization, and included suggestions of "pathology" (e.g., Moynihan, 1967). Such a focus in effect shifted the research agenda of many to defense of the African American family rather than an empirical analysis of the roots of what was actually a very recent change in black family formation patterns. As these patterns now characterize the American population generally, although they are more strongly evident among blacks, we have chosen to examine the correlates of marital behavior and expectation in distinct sociocultural groupings. In this manner we might discern whether the same structural conditions are associated with specific marital trends in all groups, whether culturally distinctive attitudinal patterns are more prominent correlates, or whether a combination of structural and sociocultural factors is associated with marital behavior and expectation. We might also determine whether such groups differ in their perceptions of marital opportunities and constraints.

In both the earlier and the current periods of focus, theory and research on the problem have been dominated by a focus on macro-level constructs. Key issues have been the economic viability of men (which, if unhealthy, would discourage marriage) and the demographic characteristics of groups (in particular a shortage of men most pronounced at middle and older ages)—both directly influencing the marriage market. It remains to be seen, however, whether individual perceptions of the marriage market bear a relationship to structural assessments. Also unknown is the extent to which determinants of marital behavior, as perceived by the actors themselves, conform to notions developed through macro processes. Furthermore, analysis of the individual attitudinal components of marital expectations and behavior, in the context of mate availability considerations, has not been conducted. The intent of this [reading] is to address these lacunae through a focus on attitudinal level data obtained on three racial/ethnic groups through the 1989 Southern California Social Survey.

The immediate objectives are: (1) to examine the differential relationship between perceived mate availability and more objective assessments of sex ratio in broadly distinctive sociocultural groupings [and] (2) to determine whether major theoretical conceptualizations concerning the role of mate availability in family structure are differentially relevant for those groups....

Background

Changing Marital Behavior ...American patterns of family formation have undergone remarkable change in recent times. Americans now marry later, are less likely to stay married, and are less likely to marry after divorce than in previous times. Between 1970 and 1990, the percentage of women married by age twenty to twenty-four decreased from 64 percent to 37 percent, and the percentage of men married by that age declined from 45 percent to 21 percent. The divorce rate nearly tripled between 1970 and 1990 (from 60 per 1,000 to 166 per 1,000 married women) (U.S. Bureau of the Census, 1991a). Notably, the proportions of both women and men who *ever* married (that is, married at least once) did not change substantially over this twenty-year period. The cumulative data, therefore, suggest that in recent years Americans have been as likely to marry eventually as before—but spend significantly greater portions of their lives as unmarrieds, due to later age of first marriage and a greater likelihood of divorce. Still, based on current trends in marital incidence, some demographers and economists have predicted substantial decreases in the likelihood of being ever married for American women (Rodgers and Thornton, 1985; Bloom and Bennett, 1985).

Group-Specific Trends Although American trends in family formation are pervasive, reflecting (and perhaps driving), in part, worldwide changes, these transitions have not been equally experienced across ethnic groups. In particular, the transformations evident in the family formation patterns and living arrangements of African Americans, and some Latino groups, have been more substantial along certain dimensions. Prior to 1950, African Americans displayed a long-established tendency to marry earlier than whites. That pattern is now reversed (Cherlin, 1981). Between 1970 and 1990, the proportion of women who had ever married declined sharply from nearly 83 percent to 63 percent among blacks, while there was virtually no change in the extent of being ever married among either whites or women of Spanish origin (Norton and Moorman, 1987; U.S. Bureau of the Census, 1991a—Asian breakdown not reported and figures for 1970 unavailable). [... However, marriage among younger cohorts of white women has declined very dramatically, indicating either greater marital delay or real generational differences in the tendency to marry.] It has been estimated that only 70 percent of black women born in 1954 will ever marry, compared to 86 percent of black men and 90 percent of whites of both sexes (Rodgers and Thornton, 1985). Of black women born in the 1930s, 94 percent eventually married.

Although Latino marriage patterns overall had been relatively stable through 1985, a nearly nine-point drop in marital prevalence occurred over the next five-year period—at a time when African American and white patterns had begun to stabilize. Furthermore, for certain indicators, within-group differences among Latinos are quite substantial. Although in 1990 only 16 percent of Mexican-origin households were being maintained by females without males, that figure was 21 percent for Central and South Americans, and 31 percent for Puerto Ricans (U.S. Bureau of the Census, 1991b).

Although this discussion has focused on changes over very recent decades, ... the fifties and sixties were quite distinctive in terms of family formation patterns. When trends over the last fifty years are considered, some changes no longer seem as dramatic, due to substantial fluctuations in marital patterns over time. Nevertheless, later marriage and more divorce are two very significant changes evident among all Americans, and the decline in African American marital prevalence is substantial, regardless of whether the comparison point for current behavior is 1970 or 1940. These trends raise questions about the underlying attitudinal components, as well as the structural underpinnings, of marriage behavior and marital expectations. Do Americans still want to marry? Do they expect to marry? Are the desire and expectation to marry among African Americans significantly less than that observed in other American populations? What are the perceived constraints on marriage? What are the attitudinal correlates of marriage and marital expectations and do they differ among the groups that evidence differential marriage patterns?

Conceptual Perspectives

... A number of ... broad conceptual perspectives ... have been offered to explain current marital trends.... In our research program, we have focused specifically on the demographic and economic arguments because such causes, in our view, offer the greatest potential for influence through social policy. The perspectives that have been most influential in our research are briefly outlined below.

Sex Ratio Imbalance Several theorists have argued that a shift in the availability of marriage partners affects family formation patterns and family values (Glick, Heer, and Beresford, 1963; Guttentag and Secord, 1983; Rodgers and Thornton, 1985; Schoen, 1983). Because the number of males, relative to females, in the African American population has declined steadily since the 1920s, and has become increasingly divergent from white sex ratios, some believe that sex ratio imbalance is a significant factor in black marital decline (Guttentag and Secord, 1983; Staples, 1981a,b). Epenshade (1985) has argued that since decreases in black marriage have only been evident since the 1960s, sex ratio declines could not be the primary cause of this change. However, as demonstrated in Table 1, the most dramatic sex ratio declines have been evident only in recent decades. It is also possible that the inconsistencies noted by Epenshade (1985) stem from Guttentag and Secord's (1983) narrow focus on the demographic variable of sex ratio, rather than the range of issues related to marital opportunity.

TABLE 1 *Sex Ratios from Raw Census Data for Blacks and Whites: 1830–1990*

	Race	
Year	Blacks	Whites
1990	88.2	96.0
1980	89.6	95.3
1970	90.8	95.3
1960	93.4	97.4
1950	94.3	99.1
1940	95.0	101.2
1930	97.0	102.9
1920	99.2	104.4
1910	98.9	106.7
1900	98.6	101.5
1890	99.5	105.4
1880	97.8	104.0
1870	96.2	102.8
1860	99.6	105.3
1850	99.1	105.2
1840	99.5	104.6
1830	100.3	103.7

Sources: U.S. Bureau of the Census (1960, 1964, 1973, 1982, 1983, 1992).

Male Economic Viability Economic factors have been cited by a number of theorists as central contributors to marital decline (e.g., Darity and Myers, 1986/87; Ross and Sawhill, 1975; Wilson, 1987; Wilson and Neckerman, 1986). These arguments have taken various forms and focus on different economic processes including declining male economic power, increasing female economic power, and a declining differential between male and female economic power, among others. Testa and Krogh . . . outline a number of the major competing economic hypotheses concerning marriage among inner-city minorities. Also included among these theories is Oppenheimer's (1988) assertion that increasing marital delay is a function of an elongated search strategy in which more time is required to assess male economic prospects. All of these arguments suggest that societal shifts in marriage patterns are related to the fact that the economic incentives to marry and remain married have undergone change.

Other Explanations of Marital Change Clearly, the above conceptualizations cannot fully explain the substantial shifts in family formation and living arrangements that characterize either the American population generally or African Americans. . . . [I]t seems clear that some ideological shifts are factors, including significant shifts in views of the roles of women and greater acceptance of cohabitation and childbirth out of wedlock. Also, significant improvements in contraception have certainly provided a freedom of choice for women unparalleled in human history.

Toward Synthesis Through Microanalysis ... [A] number of studies have attempted to compare and test various explanations of marital change and ethnic differences in marriage behavior. The results have been varied and inconclusive, although it seems clear that both demographic and economic factors are implicated in recent trends in family formation.

We propose theoretical integration in several respects. First, the economic and demographic theories are not incompatible. As separable issues, each is more or less salient depending on the circumstances of the population under study. In earlier work, we presented a detailed discussion of the factors affecting mate availability (Tucker and Mitchell-Kernan, in press). Depending in part on geographic location, the marriage pool may be influenced to a lesser or greater extent by true demographic imbalances (i.e., an absence of "warm bodies"), compromised eligibility (which could be due to economic limitations or institutionalization, as two examples), or sociocultural unavailability (e.g., persons of other races have not normally been available as potential partners). Locational variations of a related sort would include differential inclinations to out-marry (e.g., between 1970 and 1980, one out of six black men in the western states who married for the first time wed a woman of another race, although the black female tendency was only one-quarter that rate; during the same time period in the South only 2.5 percent of black men out-married compared to less than 1 percent of black women—U.S. Bureau of the Census, 1985).

Using the demographic and economic factors as anchoring points, it seems plausible that different groups are differentially affected by these factors. In particular, we would argue that white American marital behavior and expectations are influenced to a greater extent by the demographic forces cited by marriage squeeze theorists (per the "baby boom" effect); that local Latino marital behavior and expectations are more influenced by economic forces than by demographic forces (i.e., a shortage of economically viable men, but no shortage of men per se); and that African American marital behavior and expectations are a function of both economic and demographic phenomena. If the impact of these forces is at least additive (i.e., first fewer men, then those remaining men are economically compromised), we would therefore expect black marriage rates to be more constrained than either of the other groups....

Results and Discussion

Perceived Mate Availability

Table 2 presents the 1990 Los Angeles–Long Beach PMSA (Primary Metropolitan Statistical Area) sex ratios for persons aged eighteen and over for blacks, non-Hispanic whites, and Latinos, as well as two indicators of perceived mate availability by gender: the perceived number of men for every ten women (sex ratio), and whether there were "not enough," "enough," or "more than enough" women/men for "people like yourself." [We recognize the limitation of census data and the fact that black men and Latinos in general were undercounted. This would only indicate, however, that the shortage of black men is less than indicated by these figures. Furthermore, as we have argued elsewhere, many of the men who are not counted by the census are essentially out of the marriage market anyway because of severely compromised economic situations—the homeless, for example (Tucker and Mitchell-Kernan, in press).]

The results indicate that both black men and women perceive a substantial shortage of men. Whites and Latinos also see shortages, but to a lesser extent than blacks. A fair degree of consistency across gender for all races exists. The perceived mate availability item reflects the same trend with black women feeling the most substantial limit on partner availability (nearly 70 percent of black women believed that not enough men are available). Black men confirmed this perception with 60 percent of them believing that there are more than enough women. A similar but less dramatic pattern was observed among whites, but the pattern was much less extreme among Latinos.

Overall, black and white perceptions tended to reflect their relative "realities" as indicated by the census-derived sex ratios for the two races. Although the Latino perceived sex ratio did not reflect the census figure, fewer Latinas (than observed with either

TABLE 2 *Family/Relationship Characteristics and Attitudes, by Ethnicity and Gender*

	Blacks		Whites		Latinos	
L.A.–Long Beach PMSA sex ratio (1990 census)	85.2		96.0		107.4	
	M	F	M	F	M	F
Perceived sex ratio (ten-point scale)	4.7	4.8	7.0	6.0	5.2	4.8
Perceived mate availability (%)						
Not enough	15.4	69.4	20.0	59.0	14.1	54.3
Enough	26.4	20.0	38.5	32.5	46.5	32.6
More than enough	58.2	10.6	41.5	8.5	39.4	13.0
Marital status (%)						
Married	26.9	23.4	48.1	47.6	32.4	43.0
Separated	6.5	13.7	2.3	1.8	2.9	8.0
Divorced	15.1	17.1	9.7	13.2	10.8	13.0
Widowed	3.2	11.4	4.6	13.2	0.0	8.0
Never married	48.4	34.3	35.2	24.2	53.9	28.0
Married or romantically involved (%)	61.4	53.0	65.0	67.8	73.4	62.0
Want to marry/remarry[a] (%)	81.3	72.4	79.0	71.8	93.8	76.0
Importance of long-term involvement[b]	7.9	7.6	8.3	8.3	8.0	8.3
Importance of marriage[b]	7.0	7.1	7.3	7.1	7.3	7.6
Marital expectations[a,b]	6.6	6.3	6.7	5.7	8.1	6.1

[a]Asked of single persons only.
[b]Ten-point scale.

black or white women) felt that there are "not enough" men and fewer Latino men believed that there are "more than enough" women. At least in relative terms, then, individual mate availability perceptions of these distinct sociocultural groupings did indeed reflect the census-derived "reality."

Notably, however, despite their demographically favorable position, the Latina perspective on mate availability seemed more similar to that of black and white women. The fact that their perceptions were somewhat removed from the objective "reality" as measured by the sex ratio lends support to the notions we have presented in earlier writings (Tucker and Mitchell-Kernan, in press). That is, personal perceptions of mate availability are not determined solely by the existence of "warm bodies," but by the perceived availability of potential partners with socially desired features. Latina assessments of availability were most likely a function of their perceptions of the economic viability of potential Latino mates—an area in which their eligibility is greatly compromised—rather than numerical representation.

Relationship Involvement

Marital Status As shown in Table 2, marital status differed substantially among the three groups, with the greatest extent of marriage evident among white men and the lowest among black women. Table 2 also shows that the greatest source of nonmarriage for black women is marital dissolution through separation, divorce, or widowhood (42.2 percent), rather than never having been married (34.3 percent). Also, despite the fact that a substantial proportion of the Latino sample was Catholic, divorce and separation were fairly substantial among Latinas (21 percent). Marital dissolution is therefore highest among women married to the most economically disadvantaged groups of males. (Since all men are more likely to remarry, their dissolution rates are *artifactually* deflated.) It must also be noted, however, that black women and Latinas remain the *most* economically disadvantaged of all groups and that marital dissolution is often the source of further economic decline (U.S. Bureau of the Census, 1991c). It therefore seems unlikely that the primary source of marital breakdown is the increased economic fortunes of these women. A more likely source is the strain caused by financial difficulties, which is amplified among blacks by the actual shortage of men. Previous research has provided fairly unambiguous evidence of a relationship between husband-specific economic factors and black separation and divorce (Hampton, 1979; Ross and Sawhill, 1975).

Latino men were the most likely to be never married, which is probably the result of the age structure differences among the groups. The dominance of young men in the Latino population in particular, but across all groups, inflates the never-married rates among men; while the relatively larger number of older white women and relative absence of young white women inflates the widowed category.

Nonmarital Romantic Involvement When those who were involved in "long-term" romantic relationships were added to those who were married, the extent of relationship involvement across groups becomes more approximately equal with most individuals in all groups involved in a long-term relationship of some kind. Nevertheless, black female relationship involvement is lowest. Taken together with the marital status figures, the dominant form of relationships for both black and Latino men is *nonmarital* romantic involvement.

This last point is particularly notable in terms of the economic eligibility framework. In the two clearly economically compromised groups of men, nonmarital romantic involvements predominate. Although it could be suggested that the youthfulness of the Latino male sample explains the absence of marriage among them, over 70 percent of the Latinas (who were nearly as young as the men) were either currently or previously married. These findings are supported by an earlier study conducted by Tucker and Taylor (1989). In that study, National Survey of Black Americans data showed that marriage as opposed to nonmarital romantic involvement was related to male personal income.

Desire to Marry or Remarry

Given the great variation in the extent of marriage among these groups, is "desire for marriage" similarly distinctive? When asked whether "you ever want to marry or remarry," respondents across all races indicated a strong willingness to become legally attached. Although gender differences characterized all races, with men consistently being *more* desirous of marriage, the greatest discrepancy between men and women was evident among Latinos (with Latino men being exceptionally desirous of marriage relative to all other groups). Also, black women were least likely to express a desire to marry—a view that may be tempered by the reality of their circumstances. Still, there existed relative consistency within race, with Latino men being more desirous of marriage than other men, and Latinas wanting to marry more than other women. Whites were least likely to want to marry or remarry.

Importantly, these findings contradict the commonplace stereotypical implication that blacks simply do not want to marry. We do not believe that "social desirability" any longer demands a positive response to the question. It would seem that Americans are freer than ever before to reject marriage. The fact that blacks were more likely to express a desire to marry than whites strongly suggests that the reasons for not marrying may be structural, rather than the result of a fundamental change in the way they value the institution of marriage. This result also conflicts with the Guttentag and Secord (1983) predictions concerning the consequences of male shortage, namely, that when men are in short supply they devalue the institution of marriage.

Marital Values

When asked directly about the importance of long-term relationships and marriage, there was again considerable similarity in the responses supplied by the

TABLE 3 *Multiple Classification Analysis: Adjusted Means for Marriage and Family Values, by Ethnicity (controlling for age, gender, and income)*

	Blacks (*n* = 247)	Latinos (*n* = 192)	Whites (*n* = 418)	*Eta*
Importance of long-term involvement	7.71	8.06	8.28	.10
Importance of being married	7.17	7.55	7.10	.05
Importance of similar sociocultural background for successful marriage	6.96	7.39	6.62	.07
Importance of same religion for successful marriage	6.71	7.52	6.02	.21
Importance of love for successful marriage	9.17	9.21	9.30	.03
Importance of "being faithful" for successful marriage	9.36	9.36	9.59	.06
Importance of lifelong commitment for successful marriage	9.07	9.29	9.50	.11
Importance of having children for successful marriage	7.77	8.39	6.40	.29
Importance of being of same race/ethnic group for successful marriage	5.55	6.46	5.99	.07
Importance of similar likes/dislikes for successful marriage	7.64	7.47	7.41	.07
Importance of adequate income for successful marriage	8.70	8.22	7.96	.18
Importance of good sex for successful marriage	8.80	8.94	8.62	.11
Importance of being good friends for successful marriage	9.56	9.68	9.59	.04
Importance of having children	8.22	8.57	7.76	.11
Importance of being married when you have children[a]	8.50	8.50	8.44	.01

Note: All value scores are based on ten-point scales, with one being "not very important at all" and ten being "extremely important."
[a]Asked only of childless respondents.

three ethnic groups. As shown in Table 2, responding on a scale of one to ten, with ten being extremely important, means for all three groups averaged between 7 and 8.3. There were no significant differences on the basis of either ethnicity or gender.

In order to determine whether ethnic similarities and differences in perspectives on marriage and family might be due to differences in the compositions of the three populations (e.g., on age), we used Multiple Classification Analysis (Andrews et al., 1973) to compute adjusted means for all of our marriage and family value items—controlling for age, gender, and neighborhood income level. Table 3 shows the results of those analyses. It is clear that there are very few differences between the ethnic groups in terms of the marriage and family values assessed here. The only clear patterns of distinction are: (1) the tendency for Latinos to view children as more important for a successful marriage and for whites to see them as substantially less important, and (2) the tendency for blacks to view adequate income as more central for marital success than other groups. These responses provide further indication that, despite current distinctive African American marital patterns, the underlying values concerning marriage held by blacks do not differ from those held by groups with greater proportions of intact marriages.

Reasons for Not Marrying

All single persons (i.e., those never married, separated, divorced, or widowed) were asked why they had not married or remarried (see Table 4). Although, as was expected, all groups indicated that having "never found the right person" was the primary reason (and for that reason, this response category was presented last in the interview), the next most dominant reasons are more instructive. Women in general believed, more than men, that there are not enough persons of the opposite sex who meet their standards, although

TABLE 4 *Reason for Not Marrying Among Unmarried Persons, by Ethnicity and Gender: Percentage Indicating Reasons*

	Blacks		Whites		Latinos	
	M	F	M	F	M	F
Devoted energies to school or work	40.3	22.9	57.1	31.3	63.1	37.5
Not enough men/women who meet standards	47.5	55.6	27.6	51.4	18.5	46.7
Do not want to lose freedom	25.8	26.9	30.7	23.9	41.5	32.6
Do not believe in marriage	6.5	7.3	12.5	5.3	16.9	19.1
Not ready to settle down	48.3	34.6	53.9	41.6	76.9	39.6
Having fun playing the field	16.1	10.1	20.6	15.9	44.6	10.4
Not enough money to support a family	41.9	25.7	36.5	15.9	52.3	18.8
Never found the right person	68.9	66.7	77.0	73.6	55.6	77.1

Note: Since respondents could indicate more than one reason, the columns do not sum to 100 percent.

black women were most likely to feel this way. Latinas (who enjoy a favorable sex ratio) were least likely among women to report availability concerns as reasons for not marrying. Interestingly, however, nearly half of black men also reported that there were not enough women who met their standards.

"Not ready to settle down" was cited more by men than by women, but may be more a reflection of the fact that the men in this sample were younger than the women (since this reason would seem on the face to be more associated with youth). Men, overall, were also more likely to indicate that not having enough money to support a family was a reason for not getting married, with over half of all Latino men and over 40 percent of black men giving that response. Over a quarter of black women also felt that financial concerns were reason for not marrying, compared to much lesser percentages for white women and Latinas. Significant percentages of men in all groups also cited concentration on school or work as factors in their staying unmarried, with women in each group only half as likely to indicate the same. Latinos were most likely and blacks were least likely to give this reason.

Importantly, only relatively small percentages of all groups did not believe in marriage; but those percentages were lowest overall among blacks and highest among Latinos. With the exception of the Latino men (who are relatively young), few persons indicated that "having fun playing the field" was a reason for not marrying.

In general then, in this sample, nonmarriage is not a function of "not believing in marriage." Economic factors figured prominently for all groups of men, more so than for the women in those groups—and particularly so among Latino men. Women's concerns in all groups centered on the issue of availability, although availability was an important dimension of nonmarriage among black men as well.

These findings lend support to our notions about the salience of the different structural constructs for different groups—that is, economic versus demographic/availability factors. Latinos, among men, were most concerned about financial readiness for marriage but basically believed that there were enough available women who met their standards. In contrast, although black men were also concerned about economic readiness, they did not believe that there are enough suitable women. Stated in theoretical terms: numerical scarcity inflates male value while economic deficiency deflates male marital capabilities. As seen in this sample, black men whose marriage market value has been enhanced by their scarcity apparently believe that they can be "choosier" in terms of mate selection [which would fit the Guttentag and Secord

(1983) social exchange model predictions—in situations of scarcity, the alternatives available to the sex in short supply increase and their selection standards are raised]. Because of their economic constraints, black men are less able to enter into marriage, but also due to the greater availability of alternatives for them, black men would have less need to commit to marriage. On the other hand, Latino men are not a scarce commodity and could therefore not afford to be too "choosy," but were also not financially "ready" for marriage.

On the basis of these frequencies alone, it would appear that black men are less likely than Latino men to view economic factors as marital constraints. However, we believe that this represents a theoretically consistent pattern of "exchange." In an objective sense, black men's overt concerns about economics as a marital inhibitor may be diminished by their numerical advantage....

Conclusions

This study has provided evidence that certain global conditions, determined on the basis of aggregate-level assessments, are accurately perceived by individuals. For the most part, the assessments of mate availability by men and women in the three ethnic groups reflected their particular demographic and socioeconomic realities.

Overall, this research has demonstrated that economic and mate availability concerns are differentially related to marital behavior and marital expectancy among socioculturally distinct groups. As predicted, economic as well as availability indicators were salient for blacks (both structurally and attitudinally). Economic factors were not as salient for whites, but perceived availability played a significant role as a correlate of both marital behavior and expectation....

We conclude that economic and demographic factors separately and jointly affect marital status and marital expectations, acting alternatively as constraints and incentives at different historical moments. It should be kept in mind, however, that the impingement of such structural factors as marital attitudes and behavior also requires examination within the context of sociocultural background factors which may differ for the groups under study and which may produce internal dynamics that result in the changes in marital behavior observed today....

These findings must be interpreted within the context of change. We have placed what may be an artificial constraint on these analyses by assuming that mate selection and marriage occur primarily within ethnic groups. Certainly, intermarriage data for whites, African Americans, and Latinos presently support such an assumption. However, there has been significant change in this regard in recent decades (Tucker and Mitchell-Kernan, 1990). Two-thirds of the respondents in this survey indicated that they had dated persons of another race; the same proportion stated that they would be willing to marry a person of another race. Although interracial marriage rates in the West are higher than those observed in other regions, the increase in interracial unions is a nationwide trend (Tucker and Mitchell-Kernan, 1990). If this is indicative of a fundamental change in societal attitudes concerning intermarriage, it may be that persons who find themselves in severely constrained marriage markets will broaden their eligibility criteria to include persons of other ethnic groups.

A related issue is the function of changing sex-role ideology. Jessie Bernard (1981) argued that the "good-provider" role has undergone change in American society generally, resulting in greater emphasis on the nonprovider roles of husband. Secord and Ghee (1986) have discussed the implications of these notions in the context of the black marriage market, arguing that in situations where men are constrained in attempts to act as provider, marital roles must be reevaluated. To be sure, the basis of marriage is not solely economic. Some women choose to marry poor men; some poor men choose to marry.... Economic uncertainty and changes in economic incentives not only affect women and men differentially, they may entail further sequelae by elevating other mate-selection criteria to new levels of importance. In particular, black women and Latinas who are less likely to expect economic benefits from marriage may begin to give greater emphasis to other eligibility factors, such as the ability to care for children and emotional supportiveness. This suggests that marriage as an institution may be taking on a different meaning for such women—although still highly valued....

This study was undertaken in part to examine the relative explanatory significance of two dominant theoretical paradigms. However, an additional interpretive task demanded by the study's results is that of explaining the gap between marital attitudes and behavior in the African American population. The theories that form the basis of this study would suggest that several factors—in particular, economic and mate availability concerns—constrain behavior that would conform with the stated values....

Future research on African American marriage and family formation must go beyond the necessary focus on economic factors and consider other marital facilitators and inhibitors. Work in this area must also consider individual-level decision making and its relationship to macro phenomena in a more specific manner. While we would hope for a new economic order that provides black men as well as women with the means to provide an emotionally as well as materially supportive family atmosphere, in the short run, we must bolster those mechanisms that support marital bonds despite financial constraints.

REFERENCES

Ajzen, I., and Fishbein, M. 1980. *Understanding attitudes and predicting social behavior.* Englewood Cliffs, NJ: Prentice-Hall.

Andrews, F. M., Morgan, J. N., Sonquist, J. A., and Klem, L. 1973. *Multiple classification analysis: A report on a computer program for multiple regression using categorical predictors.* 2nd ed. Ann Arbor: University of Michigan.

Bernard, J. 1981. The good-provider role: Its rise and fall. *American Psychologist, 36* (1):1–12.

Bloom, D. E., and Bennett, N. G. 1985. Marriage patterns in the United States. National Bureau of Economic Research Working Paper Series. No. 1701. Cambridge, MA: National Bureau of Economic Research.

Cherlin, A. J. 1981. *Marriage, divorce, remarriage.* Cambridge, MA: Harvard University Press.

Darity, W., and Myers, S. L. 1986/87. Public policy trends and the fate of the black family. *Humboldt Journal of Social Relations,* 14:134–164.

Entwistle, B., and Mason, W. M. 1985. Multilevel effects of socioeconomic development and family planning programs on children ever born. *American Journal of Sociology,* 91:616–649.

Epenshade, T. J. 1985. Marriage trends in America: Estimates, implications, and underlying causes. *Population and Development Review,* 11:193–245.

Fishbein, M., and Ajzen, I. 1975. *Belief, attitude, intention and behavior.* Reading, MA: Addison-Wesley.

Glick, P.C., Heer, D. M., and Bereseford, J. C. 1963. Family formation and family composition: Trends and prospects. In M. B. Sussman, ed., *Sourcebook in marriage and the family.* New York: Houghton Mifflin.

Guttentag, M., and Secord, P. F. 1983. *Too many women: The sex ratio question.* Beverly Hills: Sage Publications.

Hampton, R. L. 1979. Husband's characteristics and marital disruption in black families. *The Sociological Quarterly,* 20:255–266.

Kiecolt, K. J. 1988. Recent developments in attitudes and social structure. *Annual Review of Sociology,* 14:381–403.

Liska, A. E. 1984. A critical examination of the causal structure of the Fishbein/Ajzen attitude-behavior model. *Social Psychology Quarterly,* 47:61–74.

Liska, A. E. 1990. The significance of aggregate dependent variables and contextual independent variables for linking macro and micro theories. *Social Psychology Quarterly,* 53:292–301.

Moynihan, D. P. 1967. The Negro family: The case for national action. In L. Rainwater and W. L. Rainwater, eds., *The Moynihan report and the politics of controversy.* Cambridge, MA: MIT Press.

Norton, A. J., and Moorman, J. E. 1987. Current trends in marriage and divorce among American women. *Journal of Marriage and the Family,* 49:3–14.

Oppenheimer, V. K. 1988. A theory of marriage timing. *American Journal of Sociology,* 94:563–591.

Rodgers, W. L., and Thornton, A. 1985. Changing patterns of first marriage in the United States. *Demography.* 22:265–279.

Ross, H. L. and Sawhill, I. 1975. *Time of transition: The growth of families headed by women.* Washington, DC: The Urban Institute.

Schoen, R. 1983. Measuring the tightness of the marriage squeeze. *Demography,* 20(1): 61–78.

Secord, P. F., and Ghee, K. 1986. Implications of the black marriage market for marital conflict. *Journal of Family Issues,* 7:21–30.

Staples, R. 1981a. Race and marital status: An overview. In H. P. McAdoo, ed., *Black families,* pp. 173–175. Beverly Hills: Sage Publications.

Staples, R. 1981b. *The world of black singles.* Westport, CT: Greenwood Press.

Tucker, M. B., and Mitchell-Kernan, C. 1990. New trends in black American interracial marriage: The social structural context. *Journal of Marriage and the Family,* 52:209–218.

Tucker, M. B., and Mitchell-Kernan, C. In press. Mate availability among African Americans: Conceptual and methodological issues. In R. Jones, ed., *Advances in black psychology.* Hampton, VA: Cobb and Henry.

Tucker, M. B., and Taylor, R. J. 1989. Demographic correlates of relationship status among black Americans. *Journal of Marriage and the Family,* 51:655–665.

U.S. Bureau of the Census. 1960. *Historical statistics of the United States, colonial times to 1957.* Washington, DC: U.S. Government Printing Office.

U.S. Bureau of the Census. 1964. *U.S. census of population: 1960,* Vol. 1, *Characteristics of the population,* Part I, United States Summary. Washington, DC: U.S. Government Printing Office.

U.S. Bureau of the Census. 1973. *U.S. census of population: 1970,* Part 1, United States Summary—Section 2. Washington, DC: U.S. Government Printing Office.

U.S. Bureau of the Census. 1982. *Statistical abstract of the United States: 1982–83.* 103rd ed. Washington, DC: U.S. Government Printing Office.

U.S. Bureau of the Census. 1983. *U.S. census of population: 1980,* Vol. 1, *Characteristics of the population,* Chapter B: General Population Characteristics, Part 1, United States Summary. Washington, DC: U.S. Government Printing Office.

U.S. Bureau of the Census. 1985. *U.S. census of population: 1980,* Vol. 2, *Subject reports: Marital characteristics.* Washington, DC: U.S. Government Printing Office.

U.S. Bureau of the Census. 1991a. Marital status and living arrangements: March 1990. *Current Population Reports,* series P-20, no. 450. Washington, DC: U.S. Government Printing Office.

U.S. Bureau of the Census. 1991b. The Hispanic population in the United States: March 1990. *Current Population Reports,* series P-2, no. 449. Washington, DC: U.S. Government Printing Office.

U.S. Bureau of the Census. 1991c. Family disruption and economic hardship: The short-run picture for children. *Current Population Reports,* series P-70, no. 23. Washington, DC: U.S. Government Printing Office.

U.S. Bureau of the Census. 1992. Marriage, divorce, and remarriage in the 1990s. *Current Population Reports,* series P-23, no. 180. Washington, DC: U.S. Government Printing Office.

Wilson, W. J. 1987. *The truly disadvantaged.* Chicago: The University of Chicago Press.

Wilson, W. J., and Neckerman, K. M. 1986. Poverty and family structure: The widening gap between evidence and public policy issues. In S. H. Danziger and D. H. Weinberg, eds., *Fighting poverty: What works and what doesn't.* Cambridge, MA: Harvard University Press.

CHAPTER 2

Questions for Writing, Reflection, and Debate

READING 6

Values, Attitudes, and the State of American Marriage • *Norval D. Glenn*

1. Glenn concludes that Americans still value marriage, have high expectations for what they want out of marriage, yet don't regard marriage as permanent. What evidence do you find to support this finding among your family, friends, and neighbors?
2. In his study, Glenn did not distinguish unmarried people living alone from those living in committed relationships. Does this omission affect his results? How?
3. Do you think that lowering or raising the barriers to divorce is related to marital happiness? What does this article suggest? Do you agree? Why or why not?

READING 7

Marital Status, Gender, and Perception of Well-Being • *Harsha Mookherjee*

4. Some studies indicate that men have more to gain from marriage than women. How could this be true?
5. How has Mookherjee measured well-being? What are other aspects of well-being not measured in this study?
6. How can you explain the conclusion of this study, that marriage enhances the well-being of both men and women, but women are more satisfied than men with their lives regardless of their marital status?

READING 8

A Comparison of Marriages and Cohabiting Relationships • *Steven L. Nock*

7. In what ways are the cohabitating relationships described in this study less successful than the married relationships?
8. Nock suggests that the lack of institutional sanction for cohabitation may be at the heart of some of these differences. Do you think Americans will ever place the same value on cohabitation that they do on marriage? What would facilitate or prevent this change in norms?

READING 9

Marital Behavior and Expectations: Ethnic Comparisons of Attitudinal and Structural Correlates • *M. Belinda Tucker and Claudia Mitchell-Kernan*

9. Does this study find racial differences in the desire to marry or in the importance placed on marriage? Explain your answer.
10. How does this study explain different marriage rates for different racial groups? Can any of these factors be altered to promote marriage? Should they be? Explain.

3

Childbirth and Infertility

Is parenthood biological or social?

On June 12, 1998, I heard a radio story by Stephen Smith, a Minnesota Public Radio reporter, about two male homosexual life partners who were becoming parents. Their child was created from sperm from one member of the couple and an egg from the other partner's sister, which was fertilized with the help of a clinic and implanted in a surrogate mother. This child will have at least four adults with potential claims to parenthood: the two male partners, the egg donor, and the surrogate mother. Who will the child's parents be? How you answer this question reveals whether you think of parenthood as a biological or a social phenomenon.

When the radio reporter interviewed one of the expectant gay parents, he asked how the couple intended to tell their child about his or her conception and birth. The man replied that they intended to be honest with their child. The men have a book that they will use to reveal gradually to the child, when he or she is ready to understand, the child's complex biological history. However, this couple also felt their child would have two fathers, both fulfilling the social role although only one had a biological connection.

The men in this story were able to become parents because of recent developments in reproductive technology. Some people argue that homosexuals should not benefit from the research undertaken to assist the increasing numbers of couples seeking medical treatment for infertility. Regardless of your ethical position on access to these technologies, the fact remains that both homosexual and infertile couples are increasingly seeking alternative routes to becoming parents.

Ironically, the pursuit of fertility has arisen from the desire to prevent or control fertility. During the transition from an agricultural economy to industrialization, when children became more of an economic burden than an asset, each new baby could be viewed as another mouth to feed rather than more hands to work the land or the machines. Limiting family size thus became an issue, especially for families in poverty. However, contraceptive methods available at this time were ineffective.

The development of more efficient and available contraceptives had at least two important outcomes. First, it provided people with the technology to choose whether and when to become parents with a degree of certainty previously unavailable. Second, it created a separation in people's minds between engaging in sexual intercourse and conceiving a child. As a result of these advances, many contemporary couples have chosen to have fewer children or remain "childfree," a term implying choice. Meanwhile, "childless" couples (an older term connoting infertility and a desire to have a child) have found that fewer children are available for adoption. The result has been a change in the "market value" of children, as Viviana Zelizer explains in "From Baby Farms to Black-Market Babies: The Changing Market for Children."

In their desperation, some parents have turned to other less developed countries to adopt children. However, overseas adoptions are long and costly. They are also controversial, as Mary Ellen Fieweger reveals in "Stolen Children and International Adoptions." This article raises concerns regarding poverty as a motivation for relinquishing parental rights and documents the coercive or illegal tactics used in developing countries to obtain babies for adoption by parents in wealthier countries. Given these issues, most infertile couples today opt to try to produce a child to whom they have some biological connection. Reproductive technology is advancing rapidly, but investment in time, money, discomfort, and disappointment remains formidable.

A number of social and demographic factors have influenced the pursuit of less traditional routes to parenthood— for example, the trend for people to marry at an older age than in the past and the rise of untreated sexually transmitted diseases, which can decrease fertility. These seem intensely private circumstances. Yet many infertile couples and gay couples who seek to have a child through means other than heterosexual intercourse find that their decision to become parents is a public concern. The notorious adoption screenings provide a good example. No one makes decisions about the suitability of people to become parents before they are allowed to have sex, despite the definite potential to produce a child. Yet adoptive parents are submitted to intense scrutiny to determine their worthiness. Infertile couples seeking high-tech methods of conception are also screened using such criteria as health, age, genetics, and motivation. Once "approved" for treatments, couples must be willing to have their bodies and their bodily functions examined, monitored, and assessed by others. Elaine Tyler May's article, "Designer Genes: The Baby Quest and the Reproductive Fix," gives us insights into the potential impact of such intrusions on individuals and couples. Ellen Lewin explains in "Lesbian Mothers: This Wonderful Decision" how the situation is even worse for homosexuals. Lesbians who choose artificial methods of conception, such as in vitro fertilization, are subject to the same loss of privacy as anyone else attempting pregnancy this way, but in addition they face public concern with their sexuality. However, sexual orientation is not the same as infertility. Lewin discusses many additional options open to fertile homosexuals who want to become parents.

In these readings, you will learn about the social inequalities that pervade the issue of becoming a parent noncoitally. The new reproductive technologies discussed in several of these readings are available only to those with a reasonable income or substantial health insurance. Leslie King and Madonna Harrington Meyer, in "The Politics of Reproductive Benefits: U.S. Insurance Coverage of Contraceptive and Infertility Treatments," provide an insightful analysis of how public policy and insurance practices work in tandem to promote parenting among those of economic means while discouraging those who live in poverty. This inequality mirrors the social inequalities in the adoption market.

As you enjoy these readings, keep in mind the role of society in determining the biological and social aspects of parenting. In America, access to new ways of becoming a parent is challenging our traditional policy preference to preserve the rights of the biological parents. In the radio broadcast I mentioned earlier, the man who was interviewed said he saw similarities between the social acceptance of these new routes to parenting and of interracial marriage. Not too many years ago, interracial marriage was a rare thing in America. In fact, it was against the law in many states. Those who challenged this law found themselves open to negative public sanction. But today interracial marriage is both more widely practiced and more widely accepted. Maybe in another 40 years our society will find it both easier and more acceptable for children to have a complex biological history. What do you think?

From Baby Farms to Black-Market Babies: The Changing Market for Children

Viviana A. Zelizer

In the 1870s, there was no market for babies. The only profitable undertaking was the "business of getting rid of other people's [unwelcome] babies." For about ten dollars, baby farmers took in these generally illegitimate children. Yet some fifty years later, adoptive parents were eagerly paying $1,000 or more to purchase an infant....

Legal adoption, rare in the nineteenth century, became increasingly popular in the twentieth century. A judge from the Boston Probate Court remarked in 1919, "the woods are full of people eager to adopt children—the number appearing to be in the increase."[1] By 1927, the *New York Times* reported that the new problem in adoption "has become one of finding enough children for childless homes rather than that of finding enough homes for homeless children." Despite greater regulation, and more thorough screening of adoptive parents, legal adoptions increased threefold between 1934 and 1944, finally breaking a long-standing monopoly of institutional care for dependent children.[2]

The quest for a child to love turned into a glamorous and romanticized search as a number of well-known entertainment and political figures proudly and publicly joined the rank of adoptive parents. Minnie Maddern Fiske, a respected stage actress, told about her adoption of a thirteen-month-old baby, who made his stage debut in one of her tours. Al Jolson explained to the press his decision to adopt a child: "I think it is selfish to go through life without children." In the 1930s, Gracie Allen and George Burns, Mayor La Guardia, Babe Ruth, and Eddie Rickenbacker, among others, similarly announced their decision to adopt a child.[3]

The fairy-tale dimension of adoption was further magnified by many stories of poor waifs taken into the homes of generous, wealthy foster parents. A 1905 article in *Cosmopolitan* had already noted the fantastic prospect of transforming a little "plebeian" into a "lord"; "[the] little ones go from the ... doorsteps and sewers, to comfort always, and sometimes to luxury."[4] In 1925, Edward W. Browning, a wealthy New York real estate operator, made front-page headlines when he advertised for a "pretty, refined girl" to adopt, thus opening up "the gates of fairyland for many a poor child." Browning allegedly received over 12,000 applications from all over the country. Each girl he interviewed was sent home in Browning's automobile, "with a chauffeur in livery ... and a footman to help them in and out of the coach, just like Cinderella."

The Browning case ended in a scandal after Mary Spas, the girl he adopted thinking she was sixteen years old, confessed to being twenty-one. Mary left Browning lured by an agent's offer to write her story for publication and the promise of a movie job. Browning, who claimed to have spent $20,000 for the adoption (including forty dresses for Mary), had the adoption nullified.[5] While the incident was exceptional, the social class of adoptive parents was indeed undergoing change. A comparison by the New York State Charities Aid Association of the occupations of 100 foster fathers between 1898 and 1900 with the same number of fathers in the period between 1920 and 1921, found that nearly three-quarters of foster fathers in the first period were in skilled, semi-skilled, or unskilled labor, or in farming, while in the latter period there was a predominance of men in business and office work. Sophie Van Senden Theis, author of the report, recalled how "Many of the plainest homes were used for the first children placed, for in those days the Agency had to take what it could get in the way of foster homes."

The shift in social class, detected by Theis, was tied to the sentimentalization of adoption. A study of adoptive parents in Minnesota during the period from 1918 to 1928 found that adoptive fathers surpassed the proportion of males in the general population in the higher occupational levels (professional, semi-professional, and managerial). But adoptive fathers of older, and therefore potentially useful, children were more likely to belong to a lower occupational category, in particular farming.[6]

Sentimental adoption created an unprecedented demand for children under three, especially for infants. In 1910, the press already discussed the new appeal of babies, warning, "there are not enough babies to go around."[7] The Home-Finding Committee of the Spence Nursery, an agency organized for the placement of infants, was surprised to discover that, "instead of our having to seek these homes, they have sought us, and so great is the demand for babies that we cannot begin to meet it." In 1914, Judd Mortimer Lewis, a Texas poet and humorist, achieved national notoriety by working as a one-man baby bureau, using his column in the Houston *Post* to find infants for "baby-hungry" parents. Babies, observed the *New York Tribune* in 1923, "are being taken into homes in numbers and for reasons that mark a new era in the huge task of caring for parentless children." By 1937, infant adoption was being touted as the latest American fad: "The baby market is booming.... The clamor is for babies, more babies.... We behold an amazing phenomenon: a country-wide scramble on the part of childless couples to adopt a child." Ironically, while the economically "useless" nineteenth-century baby had to be protected because it was unwanted, the priceless twentieth-century baby, "needs protection as never before ... [because] too many hands are snatching it."[8]

The priceless child was judged by new criteria; its physical appeal and personality replaced earlier economic yardsticks. After talking to several directors of orphan asylums, the *New York Times* concluded that "every baby who expects to be adopted ... ought to make it a point to be born with blue eyes.... The brown-eyed, black-eyed, or grey-eyed girl or boy may be just as pretty ... but it is hard to make benevolent auxiliaries of the stork believe so."[9] But the greatest demand was for little girls. Soon after launching its popular Child-Rescue Campaign in 1907, promoting foster home care, the *Delineator* commented that requests for boys were half that for girls; "a two-year old, blue-eyed, golden haired little girl with curls, that is the order that everybody leaves. It cannot be filled fast enough."[10] Similarly, in its first thirty years of work, the New York State Charities Aid Association received 8,000 applications for girls, out of a total of 13,000. While working homes sought older girls for their domestic labor value, adoptive homes wanted little girls for their domestic sentimental value: "a doll on which they could tie pink sashes." In the 1920s, wealthy Americans even imported their "English-rose" golden-haired baby girls from London. Jews were apparently an exception. According to an interview with the assistant superintendent of the Hebrew Sheltering Guardian Asylum in 1910, three-year-old boys were in much greater demand among Jewish adoptive parents than little girls.[11]

Considering the widespread parental preference for a male first-born child, the popularity of adopted daughters was puzzling. As Hastings Hart observed in 1902, "When people pray for a child of their own, they are apt to pray for a boy; when they want it for adoption, they want a girl. It is an unexplainable fact that every one who is engaged in placing out children is familiar with."[12] Parents, suggested one adoption agency in 1916, "seem to feel that a girl is easier to understand and to rear, and they are afraid of a boy...."[13] Twenty years later, the *Canadian Magazine* linked the persistent preference for girls to parents' fear of a lonely old age: "Girls do not break the home ties so early as boys and outside interests do not play so large a part in their lives." Why do "pretty little picture-book girls, go like hot cakes," speculated rhetorically a writer in the *Saturday Evening Post,* because they are "grand little self-advertisers and they know instinctively how to strut their stuff.... They stretch out their dimpled arms, gurgle some secret baby joke, smile a divine toothless smile ... and women and strong men go mad, become besotted with adoration...." Boys, on the other hand were promotional failures, "slower, more serious and aloof."[14]

The sex and age preferences of twentieth-century adoptive parents were clearly linked to the cultural revolution in fostering. While the earlier need for a useful child put a premium on strong, older children, preferably male, the later search for a priceless child led to babies and particularly, pretty little girls. It was not the innate smiling expertise of females, but established cultural assumptions of women's superior emotional talents which made girls so uniquely attractive for sentimental adoption. The new appeal of babies was further enhanced by the increasing acceptance, in the 1920s, of environmental theories of development. Couples considering adoption were now reassured that "heredity has little or nothing to do with our characters. It is the environment that counts. ..."[15] Intelligence tests and improved methods of determining children's physical health, reduced the "old prejudice against thrusting one's hand in a grab-bag, eugenically speaking, and breeding by proxy."[16] Even the stigma of illegitimacy was turned into an asset by suggestions that "love babies" were particularly attractive and desirable.[17]

Ironically, as the priceless child displaced the useful child, the dangers of adoption shifted from economic to emotional hazards; the previously exploited little laborer risked becoming a "pretty toy." Prospective adopters were warned: "If you are planning to have a plaything to cuddle and pet and dress prettily, don't do it!"[18] Parents were also advised against seeking an emotionally or psychologically "useful" child. Experts now wrote about the dangers of "seeking compensation in children for frustrated affections," or unfulfilled ambitions.[19] If child placing agencies were less often confronted by requests for a sturdy working child, they now faced new expectations, as the couple who applied to the New York State Charities Association for a three-month-old baby, "who could eventually go to Princeton."[20]

Black-Market Babies: The Price of a Priceless Child

The sentimentalization of adoption had an unanticipated and paradoxical effect. By creating a demand for babies, it also stimulated a new kind of baby market. While nineteenth-century mothers had paid baby farmers to accept their unwanted baby, twentieth-century adoptive parents were willing to pay to obtain an infant. "Baby traffickers" thus found an additional line of business—making money not only from the surrender of babies, but doubling their profits by then selling them to their new customers.[21] As a result, the value of a priceless child became increasingly monetized and commercialized. Ironically, the new market price for babies was set exclusively by their noneconomic, sentimental appeal.

The traditional baby market continued into the second decade of the twentieth century. An influential investigation conducted by the Chairman of the Maryland State-Wide Vice Commission in 1914, reported that in many cases the adoption of children remained a "means of earning money; the child was entirely secondary and was taken simply for a price." Respondents to several decoy newspaper ads seeking a home for a one-week-old baby, requested sums ranging from $100 to $7,000 to take the child permanently. As one woman explained, "I love to care for a baby ... but I would expect to be paid a fair price for my services." Two established maternity hospitals in Baltimore actively participated in this "commercialized traffic." Parents were asked to pay between $100 to $125 to be "relieved of all responsibility and relinquish all right and claim to the child." If a woman was too poor to pay with money, she paid with labor, working for one year as a maid in the hospital. The report uncovered the routine collaboration of physicians, nurses, midwives, and even clergymen who were willing, usually for a fee, to help a mother dispose of her infant.[22]

But by 1910, there had been signs that the structure of the baby market was changing and expanding. An article in *Cosmopolitan* referred to a "desultory and elusive traffic," with babies being sold for as much as one thousand dollars. A speaker at the 1913 National Conference of Charities and Correction remarked on the developing double baby market in

California. Maternity homes and lying-in hospitals were finding that unwanted babies were a new source of revenue: "There are enough childless marriages to create a demand for promising babies, and therefore a market." The going price: $200 per baby. In Chicago, the Juvenile Protective Association uncovered a "regular commercialized business" in 1917. As before, the unmarried mother "willing to pay any amount of money to dispose of her child," was charged from fifteen to sixty-five dollars by maternity hospitals or individuals to dispose of her baby. But the study also reported a different type of baby market. An "unusually attractive" infant sold for fifteen to one hundred dollars, paid in cash or installments. The new trade slogan of one baby seller was, "It's cheaper and easier to buy a baby for $100 than to have one of your own."[23]

In 1922, the dramatic findings of "A Baby a Day Given Away," a study conducted by the New York State Charities Aid Association, put commercialized adoption directly in the national public spotlight. The six months investigation of newspaper advertisements offering and requesting children for adoption, revealed an "indiscriminate exchange of children." An average of a baby a day was being disposed of in New York, "as casually as one would give away a kitten"; many sold at "bargain-counter" prices. It was not a peculiar New York arrangement.[24] In the classified advertisement column of almost any Boston newspaper, "together with items relating to automobiles, animals, amusements . . . may often be found the child offered for adoption."[25]

Three years later, the notorious prosecution of a New York baby farmer shocked the nation and further raised the visibility of commercial child placement. Helen Augusta Geisen-Volk was charged and indicted for child substitution and for starving infants to death. The young wife of a well-to-do manufacturer added fuel to the scandal by publicly confessing that, unknown to her husband, Mrs. Geisen-Volk had sold her an infant for 75 dollars. None of the crimes committed by Geisen-Volk were new to the baby farming business; similar accusations were made as early as the 1870s. More unusual was the severity of the reaction and the degree of public interest in the case. Never before had a baby farming case, for instance, made several front-page headlines in the *New York Times*.[26]

Commercial child placement emerged as a significant social problem in the 1920s in large part because it violated new professional standards in adoption. Without proper supervision by a licensed child-placing agency, adoption could be dangerous both for children and their adoptive parents. The 1922 report by the New York State Charities Aid Association found many babies given away to "immoral and unfit homes . . . in some instances a baby was 'tried' in a new home every week for a period of six or seven weeks." Besides endangering children, such practices discredited "conscientious and intelligent home finding done by competent child-placing agencies."[27] But selling children undermined not only professional adoption; it also betrayed the new standards of sentimental adoption. It was a sacrilege to price a priceless child. Worse than a criminal, Mrs. Geisen-Volk was indicted by the judge as a "fiend incarnate." As a probation officer told reporters, ". . . the woman . . . has no maternal affections . . . [Babies] to her . . . are articles of merchandise to be bartered or exchanged. The defendant represents a revolting anomaly in humankind."[28]

Yet baby sales did not stop. Despite increased public regulation of childcare and the multiplication of adoption laws, including stricter licensing of boarding homes and new laws against adoption by advertisement, informal child placement persisted. A study of 810 children adopted in Massachusetts between 1922 and 1925 showed that two-thirds were adopted without assistance from social agencies. Similarly, of 1,051 adoptions in New Jersey during 1928, only 289 were sponsored by an agency.[29] Independent adoptions were often arranged by well-intentioned intermediaries, without involving profit. But in many cases, middlemen built a lucrative business by "bootlegging" babies.

Harshly denounced as an "iniquitous traffic in human life," and a "countrywide shame," the black market in babies flourished in the 1930s and 1940s.[30] As demand for adoptable children grew, the "booming" traffic in infants reached a new, third stage. It was now a seller's market. Therefore, the mother of an unwanted child no longer needed to pay to dispose of her baby. Instead, entrepreneurial brokers approached her, offering to pay medical and hospital expenses and often a bonus in exchange for her baby. Even in independent placements arranged without profit, it became common practice to pay the hospital and medical expenses of the natural mother.[31]

In 1955, a Congressional investigation conducted by Senator Estes Kefauver officially pronounced baby selling a national social problem. Its exact magnitude was unclear. While an estimated three out of every four adoptions were made independently of any agency, the percentage arranged for profit was unknown. Unquestionably, however, baby selling was no longer a small local business. In Memphis, a Mrs. Georgia Tann, executive of the Tennessee Children's Home Society, was found guilty of intrastate black marketing. Between 1930 and 1950 she placed over 1,000 children in some fifteen states, making more than a million dollars profit. In another case, Marcus Siegel, a Brooklyn attorney and baby broker, collected about $160,000 in only eighteen months of business. The price tag of black market baby rose from an estimated $1,000 in the 1930s to $5,000 in the late 1940s. By 1951 some babies sold for as much as $10,000.[32]

The money value of infants was partly determined by a reduced supply. As the dramatic decline in the national birth-rate, which began early in the nineteenth century, continued into the 1930s, fewer babies were available for adoption. Contemporary observers also suggested that the increased demand for babies was partly the result of higher rates of childlessness among American couples. In *The Conservation of the Family,* Paul Popenoe blamed higher infertility on a "meat diet, imperfect ventilation of houses, nervous strain of city life, spread of twin beds, [and] wearing of corsets."[33]

Growing concern with the preservation of the family unit further contributed to the baby shortage. After 1911, the mothers' pension movement allowed widows, and in some cases deserted wives or mothers whose husbands were physically or mentally handicapped, or in prison, to keep their children. Reformers also encouraged unmarried mothers to keep their babies. As a result, the supply of adoptable infants shrunk, and the waiting lists of adoption agencies grew longer. Unwilling to wait two or more years for a child, and impatient with the increasingly restrictive standards set by agencies, parents turned to the black market. As one 1951 exposé of baby selling noted, "Babies . . . are on the auction block because there are ten or more potential adopting couples in the country for every child available for adoption."[34]

But scarcity alone cannot determine value. A reduced supply raised the price of babies only because there was a growing number of enthusiastic buyers for white, healthy infants. The market exploited, but did not create the infatuation with priceless babies. In sharp contrast, older children found few customers. Deprived of their former labor value, they were excluded from the new emotional market. Therefore, while the agencies' waiting lists for babies had the names of hundreds of impatient parents, it was virtually impossible to find homes for children older than six, who had become both economically and sentimentally "useless."[35]

Pricing the Priceless: The Special Market for Children

The sentimentalization of adoption in the twentieth century, thus, led paradoxically to a greater commercialization and monetization of child life. As the market for child labor disappeared, a market price developed for children's new sentimental value. In 1975, a second Congressional hearing on black-market practices estimated that more than 5,000 babies were sold each year in the United States, some for as much as $25,000. Sellers retained bargaining leverage. As one black-market lawyer told a prospective customer, "Take it or leave it. I have five other couples."[36] The capitalization of children's value extended into legitimate child-placement. Reversing a long-standing policy by which "the question of a money transaction never comes up in negotiations for a child,"[37] many agencies in the 1940s introduced adoption fees.

An apparently profound contradiction was thereby created, between a cultural system that declared children priceless emotional assets, and a social arrangement that treated them as "cash commodities."[38] In the view of some economists, this persistent conflict between social values and structure should be resolved in favor of the market. Landes and Posner, for instance, advocate legalized babyselling: "The baby shortage and black market are the result of legal restrictions that prevent the market from operating freely in the sale of babies as of other goods. This suggests as a possible reform simply eliminating these restrictions." An undiluted price system, they argue, would match adoptive parents with adoptable children more efficiently than agencies. In fact, studies comparing independent adoptions with agency placements find little, if any, difference in outcome. Landes and Posner dismiss "moral outrage" or "symbolic" objections against baby sales, as antiquated and impractical.[39]

Others strongly defend substitute parenting as a "gift" of love that should be regulated exclusively by altruism, not profit. As Senator Jennings Randolph told the Subcommittee on Child and Youth during the 1975 black-market hearings, "I cannot conceive of someone coldly and calculatingly selling another human being.... Many thousands of Americans want to provide parental love. It is certainly immoral ... for individuals to profit from that desire." In this ideological context, adoption fees are no more justified nor less venal than black-market purchases. For instance, in testimony presented to the Congressional Subcommittee, the director of an organization of adopted adults, rejected the claim that "monies collected by agencies are respectable while monies collected by independent agents are not:" "It ... doesn't matter to the people involved ... whether the fee was $5,000 or $25,000 and whether it was paid to an agency or to an independent agent.... No rationale of fees will relieve adoptive parents of the certain knowledge that they have bought a human being...." From this perspective, the tension between values and structure can only be resolved by drastically eliminating every opportunity for a "system of merchandising adoptive babies."[40] In fact, some states have already banned all independent placements, even those not involving profit, in order to deter commercial baby markets.[41]

Ideological defense or rejection of the market equally ignores the interrelationship between market systems and values. Both positions presuppose, for better or for worse, the inevitable and unilateral power of the market. But a "free" independent market for babies is a theoretical illusion; cultural constraints cannot be simply dismissed as obsolete. On the other hand, denying the market any function at all ignores distinctions between types of markets. Not all markets are equal. As Bernard Barber notes, "As a result of ... interdependencies with, or constraints from, both values and other institutional structures, economic exchange can be patterned in different ways."[42] From the start, the baby market was shaped by the cultural definition of children as priceless. It was not the contradiction, therefore, but the interaction between notions of children's pricelessness and pricing arrangements that resulted in the differentiation of legitimate and illegitimate baby markets.

The black market is unacceptable because it treats children in the same impersonal, economizing manner used for less sacred commercial products. For baby brokers, price and profit are dominant considerations: "If they were not selling babies, they would be selling whatever else was hot and produce [*sic*] a profit."[43] Black-market practices are not only illegitimate but also illegal. In most states, it is a crime to accept payment for placing a child for adoption. Yet a different kind of market exists which is, in most cases, legal and compatible with sentimental adoption. In this "gray-market," placements are arranged "without profit by well meaning parents, friends, relatives, doctors and lawyers."[44] Within this context, professional fees for legal or medical services are acceptable. Justifying such payments during the 1975 Congressional hearings, the executive director of the Child Welfare League of America explained, "Money exchanges hands, but it is only to pay for actual costs. There is no thought of profit."[45] Not only do most adoption experts support the right to collect "reasonable fees for professional services," but certain statutes specifically allow legal fees and compensation for the mother's medical expenses. Thus, while the black market is defined as a degrading economic arrangement, a modified, legitimate market exists for the exchange of children. To be sure, the boundary between a legitimate market and a "dangerous" sale is not always easy to maintain. As a means of market control, parents in California are required to submit a detailed itemized account of their expenditures for a private adoption. Florida, on the other hand, limits lawyers' fees to $500.[46]

Adoption fees also constitute a separate market. From the start, agencies sought to define their work as consistent with sentimental adoption. Until the 1940s, only "gratitude donations" were accepted from adoptive parents. The Children's Home Society of Virginia, for instance, told parents, "that a gift from them in such an amount as they choose will be gratefully received, but that it must be made as a gift and not as payment for services."[47] The Society's directors refused to even discuss any definite sum with foster families. The boundary between adoption and purchase was preserved by defining the money as an elective gift and a symbol of gratitude, not a price. As Simmel points out in his discussion of marriage by purchase, "the gift contains something more personal—because of the indeterminateness of the gift's value and the individual freedom of choosing even if governed by conventions—than a definite sum of money with its uncompromising objectivity."[48]

The shift from donations to fees, was therefore, a sensitive matter. After all, as late as 1939, prospective adopters were warned, "Never pay anybody any money for a child—reliable agencies never ask fees."[49] Yet the system was accepted; the number of agencies charging a fee increased from 18 in 1949 to 105 in 1954. Despite opposition and predictions that fees would "degenerate into a price for placement,"[50] the adoption market retained its distinctive structure. Agencies did not turn into "efficient profit-maximizing firms," but still operate as nonprofit organizations. Their price is restricted to the costs of the services provided. Often, only a nominal fee is requested. Some agencies employ a sliding scale, charging a token fee for lower-income families and a larger sum for those who can afford it.

In large measure, the differentiation between an adoption fee and a purchase price hinged on defining the payment as compensation for professional services. But a fee was also legitimized as a symbolic payment: a more efficient expression of gratitude than the traditional donations. "We believe that a financial payment is one of the ways that applicants to adopt children can fulfill their need to pay ... many adoptive parents ... fretted a good deal whether to give and how much ... What we are now doing defines a tangible and specific requirement that is much fairer for applicants and for us."[51] Adoption fees were usually portrayed as a psychological crutch for parents, rather than a commercial device for agencies: "For any human being to be in the position of asking another ... for a child ... is to admit inadequacy. ... Payment of the fee may ease some of the discomfort arising from this deeply humiliating experience."[52] Parents' voluntary contributions of additional monies to the agency, beyond the stipulated fee, further reinforced the boundary between the adoption market and other forms of economic exchange. Their elective gift of money served as a symbolic reminder that adopting a child is not an ordinary business deal.[53] ...

NOTES

1. Robert Grant, "Domestic Relations and the Child," *Scribner's Magazine* 65 (May 1919): 527. Legal adoption did not exist in common law. The first adoption statute in the United States was passed by Massachusetts in 1851 and it became the model for other states. See Jamil S. Zainaldin, "The Emergence of a Modern Adoption Law: Child Custody, Adoption and the Courts, 1796–1851," 73 *Northwestern University Law Review* 1038–89 (1979) and Stephen B. Presser, "The Historical Background of the American Law of Adoption," 11 *Journal of Family Law* 443–556 (1971).
2. *New York Times*, May 8, 1927, VII, p. 14. See "Moppets on the Market: The Problem of Unregulated Adoptions," 59 *Yale Law Journal* 716 (1950). The increase was not only in adoptions by unrelated persons but also in adoptions by relatives, particularly step-parents. By 1962, the 1923 statistics on the substitute care of dependent children were reversed; 69 percent of these children were in family care (adoptive and boarding homes), and only 31 percent in institutions. Wolins and Piliavin, *Institution or Foster Family*, pp. 36–37.
3. *New York Times*, Mar. 17, 1923, p. 9; Jan. 20, 1925, p.19; Dorothy Dunbar Bromley, "Demand for Babies Outruns the Supply," *New York Times Magazine*, Mar. 3, 1935, p. 9.
4. Ada Patterson, "Giving Babies Away," *Cosmopolitan* 39 (Aug. 1905):411.
5. *New York Times*, July 7, 1925, p. 1; Aug. 5, 1925, p. 1; Aug. 10, 1925, p. 1.
6. Alice M. Leahy, "Some Characteristics of Adoptive Parents," *American Journal of Sociology* 38(Jan. 1933): 561–62; Sophie Van Senden Theis, *How Foster Children Turn Out* (New York: State Charities Aid Association,

1924), pp. 60–63. On the increase of upper-class adoptive parents in England between the 1920s and 1940s, see Nigel Middleton, *When Family Failed* (London: Victor Gollancz, 1971), p. 240.

7. Arno Dosch, "Not Enough Babies To Go Around," *Cosmopolitan* 49 (Sept. 1910): 431.
8. Spence Alumnae Society, *Annual Report,* 1916, p. 37; Judd M. Lewis, "Dealing in Babies," *Good Housekeeping* 58(Feb. 1914): 196; *New York Tribune,* cited in "Cradles Instead of Divorces," *Literary Digest* 77 (Apr. 14, 1923), p. 36; Vera Connolly, "Bargain-Counter Babies," *Pictorial Review* 38 (Mar. 1937), p. 17. The Spence Nursery, as well as the Alice Chapin Adoption Nursery, organized respectively in 1909 and 1910 in New York, became leading agencies for the placement of infants.
9. "Blue-Eyed Babies," *New York Times,* Jan. 17, 1909, VI, p. 7.
10. Mabel P. Daggett, "The Child Without a Home," *Delineator* 70 (Oct. 1907): 510.
11. Dosch, "Not Enough Babies to Go Around," p. 434; Carolyn C. Van Blarcom, "Our Child-Helping Service," *Delineator* 95 (Nov. 1919): 34; *New York Times,* Mar. 12, 1927, p. 3.
12. Hastings H. Hart, *Proceedings of the 29th National Conference of Charities and Correction,* 1902, p. 403.
13. Spence Alumnae Society, *Annual Report,* 1916, p. 38.
14. Frederick A. Given, "Bargains in Babies," *Canadian Magazine* 83 (Apr. 1935): 30; Frazer, "The Baby Market," pp. 25, 86. For an analysis of daughter preference in adoption, see H. David Kirk, "Differential Sex Preference in Family Formation," *Canadian Review of Sociology and Anthropology* I (Feb. 1964): 31–48, and Nancy E. Williamson, *Sons or Daughters* (Beverly Hills, CA: Sage, 1976), pp. 111–15.
15. Josephine Baker, "Choosing a Child," *Ladies' Home Journal,* 41 (Feb. 1924): 36.
16. Grant, "Domestic Relations," p. 527. See Tiffin, *In Whose Best Interest?,* pp. 269–70.
17. Gatlin, "Adopting a Baby," p. 84; Ida Parker, *"Fit and Proper"? A Study of Legal Adoption in Massachusetts,* (Boston, MA: Church Home Society, 1927), p. 18. A more lenient view of illegitimate children was consequential, since most adoptable children were born out of wedlock.
18. Honore Willsie, "When Is a Child Adoptable?" *Delineator* 95 (Dec. 1919): 35; Honore Willsie, "Not a Boy, Please!," ibid., (July 1919): 33.
19. Mary Buell Sayles, *Substitute Parents* (New York: Commonwealth Fund, 1936), p. 17.
20. *New York Times,* May 8, 1927, VII, p. 14.
21. George Walker, *The Traffic in Babies,* (Baltimore, MD: Norman Remington Co., 1918), p. 151.
22. Ibid., pp. 130, 136, 153. See also Carrington Howard, "Adoption by Advertisement," *Survey* (Dec. 11, 1915): 285–86.
23. Dosch, "Not Enough Babies To Go Around," p. 435; W. Almont Gates, "Caring for Dependent Children in California," *Proceedings of the 40th National Conference of Charities and Correction,* 1913, p. 309; Arthur Alden Guild, *Baby Farms in Chicago* (Chicago: Juvenile Protective Association, 1917), pp. 24–25.
24. See *New York Times,* Apr. 9, 1922, IX, p. 12; Apr. 16, 1922, II, p. 8; Mar. 11, 1923, VIII, p. 14, and NY State Charities Aid Association, *Annual Report,* 1922, p. 20.
25. Parker, *Fit and Proper,* p. 31. See also Arlien Johnson, *Public Policy and Private Charity* (Chicago: University of Chicago Press, 1931), p. 73.
26. See Ernest K. Coulter, "The Baby Farm and Its Victims," *National Humane Review* 14 (Jan. 1926): 3–4; *New York Times,* May 8, 1925, p. 1; May 9, 1925, p. 1; May 21, 1925, p. 1; July 16, 1925, p. 21; July 23, 1925, p. 1.
27. Ibid., Apr. 9, 1922, IX, p. 12; Apr. 16, 1922, II, p. 8. For a good overview of the professionalization of child-placing, see Tiffin, *In Whose Best Interest?,* pp. 253–80.
28. *New York Times,* July 23, 1925, p. 1.
29. Parker, *Fit and Proper,* p. 29; Josephine Nelson, "Would You 'Bootleg' a Baby?" *Independent Woman* 15 (Feb. 1936): 43. On regulation of child-placing and adoption laws, see Abbott, *The Child and the State,* II, pp. 17–21; Emelyn Foster Peck, *Adoption Laws in the United States,* U.S. Children's Bureau Publication No. 148, 1925.
30. Connolly, "Bargain-Counter Babies," p. 96.
31. Francis Lockridge, *Adopting a Child* (New York: Greenberg, 1947), p. 7; "Moppets on the Market," p. 715, fn. 2; *New York Times,* Jan. 2, 1945, p. 22.
32. Hearings before the Subcommittee to Investigate Juvenile Delinquency of the Committee on the Judiciary. United States Senate. 84th Congress. First Session, 1955, pp. 9, 153.
33. Paul Popenoe, *The Conservation of the Family* (Baltimore, MD: Williams & Wilkins, 1926), p. 95. Wilson H. Grabill, Clyde V. Kiser, and Pascal K. Whelpton, "A Long View," in Michael Gordon, ed., *The American*

Family in Social-Historical Perspective (New York: St. Martin's Press, 1973), pp. 393–94, note a gradual increase in involuntary sterility after 1910. See also Nancy J. Davis, "Childless and Single-Childed Women in Early Twentieth-Century America," *Journal of Family Issues,* 3 (Dec. 1982): 431–58.

34. Henry F. and Katharine Pringle, "Babies for Sale," *Saturday Evening Post* 224 (Dec. 22, 1951); p. 11. The secrecy and speed of black-market sales also appealed to the unmarried mothers. On the mothers' pension movement, see David M. Schneider and Albert Deutsch, *The History of Public Welfare in New York State, 1867–1940,* pp. 180–99. A new awareness of high infant mortality rates among illegitimate children further encouraged programs to prevent the separation of babies from their unwed mothers. See *The Welfare of Infants of Illegitimate Birth in Baltimore,* U.S. Children's Bureau Publication No. 144, 1925; A. Madorah Donahue, *Children of Illegitimate Birth Whose Mothers Have Kept Their Custody,* U.S. Children's Bureau Publication, No. 190, 1928.
35. Handicapped and minority children were also excluded from the adoption market. Only recently have agencies begun seriously considering adoptions for such children. Barbara Joe, *Public Policies Toward Adoption* (Washington, D.C.: Urban Institute, 1979), p. 6. See also David Fanshel, *Study in Negro Adoption* (New York: Child Welfare League of America, 1957).
36. Hearings before the Subcommittee on Children and Youth of the Committee On Labor and Public Welfare, 94th Congress, 1st Session (1975), pp. 142–45.
37. *New York Times,* Nov. 11, 1934, IV, p. 5.
38. Mona Gardner, "Traffic in Babies," *Collier's* 104 (Sept. 16, 1939): 43.
39. Elisabeth M. Landes and Richard A. Posner, "The Economics of the Baby Shortage," 7 *Journal of Legal Studies* 339 (June 1978). On the similarity of outcome between agency and independent adoptions, see Joe, *Public Policies Toward Adoption,* pp. 48–49; Daniel G. Grove, "Independent Adoption: The Case for the Gray Market," 13 *Villanova Law Review* 123–24 (1967).
40. Hearings Before the Subcommittee on Children and Youth (1975), pp. 2, 3, 580.
41. *See* Margaret V. Turano, "Black-Market Adoptions," 22 *Catholic Lawyer* 54–56 (1976); "Moppets on the Market," pp. 732–34.
42. Bernard Barber, "The Absolutization of the Market: Some Notes on How We Got From There to Here," in G. Dworkin, G. Bermant, and P. Brown, eds., *Markets and Morals* (Washington, D.C.: Hemisphere, 1977), p. 23.
43. Statement by Joseph H. Reid, Executive Director, Child Welfare League of America, in Hearings Before the Subcommittee on Children and Youth (1975), p. 19.
44. "Moppets on the Market," p. 715.
45. Hearings Before the Subcommittee on Children and Youth (1975), p. 4.
46. "Survey of New Jersey Adoption Law," 16 *Rutgers Law Review* 408 (1962) fn. 34; Grove, "Independent Adoption," p. 127. Independent, "gray-market" adoptions are justified as a necessary alternative to often highly bureaucratized and overworked agencies, Robert H. Mnookin, *Child, Family and State* (Boston: Little, Brown, 1978), pp. 621–22. The number of independent placements has significantly diminished. While in 1945 only about one-fourth of nonrelative adoptions were made by authorized child-placing agencies, in 1971, almost 80 percent of all nonrelative adoptions were arranged by agencies. Joseph L. Zarefsky, "Children Acquire New Parents," *The Child* 10 (Mar. 1946): 143.
47. Letter to the Editor, *Child Welfare League of America Bulletin* 20 (Dec. 1941): 9.
48. Georg Simmel, *The Philosophy of Money* (London: Routledge & Kegan Paul, 1978), p. 373.
49. Dorothy Canfield, "Children Without Parents," *Woman's Home Companion* 66 (May 1939): 48.
50. *Child Welfare League of America Bulletin* 20 (Nov. 1941): 9.
51. C. Rollin Zane, "Financial Practices of Children's Agencies," *Child Welfare League of America Bulletin* 25 (Oct. 1946): 5.
52. Sybil Foster, "Fees for Adoption Service," *Child Welfare League of America Bulletin* 26 (May 1947): 11.
53. See Michael Shapiro, *Fees in Adoption Practice* (New York: Child Welfare League of America, 1956), p. 12; Eilene F. Corsier, "Fees for Adoption Service," in I. Evelyn Smith, *Readings in Adoption* (New York: Philosophical Library, 1963), pp. 381–82. The average adoption fee charged by a public agency in 1975 was $200–400, and $450–900 for private agencies licensed by the state. Turano, "Black-Market Adoptions," p. 51, fn. 17.

Stolen Children and International Adoptions

Mary Ellen Fieweger

International adoption is a controversial matter in Ecuador and other Latin American countries. At the end of 1988, the Ecuadorean press reported the discovery of a ring of individuals and private institutions involved in the adoption business, and the Ecuadorean office of Defense for Children International (DCI) investigated further. Members of the organization believed that the subjects of these newspaper articles were not the only ones involved in this type of activity. The director of the Ministry of Social Welfare's Office for the Protection of Minors cooperated with the investigation, making available files of children who had been adopted by foreign couples. The study covered 1987 and 1988. The following findings are available at this writing.

Study Findings

In 1987, 132 children, between the ages of one month and 12 years, were adopted by foreign parents [DNI 1981]. The majority of these, 73, were from Pichincha, an Andean province in which Quito, Ecuador's capital, is located. Forty-seven of the adoptees went to Italian homes and 31 to homes in the United States. Twenty-one were adopted by Norwegian parents, 12 by English couples, and all but three (one to Canada and two to Israel) went to homes in other European countries. In 1988, the number of foreign adoptions increased significantly, to 166, approximately 25% greater than the number of adoptions the previous year. Once again, a majority of the children were from the province of Pichincha (103). Italy, with 58, and the United States, with 39, were again the destinations of most adoptees. One of the more striking findings was that a small group of lawyers apparently specialized in international adoptions. For 1988, one lawyer handled 33 adoptions, and another, 16. Eleven other lawyers handled five or more.

This information, along with additional information that suggested improprieties in the adoption procedure, led DCI to take its investigation into the field. DCI members chose a few cases with a number of apparent irregularities and made contact with social workers and other individuals directly involved. This phase of the study brought to light highly disturbing information.

In some cases, the lawyers and institutions involved had actually kidnapped the children they later supplied to foreign couples. Public markets were a favorite area of operation. Salespersons in the markets tend to be women from the lower social strata. The vast majority take their children to work with them because other child care options are not available. Individuals on contract to the ring (often other women from the lower classes who were paid small sums for their services) would take advantage of busy moments to spirit away the vendor's infant or toddler when the vendor was involved in making a sale.

In other cases, poverty-stricken parents were persuaded by economic enticements to put their children up for adoption, sometimes with explanations as to how much better off the child would be living with foreign parents. If these arguments did not work, threats were sometimes used. Evidence suggests that one child care institution involved in illegal adoptions resorted to macabre procedures in at least one case. A couple temporarily left their son in the care of personnel at this orphanage. When they returned to claim him, they were told to come back another time. The child, it was claimed, had been sent to a hospital for medical treatment. When the parents returned a second time, they were told that he had died. There are reasons to believe that a false death certificate was filed, and DCI has asked the government to exhume the body—if, indeed, there is a body—to verify the claims of the institution's director.

Another interesting fact uncovered by DCI is the manner in which members of these rings "complied" with adoption requirements. When a child has been kidnapped, the biological mother cannot be asked to sign the necessary documents. Consequently, others—women in need—were pressed into service to function as "mothers" of the children up for adoption. It is not clear how widely this has been done in Ecuador. Members of adoption rings commonly resort to this tactic in Central American republics. DCI has learned that, in at least one case, the woman involved was initially not fully aware of the implications of her act. When she did realize that what she was doing was illegal, she refused to cooperate further. The brother of one of the lawyers involved went to her home and threatened to kill her if she did not sign the final documents. Under the circumstances, she signed, but she also subsequently testified in court against this particular lawyer. The woman continues to receive threats. There is further evidence that social workers in the Ministry of Social Welfare have facilitated adoptions they have known to be irregular, and there is reason to believe that individuals in the courts and the civil register—the agency that provides birth certificates—have also cooperated with adoption rings.

Through the efforts of DCI and the local press, three lawyers and a number of accomplices, including the director of the orphanage mentioned earlier, have been charged with criminal offenses related to these cases. Nevertheless, few believe that the matter will end here.

The Present Picture

According to sociologist Francisco Pilotti [1983, 1985], international adoptions—both legal and illegal—can best be understood in terms of supply and demand. International adoptions began in the aftermath of World War II, when thousands of European children were left homeless. In the months and years following the war, many of them were placed with adoptive parents in European countries and the United States.

The demand from couples in these same countries has steadily increased in response to profound social changes. Since the 1960s, the number of babies available for adoption locally in the countries in question has diminished drastically. Birth control and abortion are widely practiced and unwed motherhood no longer carries the social stigma it once did. As a result, fewer unwanted babies are being born. In Paris, for example, approximately 5,000 children were abandoned in the year 1990. By 1976 that number dropped to 144. About 1,000 Swedish babies per year were available for adoption in that country in the 1950s. Today, fewer than one hundred are placed for adoption annually. In the United States, the number of foreign-born babies adopted between 1972 and 1973 increased by 33% from the previous 12-month period, a reflection of the diminishing number of children born within the country who were available for adoption.

Although the demand has increased significantly, it continues to originate from the same nations whose citizens initially adopted foreign children. The supply side, however, has undergone a radical transformation as regards the origin of adopted children.

The first international adoptions, right after World War II, involved children and parents of similar ethnic and socioeconomic backgrounds. Since the 1950s this has no longer been the case. With the decrease in the number of same-nationality children available, prospective parents and adoption agencies began to look elsewhere. The Korean War and the Chinese Revolution produced a supply of homeless Asian children in the 1950s. Many were adopted by European and U.S. couples. From that decade to the 1970s, Asian countries, including Thailand and Vietnam as well as Korea, were the major countries supplying children for adoption. In 1974, Asian children accounted for 90% of all those adopted by foreign parents. But in the years immediately following, the situation began to change dramatically. By 1980, Asian children accounted for less than 64% of all international adoptions. Asian countries, formerly major sources of adoptees, are undergoing radical changes. With economic well-being comes a marked decrease in the number of parents unable to care for their children. Simultaneously, governments in these countries have mounted aggressive birth control and sex education campaigns; with fewer children to be cared for, fewer are available for adoption. The demand on the part of the United States and European parents, however, continues to grow, and a new supply has been found since the mid-70s, and to an increasing degree in subsequent years, in Latin American countries. Whereas in 1977, 23.2% of all children adopted by foreign parents were from Latin America, by 1980, only three years later, that figure had increased to 32.1%. In the case of Ecuador, the juvenile courts processed 21 foreign adoptions in 1973. In 1988, the courts in a single province, Pichincha, heard 166 cases (DNI 1989). Figures for other countries, though not as recent, reveal a similar trend. The number of Colombian children adopted by foreign parents in 1973 was 107. In 1980, it reached 659. The 1973 figure for Peru was seven; the 1980 figure, 72. For Latin America and the Caribbean in general, in the seven-year period between 1973 and 1980, the increase was greater than 25% [Pilotti 1985].

The reasons why Latin American nations are replacing Asian countries as a supply source for children are not difficult to identify. Unlike birthrates in Asia, those in Latin America are high. In Ecuador, for example, at the present growth rate the population doubles every 27 years. Most of these children are born to Ecuador's poor, a vast majority of the country's population. Due to the current debt crisis, for which there are no genuine solutions on the horizon, the money available for social services diminishes annually. Approximately 70% of Ecuador's children under the age of 15—and this age group makes up 52% of the total population—are at risk, according to government statistics [INNFA 1986, 1989; CEPAR 1986]. They suffer from malnutrition and lack of access to schools and health care. In some rural provinces, up to 70% of the population does not have access to potable water. In those same provinces, from 75% to 87% of the population, depending on the province, have no sewage facilities. Of the 70% of Ecuador's children who are at risk, only 2% benefited from governmental social services in 1984. Clearly, the number of children in need—the adoption supply—is rising in Ecuador.

But this sad fact is coupled with a slow, often inefficient bureaucracy responsible for processing adoptions. Prospective foreign adoptive parents must travel tremendous distances and stay, often for months, in the country while the adoption is being processed. On occasion, after large expenditures of time and money, they are denied the right to adopt. Under the circumstances, it is no surprise that enterprising individuals are offering alternatives to the official adoption procedures. Some people wonder what all the fuss is about. Applicants desperate for a child willingly pay the 10 to 15 thousand dollars charged by lawyers who promise and deliver a quick, troublefree adoption. The adopted child now has an opportunity to develop in material circumstances far superior to those he or she was born into. The only apparent losers are the biological parents who have been pressured or talked into giving up their child, or whose child has simply disappeared.

From an ethical point of view, the situation is very serious. We cannot condone the means—kidnapping or undue pressure—or the motivation of individuals involved in these rings at the highest level—greed—or the consequent suffering of the biological parents deprived of their children in this fashion, no matter how laudable the ends in terms of the adopted children and their adoptive parents.

But the issue is more complex. Illegal international adoptions are, by definition, beyond the reach of the law of either the sending or receiving country. Therefore these adoptions are beyond the reach of agencies and institutions responsible for assuring the fitness of the adoptive parents and the well-being of the adopted children. Cases of serious abuse of these children at the hands of their foreign adoptive parents have been documented.

Another serious objection to this process is the absence of physical and psychological evaluation of these children. All are born to low-income parents. They are children at risk, likely to have suffered malnutrition from the moment of conception, and may well present the new parents with a whole set of unexpected difficulties.

The ethnic origin of the child is another complication. These children are being sent to a new culture and to social and economic circumstances markedly different from those of their place of birth. Particularly for older children, learning a new language and cultural norms will require severe adjustments. Children from Central American and Andean countries, with large indigenous populations, adopted by white middle-class European and North American families, may be subject to racial discrimination as well [DCI 1989].

Finally, questions of national sovereignty and national pride are at stake. Since the Spanish Conquest, the relationship between Latin American countries and those of the developed world has not been one of equality. Traditionally, these third-world republics have been providers of natural resources, purchased at bargain prices by the developed world, first Spain, then England, and today, the United States. Many Latin Americans object to international adoptions because, as they see it, Latin American children have become another natural resource in demand in the developed world. This point of view has taken on additional credibility as a result of the apparent unwillingness of many European countries, as well as the United States, to sign bilateral agreements to assure that proper care is taken when placing a child and that adequate followup is provided.

Despite these objections, however, most people in Ecuador are not opposed to international adoptions per se as a solution to the problem of abandoned children, if they are viewed as a last resort, to be applied only in cases where the controls described above are adhered to.

National Adoptions

Efforts are now underway to encourage national adoptions, the ideal solution to the problem of abandoned children, but increasing the number of adoptions by Ecuadorean couples of Ecuadorean children will not be an easy task. To begin with, adoption is not a widespread practice in Ecuador. Rigid social stratification is pervasive. Among the upper- and upper-middle classes, ancestry is vital. Though the vast majority of Ecuadoreans who think of themselves as white are, in fact, mestizo, the Indian is held in contempt by broad sectors of nonindigenous Ecuadorean society. Children in need of homes are from the lowest social strata, that is, the most obviously Indian. An increase in adoptions on the part of the middle strata is also problematic, because some members of this class share the prejudices of the upper socioeconomic sectors. Probably more important, however, is the economic factor. The foreign debt has created tremendous hardships, not only for the very poorest but also for the middle class. During the first part of 1989, inflation reached 100% per annum. Although the government currently maintains that it has fallen to a more manageable level—70% to 80%—no one really believes that. Many members of the middle class work at two or three jobs just to make ends meet.

More and more women are going to work out of necessity, not because feminism has made great strides in Ecuador. The Ministry of Health recently advertised in the newspapers for a doctor to direct a neighborhood health center. The salary offered was 30,000 sucres a month, about $50. An Ecuadorean family currently requires a minimum of 70,000 sucres a month merely to subsist. There is also a question of size: Ecuador's middle class is small and the number of homeless children is large. The likelihood that vast numbers of Ecuador's middle class will find themselves able, any time soon, to accept the financial burden of an adopted child seems very dim.

Conclusion

For the time being, that last resort—international adoption—may be the only real hope for a portion of Ecuador's abandoned children. But the adoptions must take place only when the children have been voluntarily given up by their parents, and the adoptions must be strictly controlled by responsible institutions in both the sending and receiving countries.

REFERENCES

"Hacia una movilización nacional por los niños." Quito, INNFA, 1986.

"Informe parcial de la investigación sobre adopciones internacionales." (Documento de la DNI, Ecuador, 1989).

"International Investigation into the Rights of Abandoned Children." Geneva, Defence for Children International (DCI), 1989.

"Maltrato al niño: Un problema multidimensional." Quito, INNFA, 1989.

"Mortalidad infantil y de menores de dos años." Quito, CEPAR, 1986.

Pilotti, Francisco. "Las adopciones internacionales en América Latina: Antecedentes sociales, psicológicos e históricos." Montevideo, Instituto Interamericano del Niño, OEA, 1983.

Pilotti, Francisco, "Intercountry Adoptions: A View from Latin America." Child Welfare, 1985.

Designer Genes: The Baby Quest and the Reproductive Fix

Elaine Tyler May

... Patricia Painter played by the rules. She worked hard, saved money, put herself through school, became a teacher, got married, and looked forward to living the American dream, which for her included children. When the expected reward eluded her, she felt betrayed. She had planned everything perfectly and never imagined that her body would malfunction. She and her electrician husband lived comfortably in a Los Angeles suburb and used birth control to prevent conception for the first five years of their marriage. When they decided to go off birth control, she got pregnant right away. To her astonishment, she miscarried.

> I absolutely could not believe that I miscarried. . . . It was really a horrible experience and just devastating. . . . I had always been the kind of person that got anything I wanted. I wanted to own a house so I worked two jobs and I went to school full time, and was able to save up enough money to buy a house. I wanted a career and a degree, and . . . I ended up with three teaching credentials and a master's degree. I wanted a husband and got a husband. . . . And there just seemed nothing that I couldn't do. I was living the American dream. And then, what a surprise, you know. Miscarriages do not happen to people like me. . . . I could not believe that this had happened.

When she miscarried a second time, "we got kind of crazy at that point."

According to the American ethos, if you work hard, earn a decent living, delay gratification, plan, and live by the rules, your life should be in your control. In the twentieth century, the work ethic has fed into the consumer ethic: Even if the job has few inherent satisfactions, doing the job should bring its reward in a home, leisure, consumer goods, and a fulfilling private life. Reproduction is linked to both the consumer and producer ethic. Children are both a reward for hard work *and* one of the few products that can still be created by the labor of one's own body. For most Americans, children are central to their vision of the good life; most polls show that children are still highly desired. In spite of the increasing visibility of the childfree, over 90 percent of American women who were surveyed in the 1980s considered childlessness to be undesirable for them and others.[1] By the late 1980s, the rate of childlessness had leveled off, the birthrate inched up, and a renewed interest in babies permeated the popular culture.

The New Pronatalism

On the heels of the childfree movement of the 1970s came what several observers identified as the new pronatalism of the 1980s. According to feminist author Susan Faludi, the renewed push toward parenthood took the form of a media blitz aimed at educated career women, warning them that if they delayed childbearing, they were likely to find themselves infertile. Few of the alarmists who pointed to a new "infertility epidemic" took note of studies showing a troubling trend in male fertility: the decline in the average sperm count by more than half in the past thirty years. Nor did they mention the fact that less-educated poor women were more likely than professional women to be infertile, as a result of pelvic inflammatory disease caused frequently by sexually transmitted diseases. Rather, many articles claimed that the alleged increase in infertility resulted from women postponing motherhood until they were in their thirties, when it might be too late to conceive.[2]

In 1987, NBC correspondent Maria Shriver called childlessness "the curse of the career woman." In the same year, *Life* published a special report entitled "Baby Craving." Headlines warned against "Having It All: Postponing Parenthood Exacts a Price" and bemoaned "The Quiet Pain of Infertility: For the Success-Oriented, It's a Bitter Pill." A columnist for the *New York Times* described the infertile woman as "a walking cliché" of the feminist generation, "a woman on the cusp of forty who put work ahead of motherhood." *Newsweek* noted the "trend of childlessness," and *Mademoiselle* warned, "Caution: You Are Now Entering the Age of Infertility."[3]

This media blitz of the 1980s was not lost on the childless. Margaret Lewis noticed a marked change in movie themes during the decade:

> The movies used to end when the couple fell in love. Now they end when the couple achieves parenthood. Anyone who doesn't have kids is portrayed as selfish, cold, deprived, or pitiful. It seems as if all of Hollywood is having babies. I mean, Warren Beatty expounding on the virtues of fatherhood?! Really.

She observed that not only happy endings but also advertisements suddenly began to link children to consumerism and the good life: "The newspaper is filled with cute little 'Gap Kids' with little denim jackets."

Reminiscent of the early baby-boom days, babies, children, and parenthood began to permeate the nation's popular culture. Plots of movies and television shows and even popular songs revolved around the baby quest. Bonnie Raitt sang, "A friend of mine, she cries at night and she calls me on the phone/ Sees babies everywhere she goes and she wants one of her own."[4] Films like *Parenthood, Three Men and a Baby, Look Who's Talking, Mrs. Doubtfire, Made in America, Immediate Family,* and countless others focused either on babies and their charms or adults' preoccupations with having children and child rearing. The trend culminated in 1994 when Arnold Schwartzenegger, in *Junior,* gave birth to a baby himself! Unlike the earlier formulas, however, the traditional links between marriage, sex, and procreation no longer prevailed in these films. Single men raising babies, divorced fathers scheming to be with their children, single women having children, couples struggling with infertility, and even artificial insemination by donor became plot devices for comedies. The television character *Murphy Brown* sparked a national political debate when she became an unwed mother. Most striking about this new infatuation with having and raising children was that the marital imperative faded into the background. Whereas the romantic dyad for the childfree was the married couple, the romantic dyad featured in the new pronatalism was the parent and the child.

The new pronatalism took other forms as well. Almost as quickly as the population-control movement gained momentum, critics began to attack it. As early as 1977, James A. Weber complained about the "barrage of population control propaganda." Noting that the birthrate had already dropped below replacement level, he argued, "If the ZPGers [members of Zero Population Growth] were right, we should now be on the threshold of a new era of unprecedented plenty. In fact, the opposite is the case." Pointing to recession, pollution, and social problems, he blamed "a suicidal population theory" for the fact that "our posture in the world sinks lower." Sounding a theme reminiscent of Manifest Destiny, he hoped to "rekindle the instinctive appreciation and approval of population growth that has infused the country since its founding by showing how population growth is vital to the coming health and well-being of the American people."[5]

Ten years later, in 1987, Ben Wattenberg published his controversial pronatalist tract, *The Birth Dearth.* Sounding a eugenic theme, he warned, "I believe the demographic and immigration patterns inherent in the Birth Dearth will yield an ever smaller proportion of Americans of white European 'stock'... and this will likely cause more ethnic and racial tension and turmoil than would otherwise occur." He continued, "It will make it difficult to promote and defend liberty in the Western nations and in the rest of a modernizing world." Wattenberg's warning sounded a cold-war theme: that the communist nations were propagating at a higher rate and the "free world" would soon be reduced to a tiny minority of the world's peoples. At the same time he harked back to earlier national concerns: "'Manifest Destiny' was not the cry of a no-growth continent of old people." Wattenberg called attention both to the declining birthrate, which had dipped to fewer than two children per woman by the mid-1970s, and to the increase in childlessness. "I believe that most people—men and women—who freely decide not to have children will probably live to regret it." Critics claimed that Wattenberg raised "the 'yellow peril' argument all over again," and expressed deep-seated "collective fears of being swamped by external and internal enemies." Nevertheless, *The Birth Dearth* joined a chorus of voices calling for more American babies.[6]

As the politics of reproduction sparked public debate, articles in the popular press began to trumpet "The Coming Baby Boom." Demographers predicted a new epidemic of "fertility fever." As early as the 1970s, when the birthrate dipped to its lowest level ever, commentators began to predict that "pregnancy will once again become chic." For some, babies were already chic. As one young college-educated mother explained, "I married a man I thought was rather boring at first, but decided that having a family really sounded fascinating. I expected my friends to scoff but actually they think I've done something original, even quaint." Having a baby marked her as "the class bohemian."[7]

For this young woman, marriage was a means to an end. She married her "rather boring" husband to have a child, reversing the earlier romantic notion that the baby was the expression of the love of the couple. Doreen Welsh also considered marriage as a means to an end, the end being motherhood. Describing herself as divorced, "of Jewish descent . . . a professional classical symphonic musician," Doreen was looking for a man to achieve her goal.

> It is very difficult to find men interested in dating a 40 year old woman who wants to start a family. It puts me in a desperate condition. It is hard to be natural in meeting men, because I am on some kind of crash course deadline in creating a relationship and getting pregnant. . . . I still have a strong desire to have my own, natural child. I do not feel good about the thought of artificial insemination. I must be a traditionalist in wanting to have a partner to raise the child, and to share in my life. In addition, esoterically speaking, I want to choose the father of my child. . . . A desperation has overtaken my rational mind. I get strong urges to get pregnant during the middle of the month.

A few short decades ago, the desire for a partner and the ability to choose the father of one's child would not have seemed "traditionalist," "esoteric," or "irrational"; it would have simply been expected. But not any more.

In recent years, marriage has become a desirable but not necessarily a required precondition for motherhood. In the Black community, more children are born to single than to married women. In the White community, an increasing number of women are pursuing single parenthood through adoption, artificial insemination, and bearing children out of wedlock. The past decade also witnessed "a virtual baby boom in the lesbian and gay community," according to one observer. One study estimated that by 1988, between 1,000 and 3,000 children in the United States and Europe were born to lesbians using donor insemination.[8] Institutions responded to these trends. Adoption agencies began to allow single individuals to adopt, as they had in the early part of the century. And many infertility clinics and sperm banks began to inseminate single heterosexual women as well as lesbians. By the 1980s, single men and women, both gay and straight, found ways to become parents that previously had not been available.

The new pronatalism brought together advocates from the Left and the Right. Wattenberg's conservative warning about the "birth dearth" included a call for measures to enable more women to have children, such as on-site day care centers, parental leaves, and flexible work schedules—ideas that feminists had promoted for many years. And although the calls for "family values" emanated largely from the Right and condemned all nontraditional forms of sex and procreation, the lesbian and gay community, as well as single parents, drew upon similar rhetoric in their claims for legitimacy as parents and as families.

The movement for reproductive choice, although focused largely on the issue of abortion rights, also fed the new pronatalism and drew attention to the plight of the infertile. Resolve, a national organization serving infertile women and men, offered both support and advocacy for childless adults hoping to become parents. Feminists argued that women should not be forced to choose between parenthood and careers, claiming that women, like men, should have the opportunity to do both. On the opposite end of the political spectrum, the antichoice activists also heightened the public rhetoric around reproduction and sentimentalized babies anew. It is no wonder that infertile women like Margaret Lewis, surrounded by images of "power motherhood," felt that every political and cultural message, from the Left as well as the Right, urged her not to accept her childlessness but to fight against it tooth and nail.

Infertility and Reproductive Control

In this context, the old problem of infertility became a new kind of frustration. Many observers assume that infertility is on the rise. But there is no evidence that the proportion of infertile Americans has increased. There is evidence, however, that the number of people who are seeking treatment has risen dramatically. The number of visits to physicians for infertility treatment rose from 600,000 in 1968 to 1.6 million in 1984. The increase has been due, in part, to the huge baby-boom generation; the infertile among them are a large and visible group. But there are other reasons as well. Even if the chances for successful treatment are not much better than they were a half century ago, dramatic new technological interventions are now available. High-tech approaches, such as in-vitro fertilization (IVF, fertilization of the egg in a laboratory petri dish and then its insertion directly into the uterus), first successfully used in the birth of Louise Brown in England in 1978, appear to offer "miracle babies" to the childless. Treatments using assisted reproduction techniques jumped 30 percent from 1990 to 1991, even though the chance of ending up with a "take-home baby" from these procedures was only about 15 percent.[9]

The promise of a technological fix, combined with a faith in medical progress, led many Americans to believe that they could triumph over most physical limitations. Physicians have responded to the demand. Studies have shown that American physicians are more likely than British practitioners to resort to heroic measures for treating infertility, probably because their patients request such intervention.[10] But reproductive medicine, despite its many advances over the past century, remains an imperfect art, available only to those who can afford it. Nor does it guarantee success. Infertility treatment is a high-stakes gamble: It is possible to lose all the money, time, and effort invested and gain nothing in return. If all the efforts of modern science, human struggle, and economic sacrifice do not result in the desired child, the rage, desperation, and anguish can be overwhelming.

Because birth control and reproductive choice are widely taken for granted, the infertile experience extreme frustration. Reproductive choice is much easier to achieve if the goal is to avoid pregnancy. Contraceptive technologies offer a success rate of nearly 100 percent, and legal abortion provides a backup when birth control methods fail. But infertile couples who seek treatment have only a 50 percent chance for success in the 1990s, odds that have not dramatically improved since the 1950s.[11] The inability to "control" one's reproductive fate is among the most exasperating experiences of infertility, especially for those who have put so much effort into the struggle. As the reporter Susan Sward wrote of her struggle with infertility, "As an organized, energetic person, I was used to getting what I wanted in life most of the time. To a major extent, I was also used to feeling in control of my life and knowing what I did would produce results if I tried hard enough. When it came to making babies, I found I had a lot to learn."[12]

Those who become pregnant while using birth control tend to blame the technology. But infertility patients who do not conceive often blame themselves. They feel unable to control their bodies or their destinies, even with medical intervention. For Roberta

O'Leary, "It gets more and more difficult to pick up the pieces after each failure. I also don't like the feeling of having no control over what happens." Amanda Talley "felt like a freak of nature . . . embarrassed and shameful. . . . I felt as though my body betrayed me." Dierdre Kearney explained, "My feelings of helplessness have been hard to handle. We humans like to have control over our own lives and the one thing we think we can control is our body." She has done everything to have a baby and

> still my body betrays me and deprives me of one of the things I want most in life. I cannot make my body do what I want. . . . I've heard some women say that being infertile makes them feel less like a woman. I've never really felt this. I guess, this has made me feel all too much like a woman because it's what makes me a woman that has caused my problem—PERIODS and HORMONES! I just feel helpless in determining my own future. Sometimes I feel like a ship at sea and just when I am close to land, a huge wave washes me out to sea again.

The inability to control one's reproductive functions often leads to feelings of shame and worthlessness, especially for women. Maureen Wendell explained, "I began to feel defective, ashamed. I can't do a 'normal' biological function that most anyone else could do. I had to re-evaluate my life, my hopes, my dreams and my identity as a woman. I am blessed to have a very supportive husband but even with that I felt inadequate as a wife." Feelings of inadequacy were magnified by the association of fertility with sexuality. In a taped message, Patricia Painter used the language of sexual potency when describing her husband's healthy sperm. "My husband has this, you know, magnificent, I guess he's extremely virile. He has like super sperm. . . . Everybody from the lab technicians to the receptionist at the doctor's office was always so amazed at the amount and the virility of this sperm. It's like super-human sperm." When they accidentally spilled some of the semen sample, a physician replied:

> "It doesn't matter. He could impregnate the whole block with what's left in here. It's amazing." Which made me feel absolutely horrible because he couldn't impregnate me. Well, it was real obvious who had the problem in this relationship, as far as who was the one responsible for us not getting pregnant, and that was me. So I felt extremely terrible about that. This resulted in my being very embarrassed around people. I felt very defective. . . . It was just really such a blow. . . . I would get physically ill . . . 'cause I felt so defective and so embarrassed.

Laura Lerner also felt "abnormal," even though she was not infertile. But she was single, and her singleness deprived her of the opportunity to become a mother. "I am a woman. I am supposed to have children, right? What am I if I don't produce children?" She considered adopting as a single parent or trying donor insemination. But she could not bring herself to do it: "Withdrawing some sperm from the sperm bank sounds so cold and mechanical." Without children, she felt "unnatural. . . . I have had these damn menstrual cycles since age 11 and I have nothing to show for it. . . . I get so I hate the cycle when it comes. . . . I have the most trouble trying to determine why I am here. I feel very incomplete, and very abnormal."

Before she discovered she was infertile, Leila Ember felt that "Life was good! Most importantly I was in control of it!" When she did not get pregnant, however, she began treatment, even though she recoiled at the invasive procedures: "For a person who had never had so much as a band-aid applied to any part of their body I found it quite difficult to endure the poking and prodding and exploratory procedures which were both financially and emotionally expensive." But her body remained uncooperative. Infertility destroyed her peace of mind and self-confidence: "I remember sitting on the floor of my bathroom for what seemed like hours and sobbing, I'd look at my husband and begin to scream how sorry I was that I 'messed up again!'" Blaming herself, she wrote, "My biological clock isn't ticking; it isn't working at all!"

Many childless women who wanted children questioned their own womanhood. Suzanna Drew felt "less of a woman—somehow not complete." Kate Foley felt "barren." Paula Kranz described feelings of "failure . . . it's like an empty space within yourself that you can not fill." Marie Gutierrez blamed herself when her husband's semen analysis

> came back ok, then O Boy! All fingers pointed to me, *wow* was I ever so unhappy, people don't know what it is like to try and try and never succeed.... I told no one... we were both embarrassed, marked, hurt. My husband is a very supportive husband, a good man and tells me that he accepts whatever happens, but... I can't accept the fact that I feel like some sort of alien, all women who are "normal" have children.

Along with at least four other women and men who wrote, Marie offered to release her spouse from the marriage so that he could find a fertile partner and have children.

Not every infertile woman felt these feelings of inadequacy, but the sentiment was widely shared. Talia Herman was shocked that in her infertility group of ten women, all but she said they felt less complete as women.

> I have come to understand how far women *haven't* come in their acceptance of themselves. We still define our worth by the size, beauty, health of our families.... Women of past generations have had to break down societal stigmas. Our grandmothers perhaps went to work. Our mothers may have brought home more money than our fathers. How can women of my generation feel less complete because of societal pressures...? Do men feel less complete? Do we love any differently?

Sheila Turner-Cohn noticed that physicians can contribute to the way women understand their infertility even by the language they use:

> It's interesting to me that my new gynecologist refers to my uterus and childbearing inability as something separate from myself. When he examined me, he made the statement "I don't know if *this uterus* could carry a baby." I thought that was so strange, for him to put it that way. But when I thought about it, I decided that it was very enlightened of him to make it clear that it was my organs and not me as a person who lack this "ability." That's a key distinction, I think, because of how an infertile woman tends to globally perceive herself as a failure because she lacks this basic biological ability.

Her previous gynecologist infuriated her when he referred to "'some hapless soul who can't produce eggs of her own.' All I could do was repeat, 'Some hapless soul!' in disbelief. I really don't think he had a clue about why I was offended."

Although the women expressed more emotional pain over infertility, the men who wrote experienced similar feelings of inadequacy and loss of "manhood."[13] Dave Crenshaw felt like a failure, "alienated from the rest of society, as I knew I never could be one of them." Nothing could ever "replace the one thing I really wanted to be, a father to my own child." He questioned "what my function as a man was.... I felt I was no longer of any use to anyone for any reason ... of no value to society as a whole, because we, childless people don't contribute to society what society holds so dear ... children." Although his sterility resulted from a case of childhood mumps, he could not avoid blaming himself. He felt "bitterness.... I cannot get over the feelings that somehow I failed, and life passed me by.... I still have the guilty feelings of being inadequate."

Among the men who wrote, few actually discussed the question of masculinity, but those who did acknowledged that it was an issue. Joanne and Tom Paisley sent a tape in which Joanne described their response to the news about his negative semen analysis. It "did affect his image of manhood.... It was very difficult for Tom to accept that." Tom said simply, "We try to look at the bright side of things" and praised his wife for being so supportive. Lester Bernstein, on the other hand, acknowledged the blow to his masculinity with a sense of humor. When he was diagnosed as infertile, "that one term transformed this budding stud into a wimp." And when he learned that the quality of his semen was "borderline, I was shattered. My voice went up two octaves."

Parenthood and Identity

The comments of these infertile women and men indicate that for many, infertility meant more than an inability to produce children; it was also a blow to their sense of manhood or womanhood. In a society that often equates adulthood with parenthood, infertility affects personal identity as well as reproductive behavior. Childlessness involves at least two fundamental realities that are at odds with the adult norm: not having children and not being a parent. The two conditions are connected, but they are not the same.[14] Bearing and raising children pertain to behavior and private life; the status of being a parent has more to do with identity and public life. Although the voluntarily and involuntarily childless have different ideas about how to achieve happiness in private life, they share similar experiences as nonparents in the wider world. According to a 1979 study, most Americans ranked becoming a parent as the most significant marker of adulthood, more than marriage or getting a job.[15] Whether or not childlessness is chosen, it means the lack of a certain privileged adult status.

For some of the infertile, childlessness was more painful in public than in private. Margaret Lewis said, "So, how is a life without children? Honestly, not at all bad and actually pretty nice." But "it can be lonely," she admitted, not at home but in the social world where she and her husband felt "out of the 'mainstream' of the life of our peers.... This kind of exclusion has broken my innocence and my sense of trust in belonging in the world. I never felt like an outsider in any group and was always able to feel accepted on some level until now."

Many complained that the parents in their midst talked incessantly about their children. Margaret remarked, "At a cocktail party, the women will engage in baby talk the entire time, excluding the one (me) who has no children. Make no mistake: I like children and I like to hear about them. This is life, and children are a wonderful part of it. But I don't want to hear about that and only that for three hours." Stella Sims and her husband wondered

> if people lose their brains when they have children. It's as if these parents are incapable of discussing anything outside of their children. People who were formerly bright, charming and intelligent are now rendered conversationally incompetent when it comes to sports, politics or religion.

For many respondents, feeling "left out" was the worst part of being infertile. Sharon Stoner and her husband felt a profound sense of "isolation.... I just wish our society would not look upon us as incomplete people because we did not have children." Lydia Sommer wrote, "I think I am becoming more isolated each year—I have no connection with all these childbearing women. I almost have no connection with myself anymore—I can't find a purpose in life.... It really gets depressing sometimes!" Tamara Afton found family get-togethers and holidays to be the most difficult. "Holidays only make me feel worse and more like a misfit." The Aftons now spend holidays with friends they met through Resolve, a national organization for infertile women and men. "We joke about a 'Misfit's Christmas' party."

This lack of connection, not "fitting in" as many described it, has little to do with children. It has much to do with adult identity and one's place in the world.[16] American society places a premium on parenthood. "What am I if not a parent?" wondered Corinne Gary. Tammy Bolen wrote that people treat her as if "I'm not grown up. Not mature, that's what people do. Grow up get married, have kids. Not me. I must not be grown up. What's wrong with me?" Not only full adulthood, but a measure of respectability is attached to parenthood. Daniel Steiner noticed a change in status when he and his wife had their first child after thirteen years of marriage. Their childlessness had marked them as "radical"—now they are "safely in the bosom of middle America!" He explained:

> Both of us are active in various social causes such as AIDS, homelessness, health care access, gay and lesbian issues, etc.... I sensed that people just sort of wrote us off as activist nuts. Now by deciding to have a child, people are all of a sudden very friendly towards us. "Ah-ha," I can hear them think, "Now you've decided to become a 'real' couple." What bullshit!!

...

The Cost of the Baby Quest

The baby quest is largely driven by demand and supply, like any other market system. Although the "market" is an odious concept when human lives are concerned and baby selling is almost universally considered to be an abomination, there is no avoiding the fact that babies—or, more precisely, reproductive and adoption services—are bought and sold. The infertility industry is a $2 billion a year business.[17] Each try at IVF, for example, can cost anywhere from $6,000 to $13,000. Susan Sward spent $15,000 on more than forty visits to the fertility clinic, plus three unsuccessful IVF attempts and five unsuccessful artificial inseminations before she finally became pregnant. Annie and Tony Hillman spent over $20,000 for their IVF baby.[18] Vivian Johnson and her husband finally gave birth to twins conceived by IVF, after spending four months and $5,000. Vivian was luckier than Rose Norsika, a lesbian who, with her partner of twenty-two years, "spent twelve years and several thousands of dollars trying to get pregnant with artificial insemination. It was a long hard journey for both of us," and they never managed to conceive.

Angela Carter and her husband spent three hundred dollars per month on treatment for several years, but they stopped short of IVF, which they could not afford. Marie Gutierrez and her husband were unable to follow through with treatment because the clinic required payment "up front" at each visit. She spoke for many when she complained that

> already this year alone my husband and myself have paid dearly with our savings and simply ran out of money, which really disappoints both of us, we both have full time jobs but the cost of infertility treatments are so costly we just can't afford it, and both of our clocks are ticking soon to be 30 years. And it hurts terribly. . . . I will strive with all my heart and Soul to make this dream my reality. My husband and I did not chose this pain, but maybe some of the Clinic's can understand that we *infertile people* need more affordable care. . . . I could get treatments I need and deserve if they were affordable for all.

Like Marie Gutierrez, Kendra Groneman and her husband were unable to afford the costly treatment. Because their insurance would not cover the treatment, each insemination would cost them $150, plus $1,500 for the Pergonal, in addition to the fees for all the tests. "We discontinued treatment terrible hurt and discouraged." Although they were still young, they feared that they would never achieve parenthood. "I'm 25 and my husband will be 24, my clock won't tick forever. Then people say you can always adopt," but she worried about the cost and the shortage of healthy White infants. For White couples who turn to adoption for the same reasons that they seek infertility treatment—their desire for a healthy newborn who resembles them—the costs of adoption can be as prohibitive as infertility treatment. Kendra Groneman believed that the system discriminated against people of modest resources and exploited the desperation of the infertile: "I feel like infertile couples are being punished for not being able to have children by charging outlandish fees for us to adopt a baby that needs a good home with people that will love them. Tell me who is making out here? These agencies are."

Sometimes couples chose between infertility treatment and adoption by considering the costs and risks of both strategies. Most infertile couples are not wealthy, and most insurance companies only partially cover treatments. High-tech interventions like IVF can cost thousands of dollars per attempt, and pursuing a course of treatment can quickly become a $40,000 decision.[19] Dierdre Kearney and her husband were a White middle-class couple in their early thirties with college degrees who were living in a small town in Tennessee. Their opportunities for pursuing parenthood were limited by their resources. They tried to conceive for four of their five years of marriage. "I think we have experienced every emotion and feeling one can in dealing with this situation. We have also been through every fertility test there is." Her husband was on medication for three years, and his problem was corrected. She had four surgeries, medication, fertility drugs, and artificial insemination using her husband's semen. "The only option left for us is IVF. We do not have the money; what savings we have is going toward adoption." They hope to adopt within a year. "It is very expensive to adopt plus the strain and waiting are a continuation of the strain and waiting we've already endured." Sue Kott echoed the same sentiment when she wrote that she and her husband have tried everything except IVF, which they cannot afford. "At $5,000 a try with less than 30% success rate you have to be rich to have a baby that way."

Adoption was not always an option. For Dave Crenshaw, who is sterile, infertility treatment would not help and adoption was beyond his means. He found the adoption process not only expensive, but humiliating and invasive. "The poorest person in the world can give just as damn much love, caring, etc. to an adopted child, as someone whom can buy their way through the system.... Some of us quit as we feel our Rights as Human beings are being violated." Money was not the only obstacle to adoption. Janet Dewey and her husband also wanted to adopt, but "no adoption agency in the U.S. would consider us due to our age and religion (excess of the former and absence of the latter)."

Institutional policies, especially insurance coverage, often made the difference between infertility treatment, adoption, or permanent childlessness. Roberta O'Leary and her husband had insurance that covered only four tries at any one procedure, so they tried Clomid, then Pergonal, and artificial insemination with the husband's semen four times without success. They finally decided to try IVF before their insurance expired. The grueling routine required a great deal of time and a willingness to go through extremely invasive procedures. In addition to the physical and emotional impact of hyperactive ovaries caused by the hormone supplements that are used to stimulate the ovaries to produce eggs, there were "three injections a day.... Time became an issue as I would need daily Ultrasounds to monitor the progress of the eggs. I wasn't real excited but decided to try anyway." At the time of the surgery, there were thirty-one eggs to harvest, but twenty-eight were immature and one was damaged; only two were mature, and neither one was fertilized. As the O'Learys prepared for their second attempt, they noted that their treatment would have been impossible without good insurance coverage and jobs with flexible hours.

For couples whose insurance covered unlimited infertility treatment, money was not an issue, and medical treatment was less costly for them than was adoption. But pursuing infertility treatment with unlimited insurance coverage did not guarantee success. Patricia Painter still had "unexplained infertility" and no children after years of treatment that cost her insurance company $300,000 before she and her husband decided to stop treatment. Similarly, Penny Singer and her husband considered themselves fortunate because their "insurance has paid for everything at about $100,000 a year (three years now)." In spite of the strain on their marriage, which became "shaky to say the least," and the torture to Penny's body, "a thousand tests, some have been so painful," the Singers felt that continuing treatment was their only option, given their financial situation. "We tried adopting, it comes down to you pay about $15,000 or wait about 12 years—they visit your house several times to check a million things—we both decided not to pursue it." Because they did not have the $15,000 to adopt, and adoption was not covered by health insurance, their insurance company paid twenty times that amount for treatment that was ultimately unsuccessful.

Maureen Wendell did not have comprehensive insurance that covered the costs of treatment, "so materially we've sacrificed much." She and her husband paid for several expensive procedures, including two attempts at gamete intra-fallopian transfer (GIFT), in which "harvested" eggs and sperm are inserted directly into one of the fallopian tubes, before she gave birth to a healthy son, who she described as "a miracle." She also understood that by pursuing costly treatment, they made a consumer choice: "He's the exotic vacation I'll never take. The new house I'll never build. The jewels I'll never wear.... But for him I'd do it all again—gladly."

Infertile people like Maureen invested their money in medical treatment that held out the promise of achieving a new status—parenthood—and the happy domestic life with children they desired. Maureen's child was an expensive proposition, even before he was conceived. But ultimately she made the same sacrifices, or consumer choices, that all parents make to one extent or another. For the infertile who poured their resources and energies into treatment, the desired child became the source of personal pleasure, family status, leisure enjoyment, even consumer indulgence: the "cute little 'Gap Kids' with little denim jackets" that Margaret Lewis described. The childfree made a different choice, but for essentially the same reasons. They, too, used their resources to achieve the status and private lives they wanted. They chose to spend their time and money on other sources of personal enjoyment, such as travel, nice homes and furnishings, leisure pursuits, and other pleasures. As different as their reproductive goals appeared, they all invested their resources and energies in creating a personal life that would provide them with the domestic intimacy, status, leisure pursuits, and rewards they desired.

Although children are costly sources of satisfaction, like consumer goods and leisure pursuits, the comparison ends there. Whether or not children are commodified, they are certainly not commodities. No disappointment caused by the inability to purchase a particular item or adventure can compare with the heartache and anguish of infertility for those who are desperate for a child. It is not simply that the infertile often pay and pay but never get the desired goods, it is also that the pain they experience is unlike any other. Because the pain and frustration continue as long as the possibility for parenthood exists, many of the infertile described their quest for a child as an "obsession." Both the nature of infertility treatment and the adoption system contribute to their obsession because of the huge amount of time and energy required in the effort. Like many other respondents, Lester Bernstein discovered quickly that once he and his wife began the grueling routine of treatment, "infertility had taken over our lives."...

Third Parties to Conception

Along with the decision about whether to pursue infertility treatment and when to stop, infertile couples had to decide whether to pursue conception with the help of a donor or to try to adopt a baby. Reproduction achieved with the sperm or egg of another person is usually considered part of reproductive "medical technology," but in many ways it is closer to adoption. Sperm donation is simpler than egg donation and is usually anonymous. Egg donation can also be anonymous, ... [b]ut it can also involve a "surrogate" (a woman who provides the egg and carries the pregnancy), as in the notorious case of "Baby M." Controversies over the rights of surrogates versus "contracting parents" have stymied courts and traumatized parents on all sides of the issue.

AID [artificial insemination by donor] became part of the medical repertoire when Dr. William Pancoast and Dr. Addison Davis Hard collaborated to inseminate an unconscious woman in the 1880s. But it first became widely available and controversial in the 1930s. The process carried such a stigma of dubious legitimacy that it remained a highly secret matter. In the 1950s, a judge ruled in a custody case involving a child conceived by AID that the woman had committed adultery because she had given birth to a child whose father was not her husband, even though she never had sex with any man except her husband.[20] Although that was an extreme case, until fairly recently most clinics would not perform insemination unless a legally married couple requested it and the husband signed forms agreeing to be the child's legal father. These policies not only reflected the prevailing morality, they also protected infertility physicians, who feared they would be sued for child support if the husbands ever denied paternity.

The procedure has become much more respectable, and it is now available to single women, whether heterosexual or lesbian. But its most common use is among married couples, and it still involves secrecy and controversy over whether children should be informed; about the screening of donors; about who should have access to insemination; and about whether the donor's semen should be mixed with the husband's, a practice some physicians use so that husbands can think the child may be their own biological offspring. Although some practitioners think that semen mixing offers a psychological benefit, critics claim that it is dishonest, does not allow a man to come to terms with his infertility, and encourages unnecessary fantasies about biological parenthood. According to one study, secrecy became an "obsession" for couples using AID. The authors concluded that efforts to hide AID were unhealthy and detrimental to the parents and children alike.[21]

The selection of donors also perpetuates certain eugenic theories. Whereas donors are usually screened for inheritable diseases, most are selected on the basis of their academic achievement. They are usually medical students, or at some centers, undergraduates with B (3.0) or better grade-point averages. Since the genetic basis of grade-point averages is questionable at best, the sociologist Barbara Katz Rothman sarcastically dubbed these specimens the "3.0 sperm." Nevertheless, the qualities of sperm donors are a matter of controversy. The notorious commercial "genius sperm bank" in California offers to sell the sperm of Nobel laureates and noted scientists to parents who are hoping for genius offspring. Some sperm banks offer forty dollars per donation from male students who do well on standardized tests, with bonuses for extremely high achievers.... These methods of selection do not guarantee genetic "superiority," and the system is vulnerable to abuse. One physician so abused the system that he used his own semen to inseminate dozens of women at his clinic, creating a community filled with genetic half siblings.[22]...

Continuity and Change

The pursuit of happiness over the past century has taken a decidedly private turn. Although the family has always been central to American national identity, it originally held a place in political ideology as the place where citizens were nurtured. Private life served public life by sending forth productive individuals to build the nation. In the twentieth century, that process has reversed. Now public life largely serves the needs of private life. The family no longer produces goods or even citizens; rather, it consumes. Although society remains vastly interested in the reproduction of its citizens, its citizens have largely given up on public life. Reproductive goals now reflect individual desires for the good life, and Americans expect the nation's institutions to help them achieve those goals.

Although this intense quest for private fulfillment represents a trend that has accelerated since World War II, it has its roots deep in the nation's past. From the era of early nationhood, when George Washington described "the sequestered walks of connubial life" as the place where "permanent and genuine happiness is to be found,"[23] Americans have directed their pursuit of happiness largely toward private life. But many have balanced that quest with an equally powerful engagement in community and civic enterprises. That is why, until quite recently, childless Americans who dedicated themselves to public pursuits found a respected place in the cultural landscape. Today the voluntarily childless have gained a different kind of legitimacy, based on the widespread belief in the individual choice of personal lifestyles. Even that legitimacy is offered grudgingly in a pronatalist society that equates adulthood with parenthood. The childless who carve out active public lives in lieu of parenthood do so in spite of, rather than because of, the cultural norms and expectations that surround their endeavors. At the same time, the involuntarily childless have gained a new kind of sympathy, grounded in the idea that only the family life of one's choosing can confer happiness. In accord with that belief, personal and societal resources are poured into the effort to enable the infertile to procreate.

The emphasis on private life is not so much a rupture with the past as it is an intensification of a trend. As public life in the United States has atrophied, the private side has loomed larger and larger. Reproduction, accordingly, has become a matter of increasingly high private stakes. But the public stakes in reproduction have not diminished. With private life so evidently detached from civic concerns, why are public institutions and politicians still so preoccupied with individual reproductive behavior? In this matter, too, the historical record is instructive. Since the early national period, reproductive manipulation has served public as well as private ends. Americans of today may either cringe or cheer at the news that postmenopausal women can now give birth to babies; that grandmothers bear their grandchildren for their infertile offspring; and that scientists inject sperm into eggs, "harvest" ova, and clone gametes. But the "brave new world" of reproductive engineering must have raised as many eyebrows over a century ago with such experiments as artificial insemination or the eugenic breeding practiced by Oneida Perfectionists. And the same thin odds that sparked hopes a century ago in the childless who purchased Lydia Pinkham's Vegetable Compound, with its promise of "a baby in every bottle," now lure today's infertile to the IVF clinics, with their dismal 80 percent failure rates.[24] The means and technology have changed, but the belief that scientific progress will enable individuals, as well as the society, to manipulate procreative behavior is more powerful than ever.

It is small wonder, then, that as Americans look to the home for personal happiness, they turn to experts to help them achieve their reproductive desires. These expectations are embedded in our national ideals. What has evaporated, however, is the optimism and civic engagement that once connected private to public life. Today, individuals look to social institutions to help them achieve private goals, while social institutions look to the family to fix the nation's ills. Parenthood looms larger and larger as a force for good or evil. Ideas about parental worthiness still overshadow the true needs of children. For more than a century, institutional and political policies and practices have operated on the assumption that if the "right" people became parents—those with the "right" genes, background, race, or income—children's needs would be met automatically. That presumption has proved tenacious, even though it has never been supported by evidence.

Along with the desire to control and manipulate human reproduction, beliefs about parental worthiness also have a long history. Manifest Destiny no longer permeates political rhetoric or explicitly provides the foundations for social policies concerning reproduction. But eugenic ideas are still salient, as is evident in the media blitz surrounding the publication of an old-fashioned eugenic polemic like *The Bell Curve,* by Richard J. Herrnstein and Charles Murray, in 1994. Parental worthiness, based on alleged genetic superiority or economic status, affects decisions ranging from the selection of donor sperm for artificial insemination to custody cases of children with more than one set of "parents." Political debates continue to rage over whether welfare policies should cap benefits after two children and whether poor women should be coerced into using some form of birth control. Parents are still expected to solve the nation's ills by instilling "family values" in their children, while legislation geared toward providing support for families, such as parental leaves and government-sponsored day care, continues to meet with defeat. The nation still expects the family to operate as an autonomous unit that builds society from the bottom up by rearing its citizens, even though it has never functioned in that capacity and cannot do so without institutional support.[25] Today, the family is at the receiving end of the nation's woes, but it is still expected to provide the fix.

The lion's share of that burden falls on women. Although the majority of all women hold jobs outside the home and much lip service is paid to shared reproductive and child-rearing responsibilities, procreation still falls largely within the female domain. Women are still much more likely to be infertility patients than are men (even though men are just as likely to be infertile); there are more available treatments for female reproductive problems; women's bodies are likely to be the ones that are poked, prodded, and pumped with hormones, even if they are fertile and the problem resides with their male partners. They are also more likely to suffer the stigma of childlessness and to be driven to distraction by their inability to achieve their reproductive goals. Women are both more tormented by infertility and more vehemently committed to voluntary childlessness than are men. The fact that reproduction remains primarily a woman's problem is due less to women's biological capacity for pregnancy and more to the cultural norms that still place motherhood at the center of female identity.[26]

The childless do not have different values than the rest of the society. But they articulate their hopes and dreams because they are so often frustrated in their pursuit of happiness. Increasingly, parents and nonparents alike hope to find meaning, status, and happiness in their private lives, removed from the civic and economic arena. The particular reproductive culture that has emerged in the late twentieth century is grounded in a society that looks to the family to solve its problems, but offers the family little in the way of support. Yet many Americans believe that the only place they can find true happiness is in private life. This belief places an overwhelming burden, as well as an overwhelming sense of expectation, on individual procreative behavior. Because the larger society appears out of control and beyond their ability to influence, Americans pursue their reproductive goals with fierce determination. Now that medical, legal, technological, and cultural developments hold out the promise of reproductive self-determination, procreation appears to be one of the few areas of life within a person's control. As long as public life appears bankrupt and alienating, Americans with or without children will continue their pursuit of happiness in the most private areas of life. And as long as Americans care more about each other's reproductive behavior than about each other's children, our private obsession is likely to remain.

NOTES

1. See Arthur L. Greil, *Not Yet Pregnant: Infertile Couples in Contemporary America* (New Brunswick, N.J.: Rutgers University Press, 1991), 51.
2. Susan Faludi, *Backlash: The Undeclared War Against American Women* (New York: Crown, 1991), 24–27. On the declining sperm count, see Amy Linn, "Male Infertility: From Taboo to Treatment," *Philadelphia Inquirer,* May 31, 1987, AI, cited in *Backlash,* 31–32. On the new pronatalism, see also Margarete J. Sandelowski, *With Child in Mind: Studies of the Personal Encounter with Infertility* (Philadelphia: University of Pennsylvania Press, 1993), 9.
3. Articles cited in Faludi, *Backlash,* 104–10.
4. See Mary Ellen Barrett, "Wanted: Baby," *USA Weekend,* October 6–8, 1989, 4–5.
5. James A. Weber, *Grow or Die!* (New Rochelle, NY: Arlington House, 1977), 9–11. Observers noted the trend even earlier. See, for example, "Population Slowdown—What It Means to U.S.," *U.S. News and World Report,* December 25, 1972, 59ff.
6. Ben J. Wattenberg, *The Birth Dearth* (New York: Pharos Books, 1987). See also Art Levine, "The Birth Dearth Debate," *U.S. News and World Report,* June 22, 1987, 64–65.
7. Linda Wolfe, "The Coming Baby Boom," *New York Magazine,* January 10, 1977, 38–42.
8. Cheri Pies and Francine Hornstein, "Baby M and the Gay Family," *Out/Look* 1 (Spring 1988): 79–85.
9. On the proportion of the infertile, see Greil, *Not Yet Pregnant,* 27–28; data on physicians visits from Office of Technology Assessment, in Philip Elmer Dewitt, "Making Babies," *Time,* September 30, 1991, 56–63; see also David Perlman, "The Art and Science of Conception: Brave New Babies," *San Francisco Chronicle,* March 3, 1990, B3; on the success of IVF, see Nancy Wartik, "Making Babies," *Los Angeles Times Magazine,* March 6, 1994, 18ff.
10. Study cited in Greil, *Not Yet Pregnant,* 11.
11. Most estimates gave infertile couples a 50 percent chance, as they did in the 1950s and 1960s, although some physicians were more conservative. One physician in 1962, for example, gave infertile couples a 40 percent chance of a cure, saying that "more could be helped if husbands would cooperate completely with medical examination and treatment." See Grace Naismith, "Good News for Childless Couples," *Today's Health* 40 (January 1962): 24ff. For 1990 data, see Greil, *Not Yet Pregnant,* 11.
12. Susan Sward, "I Thought Having a Baby Would Be Easy," *San Francisco Chronicle,* March 5, 1990, B4. See also Miriam D. Mazor, "Barren Couples," *Psychology Today,* May 1979, 101–12.
13. Greil, *Not Yet Pregnant,* chap. 3, also found this to be true.
14. See Ibid., esp. pp. 51–52.
15. See Lois Wladis Hoffman and Jean Denby Manis, "The Value of Children in the United States: A New Approach to the Study of Fertility," *Journal of Marriage and the Family* 41 (August 1979): 583–96. The exceptions were nonparents, who gave other markers, and Black men, who ranked supporting themselves higher.
16. Tim Carrigan, Bob Connell, and John Lee, "Toward a New Sociology of Masculinity," in Harry Brod, ed., *The Making of Masculinities: The New Men's Studies* (New York: Routledge, 1987), 63–100.
17. Wartik, "Making Babies." See also Jerry Carroll, "Tracing the Causes of Infertility," *San Francisco Chronicle,* March 5, 1990, B3.
18. Sward, "I Thought"; Jerry Carroll, "The Blessed Results," *San Francisco Chronicle,* March 5, 1990, B3.
19. Ellen Hopkins, "Tales from the Baby Factory," *New York Times Magazine,* March 15, 1992, 40ff.
20. "Test Tube Babies," *Newsweek,* December 27, 1954, 48. Commentators at the time claimed that this ruling was out of synch with contemporary practices. See "Test-Tube Test Case," *Time,* December 27, 1954, 52.
21. R. Snowden, G. D. Mitchell, and E. M. Snowden, *Artificial Reproduction: A Social Investigation* (London: Allen & Unwin, 1983).
22. Paul McEnroe, "The Genius Bank," Minneapolis *Star Tribune,* August 28, 1Aff; Sabra Chartrand, "Parents Recall Ordeal of Prosecuting in Artificial-Insemination Fraud Case," *New York Times,* Sunday, March 15, 1992, 10.
23. Miriam Anne Bourne, *First Family: George Washing-*

ton and His Intimate Relations (New York: Norton, 1982), 106. . . .

24. Greil, *Not Yet Pregnant,* 40.
25. Richard J. Herrnstein and Charles Murray, *The Bell Curve: Intelligence and Class Structure in American Life* (New York: Free Press, 1994). See also Stephanie Coontz, *The Way We Never Were: American Families and the Nostalgia Trap* (New York: Basic Books, 1992); and Leila Zenderland, *Measuring Minds: Henry Herbert Goddard and the Origins of American Intelligence Testing* (New York: Cambridge University Press, forthcoming).
26. Many feminist theorists have commented on the implications of the normative connections between womanhood and motherhood. See, for example, Peter Osborne and Lynne Segal, "Gender as Performance: An Interview with Judith Butler," *Radical Philosophy* 67 (Summer 1994): 32–39, for one recent articulation of this phenomenon. See also Carolyn M. Morrell, *Unwomanly Conduct: The Challenges of Intentional Childlessness* (New York: Routledge, 1994).

Lesbian Mothers: "This Wonderful Decision"

Ellen Lewin

Being a single mother is difficult, often very difficult. As the lesbian mothers who speak in these pages make clear, and as other research on single mothers documents abundantly, motherhood without a husband brings with it a range of problems. Single mothers, whether they are lesbian or heterosexual, are likely to face financial pressures, the stresses of bearing sole or primary responsibility for their children's welfare, problems in their efforts to manage time and energy, discrimination in housing, difficulties in finding adequate child care, and the varied stresses that derive from the need to orchestrate children's links with their fathers.[1] Though no reliable figures are available, a substantial number (and possibly the majority) of lesbian mothers seem to resemble single heterosexual mothers in the pathways that led them to their current situation: they had their children during a marriage or a long-term heterosexual relationship and through various circumstances have made a transition to single/lesbian motherhood.

But increasing numbers of lesbian mothers present a very different picture. Like some heterosexual women who find themselves wondering what their lives will be worth if they never have children, more and more lesbians are deciding that conventional circumstances are not the only context in which a child can be born. They are having children on their own, becoming "intentional" single mothers. Though we have no way of knowing how many, or what proportion of lesbians are taking this path, we do know that the visibility of intentional motherhood among lesbians is increasing. Groups for lesbians considering parenthood are thriving in major cities; books and manuals have been written for women who want to become pregnant or adopt children;[2] documentary films have sought to present positive images of lesbian families;[3] and even the mainstream press is giving significant attention to the phenomenon of artificial insemination among lesbians.[4] Media treatment, not surprisingly, is superficial, tending toward either sensationalism or blandness.

But there can be little doubt that intentional motherhood through donor insemination or, less often, through adoption is becoming a common feature of life among lesbians. Gay media are making more frequent mention of children and family issues, and child care, once rarely even thought about in connection with lesbian or gay cultural and political events, has become a routine feature of such functions at least of those expected to draw women. San Francisco's lesbian and gay synagogue, for instance, has so many members with children that a religious school has been opened to provide several levels of instruction, including preparation for Bar and Bat Mitzvah. The coincidence of these developments with the AIDS epidemic and its devastating impact on the gay community in the San Francisco Bay Area cannot be ignored; synagogue members suggested to me that the enthusiasm for activities involving children now evidenced by the men in the congregation seems to parallel their weariness with disease and death.

The Link with Technology: Artificial/Donor Insemination

Artificial insemination has joined in vitro fertilization, embryo transfer, and sex predetermination among the "new" reproductive technologies commonly being talked about. But artificial insemination, the introduction of sperm into the vagina by means other than sexual intercourse, is in fact an ancient procedure. According to Jalna Hanmer, the earliest recorded mention of artificial insemination is in the Talmud, reflecting its practice in the third century A.D.[5] Originally applied to animal husbandry, as of course it still is today, it was first successfully applied to humans in 1790 by the Scottish anatomist and surgeon John Hunter.[6] For nearly a century only the husband's sperm was used (AIH, or artificial insemination by husband), but after experiments proved successful in 1884, artificial insemination by donor (AID) slowly came into use for wives of infertile men.[7] By 1979, AID conceptions were thought to account for between 6,000 and 10,000 births in the United States annually.[8]

Aside from mastery of the procedure itself, by which sperm is introduced into the vagina with a needleless syringe at a time calculated to correspond to the woman's ovulation, the ability to freeze sperm, perfected in 1949, created the basis for expanded use of artificial insemination, both in animals and in humans.[9] Some controversy has surrounded Herman Muller's suggestion that artificial insemination be used for eugenic purposes in humans; the infamous plan to store the sperm of Nobel Prize winners for this purpose is only the most publicized of such efforts.[10]

Less well reported is the lack of regulation governing the operation of existing sperm banks, which are under the control of physicians. Not only is medical screening of donors not consistent, but doctors appear to use their personal values as a way of deciding who may use their costly services.[11] As a result, unmarried women as well as low-income patients may not have the same access to artificial insemination afforded affluent married couples.[12] Meanwhile, debates over the paternal status of the donor and the legitimacy of the offspring continue to rage, inflamed by the application of the technology to so-called surrogate motherhood.[13]

Despite these obstacles, the low-tech nature of artificial insemination and the existence of alternatives to established sperm banks have permitted women to exercise some control over the procedure. At the same time that the women's health movement and self-help gynecology were changing women's views of their reproductive options, women were beginning to circulate information about how to achieve insemination outside the medical establishment.[14]

Adoption and Other Options

The right of lesbians and gay men to be adoptive or foster parents is highly contested, and so far efforts to establish the principle of equal treatment in this area have only occasionally been successful. Unmarried adults, even if there is no question about their sexual orientation, are not preferred as placements for children, particularly for the much-desired healthy Caucasian infants who seem to be in such short supply. Such people are likely to find themselves at the end of a long waiting list with little hope of even having a child placed with them. Their chances increase, of course, if they are willing to adopt so-called hard-to-place children—those who are older, are physically or mentally disabled, are of mixed racial backgrounds, or have not fared well in previous placements.[15] But only in a few areas of the country are agencies willing to consider the possibility that a lesbian or gay applicant might offer a suitable home for a child. Foster placements to lesbian and gay applicants have been increasing in recent years, however, particularly for teenagers who have been identified as homosexual.[16]

Lesbians and gay men who wish to adopt seem to do somewhat better when they make a private arrangement directly with the biological parent or parents, bypassing agency waiting lists. The adoption must still be approved by a state or private agency, but since the evaluation is carried out after the placement, a positive recommendation is more likely. Still in contention, however, is the status of the second parent. Since all states require that unmarried persons adopt only as single individuals, it is difficult to achieve legal recognition for a second parent, either at the time of the adoption or when a biological parent seeks to have the relationship between her partner and her child validated. Some adoptions of this type have been approved, nevertheless, though legal scholars generally doubt that many will follow.[17]

There is no way to gauge how often women undertake relationships with men in order to become pregnant; certainly instances in which men's personal qualities are secondary to women's primary reproductive goals may be far more common than is generally acknowledged.... [S]ome formerly married mothers, both lesbian and heterosexual, tend to view their marriages as arrangements that permitted them to have children under culturally sanctioned circumstances. Not a few of these women, now that their marriages are over, go further and perceive single motherhood as having improved their situation in various ways. These women may see single motherhood as more desirable than motherhood in a marriage at the same time that they are constantly concerned with a range of financial and social problems exacerbated by their status as heads of households.

Though it appears that "intentional" mothers may still be in the minority among lesbian mothers, they are the most visible element of the so-called lesbian baby boom, or at least the one that attracts the most opprobrium. But these mothers afford us special insight into the underlying meaning of motherhood in the wider culture; as levels of social approval are stripped away, we are left with a view of the goals that lesbians and other unmarried women seek to achieve through motherhood and the strategies they employ in their attempts....

How to Have a Baby: Choosing a Method

Once a lesbian has decided to have a baby, she has to figure out how to go about it. Heterosexual and occasional lesbian women in relationships with men may seem to have a ready solution at hand, but issues of later obligation may undermine its apparent simplicity. In some situations involving a relationship with a man, his future involvement as a father may be at issue, as may the durability of the relationship itself. All of these questions may become merged with the decision to have a child, making the matter of intentionality murky at best.

Though lesbians sometimes have intimate relationships with the men who father their children, this approach is not what most prefer. A sexual entanglement with a man not only may be personally unappealing but may raise potential problems of custody or control. At the same time, insemination by a physician may represent an unpleasant intrusion into one's private life; that is, a threat to one's autonomy. A lesbian may circumvent these problems by opting for insemination outside the medical establishment, but that route may have other unwanted consequences; in particular, she may not be able to shield her identity from the donor. Finally, adoption is always difficult for a single woman, and a lesbian is likely not to qualify for adoption at all if her sexual orientation is discovered.[18] All the intentional single mothers I interviewed had to negotiate a variety of priorities in deciding to become mothers and in choosing a way to realize that goal, but lesbians had to take their stigmatized status into consideration in devising a strategy.

Relationships with Men

Some lesbian mothers I interviewed already had a relationship with a man at the time they decided to have a baby. Though some of these babies were welcome by-products of existing sexual relationships, other women turned to friends or casual acquaintances to become pregnant "the old-fashioned way."

But becoming involved with a man in an effort to conceive a child may not only lead to awkward entanglements but entail serious risks. Like formerly married women, "unwed" mothers who have ties to their child's father may find themselves either fending off attentions they consider excessive or having to compensate distressed children for their fathers' failure to show an interest in them. A few such fathers play their social role with enthusiasm, offering both time and financial support to their offspring, but most seem to feel no obligation and some even deny their role in the child's conception. Women rarely are financially or emotionally prepared to launch the kind of legal battle that must be waged to establish paternity and gain judicial recognition of the father's identity.

Before Laura Bergeron came out as a lesbian, she was in a long-term relationship with a man. As she moved into her early thirties, she decided that she wanted to have children even though her partner did not.

> He had one child already by a previous marriage and there were a lot of problems around visiting the child and he just felt that he didn't want to go through that again in case we were to split. And also he didn't want to be financially responsible for bringing up any more children.... So we had a contract that it would be OK if I had a child as long as I was willing to be totally financially responsible, and I agreed to that.

During the time the relationship continued, this agreement remained firm; the father made no financial contribution directly to the children and provided no assistance whatsoever in their care. More recently, however, he has established a regular visiting relationship with the two boys and with the daughter Laura had through artificial insemination after she came out.

More commonly, such relationships collapse soon after the women become pregnant. If the male partners are truly unwilling collaborators in parenthood, they tend to extricate themselves from the relationship as soon as they can. Lesbians have mixed feelings about these developments. They want their children to feel some sort of connection to their "fathers," but at the same time they may feel relieved to be able to avoid interference or active hostility from these men. Beth Romano puts it this way:

> I'm glad I did it that way, that I made no compromises. Just in practical matters now, there is no threat, I'm pretty free to do what I want, there is nobody saying, "I'm going to take your kid away." ... I guess it's rather egomaniacal to say, "I produced this child by myself," but that's how I feel.

These remarks are echoed by heterosexual mothers who expected or hoped for more from the men who had impregnated them. Samantha Paulson said:

> [My daughter's father] maintains very little [contact]. I think he'd like to but he feels guilty about not participating more in her growing up. So every three to six months he'll make an appearance or call, but nothing worthwhile.... I think the only reason I've survived is because I expected nothing. He told me initially he wasn't going to participate at all so I knew exactly what I was getting into before she was born.... He let me know where he was coming from; he told me he ... might not even be able to stay around because he couldn't accept the responsibility. So I knew I'd be a single parent. Although somewhere at the back of my mind I thought, he doesn't really, he's not really meaning this, he'll come around. I had some hopes, but I always knew that he wasn't willing to be a father.

Many heterosexual single mothers, like lesbians, move back and forth between a strong desire to be independent, to manage motherhood on their own, and wistful sentiments about what might have been. Many of these accounts differ little from those offered by formerly married mothers, regardless of sexual orientation, in which they attempt to rationalize (with little apparent success) constant disappointments with their children's fathers by insisting on their commitment to independence and autonomy.

Ruth Zimmerman describes the long-term relationship with her son's father before she became pregnant (and before she had come out as a lesbian) and her efforts during her pregnancy to encourage his involvement with the child. The father apparently couldn't decide how to relate to the pregnancy, and finally she gave him an ultimatum: Either stay and be a father or leave altogether. He stayed but continued to be indecisive until he was offered a job too far away to make commuting feasible. The job enabled him to avoid a definitive commitment. Over time, his professional obligations have gradually moved him farther and farther from the Bay Area, although he has continued to make regular financial contributions and to correspond with Ruth.

More commonly, lesbians try to maintain some distance from the child's father because of their concerns about possible threats to custody or to their maternal autonomy. Ruth, for example, harbors some resentment about her former boyfriend's failure to be involved with their son at the same time that she views his behavior as ultimately beneficial to her. Because he has kept his distance for so long, she reasons, it would not be in his interest to challenge her custody.

His name does not appear on the birth certificate and to claim the child he would have to establish paternity in court and become liable for the costs of AFDC during the time his child was supported by public funds, both powerful disincentives.

Some mothers carefully omit the father's name from the child's birth certificate to prevent possible custody disputes.[19] Camille Walsh made this choice mainly because she did not trust her children's fathers:

> [There is] some side of me that's very cautious and I thought if there was something on an official document that that might jeopardize my hold on [my children].

The fear of a challenge to their custody had led many lesbian mothers to distance themselves from the father and to take precautions against his discovery of their sexual orientation. Like formerly married women who have similar fears, women who have children on their own use whatever resources they have at hand to discourage their children's fathers from even considering litigation. Their strategies often are based on various ways of increasing distance from the father (and thereby decreasing his ability to scrutinize their lifestyles) as well as on more formal means to separate him from the family. Mothers reason that fathers who lack legal claims to children will not be motivated to pursue questionable custody litigation, but many of them still experience considerable anxiety about the possibility.

Some lesbian mothers, in contrast, make persistent efforts to bring the father into their children's lives. Sarah Klein conceived her daughter, now one year old, in a relationship with a man she had selected for qualities she felt would make him a good father. Although he makes no formal financial contribution, he has made a regular commitment to care for the child; the fact that he lives only a few blocks away from Sarah and her lover simplifies these arrangements. Sarah regards him as a parent and anticipates that he will play a vital role in the care of their daughter throughout her life. She apparently has no fear of his challenging her custody and in fact considers him a friend and has maintained a social relationship with him apart from the time he devotes to the child. The fact that they have continued these arrangements with minimal stress is intriguing in view of his initial resentment at being "used" for his sperm. Once Sarah became pregnant, she terminated her sexual relationship with him, and he reacted angrily.

> He was really pissed off. It was a classic, you took my sperm, you lesbians, you all plotted against me.... It was sad, he was really hurt.

Clearly the consequences of selecting a man as an "inseminator" are unpredictable. Though such premeditated conceptions may seem to preclude future ties with the father (or donor), actual relationships range from quasi-joint-custody situations such as Sarah's to total estrangement.

Some women who want to become pregnant manipulate situations in which the man is willing, or even eager, to establish a more extensive commitment. Bonnie Pereira reports, for instance, that though she was a lesbian, she embarked on a relationship with Bob because she had decided it was time to have a child. He wanted to marry her, but though she was not averse to his having some connection with their offspring, she was not interested in any legal entanglements. Bob visited often when their daughter, Tina, was small, but his involvement gradually waned and eventually disappeared entirely. Because Bonnie felt that Tina wanted a connection with her father, she made what she considers major efforts to keep channels of communication open with him. These efforts have met with no success.

> I have done what I can do: I have written, you know, and I have called and I've let him know where I've moved to and so forth and so on, and then I've made it very clear that he's never going to be turned away from here if he wants to see his daughter. He's made that decision himself, that he feels uncomfortable, I guess.

More commonly, lesbians tend to be extremely cautious in limiting their connections with the men they choose to father their children. Laura Bergeron made a written agreement with the father of her third child.

> We just wrote down a few basic things, which for one, I would never claim him as a parent for any reason. And I would never apply for welfare and give his name for the father. I wouldn't use his name on the birth certificate. I wouldn't expect any moral or financial obligations that might come up later. In other words, he was just a sperm donor, except that we were doing it in person.

Because she does know the father, however, Laura sees herself as having avoided one of the serious problems presented by donor insemination: the child's future questions about her father's identity.

Other lesbians, however, and some heterosexual women as well, perceive a relationship with the child's father as a threat, mandating secrecy and caution. Lilly Parker, for example, consciously manipulates information about her daughter's conception in an effort to prevent friends from figuring out his identity. She never told the man that he fathered her child and has no interest in any contribution from him at all.

> I don't want him to be only half-assed involved. I'd rather have no involvement at all. I don't feel like he is a father. I feel like he's a donor.

Lilly would like to have another child, but has decided to use artificial insemination if she does so to avoid the problems with secrecy she has had with her first child.

Donor Insemination

Beyond concerns about the consequences of a relationship with a "father," some lesbians can feel nothing but distaste at the prospect of having sexual relations with a man in an effort to conceive a child. In fact, many of the women I interviewed explained that they had thought biological motherhood was not a realistic goal because they were not willing to consider heterosexual intercourse. For some of those who later discovered artificial insemination, motherhood had become a remote dream, the price to be paid for living in a way that was otherwise comfortable. Joan Emerson, who now has a nine-month-old daughter, explains how she came to choose artificial insemination.

> There wasn't any decision. I didn't want to adopt, I wanted to have my own child. I didn't want to go out and pick up, I didn't want to have to sleep with a man to get her. So artificial insemination was the only way.

Maggie Walters, the mother of an eighteen-month-old daughter, had been familiar with the idea of artificial insemination since childhood and found the decision easy to make.

> I hadn't never fucked anyway, so I wasn't going to do it for that. Plus, see, when I grew up, one of my best friends . . . had been conceived that way . . . and it wasn't like it was any big thing. . . . So it was just kind of interesting, but it wasn't any big deal.

To several mothers, either specific or vague concerns about custody made artificial insemination seem the only viable option. Joan Emerson explains:

> I wanted the total responsibility of the child. . . . I guess I didn't want to take the chance of anybody trying to take her away from me.

Like many others, Joan chose to become pregnant through a medical facility, believing that this way of obtaining a donor would give her the fullest assurance of anonymity. Anonymity, however, is won at the cost of some personal control over conception, a central issue for some women.[20] These women tend to take a strong critical stance toward mainstream medical practices, particularly in regard to the increasing use of high-tech interventions in gynecological and obstetrical care; some of them are committed to various sorts of alternative or non-Western medicine.

The need to exert autonomy during the reproductive process was a central concern to Louise Green, for example. Further, because of her counterculture lifestyle, she felt so removed from mainstream medicine that she did not even consider that such resources would be available to her. Her approach was to ask men she met if they would like to be sperm donors until she found one who was willing. She was careful not to let this person know her full name and after conceiving she moved to another state. Even so, Louise still harbors considerable anxiety about the donor as a threat to custody, should he ever have the ability to trace her and make a claim.

Louise's nonmedical approach to conception met her spiritual needs as well. She carried out the procedure alone in her room.

> I had all these candles lit . . . and it was real quiet and I had this nice music playing that I really liked, my tarot cards out. It was real nice, it was real peaceful.

After doing a vinegar douche (to help conceive a girl), Louise used a menstrual sponge to keep the semen from leaking out.

> While I was lying there I was imagining . . . kind of like clouds . . . and from the cloud would . . . come like raindrops . . . [and] each one was a baby spirit. . . . It was like the perfect baby spirit was going to drop and . . . come inside me. And it did.

Louise became pregnant on her first attempt, believing that this occurred because of mystical forces. She intended to have her baby at home, where she would be able to create an agreeable spiritual environment, but after a protracted labor she was transferred to a hospital, where she had a long and difficult delivery. Despite the multiple medical interventions and considerable physical trauma she endured, she describes the birth of her daughter in mystical terms.

> It was about the best thing I ever experienced. I was totally amazed. The labor was like I had died. . . . I had just died. The minute she came out, I was born again. It was like we'd just been born together.

Louise's story reminds us of the intrinsic, and often spiritual, values women associate with motherhood. By becoming a mother a woman may achieve not only adulthood, but a glimpse of the most ultimate and stirring truths.

Like lesbians who became pregnant through relationships with men, those who resort to donor insemination are fearful of future interference by the biological father. These anxieties may conflict with the desire that many of them feel to share their child with him. Some mothers wish they could have some sort of supportive connection with their child's genitor, and others focus more on what they imagine the child later may want to know about him.

Grace Garson used a gay male friend as a go-between to obtain a sperm donation, and gave no thought to possible problems when she was setting up these arrangements. Now she wishes there had been some way to record the donor's name in case her son may later want to trace his biological roots.

Michelle O'Neill has similar feelings:

> When I did do the insemination, I deliberately did not want to know who the father was, I didn't ever want to meet him. I regret now that I did not have the name of the father put on file someplace so that if [my son] ever wants to know who the father is . . . I feel that it's his right to know.

Some mothers want to have more information about the donor but not enough to establish his identity. Kathy Lindstrom, for one, feels that it would be good to be able to tell her son about his ethnic background on the paternal side, and she suspects that he will have the same kinds of questions about his father that adopted children have about their biological parents. The underlying assumption here is clearly that "ethnic background" has something to do with biology or genetic heritage.

Maggie Walters expresses other kinds of misgivings about having a child without a known father. Her concern focuses on whether it is right to bring a child into the world with a lesbian mother and no father. This was the issue that she considered most carefully when she planned her pregnancy, and although she finally went ahead and was artificially inseminated, she still feels that the problem is unresolved.

The reactions of the families of the women who have children through artificial insemination sometimes confound their expectations. Most mothers report that after a period of confusion, the existence of a new grandchild came to overshadow the way the child had come into the world. Michelle O'Neill, who had grown up in a conservative Catholic family, feared that her mother would never accept her grandchild. When she told her mother that she was pregnant by artificial insemination, her mother was not only shocked but concerned that her daughter's "freaky" way of getting pregnant would be written up in medical journals. The actual birth seems to have eased these anxieties, however, and Michelle's mother has been consistently supportive, both emotionally and financially, since the child was born.

Though most mothers ultimately achieve some measure of acceptance by their parents, some families find out-of-wedlock pregnancy, especially by artificial insemination, simply too shocking to manage. Kathy Lindstrom's mother was very enthusiastic when her brother's wife had a baby, but she could not summon similar feelings when Kathy became pregnant. The pregnancy has apparently ended their relationship altogether. Although her mother lives in the state, she has made no effort to contact Kathy since the birth six months ago.

> Since she's known I became gay, she's maintained a visiting relationship, but that's even broken off since I had the baby.... It's just something that goes unsaid.... I guess she couldn't approve of my method of having [a baby].

Adoption

Adoption is rarely an option for a lesbian, or for any other unmarried woman, for that matter. Of all the mothers I interviewed, only five had adopted their children—four lesbians and one heterosexual. Three had found their children through public agencies and two through private adoptions.

The patterns associated with these adoptions all resemble those we might expect if we looked more generally at adoptions among single adults. The three mothers who adopted through agencies all received children who were considered "hard to place"—older, of minority or mixed race, and disabled. Those who were successful in arranging private adoptions (both lesbians) became mothers of virtually newborn Caucasian infants with no apparent disabilities. Both of these mothers, however, were employed in health-care settings and were able to learn about impending births under conditions that facilitated the adoption process.

Among the lesbian adoptive mothers, fear that their sexual orientation might undermine the adoption surfaced under a variety of circumstances. In the most benign conditions the matter had not been discussed but, the mother surmised, was suspected. Some of these women thought the possibility had not been pursued because the adoption worker didn't want to have to stop the proceedings.

Eileen Sullivan adopted two children privately and encountered no difficulties negotiating the bureaucratic aspects of the adoption process. At the same time, she feared situations in which she might be forced to answer a lot of personal questions:

> I was afraid the issue of my being gay would come up and on the basis of that they would refuse the adoption. That was the real issue. They never asked, or never had evidence enough to ask.

Janet Goldman, who also adopted privately, worked at the hospital where her daughter was born. The social worker who set up the adoption learned that she was a lesbian only after the adoption was final. She has kept in touch with Janet and her daughter, and she says now that she would not knowingly have offered a baby to a lesbian, but now that she sees how well the situation has worked out, she's glad she didn't know.

Most single women who try to adopt a child are faced with a battery of personal questions and may not be spared direct inquiries about their sexual orientation. When Emma Gibson adopted the first of two disabled minority children through an agency, she was living with a partner. The social worker asked her if the other women was her lover and she denied it. Nothing further was asked and the adoption went through. Although several years have passed, during which she adopted a second child and broke up with her lover, she continues to be extremely anxious about the possibility of being exposed. Because she lied in answer to a direct question, she fears that she has committed the equivalent of perjury and that she will lose her children if the truth is discovered. Emma's preoccupation with secrecy is reinforced by her certainty that she would be fired from her job if her lesbianism were ever revealed. For financial reasons, she recently moved to a working-class suburb far from her old neighborhood and lacks a close circle of lesbian friends. Most of her friends from the years before the adoptions are not parents, and she finds that she is out of step with their social world now.

Intentional Motherhood: What Is Intended?

American culture places tremendous emphasis on the powers of the individual, on the importance of achieving personal goals through action in one's own behalf. Lesbians who are not mothers share with other childless women a feeling of distance not only from the kinds of things "ordinary" women do but from the special relationship to the spiritual world women can derive from their connection to children. By becoming a mother, a woman can experience a moment of transcendent unity with mystical forces; by being a mother, she makes continuing contact with her inner goodness, a goodness that is activated by altruism and nurtured by participation in a child's growth and development.

By becoming a mother, a lesbian can negotiate the formation of her self; she can bring something good into her life without having to sacrifice autonomy or control. Thus the intentional single mother (whether she is lesbian or heterosexual) can achieve a central personal goal—the goodness that comes from putting the needs of a dependent being first. By becoming a mother through her own agency, she avoids the central paradox that motherhood represents to married women—a loss of autonomy and therefore of basic personhood in a culture that valorizes individualism and autonomy. Like ending a marriage, having a baby on her own allows a woman to meet her basic personal goals, and she may see it as a critical part of establishing a satisfying identity in a culture that often blocks women's efforts to be separate individuals.

Being a mother provides many benefits, but becoming a mother is a process that can be pursued in a variety of ways and can help women realize a variety of goals. The specific strategies they select—deliberate pregnancy with a man, artificial insemination, adoption—reflect not only the opportunities available to them, but the particular ends which they seek to enhance. Women who wish to distance themselves from such mainstream institutions as the medical establishment may find it difficult to maintain the anonymity of a sperm donor; those who fear the donor's intrusion on their lives may seek anonymity at the price of autonomy. There are many ways to go about becoming a mother, and they are as vital a part of women's objectives as their desire to be mothers.

Motherhood also appears to offer lesbians some resolution of the dilemmas inherent in membership in a stigmatized category. On the one hand, intentional motherhood demands specific action of some sort—a lesbian is, after all, unlikely to become pregnant by chance. On the other hand, to the extent that wanting to be a mother is a profoundly *natural* desire, and is perceived as having nothing to do with cultural or political choices, then achieving motherhood implies movement into a more natural or normal status than a lesbian can ordinarily hope to experience otherwise. But motherhood also requires planning and manipulation, and thus stands in contrast to one's natural—that is, unpremeditated—lesbian identity.

At the same time, however, a lesbian who becomes a mother has effectively rejected the equation of homosexuality with unnaturalness and the exclusion of the lesbian from the ranks of "women." In this sense, finding a way to become a mother constitutes a form of resistance to the gender limitations, and particularly to the constructions of sexual orientation, that prevail in the wider culture. Curiously, though, this act of resistance is achieved through compliance with conventional expectations for women, so it may also be construed as a gesture of accommodation.

The stories that lesbian and some unmarried heterosexual mothers tell of their ventures into motherhood, of the ways they formulated their aims and acted to achieve them, then, bring together behaviors that can be regarded simultaneously as rebellion and as compliance. For these women, negotiating motherhood consists of forging a path through these conflicting meanings and weaving them together into a rewarding definition of the self.

NOTES

1. Arendell 1986; Weiss 1979; Weitzman 1985.
2. Pies 1985; Robinson & Pizer 1985.

3. Chasnoff & Klausner 1986; Zheutlin, Reid & Stevens 1977.
4. Shah with Walters 1979; Stern 1979.
5. Hanmer 1983: 184.
6. Corea 1985.
7. Langer 1969.
8. Curie-Cohen, Luttrell & Shapiro 1979: 588.
9. Corea 1985: 36.
10. Hanmer 1983; Muller 1961.
11. Lasker & Borg 1987.
12. Curie-Cohen, Luttrell & Shapiro 1979; McGuire & Alexander 1985; Strong & Schinfeld 1984.
13. Smart 1990.
14. Achilles 1989; Hornstein 1984; Ruzek 1978.
15. Editors of the *Harvard Law Review* 1989: 135.
16. Ricketts & Achtenberg 1987.
17. Pies 1985; Ricketts & Achtenberg 1987; Zuckerman 1986.
18. Ricketts & Achtenberg 1987.
19. Some mothers omit the father's name from the birth certificate to simplify their claims to AFDC or other benefits or to protect the father from harassment.
20. See Lewin 1985.

WORKS CITED

Achilles, Rona. 1989. Donor Insemination: The Future of a Public Secret. In *The Future of Human Reproduction*, ed. Christine Overall, pp. 105–119. Toronto: Women's Press.

Arendell, Terry. 1986. *Mothers and Divorce: Legal, Economic, and Social Dilemmas.* Berkeley: University of California Press.

Chasnoff, Deborah, and Kim Klausner. 1986. *Choosing Children.* Film distributed by Cambridge Documentary Films, Cambridge, Mass.

Corea, Gena. 1985. *The Mother Machine: Reproductive Technologies from Artificial Insemination to Artificial Wombs.* New York: Harper & Row.

Curie-Cohen, Martin, Lesleigh Luttrell, and Sander Shapiro. 1979. Current Practice of Artificial Insemination by Donor in the United States. *New England Journal of Medicine* 300(11): 585–590.

Editors of the *Harvard Law Review.* 1989. *Sexual Orientation and the Law.* Cambridge: Harvard University Press.

Hanmer, Jalna. 1983. Reproductive Technology: The Future for Women? In *Machina Ex Dea: Feminist Perspectives on Technology*, ed. Joan Rothschild, pp. 183–197. New York: Pergamon.

Hornstein, Francie. 1984. Children by Donor Insemination: A New Choice for Lesbians. In *Test-Tube Women: What Future for Motherhood?* ed. Rita Arditti, Renate Duelli Klein, and Shelley Monden, pp. 373–381. London: Pandora.

Langer, G., et al. 1969. Artificial Insemination: A Study of 156 Successful Cases. *International Journal of Fertility* 14(3): 232–240.

Lasker, Judith N., and Susan Borg. 1987. *In Search of Parenthood: Coping with Infertility and High-Tech Conception.* Boston: Beacon.

Lewin, Ellen. 1985. By Design: Reproductive Strategies and the Meaning of Motherhood. In *The Sexual Politics of Reproduction*, ed. Hilary Homans, pp. 123–138. London: Gower.

McGuire, Maureen, and Nancy Alexander. 1985. Artificial Insemination of Single Women. *Fertility and Sterility* 43: 182–184.

Muller, Herman J. 1961. Human Evolution by Voluntary Choice of Germ Plasm. *Science* 134(3480): 643–649.

Pies, Cheri. 1985. *Considering Parenthood: A Workbook for Lesbians.* San Francisco: Spinsters Ink.

Ricketts, Wendell, and Roberta Achtenberg. 1987. The Adoptive and Foster Gay and Lesbian Parent. In *Gay and Lesbian Parents*, ed. Frederick W. Bozett, pp. 89–111. New York: Praeger.

Robinson, Susan, and H. F. Pizer. 1985. *How to Have a Baby without a Man.* New York: Simon & Schuster.

Ruzek, Sheryl Burt. 1978. *The Women's Health Movement: Feminist Alternatives to Medical Control.* New York: Praeger.

Shah, Diane K., with Linda Walters. 1979. Lesbian Mothers. *Newsweek*, February 12, p. 1.

Smart, Carol. 1990. "There Is of Course the Distinction Dictated by Nature": Law and the Problem of Paternity. In *Ethical Issues in the New Reproductive Technologies*, ed. Richard T. Hull, pp. 69–86. Belmont, Calif.: Wadsworth.

Stern, Susan. 1979. A Different Type of Baby Boom in S.F.: Artificial Insemination for Lesbians. *Synapse*, December 6, pp. 1, 4–5.

Strong, Carson, and Jay Schinfeld. 1984. The Single Woman and Artificial Insemination by Donor. *Journal of Reproductive Medicine* 29: 293–99.

Weiss, Robert S. 1979. *Going It Alone.* New York: Basic Books.

Weitzman, Lenore. 1985. *The Divorce Revolution: The Unexpected Social and Economic Consequences for Women and Children in America.* New York: Free Press.

Zheutlin, Cathy, Frances Reid, and Elizabeth Stevens. 1977. *In the Best Interests of the Children.* Film distributed by Iris Films, Berkeley, Calif.

Zuckerman, E. 1986. Second Parent Adoption for Lesbian-Parented Families: Legal Recognition of the Other Mother. *UC Davis Law Review* 19: 729, 731.

The Politics of Reproductive Benefits: U.S. Insurance Coverage of Contraceptive and Infertility Treatments

Leslie King and Madonna Harrington Meyer

The average woman is fertile, and therefore must attempt to control her reproductivity, for one-half of her life.[1] For most women, it is the preoccupation with preventing births that consumes their health care dollars and energies; for a small minority, it is the preoccupation with achieving a birth that dominates. The ability to control fertility is, to a great extent, linked to access to various forms of reproductive health services, including contraceptives and infertility treatments. Yet, in the United States, insurance coverage of contraceptive and infertility treatments is fragmented.

We began this research with the realization that employees of the state of Illinois have broad insurance coverage of infertility treatments but no coverage of contraceptives; yet Illinois women insured by Medicaid receive benefits in just the opposite configuration. We set out to determine whether access to benefits is similarly unequal in the United States and to discover how this unequal distribution of benefits came about. To what extent do deeply embedded values and conflicts about who should—and who should not—bear children generate class cleavages in access to insurance coverage?

Contraceptives, Infertility Treatments, and Insurance

The typical American woman spends 90 percent of her fertile period avoiding pregnancy (Gold and Richards 1994). At any point in time, about 60 percent of women ages 15–44 use contraceptives (Mosher and Pratt 1990a). The most effective forms of birth control are also the most costly; a one-year supply of oral contraceptives costs around $300 (AGI 1994). Five-year contraceptive implants average over $500, and even spermicide, necessary for use with diaphragms and condoms, costs $10 a tube. Americans spent approximately $1.5 billion for contraceptives in 1993; nonetheless, unintended pregnancies outnumber intended pregnancies every year (Gold and Richards 1994; Women's Research and Education Institute 1994).

Contraceptives have often been excluded from insurance coverage on moral grounds: Women who do not want to procreate should not have intercourse (see critique by MacKinnon 1983). Though experts repeatedly identify the limiting and spacing of births as the single greatest factor in improving physical and economic well-being among women, few in the United States define contraceptives as a key form of preventative health care for women (AGI 1994; Dixon-Mueller 1993; Gold and Richards 1994). Too often, discussions about access to contraceptives turn to pregnancy among teens or unmarried women as if they were the only women requiring contraception. But in fact, women's reproductive years include many stages; before beginning a family, between children, and once again after optimal family size has been reached, the need for safe and effective contraceptives is paramount. Fully 80 percent of pregnancies to women aged 40 and older are unwanted (Gold and Richards

1994). Regardless of their age, socioeconomic class, or marital status, most women require reproductive services to limit and space pregnancies.

At the other end of the spectrum, about 8 percent of, or 1 in 12, couples with wives ages 15–44 have difficulty conceiving or carrying a pregnancy to term (Mosher and Pratt 1990b). Couples are generally considered infertile if they fail to conceive after one year of intercourse without contraception (Mosher and Pratt 1990b). About 51 percent of couples with primary infertility (childless couples) and 22 percent of those with secondary infertility (couples with at least one biological child) seek treatment. Of those, more than 85 percent are treated with surgery, fertility drugs, and/or artificial insemination (Office of Technology Assessment 1988). Nearly 15 percent of infertile couples attempt in vitro fertilization (IVF) or similar treatments that cost between $8,000 and $10,000 per attempt. In 1987 alone, Americans spent roughly $1 billion on infertility diagnosis and treatment (Office of Technology Assessment 1988).

Like the need for contraception, infertility problems have historically been dismissed as social or moral rather than health issues, and treatments have generally been regarded as either experimental or elective. Often attributing their difficulties to emotional rather than physical problems, physicians told infertile couples to relax, accept God's will, or adopt (Stone 1991). Infertility has only recently been defined as a health concern deserving of medical treatment (Stone 1991). Untreated sexually transmitted diseases are the most common of known causes of infertility in the United States. Pelvic inflammatory disease and endometriosis have both been linked—though only partially—to early sexual activity and/or multiple partners (Stone 1991). Moreover, research indicates that the chances of infertility may increase with age: 1 of 17 women ages 30–34 are infertile compared with 1 in 5 ages 40–44 (Gold and Richards 1994). Given the links to sexual activity and delayed childbearing, some oppose coverage of infertility treatment for moral reasons (interview Bush-Joseph 1994b). But advocates of infertility treatment point out that while the choice to have a child is elective, the numerous diseases and conditions that prevent a successful pregnancy are not (interview Bush-Joseph 1994b). As the number of babies available for adoption diminished and infertility treatments became more prolific and profitable, the tendency to dismiss infertility treatments as elective waned. By 1993, about 65 percent of fee-for-service plans covered semen analysis or endometrial biopsies; 45 percent covered infertility drug treatments, and 15 percent covered various forms of IVF (AGI 1994). As insurance coverage of infertility treatment became more widespread, the number of infertility clinics rose from 4 in 1981 to at least 192 in 1990 in this multimillion dollar industry.

Locating Reproductive Benefits in the U.S. Welfare State

Despite increasing attention by welfare state theorists to issues of gender and class inequality in access to welfare state benefits, there is little analysis of access to reproductive benefits in the United States. We believe this inattention is related to two main factors. First, the literature on welfare states analyzes almost exclusively access to "public" rather than "private" benefits. In the United States, access to reproductive care is determined by access to health insurance, and the majority of U.S. residents gain access to health insurance through their jobs rather than welfare state programs per se (EBRI 1994; Short, Monheit, and Beauregard 1989). Studies that explore only the distribution of formal or "public" welfare benefits hide or ignore key class and gender differences in access to "private" benefits (Esping-Andersen 1989). Moreover, the distinction between public and private health insurance benefits is artificial given the amount of government involvement in the employment-based health insurance system (Esping-Andersen 1989). State subsidy of employment-based health insurance topped $51 billion in forgone taxes in 1994 alone (EBRI 1994). In fact, the state has a lengthy legacy of shaping "private" health care benefits; both the federal and the state government use tax breaks and other mechanisms as leverage to enforce a wide variety of health mandates including mammography screenings or immunizations.

Second, we believe access to reproductive benefits has received short shrift in the welfare state literature because the U.S. fertility policy is implicit rather than explicit. There is no unilateral program or service that distributes reproductive benefits, nor any clearly stated objectives for the outcomes of such benefits. Unlike many other nations, the U.S. policy is a de facto policy, emerging as much in the wake of inaction as action. In some instances, state policymakers have actively shaped the distribution of reproductive benefits by mandating Medicaid or employer-based insurers to cover specific treatments. In other instances, the state has taken no action; we believe failure to act characterizes implicit endorsement of the policies of reigning social institutions (MacKinnon 1983, 1989). If the state does not guarantee all women the right to a full range of reproductive treatments, regardless of social and economic class, the state may be implicitly endorsing a de facto fertility policy that encourages births among working- and middle-class women and discourages births among the poor and welfare recipients.

Class and Reproduction

The idea of encouraging births among middle- and upper-class women while discouraging births to poor women is not new in the United States. Ginsburg (1992) reports that public and private groups have long worked to limit births among poor whites, and all racial and ethnic minorities, while attempting to encourage births among middle- and upper-class whites. In the early 20th century, eugenicists promoted fertility control among poor immigrant women while accusing middle- and upper-class women of "race suicide" and admonishing them to increase their family sizes (Davis 1983). Family planning advocate Margaret Sanger became aligned with the eugenics movement, writing, in 1919, "More children from the fit, less from the unfit—that is the chief issue of birth control" (cited in Gordon 1990, 281). By 1931, 30 states had enacted eugenics laws calling for sterilization of "socially inadequate" persons. Twenty thousand sterilizations had been performed by 1935 (Shapiro 1985).

Although the eugenics movement lost prestige in the United States after World War II, health providers continued to pursue separate strategies for poor and minority women and middle-class white women. Numerous studies refer to sterilization abuse, during the 1960s and 1970s, involving poor and minority women (Davis 1983; Gordon 1990; Petchesky 1981; Shapiro 1985). Little good quantitative evidence of the extent of sterilization abuse exists, but Shapiro (1985) provides some powerful numbers to back up many of the charges. Using the 1973 and 1976 National Survey of Family Growth, Shapiro found that for those with incomes below 150 percent of poverty level, 304 women per 1,000 had been sterilized. Among other women, 209 per 1,000 had been sterilized. In addition, 366 per 1,000 Medicaid recipients were sterilized compared with 219 per 1,000 nonrecipients. Middle-class white women, on the other hand, complained of *lack* of access to sterilization during the 1970s (Rodriguez-Trias 1984).

In the 1960s and 1970s, a movement to provide federally funded family planning to the poor gained momentum, in part due to the belief that helping the underclass control their fertility would help reduce their numbers (Dryfoos 1988; Nathanson 1991). Supporters included those who "thought that the family planning program would be a cost-effective approach to cutting down on the number of people receiving public assistance" (Dryfoos 1988, 283). Littlewood (1977, 6) writes that "support for organized birth control has been bolstered in the United States by elements of the very rich who fear that the capitalist system would be threatened by lower-class overpopulation." Shapiro (1985, 18) argues, similarly, that "Members of the capitalist class, acting mainly through philanthropic organizations, articulate a strategy of population planning consistent with the needs of capitalist society" and that a "major achievement of the capitalist class has been to incorporate population policy within the welfare state." Thus, Shapiro claims, capitalists like John D. Rockefeller III and Clarence Gamble attempted, through their foundations, to lower the birth rates of the poor.

More recently, reports that fertility rates of the poor and most racial and ethnic minority groups continue to be above those for middle- and upper-class whites have sparked concerns about a "birth dearth" (Wattenberg 1987). Wattenberg (1987) voices alarm over the possibility that middle- and upper-class whites may become outnumbered by lower-class and racial and ethnic minorities. Herrnstein and Murray (1994) warn that if those with lower IQ, who are significantly more likely to be poor and/or Black or Hispanic, continue to outproduce those with higher IQ, we will witness what some have dubbed the "dumbing down" of America. Some feminists have taken a strong stand against attempts to pressure poor women to have fewer children (see Hartmann 1995; Rodriguez-Trias 1984). Others have emphasized race- or class-based inequities in access to infertility treatments or abortion (e.g., Davis 1983; Nsiah-Jefferson and Hall 1989). But no group that we are aware of has actively championed the goal of equal distribution of both infertility and contraceptive benefits to all women regardless of socioeconomic class.

In this article, we explore state involvement in the distribution of reproductive benefits. Four main questions guide this study: First, does the allocation of reproductive health benefits in Illinois create a bifurcated, de factor fertility policy that encourages births to middle- and upper-class women while discouraging births to poor women? Second, is the Illinois case generalizable to the United States? Third, what is the state's role in shaping the configuration of reproductive benefits? Finally, what assumptions guide state interventions?

Data and Methods

Our data derive from a variety of both primary and secondary sources. Much of our evidence was collected through telephone interviews with government officials and political activists. To assess access to fertility-related insurance benefits in Illinois, we interviewed spokespersons at the state board of insurance and the state Medicaid office. We also conducted telephone interviews with benefits representatives at the 25 largest employers in Illinois. To learn how the middle class in Illinois gained increased access to infertility treatments and why Medicaid recipients were excluded, we interviewed state legislators, governmental aids, activists, and lobbyists from various groups for and against mandated infertility coverage. Then, to assess the generalizability of the Illinois case to the United States, we conducted a telephone survey of state departments of insurance to find out whether insurers were required to cover contraceptive and/or infertility treatments. For information pertaining to poor women's access to infertility treatments, we interviewed supervisors in several state Medicaid offices. Finally, we interviewed members of Planned Parenthood, the National Abortion Rights League, and other key organizations active in the area of women's health.

In addition to interviews, we examined transcripts of Illinois General Assembly debates, newsletters from several key organizations, and newspaper accounts of both expansions and retractions of access to infertility and contraceptive benefits. For our discussion of poor women's access to contraceptive treatments, we were able to draw on existing literature that documents governmental attempts to encourage poor women to lower their fertility.

We were not able to interview everyone we would have liked. First, we could not reach some of the representatives to the Illinois General Assembly who spoke out about legislation mandating Illinois insurers to cover infertility. Thus, we were forced to rely on press coverage and transcriptions of general assembly debate. Second, though we interviewed leaders at national women's organizations, we did not interview those in state offices; thus, some variation in these groups' agendas has gone undetected. Finally, in our telephone survey of Illinois' largest employers, some employers were unable or unwilling to answer our questions either because they operate nationally and do not have state-specific information or because they offer many different plans and could not calculate what proportion of employees had access to particular services. Thus, we were unable to determine with precision the extent to which workers in Illinois have insurance coverage of contraceptives, but we were able to estimate the extent of coverage nationally using data from the Alan Guttmacher Institute (1994).

Access to Reproductive Benefits in Illinois

In 1991 the Illinois General Assembly passed the Family Building Act mandating that employers provide health coverage that includes infertility diagnosis and treatments of all kinds. Those unable to conceive after one year of unprotected intercourse are eligible for IVF and its variations, gamete intrafallopian transfer (GIFT), and zygote intrafallopian transfer (ZIFT). Recipients are limited to four attempts or two successful births. Though the benefits are broad, the exceptions to the mandate are numerous; those in small or religious-affiliated firms, at corporations with self-funded insurance plans, or in multistate corporations are not covered by the mandate. According to RESOLVE activist Suzanne Bessette-Smith (interview 1993), about one-half of all workers in Illinois are affected by the mandate.

Illinois has no parallel mandate requiring employers to provide coverage of contraceptives. To assess the extent of voluntary employment-based contraceptive coverage in Illinois, we conducted telephone interviews with benefit representatives at the 25 largest employers. We found that most employment-based insurance policies cover permanent sterilization such as vasectomies or tubal ligations, but few cover nonpermanent forms of contraception. In fact, refusal to cover contraceptives is so entrenched that many employer-based policies refuse to cover birth control pills even when they are prescribed for noncontraceptive purposes (interview Baker 1994; interview Roth 1995). The largest employer—the state of Illinois—reported that the one-half million state and local employees and dependents on its health insurance plans receive no coverage of nonpermanent contraceptives (Illinois Department of Central Management Services 1994). Similarly, several of the other largest employers, including Caterpillar Tractors and Osco Jewel, provide no coverage of contraceptives for any employees. Sears, on the other hand, indicated that some, but not all, of its employees had options that included contraceptive coverage.

By contrast, poor women on Medicaid in Illinois have no access to infertility treatments and nearly comprehensive coverage of contraceptives. Though the state of Illinois now requires employers to provide insurance coverage of infertility, the state does not offer that same coverage to its 1.5 million Medicaid recipients (Illinois Bureau of Research and Analysis 1995). Thus, Medicaid recipients are not eligible for either drug or surgical infertility treatments.

Illinois began providing birth control to welfare recipients in the mid-1960s, even before a 1967 federal mandate required state Medicaid programs to provide "family planning" services (Littlewood 1977). In Illinois, Medicaid covers virtually all permanent and temporary forms of contraception including birth control pills, the IUD, diaphragms, Depo-Provera, Norplant, tubal ligations, and vasectomies. The glaring exception is abortion; the Hyde Amendment, passed in 1977, bars states from using federal funds to pay for abortion unless pregnancy threatens the life of the pregnant woman. The Clinton administration now allows federal funds to be used for abortions in the case of rape or incest as well (Women's Research Education Institute 1994). Thus, low-income women on Medicaid have full coverage of contraceptives other than abortion but no coverage of infertility treatments.

Women working in part-time or low-wage work are unlikely to have employment-based health insurance or to be eligible for Medicaid. Nearly 77 percent of full-time workers, compared with only 16 percent of part-time workers, have employer-based health insurance (Perman and Stevens 1989). The working poor are only one-third as likely as other workers to receive employment-based health insurance and are five times more likely to be uninsured (Seccombe and Amey 1995). The link between wages and insurance is direct: 90 percent of adult workers with earnings over $10 per hour, compared with less than 50 percent of workers with earnings under $5 per hour, receive employer-based health insurance (Short, Monheit, and Beauregard 1989). Uninsured women must pay health care expenses out-of-pocket (Office of Technology Assessment 1988). In 1985, nearly 10 million women in their reproductive years did not have health insurance, and a much higher number did not have insurance coverage for contraceptives and/or infertility treatments (Gold and Richards 1994). To assure that poor, uninsured women receive contraceptives at reduced cost, the federal government provides funding to reproductive health clinics through block grant programs, such as Title X of the Public Health Services Act. In 1992, the federal government provided over $110 million in Title X funds for family planning services (Daley and Gold 1993).

As a result, many low-income women obtain contraceptives from public clinics or through organizations, such as Planned Parenthood, that offer contraceptives on a sliding income-based scale. At Planned Parenthood of Central Illinois, birth control pills may cost as little as $72 per year. Norplant implants, which cost about $600 elsewhere, may cost only $375 for low-income women. Even at reduced prices, these amounts are difficult to sustain on a minimum-wage budget. While federally subsidized low-cost contraceptives are widely available at public clinics, few if any of these clinics subsidize infertility treatments. Planned Parenthood in Illinois does not subsidize drug or surgical infertility treatments (interview Knaub 1995). Thus, low-income women have access to infertility treatments only if they are able to pay for them out-of-pocket.

Access to Reproductive Services in the United States

To assess the generalizability of the patterns found in Illinois, we conducted telephone interviews with officials in state department of insurance offices in all 50 states and the District of Columbia. We also surveyed selected Medicaid offices to supplement and update a *Newsweek* survey documenting Medicaid coverage of infertility treatments (Beck et al. 1994). Table 1 summarizes the results of these surveys. Columns 2 and 3 show that within the past decade, 10 of the largest and most populous states have passed insurance mandates affecting coverage of infertility treatments. Exemptions to the mandates vary significantly in different states. Nonetheless, seven states—Maryland, Hawaii, Massachusetts, Arkansas, Rhode Island, New York, and Illinois—mandate employers to provide coverage of infertility diagnosis and treatment to all employees. Three states—California, Connecticut, and Texas— have mandates requiring employers to *offer* coverage of infertility to employees, which means infertility diagnosis and treatment are not routinely covered but may be requested by employees. In these three states, employees may be required to pay higher premiums if they choose to have coverage. Mandates in the first seven states spread the costs of infertility treatments across all covered families, while mandates in the latter three states spread the costs across only those families with infertility concerns. Even in states without mandated infertility coverage, however, some companies are voluntarily offering infertility benefits to their employees. A 1991 survey showed that 38 percent of 2,409 employers offered at least some infertility coverage (Shellenbarger 1992).

Column 4 shows that there are no state mandates for employer-covered contraceptives. Even Hawaii, which has universal employer-based health insurance, has no such mandate. Lack of mandated coverage means that employers are not required to include contraceptives in health policies offered to employees. As a result, a recent nationwide survey of the 100 largest insurance companies found that only 15 percent of fee-for-service plans and 40 percent of HMOs cover all five of the most effective nonpermanent contraceptive methods—Norplant, Depo-Provera, oral contraceptives, the IUD, and the diaphragm (AGI 1994).[2] Coverage of contraceptives is so spotty that women pay at least 56 percent of the costs out-of-pocket (Women's Research and Education Institute 1994).

At one time, several Medicaid programs covered low-tech infertility treatments, as shown in column 5, but none covered high-tech infertility treatments such as IVF. Prior to 1992, at least 17 states provided Medicaid coverage for reversal of tubal ligation and vasectomy and/or drug treatments for infertility (Beck et al. 1994). Following public outcry about infertility expenditures for the poor, all states eliminated drug treatments, and only Maryland still covers reversals. Several states, including Minnesota, preclude coverage of infertility treatments but will, for example, cover surgery of endometriosis because it is a health problem in itself (interview Lindy 1996).

Though state governments are retracting coverage of infertility to poor women, funding for contraceptives remains strong. Column 6 shows that in compliance with federal mandates, every state Medicaid program provides coverage of contraceptives. In 1992, Medicaid programs spent over $300 million on nonpermanent contraceptive services. Public expenditure on contraceptives nationwide (including Medicaid and other programs) totaled nearly $650 million. In addition, state Medicaid programs spent over $125 million

TABLE 1 *State Policies Governing Private and Public Health Insurance Coverage of Contraceptives, Abortion, and Infertility Treatments*

State	Mandates Infertility Coverage	Mandates to Offer Infertility Coverage	Mandates Contraceptive Coverage	Medicaid Covers Some Infertility	Medicaid Covers Contraceptives	State Funds Medically Necessary Abortions
Alabama	No	No	No	Revoked	Yes	No
Alaska	No	No	No	No	Yes	Yes
Arizona	No	No	No	No	Yes	No
Arkansas	Yes	No	No	No	Yes	No
California	No	Yes	No	No	Yes	Yes
Colorado	No	No	No	No	Yes	No
Connecticut	No	Yes	No	Revoked	Yes	Yes
Delaware	No	No	No	No	Yes	No
District of Columbia	No	No	No	No	Yes	No
Florida	No	No	No	No	Yes	No
Georgia	No	No	No	No	Yes	No
Hawaii	Yes	No	No	Revoked	Yes	Yes
Idaho	No	No	No	No	Yes	No
Illinois	Yes	No	No	No	Yes	No
Indiana	No	No	No	No	Yes	No
Iowa	No	No	No	Revoked	Yes	No
Kansas	No	No	No	No	Yes	No
Kentucky	No	No	No	No	Yes	No
Louisiana	No	No	No	Revoked	Yes	No
Maine	No	No	No	Revoked	Yes	No
Maryland	Yes	No	No	Yes[a]	Yes	Yes
Massachusetts	Yes	No	No	Revoked	Yes	Yes
Michigan	No	No	No	No	Yes	No
Minnesota	No	No	No	Revoked	Yes	No
Mississippi	No	No	No	No	Yes	No
Missouri	No	No	No	No	Yes	No
Montana	No	No	No	Revoked	Yes	No
Nebraska	No	No	No	No	Yes	No
Nevada	No	No	No	No	Yes	No
New Hampshire	No	No	No	Revoked	Yes	No
New Jersey	No	No	No	No	Yes	Yes
New Mexico	No	No	No	Revoked	Yes	No
New York	Yes	No	No	Revoked	Yes	Yes
North Carolina	No	No	No	No	Yes	Yes
North Dakota	No	No	No	No	Yes	No
Ohio	No	No	No	No	Yes	No
Oklahoma	No	No	No	No	Yes	No
Oregon	No	No	No	Revoked	Yes	Yes
Pennsylvania	No	No	No	Revoked	Yes	No
Rhode Island	Yes	No	No	No	Yes	No
South Carolina	No	No	No	No	Yes	No
South Dakota	No	No	No	No	Yes	No
Tennessee	No	No	No	No	Yes	No
Texas	No	Yes	No	No	Yes	No
Utah	No	No	No	No	Yes	No
Vermont	No	No	No	No	Yes	Yes
Virginia	No	No	No	No	Yes	No
Washington	No	No	No	No	Yes	Yes
West Virginia	No	No	No	Revoked	Yes	Yes
Wisconsin	No	No	No	Revoked	Yes	No
Wyoming	No	No	No	No	Yes	No

[a]Covers only sterilization reversal.

Sources: O'Rourke (1992; infertility mandates); telephone survey of state departments of insurance, January 1994 (mandates for contraceptives); Beck et al. (1994); telephone survey of selected state Medicaid offices, February 1996 (Medicaid); Daley and Gold (1993; state funding for abortion).

on sterilizations (Daley and Gold 1993). Finally, though the Hyde Amendment prohibits using federal funds for abortions, some states provide abortion funding to poor women. Those states are shown in column 7. Total state expenditure for abortion totaled $80 million in 1992 (Daley and Gold 1993).

Explaining Class Differences in Access to Reproductive Benefits

The patterns represented in Illinois are pervasive throughout the United States and suggest an increasingly dualistic natalist policy. Though never formally declared as a fertility policy, state interventions have shaped the distribution of reproductive benefits in a way that encourages births among working- and middle-class women and discourages births among the poor. In this section we ask, why? We explore the set of circumstances that has led to an unequal distribution of benefits. We show how, in some states, the middle class gained access to infertility treatments while the poor did not. We then discuss why no states mandate insurers to cover contraceptives when Medicaid is mandated to provide thorough coverage.

RESOLVE and Mandated Infertility Coverage

How did working women in the state of Illinois come to have state-mandated coverage of a full array of infertility treatments? As is often the case, the expansion of the welfare state occurs in conjunction with demands by social movements (Bock and Thane 1991). RESOLVE is a national grassroots organization of 25,000 women and men, most of whom suffer from infertility problems. Since the late 1980s, they have introduced and lobbied successfully for legislation requiring health insurance coverage of infertility treatments in 10 states. In 1991, a very small group of professional, well-educated women and men stunned Illinois legislators, policymakers, and employers alike when they pushed through a mandate requiring employer coverage of infertility treatments despite opposition from every possible camp.

The passage of the 1991 Illinois Family Building Act was particularly surprising given the abundance of visible opposition and the dearth of visible support. The mood of the Illinois General Assembly had been one of retrenchment and was decidedly antimandate. In fact, 1990 legislation technically forbid the Illinois General Assembly from passing any more insurance mandates. Businesses had long opposed mandates because they were not unilaterally binding and interfered with fair competition between firms, insurers felt mandates were prompting more employers to become self-insured and thus exempt from mandates, and multistate insurers and employers feared that different state mandates would complicate their operations. Rather amazingly, sponsors of the Family Building Act circumvented these oppositions by claiming infertility coverage was not a new mandate but an extension of pregnancy-related benefits already on the books (interview Bush-Joseph 1994b).

The most powerful opponents of the infertility mandate, including the Illinois Chamber of Commerce, the Illinois Manufacturing Association, and the insurance industry, argued that coverage of infertility was prohibitively expensive (Dellios 1991; Narario 1989). Each attempt at IVF costs approximately $8,000–$10,000 (Boston Women's Health Book Collective 1992). Indeed, some RESOLVE members take out second mortgages on their homes to cover their medical bills. RESOLVE countered cost arguments by demonstrating that 85 to 90 percent of those suffering from infertility are treated without resort to IVF (Office of Technology Assessment 1988). Moreover, they showed how negligible costs are when spread across fertile and infertile couples alike: The price tag for coverage under the Massachusetts mandate was just $1.70 per family per month (Turnbill 1990). Similarly, in Delaware, the cost was just $0.60 per family per month, and in Maryland, under $1.00 (Office of Technology Assessment 1988; Picciotto 1990). Finally, RESOLVE argued, the costs of pregnancies have always been spread across fertile and infertile families; it was time to spread the costs of infertility across both groups as well. In a 1993 interview, RESOLVE president Suzanne Bessette-Smith reasoned: "I have paid for a lot of pregnancies and deliveries, paying a premium all these years. [Now] they can contribute to my pregnancy."

In explaining the insurance industry's failure to defeat the RESOLVE mandate, lobbyist Donald Pebworth (interview 1995) suggested the insurance representatives were mistaken to focus so myopically on the costs of infertility treatments. Instead, he told us, they should have focused on the need to adopt children. But the insurance industry would have made little progress on that argument either. Adoption is an increasingly problematic response to childlessness. The number of available babies has decreased due to improved contraceptive practices, increasing abortion rates, growing desire among teen mothers to keep their children, and emerging criticisms about the politics of interracial or international adoptions (McCormack 1991). As a result, potential parents often have to wait years to adopt a healthy baby, and the cost of adoption is now between $6,000 and $25,000 (Boston Women's Health Book Collective 1992).

The Catholic Church and anti-abortion activists opposed the infertility treatments because they go beyond "natural" family planning, and even more so because IVF occasionally involves aborting one or more of the implanted embryos. In response, RESOLVE rewrote the bill to exempt specific procedures such as IVF for employers who were opposed due to religious beliefs (interview Bush-Joseph 1994b). Moreover, RESOLVE and its supporters in the general assembly presented the Family Building Act as a "pro-life" bill. The result was a split among pro-life legislators (interview Cullerton 1995; interview Currie 1995; Illinois House of Representatives 1991; Illinois Senate 1991). Several of the traditionally pro-life legislators eventually became supporters of the mandate because it would potentially help people create families (interview Cullerton 1995). Representative Lee Preston argued, "If, indeed, you consider yourselves, in any remote way, a pro-life legislator . . . if indeed you're for life, you have to indeed be for this bill" (June 28 transcript, Illinois House of Representatives 1991).

Opposition was so stiff that the bill was actually defeated several times in committee and on the floor of both houses. But RESOLVE members became astute lobbyists. They learned to shepherd supportive voters into chambers—and to detain opposing voters outside of chambers—in time for key votes. They convinced the author of an unrelated bill (who happened to oppose the RESOLVE mandate) to allow a supportive senator to sponsor the bill so that the RESOLVE mandate could be attached. And they reluctantly chipped away at the mandate's coverage, deleting portions that applied to individual rather than group coverage plans, excluding firms with fewer than 25 employees, and, to secure the vote of the senator representing Caterpillar's home office, exempting national firms that issue insurance policies in other states (interview Bessette-Smith 1993). Ultimately, the Family Building Act passed only because RESOLVE managed to gain some Republican support. In a 1995 interview, Illinois Senator John Cullerton remarked:

> RESOLVE members were basically suburban women, primarily white, from mostly Republican districts—not usually the type of people wanting mandates . . . [They] were good lobbyists. They had personal stories and they were very articulate. The bottom line is that they got the Republican vote.

Though Republicans generally oppose mandates, some were attracted to the "pro-family" aspect of the bill, and several felt the need to listen to their constituents; the most active RESOLVE members came from predominantly Republican districts near Chicago (interview Cullerton 1995; interview Currie 1995).

RESOLVE took full advantage of the idea that the bill was "pro-family" and relied on less traditional forms of lobbying. In an interview, the legal counsel for the insurance industry told us: "this bill was emotional. It was about motherhood . . . Happy mothers with smiling babies are a powerful statement" (interview Pebworth 1995). RESOLVE presented the "products" of infertility treatment, their children, to legislators in hopes of appealing to the legislators' emotions. They also sent Mother's and Father's Day cards to legislators and presented testimony: first, from a couple who had taken out a second mortgage on their house to pay for IVF and had twin daughters as a result, and second, from a couple who had been unable to conceive and could not afford IVF (interview Bush-Joseph 1994b;

Jacobs 1991). The tactics apparently worked. In a pivotal vote, Representative Manny Hoffman, an insurance salesman, changed his vote to "yes" after looking up at a crowd of women and children in the balcony. "'It just emotionally got to me,' Hoffman explained later. 'I wanted to give everyone else the same chance I had to raise a family'" (quoted in Dellios 1991, 1).

After the Family Building Act passed the general assembly, RESOLVE worked to persuade Governor Jim Edgar to sign the bill. Never permitted to meet directly with Edgar, they lobbied Edgar's wife, his chief of staff, and his chief staff assistant. At the signing, Edgar praised RESOLVE's tenacity, calling them one of the most effective grassroots movements he had ever encountered (interview Bessette-Smith 1993). Many sources suggest that Governor Edgar eventually signed the bill into law only because he needed a "pro-family" bill to support, having vetoed a family leave bill not long before (Johnson 1991). Others suggest that the bill's passage was greatly aided by the fact that it did not require governmental expenditures (interview Currie 1995). In the end, however, the governor seems to have been genuinely persuaded that the Family Building Act was worthwhile. In fact, when implementation rules were negotiated by the State Department of Insurance, business and insurance lobbyists tried to impose a 50 percent copayment on infertility benefits, and Edgar overruled (interview Pebworth 1995). When health care providers threatened to confine infertility treatments to specific geographical areas, the state inserted coverage of travel expenses under certain conditions. Most surprisingly, Edgar directed the state of Illinois to provide coverage of infertility treatments for its employees, even though the state was self-insured and therefore not required to follow the mandate (interview Chorle 1995; interview Sullivan 1995).

While opposition to the RESOLVE mandate was vocal, support was nearly invisible. Though they stood to profit under the mandate, neither the American Medical Association (AMA) nor the American Fertility Society (AFS) actively supported the legislation. Behind the scenes, however, they provided statistics and, when needed, "expert" definitions. RESOLVE lobbyist Vicky Bush-Joseph (1994a) suggested in an interview that if the AMA or AFS had supported the legislation publicly, it might have appeared self-serving. Similarly, RESOLVE received little help from various factions of the women's movement. Planned Parenthood and the National Organization for Women provided behind-the-scenes advice, gleaned from years of experience, on navigating the legislative system and successful lobbying tactics (interview Bush-Joseph 1994b). But given their ongoing attention to sustaining women's rights to control their reproductivity, feminist organizations were notably silent. Their inaction may be linked to a complex set of objections. First, the mandate applied only to insured, working women; therefore, constituents of family planning organizations who are unemployed or uninsured reaped no benefits from the mandate. Second, some feminists oppose the proliferation of high-tech infertility treatments because it deflects resources from other more basic forms of reproductive services that are still unavailable to millions of poor American women (Petchesky 1981; Stone 1991). In addition, many feminists oppose any actions that further define women as mothers or tie them to domestic labor (Bergmann 1982; Brenner 1987). Finally, nearly all argue that the biomedicalization of high-tech infertility treatments reflects male-dominated efforts to control women's bodies and lifestyles (Arditti, Klein, and Minden 1989; Corea 1985; Martin 1987; for review, see Petchesky 1981; Stone 1991). The process of being diagnosed and treated for infertility is enormously time-consuming, painful, and potentially dangerous for women, and success rates are only 5–10 percent (Arditti, Klein, and Minden 1989). But women are social actors responding to specific sets of opportunities and constraints (Bock and Thane 1991; Stone 1991). Thus, an alternative explanation is that many infertile women see themselves less as victims of biomedical experimentation and more as individuals forced to choose between a set of disagreeable alternatives.

Throughout the past two decades, critics have charged that high-tech, hi-cost treatments were available only to middle- and upper-class married couples and inaccessible to the poor, racial and ethnic minorities, gays and lesbians, and unmarried women. Blacks have been particularly unlikely to gain access to infertility treatments given lower incomes, higher rates of uninsurance, and lower marital rates; this is especially troubling given the higher rates of infertility among African Americans (Katz Rothman 1989; McCormack 1991; Nsiah-Jefferson and Hall 1989; Office of Technology Assessment 1988; Saluter 1992; Stone 1991). Because of RESOLVE's efforts, the state has partially democratized access to infertility treatments. To the extent that they are insured through jobs covered by the mandate, these groups now have improved access to infertility treatments.[3]

But infertility treatments were not made available to poor women, specifically women on Medicaid. In Illinois, there appears to have been no attempt to extend the Family Building Act to poor women. When Governor Edgar extended coverage to state employees, he did not suggest extending coverage to state Medicaid recipients. Years earlier, the Illinois General Assembly passed separate legislation to assure that mandated coverage of mammograms included Medicaid recipients (interview Currie 1995). But legislators avoided the issue of coverage for Medicaid recipients during the floor debate on the Family Building Act (Illinois House of Representatives 1991; Illinois Senate 1991). Only Senator Emile Jones objected to the selective application of the Family Building Act: "Many persons—the working poor ... have no insurance at all.... I think these issues should be dealt with prior to giving just a select small group [this] coverage" (June 19 transcript, Illinois Senate 1991). Even he did not openly call for the inclusion of Medicaid recipients. As a result, Illinois provides no coverage of infertility treatments to Medicaid recipients.

Indeed, behind the Illinois Family Building Act lurked the expectation that employer-based insurance coverage of infertility would encourage births only among certain types of families. Though several senators went on record as wanting to help people build families (e.g., Senator John Cullerton and Senator John D'Arco; see June 19 transcript, Illinois Senate 1991), their comments referred only to those covered by the mandate. Because Medicaid recipients would not be included in the mandate and because the working poor are often uninsured, the implicit message is that the state should encourage the birth of only working- and middle-class babies. In a letter to the general assembly, explaining why he signed the Family Building Act, Governor Edgar explained, "Many couples today experience the anguish of infertility. I fully sympathize with their desire to raise a family and support efforts such as HB 1470 that will provide them with joy and fulfillment and lead to the birth of *productive citizens of our state*" (Edgar 1991; emphasis added).

Some states have, until recently, provided some infertility services for Medicaid recipients. In states that previously provided drug treatments for infertile Medicaid recipients, Beck et al. (1994) point out, states did not actively decide to provide coverage so much as fail to delete these infertility medications from the list of covered drugs. After learning that Massachusetts had spent $46,000 on two infertility drugs, Republican Governor William Weld banned those drugs from the Medicaid coverage list (Wong 1994). In response to the negative publicity and public outcries, other states quickly repealed Medicaid coverage of infertility treatments as well. Alabama Medicaid withdrew its infertility coverage, which included only counseling and tests. Dee Lockridge (interview 1995), head of the family planning division at Alabama Medicaid, explained that the decision to cancel benefits was made because "it seems odd that we would want to help a woman get pregnant when she can't even pay her own medical bills.... It was not a good image for Medicaid." In 1994, New Jersey's legislature voted to preclude Medicaid coverage of infertility; in doing so, lawmakers in that state defined "family planning services" as only those that "prevent or delay pregnancy" (New Jersey Department of Human Services 1994).

Why was the retrenchment of Medicaid infertility coverage so immediate and pervasive? In part because many of the children born to women on Medicaid become eligible for Aid to Families with Dependent Children (AFDC). Opponents argue that Medicaid coverage of infertility encourages reliance on the state and represents an inefficient use of tax dollars. Democratic Senator Edward Kennedy states: "Our goal in using tax dollars wisely is to reduce welfare dependency, not create more of it" (quoted in Phillips 1994, 27). Beck et al. (1994, 30) echo popular sentiment: "At a time when Congress, President Clinton and state legislatures are pushing welfare reform, it hardly seems prudent to add more children to the rolls." Similarly, a Connecticut Department of Social Services official suggests that it is inconsistent to simultaneously cover infertility drugs and try to move welfare recipients toward self-reliance (Beck et al. 1994). Robert Rector of the Heritage Foundation argues that it is "morally irresponsible for someone to have a child and expect someone else to raise them" (quoted in Beck et al. 1994, 30). Even liberal columnist Ellen Goodman (1994, A4) wrote:

> The same society concerned about the number of children born in and to poverty, the same government trying to get mothers and children off AFDC, paid to help poor parents conceive more children.... In the scheme of things there wasn't much money involved.... But it was the principle of the thing.

By repealing Medicaid coverage of infertility treatments for the poor, states have vocally asserted their positions on who should have children. But the retrenchment of Medicaid infertility treatments did not go unopposed. Proponents of Medicaid coverage of infertility do not promote higher fertility among the poor, but they do promote equal access to health care benefits regardless of social class. Robert Restucia of Health Care for All told *The Boston Globe*, "If this is a mandated benefit that applies to all people, then it should apply to all people across the board. We should not stereotype Medicaid patients" (quoted in Wong 1994, 23). While RESOLVE never actively included Medicaid recipients in their state mandates, a spokesperson told *Newsweek* that secondary infertility can affect women on welfare and that their basic position is that infertility should be treated like any other medical problem. A spokesperson for the AFS argues that it is unethical to define access to services on the basis of the recipient's socioeconomic class (Beck et al. 1994). In one of the boldest statements so far, Jencks and Edin (1995) ask whether poor women have even the right—much less the means—to bear children. Much of the public discourse echoes this key ethical question: Are women who choose to bear children entitled to do so even if they cannot afford to raise them? Jencks and Edin suggest that we would have to give poor women powerful incentives to delay or forgo having children. To expect women to wait until they can afford children without any assistance from the state—or from husbands who could potentially leave them—would require jobs paying $20,000 a year (Jencks and Edin 1995). And yet, the longer women wait to procreate, the higher the chances for infertility problems (Gold and Richards 1994). Despite arguments from this camp, infertility treatments for Medicaid recipients have now been almost fully retrenched.

Contraceptive Coverage

Several states now mandate insurers to cover infertility, but no state requires insurers to cover contraceptives. Neither RESOLVE nor any of the women's movement groups, workers, or labor unions appear to have ever made contraceptive coverage for working- and middle-class women a priority. Members of RESOLVE indicated that while they would like to see women have access to contraceptives, RESOLVE activists never considered adding such coverage to their mandates. Though Planned Parenthood, the National Organization for Women, the American Civil Liberties Union Reproductive Freedom Project, feminist scholars, and many others have worked to assure access to contraceptives among poor women, activist groups have not devoted themselves to extending contraceptive coverage to working- and middle-class women. Instead, they have mainly dedicated their energies to access to abortion (Rodriguez-Trias 1995). Legal, moral, and logistical concerns surrounding abortion have eclipsed the day-to-day concerns of most women about preventing pregnancies. Sometimes begrudgingly, women's groups have battled to guarantee women the right, and the means, to abortion at the expense of a fuller slate of contraceptive options. For example, the country's leading pro-choice organization, the National Abortion and Reproductive Rights Action League (NARAL), expanded its mission in 1993 to include not only abortion but all reproductive issues, including contraception. However, a spokesperson for NARAL explained that "because of the recent Congressional elections we're having to refocus our efforts toward abortion" (interview McKee 1995).

In the absence of mandates requiring contraceptive coverage, health benefits are negotiated between employers and employees, often by unions. Historical union neglect of benefits for women has been well documented (see May 1985). The state of Illinois, which provides no coverage for nonpermanent contraceptives, negotiates benefits with the American Federation of State, County and Municipal Employees (AFSCME). During the last contract negotiation, AFSCME was successful in broadening employee health insurance coverage of dental and vision treatments as well as routine physicals (interview Scheff 1995). Yet, even though at least one-half of union members are women, the union did not ask for coverage of contraceptives (interview Scheff 1995). Health benefit negotiators for both AFSCME and the state claim that "it's just not the norm to cover birth control" (interview Sullivan 1995). Repeatedly, we were told that the type of insurance employers provide varies according to how benefits are negotiated with unions, the number of employees involved, and the employer's preferences. Usually, employers seek to minimize costs, and that often means forfeiting nonmandated "extras" such as coverage of contraceptives (interview Roth 1995; interview Sullivan 1995; interview Zehn 1995). Once the state intervenes to mandate coverage, however, benefits become part of a standard or minimum package and cease to be regarded as "additional" costs to employers.

Union and insurance representatives alike report that if employees demanded contraceptive coverage, it would be costly and highly controversial (interview Pebworth 1995; interview Scheff 1995; interview Sullivan 1995). According to Joanna Sullivan (interview 1995), health benefits negotiator for the state of Illinois, "Politically, morally, and religiously it is a hot issue. . . . If we covered abortion we would be open to intense criticism, and contraceptives are similar." A health benefits negotiator for AFSCME voiced similar sentiments regarding coverage of contraceptives:

> It's political. . . . Those who don't have to worry about politics can cover it, but in the public realm we have to be careful in negotiations. In a Republican administration, they're always watching their right flank. They get wind of this stuff and they go nuts. (Interview Scheff 1995)

By contrast, poor women have broad coverage of contraceptives because the U.S. Congress passed legislation guaranteeing it. Throughout the 1960s, several diverse groups lobbied Congress to support family planning for poor women. Dryfoos (1988) describes the "odd lot" of activists who came together to work for increased access to contraceptives for the poor. Women's groups hoped that increased access to contraceptives, especially through specialized family planning clinics, would empower women. Environmentalists felt that lowering population growth was important in the face of finite natural resources; thus, these groups supported the legislation in hopes that increased access to contraceptives for the poor would help move the country toward zero population growth. Advocates for the poor felt unintended pregnancy was an important problem for low-income families. Finally, expanding access to contraceptives for the poor received support from those who believed that family planning would decrease welfare dependency by reducing the number of people receiving public assistance (Dryfoos 1988). The Catholic Church opposed efforts to increase access to birth control but, ultimately, was no match for the coalition favoring family planning for the poor (Dryfoos 1988; Littlewood 1977). By 1967 Congress responded, altering the provisions of the Social Security Act. Federal and state funds for reproductive, maternal, and child care were expanded. In addition, states were required to offer family planning services to present, past, and potential AFDC recipients in an effort to reduce illegitimate births and corral welfare expenses (Littlewood 1977). Later, in 1970, Congress expanded access to contraceptive treatments for low-income women by passing Title X of the Public Health Services Act, which authorized funding for a national network of family planning clinics (Dryfoos 1988).

Conclusions

The distribution of reproductive benefits has been widely overlooked in the welfare state literature, both because job-related health benefits are regarded as private rather than public and because there is no explicit U.S. fertility policy. We have argued, however, that even though benefits are gained through a mix of public and private sources, and even though the fertility policy is implicit rather than explicit, reproductive benefits are part of the rubric of the U.S. welfare state.

This study contributes to a broader understanding of how the distribution of welfare state benefits creates class cleavages. Most studies conclude that benefits available to those making claims as workers are more comprehensive while benefits available to the poor making claims as dependents of the state are more meager (Acker 1988; Esping-Andersen 1989; Harrington Meyer 1990, 1996; Nelson 1989; Orloff 1993; Quadagno 1990; Ruggie 1984). For reproductive benefits, however, this is not the case. It is not that working- and middle-class women have more comprehensive benefits or that poor women have more comprehensive benefits. Rather, the benefits are distributed differentially on the basis of class. Though poor women on Medicaid have mandated coverage of contraceptives, working- and middle-class women do not. Though many working- and middle-class women have mandated coverage of infertility, the few infertility benefits that were available to Medicaid recipients have now been repealed. The effect is a de facto fertility policy that discourages births among poor women and encourages births among working- and middle-class women.

Though the practice of discouraging births among the poor is not new, the spread of state infertility mandates represents a new effort by the state to encourage births among working- and middle-class women. We believe that class bias concerning who should—and who should not—have babies is so firmly entrenched that the question of providing equitable access to all reproductive benefits seldom arises. It is difficult to support infertility mandates in the context of a medical system that leaves 30 million uninsured and millions without basic reproductive services such as prenatal and postnatal care or childhood immunizations (EBRI 1994). And few would favor a pro-natalist policy for impoverished welfare recipients. But we think it untenable to argue that location within the economic strata should define access to reproductive services. Our position is that all women, regardless of social or economic class, should have the right *and the means* to control their reproductivity; this requires access to a full range of reproductive benefits regardless of employment, marital, or socioeconomic status.

NOTES

1. Family planning and infertility issues are important for men, but women still have primary responsibility for both. Most contraceptives are designed for use by women. Similarly, though infertility may be caused by male or female subfecundity or sterility, women undergo by far most of the tests and treatments (Boston Women's Health Book Collective 1992).
2. Of private insurance plans for companies with 100 or more employees, 33 percent cover oral contraceptives, 26 percent cover IUD insertion, 21 percent cover diaphragm fitting, and 24 percent cover Norplant. HMOs provide much more substantial coverage of these services. Over 80 percent cover oral contraceptives, IUD insertion, and diaphragm fitting. Forty-four percent cover Norplant (AGI 1994).
3. In fact, the degree of democratization has proven troublesome for many health providers. One HMO representative reported that claims had been made by women who were postmenopausal, women and men who had unsuccessful sterilization reversals, and women without a heterosexual partner (interview Baker 1994). Because these cases were not specifically excluded under the mandate, local providers felt compelled to provide coverage but did so involuntarily.

REFERENCES

Acker, Joan. 1988. Class, gender, and the relations of distributions. *Signs: Journal of Women in Culture and Society* 13: 473–97.

Alan Guttmacher Institute (AGI). 1994. *Uneven and unequal: Insurance coverage and reproductive health services.* New York: Alan Guttmacher Institute.

Arditti, Rita, Renate Duelli Klein, and Shelley Minden. 1989. *Test tube women: What future for motherhood?* London: Pandora.

Beck, Melinda, Debra Rosenberg, Pat Winger, and Mary Hager. 1994. The infertility trap. *Newsweek.* 4 April, 30–1.

Bergmann, Barbara. 1982. The housewife and Social Security reform: A feminist perspective. In *A challenge to Social Security: The changing roles of women and men in American society,* edited by Richard Burkhauser and Karen Holden. New York: Academic Press.

Bock, Gisela, and Pat Thane. 1991. *Maternity and gender politics: Women and the rise of the European welfare states, 1880s–1950s.* New York: Routledge.

Boston Women's Health Book Collective. 1992. *Our bodies, ourselves.* New York: Touchstone.

Brenner, Johanna. 1987. Feminist political discourses: Radical vs. liberal approaches to the feminization of poverty and comparable worth. *Gender & Society* 1: 447–65.

Corea, Gena. 1985. *The mother machine.* New York: Harper & Row.

Daley, Daniel, and Rachel Gold. 1993. Public funding for contraceptive, sterilization and abortion services. *Family Planning Perspectives* 25: 244–51.

Davis, Angela. 1983. *Women, race and class.* New York: Vintage Books.

Dellios, Hugh. 1991. Infertility bill gives Edgar predicament. *Chicago Tribune,* 9 September.

Dixon-Mueller, Ruth. 1993. *Population policy & women's rights: Transforming reproductive choice.* Westport, CT: Praeger.

Dryfoos, Joy. 1988. Family planning clinics: A story of growth and conflict. *Family Planning Perspectives* 20: 282–7.

Edgar, Jim. 1991. Letter to members of the House of Representatives, Illinois General Assembly, 23 September.

Employment Benefit Research Institute (EBRI). 1994. Table 24: Persons aged 18–64 with selected sources of health insurance. *EBRI Issue Brief* 50: 50–1.

Esping-Andersen, Gosta. 1989. The three political economies of the welfare state. *Canadian Review of Sociology and Anthropology* 26: 10–36.

Ginsburg, Norman. 1992. *Divisions of welfare: A critical introduction to comparative social policy.* Newbury Park, CA: Sage.

Gold, Rachel Benson, and Cory L. Richards. 1994. Securing American women's reproductive health. In *The American woman, 1994–1995,* edited by Cynthia Costello and Anne J. Stone. New York: W. W. Norton.

Goodman, Ellen. 1994. Fertility treatments a low priority. *Champaign/Urbana News Gazette,* 22 March.

Gordon, Linda. 1990. *Woman's body, woman's right.* New York: Penguin Books.

Harrington Meyer, Madonna. 1990. Family status and poverty among older women: The gendered distribution of retirement income in the United States. *Social Problems* 37: 1101–13.

———. 1996. Making claims as workers and/or wives: The distribution of Social Security benefits. *American Sociological Review* 61: 449–65.

Hartmann, Betsy. 1995. *Reproductive rights and wrongs: The global politics of population control.* Boston: South End.

Herrnstein, Richard, and Charles Murray. 1994. *The bell curve: Intelligence and class structure in American life.* New York: Free Press.

Illinois Bureau of Research and Analysis. 1995, January. *Monthly data card.* Springfield: Illinois Bureau of Research and Analysis.

Illinois Department of Central Management Service, Bureau of Benefits. 1994. Personal communication with L. King, 9 March.

Illinois House of Representatives. 1991. 87th General Assembly. Transcripts of debates. 8 May, 15 May, 28 June.

Illinois Senate. 1991. 87th General Assembly. Transcripts of debates. 19, 20, 25, 26 June.

Jacobs, Tamy. 1991. Testimony before the Illinois House Insurance Committee on H.B. 0133. 9 April.

Jencks, Christopher, and Kathryn Edin. 1995. Do poor women have a right to bear children? *The American Prospect* 20 (Winter): 43–52.

Johnson, Mark. 1991. Fighting over family issues. *Rockford Register Star,* 20 September.

Katz Rothman, Barbara. 1989. *Recreating motherhood: Ideology and technology in a patriachal society.* New York: W. W. Norton.

Littlewood, Thomas. 1977. *The politics of population control.* Notre Dame, IN: University of Notre Dame Press.

MacKinnon, Catharine. 1983. The male ideology of privacy: A feminist perspective on the right to abortion. *Radical America* 17 (4): 23–35.

———. 1989. *Toward a feminist theory of the state.* Cambridge, MA: Harvard University Press.

Martin, Emily. 1987. *The woman in the body: A cultural analysis of reproduction.* Boston: Beacon.

May, Martha. 1985. Bread before roses: American working men, labor unions and the family wage. In *Women, work and protest,* edited by Ruth Milkman. London: Routledge and Kegan Paul.

McCormack, Thelma. 1991. Public policies and reproductive technology: A feminist critique. *Research in the Sociology of Health Care* 9: 105–24.

Mosher, William D., and William F. Pratt. 1990a. Contraceptive use in the United States, 1973–88. *Advance data from vital and health statistics,* no. 192. Hyattsville, MD: National Center for Health Statistics.

———. 1990b. Fecundity and infertility in the United States, 1965–1988. *Advance data from vital and health statistics, no. 192.* Hyattsville, MD: National Center for Health Statistics.

Narario, Sonia, L. 1989. Infertility insurance gains backing. *Wall Street Journal,* 5 December.

Nathanson, Constance. 1991. *Dangerous passage: The social control of sexuality in women's adolescence.* Philadelphia: Temple University Press.

Nelson, Barbara. 1989. The gender, race and class origins of early welfare policy and the U.S. welfare state: A comparison of workman's compensation and mother's aid. In *Women, politics, and change,* edited by Louise Tilly and Patricia Gurin. New York: Russell Sage.

New Jersey Department of Human Services, Division of Medical Assistance and Health Services. 1994. *Newsletter* 4 (35): 1–4.

Nsiah-Jefferson, Laurie, and Elaine J. Hall. 1989. Reproductive technology: Perspectives and implications for low-income women and women of color. In *Healing technology,* edited by Kathryn Strother Ratcliff. Ann Arbor: University of Michigan Press.

Office of Technology Assessment. 1988. *Infertility: Medical and social choices.* Washington, DC: Office of Technology Assessment.

Orloff, Ann. 1993. Gender and the social rights of citizenship: The comparative analysis of gender relations and welfare states. *American Sociological Review* 58: 303–28.

O'Rourke, Melissa. 1992. The status of infertility treatments and insurance coverage: Some hopes and frustrations. *South Dakota Law Review* 37: 343–87.

Perman, Lauri, and Beth Stevens. 1989. Industrial segregation and the gender distribution of fringe benefits. *Gender & Society* 3: 388–404.

Petchesky, Rosalind Pollack. 1981. "Reproductive choice" in the United States. In *And the poor get children,* edited by Karen Michaelson. New York: Monthly Review.

Phillips, Frank. 1994. Senator echoes GOP rivals, governor. *The Boston Globe,* 16 March.

Picciotto, John A. Blue Cross and Blue Shield of Maryland. 1990. Letter to Daniel M. Clements, 24 April.

Quadagno, Jill. 1990. Race, class and gender in the U.S. welfare state: Nixon's failed Family Assistance Plan. *American Sociological Review* 55: 11–28.

Rodriguez-Trias, Helen. 1984. The women's health movement: Women take power. In *Reforming medicine: Lessons of the last quarter century,* edited by Victor Sidel and Ruth Sidel, 107–26. New York: Pantheon Books.

———. 1995. Foreword to *Reproductive rights and wrongs: The global politics of population control,* by Betsy Hartmann. Boston: South End.

Ruggie, Mary. 1984. *The state and working women.* Princeton, NJ: Princeton University Press.

Saluter, Arlene. 1992. Marital status and living arrangements: March 1992. *Current Population Reports.* Series P20–468. Washington, DC: U.S. Department of Commerce.

Seccombe, Karen, and Cheryl Amey. 1995. Playing by the rules and losing: Health insurance and the working poor. *Journal of Health and Social Behavior* 36: 168–81.

Shapiro, Thomas. 1985. *Population control politics: Women, sterilization and reproductive choice.* Philadelphia: Temple University Press.

Shellenbarger, Sue. 1992. Infertile employees seek firms' support. *Wall Street Journal,* 12 May.

Short, Pamela F., Alan Monheit, and Karen Beauregard. 1989. A profile of uninsured Americans. DHHS Publication No. (PHS) 89–3443: 1–18. Rockville, MD: National Center for Health Services Research and Health Care Technology Assessment.

Stone, Jennifer. 1991. Contextualizing biogenetic and reproductive technologies critical studies. *Mass Communication* 8: 309–32.

Turnbill, Nancy C. 1990. Letter to Susan Crockin, 31 December.

Wattenberg, Ben J. 1987. *The birth dearth.* New York: Pharos Books.

Women's Research and Education Institute. 1994. *Women's health insurance costs and experiences.* Washington, DC: Women's Research and Education Institute.

Wong, Doris Sue. 1994. State, in about-face, bars Medicaid from funding fertility treatments. *The Boston Globe,* 5 March.

INTERVIEWS

Baker, Carol, Christy Clinic, Champaign, IL. 1994. Telephone interview with L. King, 24 February.

Bessette-Smith, Suzanne, President of RESOLVE of Illinois, Chicago. 1993. Telephone interview with M. Harrington Meyer, 17 September.

Bush-Joseph, Vicky, RESOLVE lobbyist, Chicago, IL. 1994a. Telephone interview with L. King, 18 February.

——. 1994b. Personal interview with L. King and M. Harrington Meyer, 8 August.

Chorle, Erhard, Aid to Governor Jim Edgar, Springfield, IL. 1995. Telephone interview with L. King, 16 January.

Cullerton, John. Senator, State of Illinois, Springfield. 1995. Telephone interview with L. King, 11 January.

Currie, Barbara Flynn, Representative, State of Illinois, Springfield. 1995. Telephone interview with L. King, 10 January.

Knaub, Sara, Director of Communications and Marketing, Planned Parenthood of Chicago. 1995. Telephone interview with L. King, 21 June.

Lindy, Katy, Spokesperson, Minnesota Medicaid. 1996. Telephone interview with L. King, 5 February.

Lockridge, Dee, Head of Family Planning, Alabama Medicaid. 1995. Telephone interview with L. King, 28 June.

McKee, Caroline, Legal Services, National Abortion and Reproductive Rights Action League, Washington, DC. 1995. Telephone interview with L. King, 13 January.

Pebworth, Donald, Legal Counsel to Blue Cross Blue Shield, Chicago, IL. 1995. Telephone interview with L. King, 11 January.

Roth, Shelly, Health Alliance, Champaign, IL. 1995. Telephone interview with L. King, 10 January.

Scheff, Hank, Health Benefits Negotiator, American Federation of State, County and Municipal Employees, Champaign, IL. 1995. Telephone interview with L. King, 25 January.

Sullivan, Joanna, Assistant Bureau Manager, Illinois Department of Central Management Services, Springfield. 1995. Telephone interview with L. King, 23 January.

Zehn, Dennis, Illinois Federation of Teachers, Champaign. 1995. Telephone interview with L. King, 27 January.

CHAPTER 3 Questions for Writing, Reflection, and Debate

READING 10

From Baby Farms to Black-Market Babies: The Changing Market for Children • *Viviana A. Zelizer*

1. What changes in attitudes toward parenting are discussed in this article? Do attitudes differ for biological parenting and for social parenting by adoption? If so, how?
2. What attitude toward parents who release their children for adoption dominated in the first historic period covered in the reading? How did this attitude change over time? What motivates parents to release children for adoption? How have social factors affected these motivations at various points in history? What has the effect been on the supply of babies available for adoption?
3. What race and class inequities are involved in adoption?

READING 11

Stolen Children and International Adoptions • *Mary Ellen Fieweger*

4. What motivates parents in developing countries to release children for adoption? What ethical issues arise from this situation?
5. What race and class inequities are evident in international adoptions? How do these inequalities affect the rights of biological parents?
6. Should (and could) illegal adoptions be prevented? Why or why not? Who would benefit, and who would suffer?

READING 12

Designer Genes: The Baby Quest and the Reproductive Fix • *Elaine Tyler May*

7. Do you agree that contemporary society has experienced a "new pronatalism"? What evidence do you offer to support your position?
8. The author states, "Those who become pregnant while using birth control tend to blame the technology. But infertility patients who do not conceive often blame themselves." How does this personalization increase the emotional "costs" of infertility?
9. What are the financial "costs" associated with infertility? What are the "costs" in loss of privacy?

READING 13

Lesbian Mothers: "This Wonderful Decision" • *Ellen Lewin*

10. How do homosexual couples differ from heterosexual couples in the options open to them to become parents? What are the similarities? Are the "non-biological" routes to parenting more socially acceptable for heterosexual couples than for homosexual couples?
11. Many homosexual individuals take on a social parenting role when their partner brings to the relationship a biological child from a previous heterosexual relationship. This is similar to a heterosexual person adopting the role

of stepparent. What complications are involved in stepparenting that are not present in other routes to parenting?

12. What social institutions and norms shape a homosexual couple's decision to become parents?

READING 14

The Politics of Reproductive Benefits: U.S. Insurance Coverage of Contraceptive and Infertility Treatments • *Leslie King and Madonna Harrington Meyer*

13. This article notes that poor women have social support for obtaining contraceptives, but not for infertility treatments, whereas working- and middle-class women have coverage for infertility treatment but not for contraceptives. What are the implications of this finding? What does this tell us about the way class shapes decisions about biological versus social aspects of parenting?

4

Working Families

How can men and women balance their work and family roles?

Balancing work and family roles is not a new issue. Throughout history adults have faced a balancing act, combining family and work responsibilities. However, changes in work, family, and individual expectations have altered the context in which each successive generation faces this issue. Anna Quindlen's "Men at Work" provides a tongue-in-cheek contrast between the ways fathers of the 1950s and of the 1990s differ in their involvement in work and family life.

I have always envisioned this issue as one that couples need to negotiate and renegotiate over the course of their partnership. These negotiations are of course affected by a range of personal and societal factors. Anselm Strauss is a sociologist who offers ideas about the way people in work settings negotiate to reach their goals. I think many of his ideas provide a model for looking at the way couples negotiate work and family issues.

Strauss's negotiated order theory is built on the idea that in order to accomplish work tasks, colleagues need to establish a social order. Parties develop tacit agreements and arrangements that create the order that lets them get things done. It seems to me that families also need to negotiate a social order that helps them work together as a family. I am not suggesting that couples sit down and logically work out a five-year plan, but rather that living with a partner requires give and take, compromise and negotiation. Much of this is done in informal and sometimes nonverbal ways.

The ongoing nature of these negotiations is central to the idea of negotiated order. Just when you have determined the way you want to proceed, something changes, and this change necessitates rebalancing roles and responsibilities. Changes in people and in circumstances affect the life of the average family on a daily basis. So negotiation takes place regarding very mundane things, such as who gets to sleep in tomorrow morning, whose turn it is to cook, who takes the kids to soccer, as well as more major aspects of family life, such as will you take that promotion, will we have another child, will I go back to school?

Borrowing from negotiated order theory, there are two levels of context in which these negotiations take place. In work settings Strauss labeled these the negotiation context and the structural context. The negotiation context is the more immediate of the two. It includes aspects such as personal experience with negotiation, the power balance among individuals, what each person has at stake in the negotiation, the complexity of issues, and the options participants have to avoiding the negotiation. Examples of the negotiation context are found in Scott Coltrane's study as reported in "Sharing and Reluctance." Some elements of context discussed in this study are unique to the Chicano culture, but many issues, such as power and resources, are not culture bound.

The broader or structural context needs some adaptation in applying these concepts to families. In organizations, Strauss looked at the way the work setting shapes negotiation. In terms of families, we can draw parallels with the way the larger aspects of the social structure, such as the economy, religion, and education, influence couples' negotiation. The article "Women, Work, and Family in America," by Suzanne Bianchi and Daphne Spain, conveys this concept of context in terms of cohorts. In the extreme, society can predetermine what couples negotiate. For example, as late as the 1960s a married teacher who became pregnant was forced to leave her job. However, there are many less severe influences of the structural context.

In at least three ways this larger context plays an important role in how individuals negotiate work/family balance. The most obvious is that our expectation and values regarding career and family are shaped by our experiences. As Bianchi and Spain indicate, women's participation in the labor market is related to their education level and the state of the economy. Most women who invest considerable time and money in education expect to have the opportunity to apply their training through employment. Many women today plan to combine being a wife and mother with having a job. But few understand what compromises are required in "having it all." And this is where negotiation comes in.

Closely related to the way experience with social structures shapes our negotiation is the role of cultural values. This second way in which structural context determines work and family roles concerns the images, values, and expectations of other members in our society. The readings by David Popenoe and Arlie Russell Hochschild both discuss this aspect of structural context. In "Parental Androgyny," Popenoe argues that we damage families by encouraging men and women to adopt interchangeable family roles. He uses a sociobiological argument to reason that the instability of contemporary family life is attributable to social norms pressuring men to adopt the same family roles as women. In "Understanding the Future of Fatherhood," Hochschild reminds us that messages from the structural context are not always clear and in fact can be contradictory. This article offers potential scenarios for the future that depend on which of two competing cultural values regarding fathers' roles is adopted as the cultural norm.

A third way in which the structural context affects negotiation is through restraining or promoting involvement in the labor force for family members. In "Business and the Facts of Family Life," Rodgers and Rodgers offer several examples of ways business and government can work together to encourage parental participation in the labor force. As you read this article, think about the impact some of these programs might have on a couple's negotiations regarding blending work/family responsibilities. Think also about the majority of workers, who do not benefit from such innovations, and about how their roles are affected.

Negotiating work/family balance is an ongoing process that can be fraught with conflict. We often associate this conflict with the large long-term decisions such as whether to relocate for a promotion at the risk of disrupting children's lives. However, the immediate, day-to-day lives of working parents may be just as, or more difficult, to negotiate. The readings in this chapter illustrate the impact of negotiation and structural contexts in determining the balance of work and family roles.

REFERENCE

Strauss, Anselm. 1978. *Negotiations*. San Francisco: Jossey-Bass.

Men at Work

Anna Quindlen

Overheard in a Manhattan restaurant, one woman to another: "He's a terrific father, but he's never home."

The five o'clock dads can be seen on cable television these days, just after that time in the evening the stay-at-home moms call the arsenic hours. They are sixties sitcom reruns, Ward and Steve and Alex, and fifties guys. They eat dinner with their television families and provide counsel afterward in the den. Someday soon, if things keep going the way they are, their likenesses will be enshrined in a diorama in the Museum of Natural History, frozen in their recliner chairs. The sign will say, "Here sit lifelike representations of family men who worked only eight hours a day."

The five o'clock dad has become an endangered species. A corporate culture that believes presence is productivity, in which people of ambition are afraid to be seen leaving the office, has lengthened his workday and shortened his homelife. So has an economy that makes it difficult for families to break even at the end of the month. For the man who is paid by the hour, that means never saying no to overtime. For the man whose loyalty to the organization is measured in time at his desk, it means goodbye to nine to five.

To lots of small children it means a visiting father. The standard joke in one large corporate office is that the dads always say their children look like angels when they're sleeping because that's the only way they ever see them. A Gallup survey taken several years ago showed that roughly 12 percent of the men surveyed with children under the age of six worked more than sixty hours a week, and an additional 25 percent worked between fifty and sixty hours. (Less than 8 percent of the working women surveyed who had children of that age worked those hours.)

No matter how you divide it up, those are twelve-hour days. When the talk-show host Jane Wallace adopted a baby recently, she said one reason she was not troubled by becoming a mother without becoming a wife was that many of her married female friends were "functionally single," given the hours their husbands worked. The evening commuter rush is getting longer. The 7:45 to West Backofbeyond is more crowded than ever before. The eight o'clock dad. The nine o'clock dad.

There's a horribly sad irony to this, and it is that the quality of fathering is better than it was when the dads left work at five o'clock and came home to café curtains and tuna casserole. The five o'clock dad was remote, a "Wait till your father gets home" kind of dad with a newspaper for a face. The roles he and his wife had were clear: she did nurture and home, he did discipline and money.

The role fathers have carved out for themselves today is a vast improvement, a muddling of those old boundaries. Those of us obliged to convert behavior into trends have probably been a little heavy-handed on the shared childbirth and egalitarian diaper-changing. But fathers today do seem to be more emotional with their children, more nurturing, more open. Many say, "My father never told me he loved me," and so they tell their own children all the time that they love them.

When they're home.

There are people who think that this is changing even as we speak, that there is a kind of perestroika of home and work that we will look back on as beginning at the beginning of the 1990s. A nonprofit organization called the Families and Work Institute advises corporations on how to balance personal and professional obligations and concerns, and Ellen Galinsky, its cofounder, says she has noticed a change in the last year.

"When we first started doing this the groups of men and of women sounded very different," she said. "If the men complained at all about long hours, they complained about their wives' complaints. Now if the timbre of the voice was disguised I couldn't tell which is which. The men are saying: 'I don't want to live this way anymore. I want to be with my kids.' I think the corporate culture will have to begin to respond to that."

This change can only be to the good, not only for women but especially for men, and for kids, too. The stereotypical five o'clock dad belongs in a diorama, with his "Ask your mother" and his "Don't be a crybaby." The father who believes hugs and kisses are sex-blind and a dirty diaper requires a change, not a woman, is infinitely preferable. What a joy it would be if he were around more.

"This is the man's half of having it all," said Don Conway-Long, who teaches a course at Washington University in St. Louis about men's relationships that drew 135 students this year for thirty-five places. "We're trying to do what women want of us, what children want of us, but we're not willing to transform the workplace." In other words, the hearts and minds of today's fathers are definitely in the right place. If only their bodies could be there, too.

Women, Work, and Family in America

Suzanne M. Bianchi and Daphne Spain

How the Generations Compare

Two contrasting media images—Donna Reed, wife and mother in a popular television sitcom of the 1950s and 1960s, and the much maligned Murphy Brown of the 1990s—suggest how perceptions, along with the reality, of women's lives have changed. Donna Reed did not hold a paid job outside the home but she did provide an immaculately clean house and produce cookies and wise counsel when her children returned from school, and she greeted her husband each evening with dinner on the table, seemingly effortlessly prepared. She, along with other TV moms of the 1950s like June Cleaver, epitomized the essence of a woman's adult role: To provide everything necessary for the smooth functioning of a happy family. No doubt few women achieved the level of perfection of these sitcom images, but there was widespread acceptance that women were expected to nurture families while men provided the necessary financial support.

Enter Murphy Brown in the 1990s, a successful career woman (if a bit high strung), who cannot locate a good mate and ultimately decides to have a child without one. She becomes a single mother who balances baby and career on her own. In the debate the raged over the appropriateness of Murphy Brown's choices, the changed labor force role of women (from full-time homemaker to full-time wage earner) was never an issue. Few questioned the decision of a mother of an infant to return to her full-time job. The crux of the debate was the wisdom and morality of mothering an infant without a husband.

A dramatic transformation in labor force expectations of and for women had occurred in a generation. As a nation, we are concerned with the unsettled issues arising from that revolution: What will bolster the institution of marriage as husbands' and wives' roles become more similar than dissimilar? Who will properly care for our children if all of the adults in the family are in the paid labor force?

As economist Claudia Goldin points out, the increase in women's paid labor force participation is an old story and has in fact been occurring since at least the middle of the 19th century.[1] What is relatively new is the explosion in the number and percentage of women who perform a substantial number of hours of paid work while they raise young children. Whereas poor, minority, and working-class women have long contributed economically to their families in whatever way they could, now most women, not just those with the greatest economic need, expect to work for pay outside the home.

The contrast in how adult women juggle work and family will perhaps be greatest between the women who began families between 1946 and 1964, and their daughters, the baby-boom generation (see Box 1). The baby boomers grew up in unusually expansive economic times in which, for many families, the income of one earner afforded much of the American dream. Women married early, perhaps working a year or two prior to or just after marriage, and then left the labor force to raise their children. Their daughters followed a different path. They stayed in school longer, then entered the labor force after finishing school. When they married, usually later than their mothers had, they tended to stay in the labor force. Many did not leave paid work even after the birth of their first child, though they often reduced their number of hours.

A revolution in the *pattern* of women's working lives quietly took place. Younger cohorts of women, with continuous labor force attachment similar to men's, gradually replaced older cohorts, whose lifetime work experience had been far different from the men of their generation.

The simultaneous holding of both breadwinning and caregiving roles is the hallmark of women's changed lives. Yet the images of the 1950s remain—images of the good life provided by stay-at-home moms and hard-working dads. The changed reality, the nostalgia for the past, the grappling with how to reconcile caregiving demands with paid work, and the blurring of the family roles of men and women remain salient issues high on the policy and personal agendas of millions of Americans....

BOX 1 *Cohorts and Generations*

A "cohort" is a group of individuals who share a unique set of experiences throughout life. The term usually refers to individuals born in a specified time period, although a cohort can be defined by events other than birth, such as year of marriage or graduation from high school. "Generation" has a somewhat less precise meaning, but the concepts of cohort and generation are similar and often are used interchangeably. Any portrait of women's lives at a given point in time is the amalgamation of experiences of female birth cohorts.

Generations, or cohorts, of American women can be identified according to the decade in which they made the transition to adulthood and the world of paid work. For example, women born between 1936 and 1945, labeled the World War II generation in the table, typically reached labor force age between the mid-1950s and the mid-1960s. They entered adulthood during the 10-year period leading up to the passage of the Civil Rights Act of 1964, which for the first time in American history barred discrimination on the basis of sex. Most of the World War II cohort of women therefore completed their education and began their families before the intense and widespread questioning of gender stereotypes that occurred during the 1970s.

The cohort born between 1946 and 1955, "the early baby boomers," reached adulthood between the mid-1960s and the mid-1970s. It was a large generation that created serious dislocations as it moved through school and into the labor force. First, baby boomers made elementary and high school classrooms bulge; then they flooded college campuses during the Vietnam War years, fueling the activism that became a defining marker of the late 1960s and early 1970s. These women had access to the birth control pill and to legalized abortions, factors that contributed to a radical change in sexual practices and attitudes toward marriage.

The members of the "late baby-boom cohort," those born between 1956 and 1965, were even more numerous than their older brothers and sisters. They reached adulthood and began entering the labor force in the mid-1970s. These women (and men, too) entered a labor market in which wage rates for all but the highly educated were stagnating rather than rising. This created added financial pressure for many women to work outside the home. The pressure to get a job was especially strong for less-educated women married to husbands who were most affected by the economic restructuring.

Finally, the smaller cohort born after the mid-1960s, "the baby-bust cohort," sometimes partially subsumed under other labels (such as "Generation X"), forms the group that is currently in its 20s and early 30s. The adult lives of these

(continued on next page)

BOX 1 *Cohorts and Generations (continued)*

women—shaped by choices they make about work, childbearing, and family—will be the true measure of just how profound and long-lasting the transformation of women's lives has been. They were not even born when the 1964 Civil Rights Act was passed. They have lived their entire lives during a period in which the country's stated ideology, if not always its reality, has been that men and women are equal and that discrimination on the basis of sex is not to be tolerated.

REFERENCE

Daphne Spain and Suzanne M. Bianchi, *Balancing Act: Motherhood, Marriage, and Employment Among American Women* (New York: Russell Sage Foundation, 1996): ix–xv.

Labor Force Entry of Birth Cohorts

Birth cohort	Generation	Labor force entry	Age in 1980	Age in 1995
1966–75	Baby bust	Mid-1980s through 1990s	5–14	20–29
1956–65	Late baby boom	Mid-1970s through 1980s	15–24	30–39
1946–55	Early baby boom	Mid-1960s through 1970s	25–34	40–49
1936–45	World War II	Mid-1950s through 1960s	35–44	50–59
1926–35	Parents of baby boom	Mid-1940s through 1950s	45–54	60–69
1916–25	Parents of baby boom	Mid-1930s through 1940s	55–64	70–79
1906–15	Grandparents of baby boom	Mid-1920s through 1930s	65–74	80–89

Source: Daphne Spain and Suzanne M. Bianchi, *Balancing Act: Motherhood, Marriage, and Employment Among American Women:* table 1.

Women's Changing Work Lives

The picture of women in the workplace is one of clear, albeit slow, progress toward equality with men. This bodes well for gender equality in the labor force. The more highly educated a woman, the more likely she is to work for pay, to work full-time when she is employed, and to be in a managerial or professional job where wages are higher. Hence, the rise in women's educational attainment, particularly in the 1970s and 1980s, has been a key factor in women's movement toward paid work and continuous labor force participation throughout their adult lives.

Education Trends

High school graduation is becoming the minimum level of educational achievement for men and women in the United States. Over 85 percent of adults ages 25 to 34 were high school graduates in 1994, as shown in Table 1, up from less than 60 percent in 1960. Women are somewhat more likely to graduate from high school than men. Sharper gender differences appear at the higher levels of education—and it is here that dramatic changes have occurred in the past three decades. In 1960, 54 percent of male, but only 38 percent of female high school graduates enrolled in college in the fall after their high school graduation. By 1994, the percentage of female high school graduates enrolling in college was 63 percent, slightly higher than for males (61 percent). Almost one out of four young women and men are earning college degrees in the 1990s.

Throughout the 1960–1993 period, women increased their representation among degree recipients

TABLE 1 *Selected Indicators of Educational Attainment for Women and Men, 1960–1994*

Indicator	1960	1970	1980	1990	1994
Birth cohorts (at ages 25–34)	1926–35	1936–45	1946–55	1956–65	1960–69
Percent of population (ages 25–34) who graduated from high school					
Women	60	73	85	87	87
Men	56	74	86	85	85
Percent of high school graduates who enrolled in college in fall after graduation					
Women	38	49	52	62	63
Men	54	55	47	58	61
Percent of population (ages 25–34) who are college graduates					
Women	8	12	21	23	24
Men	15	20	27	24	23
Percent of degrees conferred on women[a]					
Bachelor's	39[b]	43	49	53	54
Master's	32	40	49	53	54
Doctorates	11	13	30	36	38
Dentistry	1	1	13	31	34
Medicine	6	8	23	34	38
Law	2	5	30	42	42
Business (MBA)	4	4	22	34	36

Note: Includes degrees earned by foreign students.
[a]Data in last column are for 1993.
[b]1961.
Source: National Center for Education Statistics, *Digest of Education Statistics 1995:* tables 178, 236, 251, and 272; and U.S. Bureau of the Census, "Years of School Completed by Persons 25 Years and Over, by Age and Sex: Selected Years, 1940 to 1995" (on-line, Nov. 1996).

at all levels. Over half of all bachelor's and master's degrees were conferred on women in 1993. Women earned 38 percent of all doctoral degrees in 1993, a marked improvement over 1960, when they earned only 11 percent.

The gender differences in doctoral degrees conferred by U.S. universities in recent years have been accentuated by increasing numbers of foreign students, the majority of whom are men. While many foreign students may eventually settle in the United States, most return to their home countries. Excluding nonresident aliens, U.S. women earned 44 percent of doctoral degrees in 1993, considerably closer to parity with men. The female share of bachelor's and master's degrees conferred in 1993 increases by 1 and 3 percentage points, respectively, when nonresident aliens are excluded.

Women's share of college degrees increased in all racial and ethnic groups (see Figure 1). In 1977, for example, white women earned about 46 percent of bachelor's degrees, but by 1993, they earned a majority of degrees among whites (54 percent). The trend was even more marked for Hispanic and American Indian women. The gender gap closed for Asians as well, although women's gains were more moderate. They are the only group in which women and men earned nearly equal numbers of bachelor's degrees in the early 1990s. Among African Americans, unlike the other racial/ethnic groups, women already earned more bachelor's degrees than men by the late 1970s. The female advantage has grown. Women earned 63 percent of bachelor's degrees conferred on African Americans in 1993, up from 57 percent in 1977.

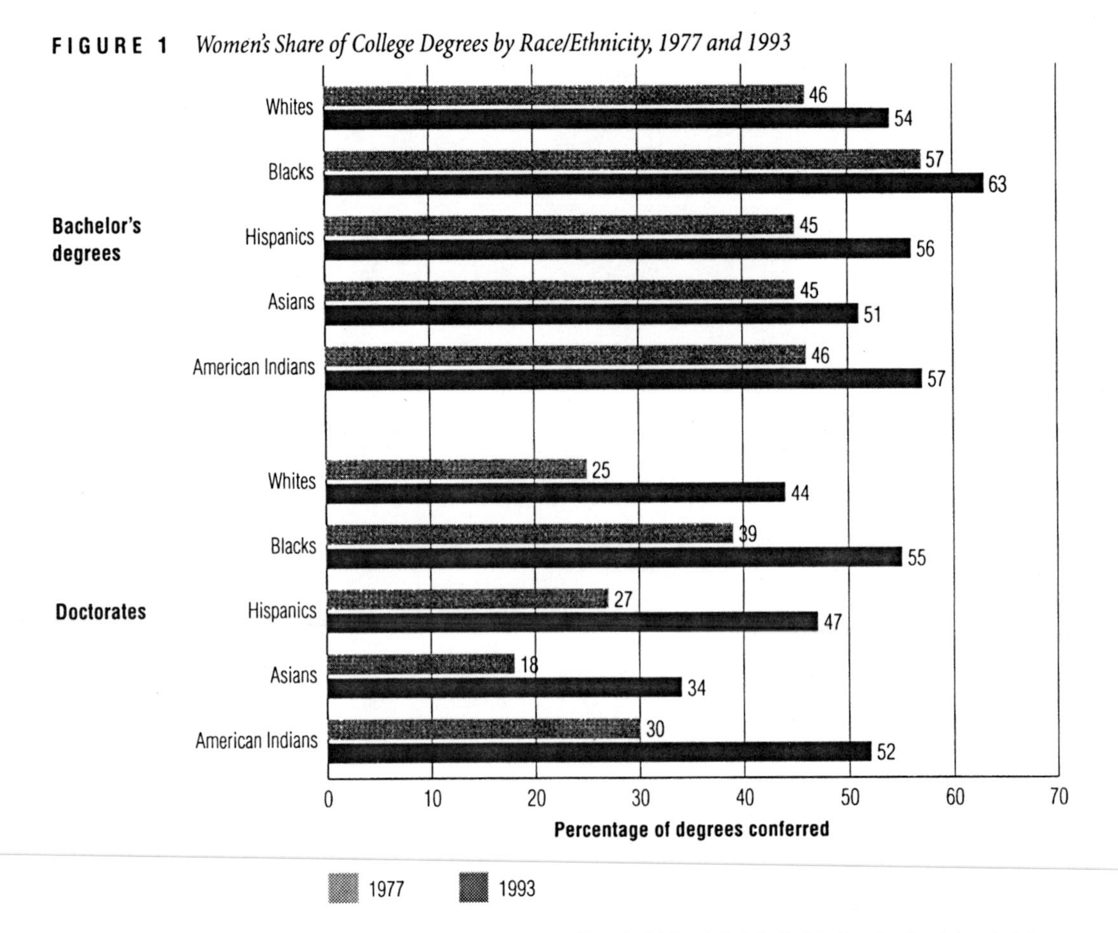

FIGURE 1 *Women's Share of College Degrees by Race/Ethnicity, 1977 and 1993*

Note: Rates for whites and blacks are for "white non-Hispanics" and "black non-Hispanics." Asians include Pacific Islanders. American Indians include Aleuts and Eskimos.
Source: National Center for Education Statistics, *Digest of Education Statistics 1995:* tables 256 and 262.

The racial and ethnic differences are wider at the higher degree levels. Women earned 55 percent of the doctorates conferred on African Americans, but only 34 percent of doctorates earned by Asians in 1993, for example. The gender gap was narrower among whites, Hispanics, and American Indians—for whom the women's share ranged between 44 and 52 percent. Women in all these racial and ethnic groups have increased their share of higher education degrees since 1977, however, as shown in Figure 1....

Increase in Women's Employment

Over the first half of the 20th century, women's participation in the labor force moved slowly upward. With each younger birth cohort, the propensity to engage in market work increased, especially at older ages, after children were reared. As Valerie Oppenheimer has shown, the demand for clerical, and later service, workers expanded rapidly after 1940 as the economy grew and changed. Many women moved into these jobs.[2] In the first four decades of the 20th

TABLE 2 *Women in the Labor Force, 1900 to 1995*

Year	Women in the Labor Force (in thousands)	Women as Percent of Total Labor Force	Women in Labor Force as Percent of All Women
1900	5,319	18.3	18.8
1910	8,076	21.2	23.4
1920	8,550	20.5	21.1
1930	10,752	22.0	22.0
1940	12,887	24.5	25.4
1950	18,389	29.6	33.9
1960	23,240	33.4	37.7
1970	31,543	38.1	43.3
1980	45,487	42.5	51.5
1990	56,554	45.3	57.5
1995	60,944	46.1	58.9

Note: Data for 1990–30 refer to employed workers ages 10 and older; data for 1940 include ages 14 and older; data for 1950–95 refer to the civilian labor force ages 16 and older. Figures for 1950–95 are annual averages from the Current Population Survey; figures for 1900–40 are based on the decennial census.

Sources: (1900–1960) U.S. Bureau of the Census, *Historical Statistics of the United States: Colonial Times to 1970,* Bicentennial Edition, part 1, series D11-25; (1970–1995) U.S. Bureau of Labor Statistics, *Employment and Earnings* 43, no. 1: tables 1 and 2.

century, the percentage of women in the labor force increased from one-fifth to one-quarter (see Table 2). In the next four decades, the increase accelerated, and by 1980, over half of U.S. women were in the paid labor force. This upward shift continued during the 1980s and 1990s, reaching 59 percent in 1995.

The composition of the labor force was transformed by the increasing numbers of women. In 1900, only 18 percent of the paid labor force was female. This rose gradually to 25 percent by 1940. By 1980, the labor force was 43 percent female, an increase of 18 percentage points. By 1995, the women's share had reached 46 percent.

From 1940 until the mid-to-late 1960s, labor force activity increased most among women who were past their prime childrearing years. During the 1970s and 1980s, as the marriage age rose, fertility declined, and women's educational attainment increased, the growth in labor force participation spread to younger women. Many women now postpone family formation to complete their education and establish themselves in the labor force. Women often curtail their commitment to paid work when they have children. Despite family obligations, however, a majority of women of all educational levels now work outside the home during the years they are raising children.

Women's labor market involvement is still lower than men's. Many mothers work part-time and some drop out of the labor force while their children are young. But there is little doubt that the economic activity of the two genders has become more similar in recent decades. As Table 3 shows, the narrowing gender gap in labor force participation is especially evident among adults in their prime working ages (25 to 54). In 1950, the participation rate of men in these ages was 97 percent, 60 percentage points higher than that of women of the same age (37 percent). By 1995, men's participation slipped to 92 percent while women's rate increased dramatically, to 76 percent. The gap separating men and women, although still sizable, fell to 16 percentage points.

Labor force participation rates (which include those employed and those who are unemployed but actively seeking work) have risen for women of all races and ethnicities. In 1995, 59 percent of white, 60 percent of black, and 53 percent of Hispanic women were in the labor force. Between 1960 and 1995, rates for white women rose the most, from 37 percent to 59 percent, until they nearly equaled the labor force rates of black women.

The gender gap in labor force participation rates was narrower for blacks than for other racial and ethnic groups in 1995, partly because participation is relatively high for black women and relatively low for black men. The greatest difference in women's and men's labor force participation rates in 1995 existed among Hispanics (nearly 27 percentage points), despite a significant rise in Hispanic women's labor force participation since 1980.

Throughout the 1970s and 1980s, labor force participation increased most rapidly for the group with

TABLE 3 *Labor Force Participation by Gender and Race/Ethnicity, 1950–1995*

Percent in Labor Force	1950	1960	1970	1980	1990	1995
Total civilian population						
Ages 16 +						
Women	34	38	43	52	58	59
Men	86	83	80	77	76	75
Difference (M–W)	53	46	36	26	19	16
Ages 25–54						
Women	37	43	50	64	74	76
Men	97	97	96	94	93	92
Difference (M–W)	60	54	46	30	19	16
Race/ethnicity (ages 16+)						
White						
Women	na	37	43	51	57	59
Men	na	83	80	78	77	76
Difference (M–W)	na	47	37	27	20	17
Black						
Women	na	na	49*	53	58	60
Men	na	na	74*	70	71	69
Difference (M–W)	na	na	25	17	13	10
Hispanic						
Women	na	na	na	47	53	53
Men	na	na	na	81	81	79
Difference (M–W)	na	na	na	34	28	27

*1972 rates. Data for those age 16 years and over were tabulated separately for blacks beginning in 1972.
Note: Hispanics may be of any race. Male/female differences were based on unrounded figures.
na: Not available.
Source: U.S. Bureau of Labor Statistics, published and unpublished tabulations.

the lowest rate of labor force participation historically—married women with children under age 18. By 1996, almost three-quarters of married women with dependent children worked in the paid labor force and almost two-fifths (38 percent) worked full-time and year-round (see Table 4). In other words, by 1996, most married mothers with children had some involvement in market work, and many, though not the majority, worked full-time....

Why Is There More Earnings Parity for Men and Women?

...Indisputably, baby-boom women greatly increased their lifetime attachment to market work and narrowed the differences in work experience between themselves and the men of their generation. Those who obtained advanced educations were poised to move into higher-paying jobs. The 1980s turned out to be a very good decade, as far as earnings, for the college educated. Women at the top of the educational distribution made strides in narrowing the gender gap even as the earnings of college-educated men improved.

The majority of persons in the labor force, however, do not have college educations. For these workers, wages stagnated or declined, especially among men, in the latter 1970s and 1980s. Less-educated women improved their earnings situation vis-à-vis men, not only because they increased the amount of time they allocated to market work but also because men's wages were *not* rising. A recent analysis of the

TABLE 4 *Married Mothers' Labor Force Attachment, 1970–1996*

Age of Children	1970	1980	1990	1996
With children under age 18				
Percentage who worked last year	51	63	73	74
Percentage who worked full-time, year-round	16	23	34	38
With children under age 6				
Percentage who worked last year	44	58	68	68
Percentage who worked full-time, year-round	10	18	28	31
With children ages 6–17				
Percentage who worked last year	58	68	78	80
Percentage who worked full-time, year-round	23	29	40	44

Note: Data are for women ages 16 and over who are married and living with their husbands.
Source: (1970–1990) U.S. Bureau of Labor Statistics, Current Population Survey, unpublished tabulations; (1996) PRB analysis of data from the March 1996 Current Population Survey.

distribution of men's and women's earnings suggests that, at least among whites, the downgrading of jobs and lowered wages of men without a college education that pushed more men into the lowest earnings brackets may have accounted for half of the narrowing of the gender wage gap in recent years.[3] Although it is good news for women that their earnings have been increasing in real terms, particularly among the college educated, and many would applaud greater equality of outcomes between the sexes, it is not necessarily good news for young women that the men they marry may do less well in the labor market than previous generations.

Indeed, a growing body of research is examining how low wages may have led to the delay in marriage among young men and women. Valerie Oppenheimer and her colleagues, for example, argue that though the desire to obtain additional education is an important motivation for delay in marriage,[4] so too are wage constraints: Men's (and women's) difficulty in becoming established in the labor market and the increased time they spend in low-wage, stop-gap jobs may also be exerting pressure to postpone marriages.

Both less- and better-educated women are filling a family breadwinner role, but for somewhat different reasons. Highly educated women have invested time and energy in obtaining skills and they are likely to have jobs with greater financial and psychological rewards, but they also tend to marry men with similar attributes. Thus, motivation to work for pay is high, but financial need may not be so high. Less-educated women may fill less-rewarding jobs, not only with regard to pay but perhaps also in terms of flexibility and autonomy.[5] Their husbands' wages alone tend to be insufficient to maintain a middle-class lifestyle, however, and they must continue to work for pay after marriage and children. Women at higher educational levels work more than less-educated women. In 1990, among young women ages 25 to 34, 87 percent of women with postgraduate educations were in the labor force, compared with 72 percent of those with a high school education; 48 percent among those with postgraduate educations worked full-time, year-round, compared with 40 percent with a high school education.[6] Regardless of educational attainment, however, the majority of women work for pay.

Historically, it has been accepted that men work for pay, with no questions asked as to what motivates their high rates of participation. The assumption has been that paid work is what adult men do—men support themselves and those who depend on them. Women may still envision more choice than men when it comes to decisions about paid work. But choice is tied to issues of economic dependence and independence. As women increasingly view themselves as independent, as responsible for themselves and frequently for their dependent children, and as less able to count on the lifelong financial support of a husband, the choice regarding whether to work for pay is disappearing. They work because that is what adult women do—they support themselves and those who depend on them.

Part of the questioning of gender roles occurred with the women's movement of the 1970s. What emerged was an increased awareness that the "separate spheres" of women's and men's lives, with women as keepers of the home and men as financial providers, were idealizations, and that this kind of arrangement worked best when marriages were highly stable and long-lasting and men earned a "family wage." What's more, it required a certain contentment with the constraints that such narrowly defined roles placed on both men and women. The economic conditions of the 1970s and 1980s, the increase in marital instability, and the increased opportunities for women that accompanied the civil rights movement and the sexual revolution all tended to discredit the separate spheres vision of adult gender roles. As a result, work and family relationships underwent a profound alteration in the 1970s and the 1980s....

The Domestic Division of Labor: Housework and Child Care

As more wives took on paying jobs over the past few decades, their proportional contribution to family earnings increased. In the 1990s, working wives contributed about 30 percent of their family's income; 40 percent when they worked full-time, year-round.[7] This development has tended to give wives greater power within their marriages.[8]

Despite their growing economic independence, however, women continue to allocate less time to market work and more time to family than do men. For example, 96 percent of married men ages 35 to 44 were employed in the year preceding the 1990 Census, compared with 77 percent of married women in that age group, a 19 percentage point difference. But wives put in almost 900 fewer paid work hours than husbands. The full-time, year-round employment rate of married women (ages 35 to 44) was 39 percentage points lower than that of married men of the same age.[9]

Most likely, these disparities reflect the continuing differences in the domestic roles of men and women in the American family and economy. To be sure, married mothers have scaled down their hours of housework—from about 30 hours in 1965 to about 20 hours per week in 1985; married fathers, however, picked up only part of the slack, increasing their household work from about 5 hours to about 10 hours per week over the period (see Table 5). By 1985, married mothers performed about two-thirds of the housework compared with 85 percent in 1965. Since 1985, men appear to have increased their share of domestic chores, but not because they are doing more housework. The hours women spend on housework have spiraled downward over the past decade, and have fallen closer to the level reported by men.[10]

Household tasks continue to differ by gender. Husbands participate most in child care and in yard and home maintenance, assuming about 40 percent of the family workload in these areas.[11] Husbands do less than 25 percent of the cooking, cleaning, dishwashing, and laundry and share just over 25 percent of the grocery shopping and paperwork associated with family finances. In families in which the wife works outside the home, however, husbands perform a greater share of domestic tasks and child care.

Although women do more household work than men, studies find that when asked, women as well as men tend to report their household division of labor as fair. Why this should be the case in the face of overwhelming evidence to the contrary continues to puzzle researchers. Sociologists and psychologists have offered several theories to explain this conundrum. The psychological theory of cognitive dissonance suggests that most people perceive their lives to be fair because of the need to reconcile expectations with reality (for better or worse).[12] If women were to dwell on the unfairness of their contribution to their household's workload, other valuable aspects of their life (for example, their psychological health or their marriage) might be threatened. They resolve the discrepancy by adjusting their expectations about the division of labor in their homes.

Social exchange theory proposes that power and dependency influence how people assess fairness, and that power depends on individual resources (such as

TABLE 5 *Hours Per Week Men and Women Spend Doing Housework, 1965, 1975, and 1985*

Gender/Characteristic	1965	1975	1985	Change 1965 to 1985
Total, ages 18–65				
Men	4.6	7.0	9.8	+5.2
Women	27.0	21.7	19.5	–7.5
Ratio (M/W)	.17	.32	.50	
Married				
Husbands	4.5	6.8	11.1	+6.6
Wives	31.6	24.2	22.4	–9.2
Ratio (H/W)	.14	.28	.50	
With preschool-age children				
Fathers	3.9	5.9	9.0	+5.1
Mothers	32.0	25.1	22.5	–9.5
Ratio (F/M)	.12	.24	.40	
With school-age children				
Fathers	5.3	7.6	10.4	+5.1
Mothers	30.3	23.9	19.9	–10.4
Ratio (F/M)	.17	.32	.52	

Note: Housework includes cooking meals, meal clean-up, housecleaning, doing laundry, outdoor chores, repairs, garden and animal care, and paying bills. It excludes childcare.
Source: John P. Robinson, "Who's Doing the Housework?" *American Demographics* 10, no. 12: 24–28.

income).[13] Neoclassical economics also proposes that differential returns to women's and men's market labor may affect perceptions of equity: Husbands who make more money outside the home may be excused from unpaid work within the home.[14] The notion that wives' economic dependence on husbands affects their sense of fairness is supported by some research evidence.[15]

Perceptions of fairness may depend on comparisons of one's status with that of others. A sense of relative deprivation is more likely to occur when peers perceive differential treatment than when actual deprivation occurs.[16] If women (and men) compare their division of labor with that of other couples, and conclude that their spouse does at least as much domestic work as other spouses, they do not perceive their own division of labor as unfair. A sense of equity may result more from a husband's *increased* efforts in the household than a wife's *reduced* efforts. A husband can be doing fewer hours of household work than the wife, but if his efforts are increasing over time, more fairness is perceived.[17]

Another explanation for women's perceptions of fairness in the face of inequality arises from gender ideology. Some women believe they *should* do most of the housework regardless of their employment status. The more conventional an employed wife's view of a woman's role in the family, the more likely she is to perceive an unequal division of household work as fair.[18]

Finally, the possibility exists that women actually have a higher tolerance for housework than men or that women's standards of cleanliness are higher than men's.[19] This may be the result of differential gender socialization, which raises girls to understand that they will be the keeper of house and kin as adults.[20] To the extent that women view housework as their job, or that they want certain standards maintained, they may rationally decide that it is better not to concern themselves too much with the fairness issue. They reduce dissonance and conflict and achieve their goals, albeit at considerable cost-of-time overload.

All this can become self-fulfilling. The greater a woman's commitment of time and energy to unpaid household work, the less her available time and energy for paid work. This can perpetuate the unequal earnings of women and men and ensure women's economic dependency on men within marriage. This earnings inequity also assures more adverse economic consequences for women than men when marriages end....

What Lies Ahead

What might the balancing act for women look like in the 21st century? Public attitudes toward women's roles and the necessity for women to work outside the home have created significant changes from which society cannot easily retreat. The unresolved issue is, what happens to the family when the role of mother undergoes such a radical transformation?

Will the stresses of our current lifestyles and our collective nostalgia for the "stability" of the 1950s push women out of the labor force, or, to frame the question differently, give them the opportunity to leave? It seems unlikely. The media often run stories about professional women who give up their careers to spend more time with their children. But these examples are atypical. And, perhaps more telling, they often involve affluent married women who probably have more economic choices than the average American mother.

There are other reasons why we are not likely to return to the idealized 1950s. As demographer Andrew Cherlin and economist Frank Levy have pointed out, the 1950s were an anomaly—socially, demographically, and economically. Never before nor since were so many families able to achieve the American dream of owning a home and a comfortable middle-class lifestyle on just one income. In the 21st century, leaving paid work to be a stay-at-home mom will be an option—but one increasingly available only to the privileged few unless families are willing to adopt a lower living standard.

Are we pushing the limit on how much gender equality women, men, and families want or can handle? Perhaps we are—although international comparisons reveal that other countries have gone even further. Scandinavian countries, for example, offer economic inducements to fathers as well as mothers to take parental leave following the birth of a child. The U.S. government has taken some steps to allow parents of both sexes to accommodate work and family life. The Family and Medical Leave Act, which took effect in 1993, guarantees unpaid leave to male as well as female workers who need to care for children or other family members.

Women's presence in the labor force could continue to increase, especially if more employers are willing to accommodate the family responsibilities of both fathers and mothers. Although we may be reaching an upper limit in the percentage of women who work for pay, most married women still work less than full-time while they are raising children.

There is considerable room for increasing the hours and years women spend in the labor force. There is also plenty of opportunity for men to change. They could increase the time spent on domestic chores and child care, allowing women to devote more energy to paid market work.

Will gender roles in work and family become more similar? For the United States as well as many other industrialized countries, the answer appears to be "yes." But the path toward gender equality is laden with anxiety for the current generations of women juggling work and family responsibilities. The movement toward gender equality will be achieved over the long run by the gradual replacement of generations—each somewhat more committed to equality than the last. If young women in the 1990s appear to be less urgent than their older sisters and mothers about the importance of establishing themselves in the labor market before beginning a family, it is because they take for granted rights in the workplace that their predecessors could not. But they also take seriously their obligation to contribute financially to their family. It is easy to misinterpret trends that seem to return us to the past because we sometimes forget that these decisions are being made by young women who are looking squarely at the future.

Our norms about the best way to balance work and family have never been universal—black women typically have started families at younger ages than whites or Asians, *and* blacks have been more committed to paid employment. The sample of black women, as well as evidence from Scandinavian countries, suggest that the rates of childlessness may decline somewhat and the timing of childbearing may shift toward the younger ages now that combining work and family is commonplace for women of all racial and ethnic groups.

Although the condition of the American family is often lamented, Americans continue to have children and to find joy and satisfaction in their family lives. They also express a strong work ethic, and dissatisfaction when economic downturns jeopardize their jobs. The need to balance work and family will not disappear. Not many men and women can really content themselves with a "separate spheres" division of labor in the home—and this will force negotiation about marriage, child care, and housework.

If we think in terms of cohorts, many baby-boom women grew up watching one model of women's appropriate roles—that mothers stay at home and nurture children. However, as economic times changed, women chose or were forced to adopt very different roles as adults. Their daughters, who grew up with mothers who work outside the home, tend not to question whether they will work. Rather, today's young women struggle with how they can achieve a balance in their lives—preferably with less angst and stress than their own parents faced. Their answers will not be exactly the same, but they will propel us forward, not backward, and they will probably lead toward even greater equality in the economic and caregiving roles of men and women.

REFERENCES

1. Claudia Goldin, *Understanding the Gender Gap* (New York: Oxford University Press, 1990).
2. Valerie Kincade Oppenheimer, *The Female Labor Force in the United States* (Westport, CT: Greenwood Press, 1970).
3. Annette Bernhardt, Martina Morris, and Mark S. Handcock, "Women's Gains or Men's Losses? A Closer Look at the Shrinking Gender Gap in Earnings," *American Journal of Sociology* 101, no. 2 (1995): 302–28.
4. Valerie K. Oppenheimer, Matthijs Kalmijn, Nelson Lim, and Vivian Lew, "Men's Career Development and Marriage Timing: Race and Schooling Differences" (Paper presented at the annual meeting of the Population Association of America, New Orleans, LA, May 1996).
5. Jennifer Glass, "The Impact of Occupational Segregation on Working Conditions," *Social Forces* 90 (1990): 779–96; and Jennifer Glass and Valerie Camarigg, "Gender, Parenthood, and Job-Family Compatibility," *American Journal of Sociology* 98, no. 1 (1992): 131–51.
6. Spain and Bianchi, *Balancing Act:* table 3.5 (New York: Russell Sage, 1996)
7. Howard V. Hayghe and Suzanne M. Bianchi, "Married Mothers' Work Patterns: The Job–Family Compromise," *Monthly Labor Review* 117 (June 1994): 24–30.
8. England and Browne, "Women's Status"; Paula England and Barbara Stanek Kilbourne, "Markets, Marriages, and Other Mates: The Problem of Power," in *Beyond the Market Place: Rethinking Economy and Society,* eds. Roger Friedland and A. F. Robertson (New York: Aldine de Gruyter, 1990); and Judith Treas, "Money in the Bank: Transaction Costs and Economic Organization of Marriage," *American Sociological Review* 58 (Oct. 1993): 723–34.
9. Spain and Bianchi, *Balancing Act:* table 6.6.
10. Jonathan Gershuny and John P. Robinson, "Historical Changes in the Household Division of Labor," *Demography* 25, no. 4 (1988): 537–52; John P. Robinson, "Who's Doing the Housework?" *American Demographics* 10 (1988): 24–28; and authors' communication with John Robinson, Nov. 1996.
11. Frances Goldscheider and Linda Waite, *New Families, No Families? The Transformation of the American Home* (Los Angeles, CA: University of California Press, 1991): chapter 7.
12. Leon Festinger, *A Theory of Cognitive Dissonance* (Evanston, IL: Row, Peterson, 1957).
13. Peter M. Blau, *Exchange and Power in Social Life* (New York: John Wiley and Sons, 1964).
14. Gary Becker, *A Treatise on the Family* (Cambridge, MA: Harvard University Press, 1981).
15. Emily W. Kane and Laura Sanchez, "Family Status and Criticism of Gender Inequality at Home and at Work," *Social Forces* 72, no. 4 (1994): 1079–102; and Mary

Clare Lennon and Sarah Rosenfield, "Relative Fairness and the Division of Housework: The Importance of Options," *American Journal of Sociology* 100, no. 2 (1994): 506–31.

16. George C. Homans, *Social Behavior: Its Elementary Forms* (New York: Harcourt, Brace, 1961); and Robert K. Merton, *Social Theory and Social Structure* (New York: Free Press, 1968).
17. Laura Sanchez, "Gender Labor Allocations and the Psychology of Entitlement Within the Home," *Social Forces* 73, no. 2 (1994): 533–53.
18. Linda Thompson, "Family Work: Women's Sense of Fairness," *Journal of Family Issues* 12 (1991): 181–95.
19. John P. Robinson and Melissa Milkie, "Back to the Basics: Trends in and Determinants of Women's Attitudes Toward Housework," Draft manuscript (College Park, MD: University of Maryland, 1996).
20. Goldscheider and Waite, *New Families, No Families.*

Sharing and Reluctance

Scott Coltrane

For these dual-earner Chicano couples, we found considerable sharing in several areas. First, as in previous studies of ethnic minority families, wives were employed a substantial number of hours and made significant contributions to the household income. Second, like some researchers, we found that couples described their decision-making to be relatively fair and equal. Third, fathers in these families were more involved in childrearing than their own fathers had been, and seven of 20 husbands were rated as sharing most child-care tasks. Finally, although no husband performed fully half of the housework, a few made substantial contributions in this area as well.

One of the power dynamics that appeared to undergird the household division of labor in these families was the relative earning power of each spouse, though this was modified by occupational prestige, provider role status, and personal preferences. When the wife earned less than a third of the family income, the husband performed little of the routine housework or child care. In two families, wives earned more than their husbands. These two households reported sharing more domestic labor than any of the others. Among the other couples who shared family work, we found a preponderance of relatively balanced incomes. In the two families with large financial contributions from wives, but little household help from husbands, couples had hired housekeepers to reduce the household workload.

Relative income thus makes a difference, but there was no simple or straightforward exchange of market resources for domestic services in these families. Other factors like failed career aspirations or occupational status influenced power differentials and helped to explain why some wives were willing to push a little harder for change in the division of household labor. In almost every case, husbands reluctantly responded to requests for help from wives. Only when wives explicitly took the initiative to shift some of the housework burden to husbands did the men begin to assume significant responsibility for the day-to-day operation of the household. Even when they began to share the family work, men tended to do some of the less onerous tasks like playing with the children or washing the dinner dishes. When we compare these men to their own fathers, or those of their mothers, however, we can see that they are sharing more domestic chores than the generation of parents that preceded them.

Acceptance of wives as co-providers and wives' delegation of a portion of the homemaker role to husbands were especially important to creating more equal divisions of household labor. If wives made lists for their husbands or offered them frequent reminders, they were more successful than if they waited for husbands to take the initiative. But remaining responsible for managing the home and children was cause for resentment on the part of many wives. Sometimes wives were effective in getting husbands to perform certain chores, like ironing, by stopping doing it altogether. For other wives, sharing evolved more "naturally," as both spouses agreed to share tasks or performed the chores they most preferred.

Economies of gratitude continually shifted in these couples as ideology, career attachments, and feelings of obligation and entitlement changed. For some main provider families, this meant that wives were grateful for husbands' "permission" to hold a job, or that wives worked harder at home because they felt guilty for making their husbands do any of the housework. Main provider husbands usually let their job commitments limit their family work, whereas their wives took time off from work to take children to the dentist, care for a sick child, or attend a parent–teacher conference.

Even in families where co-provider wives had advanced degrees and earned relatively high wages, women's work/family boundaries were more permeable than their husbands'. For example, one professional woman complained that her teacher husband was a "perpetual" graduate student and attended "endless" community meetings. She was employed more hours than he, and made about the same amount of money, but she had to "schedule him" to watch the children if she wanted to leave the house alone. His stature as a "community leader" provided him with subterranean leverage in the unspoken struggle over taking responsibility for the house and children. His "gender ideology," if we had measured it with conventional survey questions, would undoubtedly have been characterized as "egalitarian." He spoke in broad platitudes about women's equality and was washing the dishes when we arrived for the interviews. He insisted on finishing the dishes as he answered my questions, but in the other room his wife confided to Elsa in incredulous tones, "He *never* does that!"

In other ambivalent co-provider families, husbands gained unspoken advantage because they had more prestigious jobs than their wives, and earned more money. While these highly educated attorneys and administrators talked about how they respected their wives' careers, and expressed interest in spending more time with their children, their actions showed that they did not fully assume responsibility for sharing the family work. To solve the dilemma of too little time and too many chores, two of these families hired housekeepers. Wives were grateful for this strategy, though it did not alter inequities in the distribution of housework and child care, or in the allocation of worry.

In other families, the economy of gratitude departed dramatically from conventional notions of husband as economic provider and wife as nurturing homemaker. When wives' earnings approached or exceeded their husbands', economies of gratitude shifted toward more equal expectations, with husbands beginning to assume that they must do more around the house. Even in these families, husbands rarely began doing more chores without prodding from wives, but they usually did them "without complaining." Similarly, when wives with economic leverage began expecting more from their husbands, they were usually successful in getting them to do more.

Another type of leverage that was important, even in main provider households, was the existence of failed aspirations. If wives expected husbands to "make more" of themselves, pursue "more important" careers, or follow "dream" occupational goals, then wives were able to get husbands to do more around the house. This perception of failed aspirations, if held by both spouses, served as a reminder that husbands had no excuse for not helping out at home. In these families, wives were not at all reluctant to demand assistance with domestic chores, and husbands were rarely able to use their jobs as excuses for getting out of housework.[1]

The leisure gap, common among Anglo couples, is also clearly present in dual-earner Chicano families. Nevertheless, in couples where the economy of gratitude is more balanced, the leisure gap begins to shrink. It becomes much less significant, though it doesn't disappear entirely, when both spouses consider the woman's job as important as the man's.... [T]his tends to happen when wives' earnings approach those of husbands'.

The economies of gratitude in these families were not equally balanced, but many exhibited divisions of household labor that contradicted cultural stereotypes of male-dominated Chicano families. Particularly salient in these families was the lack of fit between their own class position and that of their parents. Most parents were immigrants with little education and low occupational mobility. The couples we interviewed, in contrast, were well educated and relatively secure in middle-class occupations. The couples could have compared themselves to their parents, evaluating themselves as egalitarian and financially successful. While some did just that, most compared themselves to their Anglo and Chicano friends and co-workers, many of whom shared as much or more than they did.... [C]ouples had no absolute or fixed standard against which to make judgments about themselves. Implicitly comparing their earnings, occupational commitments, and perceived aptitudes, these individuals negotiated new patterns of work and family commitments and developed new justifications for their emerging arrangements. These were not created anew, but emerged out of the popular culture in which they found themselves. Judith Stacey labels such developments the making of the "postmodern family" because they signal "the contested, ambivalent, and undecided character of contemporary gender and kinship arrangements."[2] Our findings confirm that families are an important site of new struggles over the meaning of gender and the rights and obligations of men and women to each other and over each other's labor.

One of our most interesting findings has to do with the class position of Chicano husbands and wives who shared the most household labor: white-collar working-class families shared more than upper-middle-class professionals. Contrary to findings from some nationwide surveys, the most highly educated of our well-educated sample of Chicano couples shared only moderate amounts of child care and little housework.[3] Contrary to other predictions, neither was it the working-class women in this study who achieved the most balanced divisions of labor.[4] It was the middle occupational group—the executive secretaries, clerks, technicians, teachers, and mid-level administrators—who extracted the most help from husbands. The men in these families were similarly in the middle in terms of occupational status for this sample—administrative assistants, a builder, a mail carrier, a technician—and in the middle in terms of income. What this means is that the highest status wives—the program coordinators, nurses, social workers, and office managers—were not able to, or chose not to, transform their salaries or occupational status into more participation from husbands. This was probably because their husbands had even higher incomes and more prestigious occupations. The lawyers, program directors, ranking bureaucrats, and "community leaders" parlayed their status into extra leisure at home, either by paying for housekeepers or ignoring the housework. Finally, Chicana wives at the lowest end fared least well. The teacher's aides, entry level secretaries, day care providers, and part-time employees did the bulk of the work at home whether they were married to mechanics or lawyers. When wives made less than a third of what their husbands did, they were only able to get husbands to do a little more if they were working at jobs considered "below" them—a telephone lineman, a painter, an elementary school teacher.

These interviews with Chicano couples corroborate results from previous depth-interview studies of Anglo couples and suggest that the major processes shaping divisions of labor in middle-class Chicano couples are about the same as those shaping such divisions in other couples.[5] That is not to say that ethnicity did not make a difference to the people we talked with. They grew up in recently immigrating working-class families, watched their parents work long hours for minimal wages, and understood first-hand the toll that various forms of racial and ethnic discrimination can take. Probably because of some of these experiences, and their own more recent ones, our informants looked at job security, fertility decisions, and the division of family work somewhat differently than their Anglo counterparts. In some cases, this may give Chicano husbands in working-class or professional jobs license to ignore more of the housework, and might temper the anger of some working-class or professional Chicanas who are still called on to do most of the domestic chores. If our findings are generalizable, however, it is those in between the blue-collar working class and the upper-middle-class professionals who might be most likely to share family work.

Assessing whether my findings apply to other two-job Latino or other minority couples will require the use of larger, more representative samples. If the limited sharing we observed represents a trend—however slow or reluctant—it could have far-reaching consequences. More and more mothers are remaining full-time members of the paid labor force. With the "post-industrial" expansion of the service and information sectors of the economy ... Chicanos and other minorities will be increasingly likely to enter white-collar working-class occupations. As more minority families fit the occupational profile of those we studied, we may see more assumption of housework and child care by the men within them.

Regardless of the specific changes that the economy will undergo, we can expect ethnic minority men and women, like their white counterparts, to continue to negotiate for change in their work and family roles. Economic and institutional factors will undoubtedly play a major part in the shaping of these roles, but social and personal factors will also be important. Reluctant husbands will be unlikely to accept even partial responsibility for the homemaker role unless wives are accepted as co-providers....

NOTES

1. For a discussion of the ways that economies of gratitude are shaped by past events see Karen Pyke and Scott Coltrane, Entitlement, Obligation, and Gratitude in Remarriage (paper presented at the Annual Meeting of the Pacific Sociological Association, San Francisco, California, April 1995)....
2. Stacey, *Brave New Families* [(New York: Basic Books, 1990),] p. 17.
3. Donna H. Berardo, Constance Shehan, and Gerald R. Leslie, "A Residue of Tradition: Jobs, Careers, and Spouses' Time in Housework," *Journal of Marriage and the Family* 49 (1987): 381–390; Catherine E. Ross, "The Division of Labor at Home," *Social Forces* 65 (1987): 816–833.
4. [Patricia Zavella, *Women's Work and Chicano Families: Cannery Workers of the Santa Clara Valley* (Ithaca, NY: Cornell University Press, 1987)]; Stacey, *Brave New Families.*
5. See, for example, Hochschild, *Second Shift* [(New York: Viking Penguin, 1989)]; Hood, [*Becoming a*] *Two-Job Family* [(New York: Praeger, 1983)].

Parental Androgyny

David Popenoe

Gender studies have finally come of age. Once a synonym for "female studies," they have now moved on to encompass the other half—"male studies." In the February 3, 1993, issue of *The Chronicle of Higher Education,* twenty-eight recent and forthcoming books on men and masculinity were listed and discussed. Their common theme was expressed in the headline "Scholars Debunk the Marlboro Man: Examining Stereotypes of Masculinity."

One troubled reaction to these books was that by Christopher Lasch who stated in *The New Republic:* "[They] leave us with the sinking feeling that this is only the beginning." But even more troubling than their quantity is what these works say, or do not say. In suggesting guidelines for the "new man," few of these books express a concern for men as fathers, much less have anything useful to say about the topic. Their main interest is in men as pro-feminists, men as homosexuals, men and male friendships, and men who are able to cry in the movies.

Yet, the real "masculinity crisis" today is not too many Marlboro Men, but too few fathers. All over America men are abandoning their wives and children and remain apart from family life. The widespread, voluntary father-absence from the American family today is strongly associated with two phenomena: divorce and non-marital births. It is not politically correct these days to say positive things about the 1950s era, but it is a fact that fathers participated more in the lives of their biological children then as a result of high marriage and low death rates, and divorce rates that were not out of bounds. With the startling increase in divorce today, a child's chance of making it to adulthood with a biological father in the home is only about 50 percent. About half of non-custodial divorced fathers drop out of the lives of their children, and, for those who do not, their presence is often minimal. Even more remarkable, the non-marital birth rate has jumped from 5 percent in 1960 to a current 28 percent. In most cases, the father is out of the picture; in many, he could care less, or the mother could care less about him.

These trends toward family decline strongly contribute to the deteriorating well-being of children. The evidence suggests that we may have the first generation of children and youths in our history who are less well off—psychologically, socially, economically, and morally—than their parents were at the same age. While father absence is by no means the sole cause of this deterioration, it is heavily implicated.

About the only contribution that some of these books make to this egregious national problem is to put forth the ideal of the "new father." Assuming that men wish to become fathers at all, it is said, fathers should become more like mothers. Men should become more nurturing and share homemaking activities with their working wives, including early infant care, on a fifty–fifty basis. Parental gender roles are entirely learned, we are told, and it is perfectly possible and reasonable "for daddies to become mommies."

The "new man" in the home is an exact parallel to the "new woman" in the workplace, the woman who is able to (and should) do everything in the workplace that men have always done. If women can do everything that men have done in the workplace, why can men not do everything women have done in the home? Indeed, the new father in the home is seen to be absolutely essential if women are to achieve equality in the workplace and still function as mothers.

In addition to widespread absence, the greatest difference between fathers of the present and of preceding generations is that today those fathers who are still present do participate much more in traditional female activities within the home. We have what sociologist Frank F. Furstenberg, Jr., has aptly labeled a "good dad, bad dad" phenomenon. Men who do function as fathers are more nurturing than their own fathers ever were. Some men have fully incorporated the "new father" role, even to the extent of staying home with young children so that their wives can remain in the labor force full time.

The movement in the direction of the new father is certainly strongly to be applauded. The emergence of family- and child-oriented "good" dads—dads who are nurturing and participate equally in the day-to-day lives of their children—is no doubt a positive development. However, implicit in most discussions of the "new father" is the goal of parental androgyny—fathers and mothers playing essentially the same social roles. Social androgyny may be an appropriate goal in the working world, in family life it is not appropriate. Although it is neither possible nor desirable to return to the traditional nuclear family exemplified by *Ozzie and Harriet,* we must take care not to jettison traditional mother–father roles entirely. Unlike the workplace, family organization is based on very real, biological differences between men and women. Parental androgyny is not what children need. Neither is it a good basis for a stable, lasting marriage.

The literature is overflowing with statements arguing the case against traditional gender roles, but concerns about androgynous gender roles are seldom heard. My concerns are based on evidence derived from social and biological research into requirements for optimal child development and the biological differences between men and women. And they are shaped by speculation about what is ultimately personally fulfilling for adults, and what men and women "really want" out of marriage.

Child Rearing

No one has spoken more eloquently about the requirements for optimum child development than Urie Bronfenbrenner. Two points in a recent summary of his main findings of the "scientific revolution" in the study of human development bear special significance for the present discussion.

1. In order to develop—intellectually, emotionally, socially, and morally—a child requires participation in progressively more complex reciprocal activity, on a regular basis over an extended period in the child's life, with one or more persons with whom the child develops a strong, mutual, irrational attachment and who are committed to the child's well-being and development, preferably for life.
2. The establishment and maintenance of patterns of progressively more complex interaction and emotional attachment between caregiver and child depend in substantial degree on the availability and involvement of another adult, a third party, who assists, encourages, spells off, gives status to, and expresses admiration and affection for the person caring for and engaging in joint activity with the child.

Here we have not just the "main findings of the scientific revolution," but a confirmation of a relatively traditional division of labor in marriage between husband and wife.

The key element in proposition number one is the "irrational attachment" of the child with at least one caretaker. Empirical support for this proposition has grown enormously in recent years, mostly stemming from the many psychological studies conducted by Mary Ainsworth and others that have upheld attachment theory—the theory that infants have a biosocial need for a strong, enduring, socio-emotional attachment to a caretaker, especially during the first year of life. This is what pioneering attachment theorist John Bowlby has called starting life with "a secure base." Empirical studies have shown that failure to become attached, to have a secure emotional base, can have devastating consequences for the child, and that patterns of attachment developed in infancy and childhood largely stay with the individual in adulthood, affecting one's relationships and sense of well-being.

The work on attachment theory has been paralleled by research showing some negative effects of placing infants in group care. A widely discussed finding, as yet still controversial, by psychologist Jay Belsky and others is that extensive (more than twenty hours per week) non-parental care initiated during the first year of life is likely to cause attachment problems (insecurity, aggression, and non-compliance) in children. Some recent evidence suggests that negative consequences may also derive from nonparental care during the second year of life. None of this research is conclusive, but it certainly supports what any grandmother could have told us from the outset—limited infant–parent contacts and non-parental childrearing during the first few years of life represent considerable risk.

There is little or no evidence that limited, high quality daycare has any ill effect on children after the age of three. American children have long gone to "nursery school" at three and four, and group care for children at these ages is common in most other industrialized nations, including Japan.

Why is close contact with a parent so important in the first few years of life? Because parents are motivated, as no one else is likely to be, to provide warm and supportive care for their children. The task of parenting could be, and occasionally is, successfully accomplished by a non-related caretaker, such as a full-time nanny. But attachment is much less likely in group settings with a generally high ratio of caretaker to child and a very high turnover of staff.

Yet, why should the primary parent of a young child preferably be the mother and not the father? There is now substantial evidence that fathers can do the job "if they are well-trained and strongly motivated." But it is much harder to train and motivate men than women for childcare. Most dads do not want to be mom, and they do not feel comfortable being mom. To understand why the sexes are not interchangeable in child care, it is necessary to review the biological differences between them.

Biological Differences

Nowhere in the world has there ever been a society known to exist in which men were the primary caretakers of young children. The reason has much to do with the biological nature of males and females. In recent years, any discussion of biologically influenced sex differences has been fraught with peril. As historian Carl Degler noted in his widely read work *In Search of Human Nature: The Decline and Revival of Darwinism in American Social Thought*, the idea of biological rootedness of human nature was almost universally accepted at the turn of the twentieth century. As the century wore on, however, it all but vanished from social thought in the course of a vigorous (and reasonably successful) battle against sexism and racism.

Understandably, this knowledge blackout on the discussion of sex differences was associated with the desire to challenge centuries-old stereotypes about the capacities of women, and to overcome strong resistances to a more forceful and equal role for women in economic and public life. The only sex differences academics have been willing to accept in the past are that women menstruate, that they have the ability to become pregnant, give birth, and lactate, and that men on average are taller, more muscular, and stronger. But the behavioral implications of even these differences have been left mostly vague.

Recognition of biological influences on human behavior is returning, albeit very slowly. Although the idea is still foreign, even inimical, to most social scientists, in probably no other area has the idea of biological roots to human nature become more widely discussed than in the field of sex and gender. This is evidenced by a cover story in *Time* magazine (January 20, 1993) on "sizing up the sexes" which began, "Scientists are discovering that gender differences have as much to do with the biology of the brain as with the way we are raised."

Across the world's societies, the "natural and comfortable" way most males think, feel, and act is fundamentally different from the way most women think, feel, and act. Not that biology is "determinant" of human behavior; this would be a poorly chosen word. All human behavior represents a combination of biological and sociocultural forces, and it makes little sense, as sociologist Alice Rossi has emphasized, to view them "as separate domains contesting for election as primary causes."

The case can certainly be made, in the interest of equality, that a culture should not accentuate the existing biological differences between the sexes. Cultures differ radically in this respect; consider the difference in gender roles between Arab and Nordic cultures. But an even stronger case should be presented in this time of declining family stability and personal well-being for frank acknowledgment of the very real differences between men and women. Acknowledgment by both sexes of their differences in sexual motives, cognitive styles, and communication patterns, for example, could make for stronger marriages. And recognition that the roles of father and mother are not interchangeable would probably make for better parenting. Apparently many Americans agree, if the popularity of Deborah Tannen's book *You Just Don't Understand: Women and Men in Conversation*, a *New York Times* bestseller for two years, is any indication.

Differences between men and women have universally been found with respect to four behavioral/psychological traits: (1) aggression and general activity level; (2) cognitive skills; (3) sensory sensitivity; and (4) sexual and reproductive behavior. The fact that differences are found universally does not unequivocally mean they are influenced by biology, but the implication is stronger than for most other scientific findings about human affairs. A large body of evidence also points to the fact that many universal differences are rooted in a distinct "wiring" of male and female brains and in pronounced hormonal variation between the sexes. The greatest behavioral difference is in aggression. Almost from the moment of birth, boys tend to be more aggressive and in general have a higher physical activity level than girls. To a large degree, this accounts for universal male dominance in human societies.

Differences in cognitive skills are less well known and perhaps not as great, but they have now been widely confirmed by empirical studies. From early adolescence, males tend to have greater visual-spatial and mathematical ability than females, and females tend to have greater verbal ability than males. (Spatial ability refers to the ability to form a mental picture of the shape, position, geography and proportion of physical objects.) Also females tend to be more sensitive to all sensory stimuli. They typically receive a wider array of sensory information, are able to communicate it better, and place primacy on personal relationships within which such information is communicated.

While male superiority rests with "things and theorems," female superiority rests with personal relationships. Almost from birth, girls are more interested than boys in people and faces, whereas boys "just seem as happy with an object dangled in front of them." That these differences become accentuated when they reach adolescence strongly suggests that hormones play a decisive role, specifically testosterone in men and estrogen in women. The role hormones play is further indicated by the decline of behavioral differences later in life, when hormonal levels drop. It is also worth noting that males are the best and the worst with respect to several of these traits. Males, for example, disproportionately make up math geniuses, but also math dysfunctionals.

Not all behavioral differences, however, have a direct effect on family behavior. Most important for family behavior are differences that stem from the dissimilar role of males and females in sexual activity and the reproductive process. The differential "sexual strategies" of men and women have long been noted; in crude, popular terminology, "women give sex to get love, and men give love to get sex." Sex is something that women have and men want, rather than vice versa. Relationships and intimacy are the special province of women.

Sex and Evolution

Probably the most compelling explanation for male–female differences in sexuality and sexual strategies comes from the field of sociobiology. In evolutionary terms, the goal of each individual's life is to perpetuate one's genes through reproduction and maximize the survival of all those with the same genes. Among mammals, the primary reproductive function of males is to inseminate and for females to harbor the growing fetus. Since sperm is common and eggs are rare (both being the prime gene carriers), a different sexual or reproductive strategy is most adaptive for males and females, with males having more incentive to spread their sperm more widely among many females, and females having a strong incentive to bind males to themselves for the long-term care of their offspring.

Males are more sexually driven and promiscuous while females are more relationship-oriented, thus setting up a continuous tension between the sexes. Psychologist David Buss has found that the strongest predictor of sexual dissatisfaction for American males is "sexual withholding by the wife," and for females "sexual aggressiveness by the husband."

According to sociobiologists, men tend to be more upset by their mate's sexual infidelity than vice versa because a man can never be certain that a child is really his. Women, by contrast, tend to be more upset by the loss of their mate's emotional attachment, which threatens long-term commitment and support.

Male promiscuity à la tomcat is not characteristic of humankind, however. Wide variation in male sexual strategies can be found. As anthropologists Patricia Draper and Henry Harpending have said, male sexual strategies range from the relatively promiscuous and low paternal investment "cad" approach, in which sperm is widely distributed with the hope that more offspring will survive to reproduce, to the "dad" approach, in which a high paternal investment is made in a limited number of offspring. But in every society the biological fathers are identified, if possible, and required to hold some responsibility for their children's upbringing. In fact, compared to other species, human beings are noted for relatively high paternal investment because human offspring have a long period of dependency and require extensive cultural training to survive. The nature of human female sexuality (loss of estrus, for example), too, encourages men to stay around.

Culture, of course, has a major say in which sexual strategies are institutionalized. In industrialized societies high paternal investment is the culturally expected. Monogamy is strongly encouraged in these societies (although "serial monogamy" has become the norm in many nations, especially in the United States), polygamy is outlawed, and male promiscuity is somewhat contained. Because it promotes high paternal investment, monogamy is well suited to modern social conditions.

Whatever sexual strategies are followed, our basic biological nature dictates that society face the problem of how to keep men in the reproductive pair-bond. Sex is rather ill-designed for lasting marriages, especially for males. Margaret Mead is purported to have said that there is no society in the world where men will stay married for very long unless culturally required to do so. This is not to suggest that marriage is not "good" for men, only that their innate biological propensities push them in another direction.

Biologically, male attachment to the mother–child pair is largely through the sexual relationship with the mother. Many anthropologists have noted that motherhood is a biological necessity, while fatherhood is mainly a cultural invention. Because it is not biologically based, a father's attachment to his children must be culturally fostered.

Cross-cultural comparisons show a man is likely to take active care of the children if (1) he is sure they are his; (2) if he is not needed as warrior and hunter; (3) if the mother contributes to food resources; and (4) if male parenting is encouraged by the woman. All these conditions prevail largely in modern societies. Although history is replete with stories of men who have developed very strong attachments to their children, men have almost never been closely involved in childcare in the early stages of life.

Parental Androgyny

Ample evidence suggests that men can make a significant contribution to childrearing, especially with regard to their sons, and that the lack of a male presence poses a handicap for children. The assistance men give to women in the rearing of children may be more important now than ever before because mothers have become isolated from their traditional support systems. More than in the past, it is crucial now to maintain cultural measures that induce men to take an active interest in their families. It should be recognized, of course, that the parenting of young infants is not a "natural" activity for males. To perform well in that role they require thorough training and experience, plus encouragement from their wives.

Moving too far in the direction of androgynous parenting, however, presents many difficulties both for child rearing and for the marriage relationship. First, while females may not have a "maternal instinct," hormonal changes occur during and after childbirth that strongly motivate a woman to care for her newborn. These hormonal changes are linked, in part, to the woman's capacity to breastfeed. Also, several of the sex differences already noted are directly related to this stage of the reproductive process. "In caring for a nonverbal, fragile infant," sociologist Alice Rossi has noted, "women have a head start in reading an infant's facial expressions, smoothness of body motions, ease in handling a tiny creature with tactile gentleness, and soothing through a high, soft, rhythmic use of the voice."

Men seem better able to perform the parental role after children reach the age of eighteen months. By then children are more verbal and men do not have to rely on a wide range of senses. Yet, even at that age men interact with children in a different way than do women. The father's mode of parenting is clearly not interchangeable with the mother's. Men tend to emphasize "play" over "caretaking," and their play is more likely to involve a "rough-and-tumble" approach.

Reasonably sex-typed parenting in which mothers are "responsive" and fathers are "firm" also seems to have its value. One research review determined that "children of sex-typed parents are somewhat more competent than children of androgynous parents." Social psychologist Willard W. Hartup concluded: "The importance of fathers, then, may be in the degree to which their interactions with their children do not duplicate the mother's and in the degree to which they support maternal caregiving rather than replicate it."

Less widely discussed, certainly much more speculative but probably no less important, are the effects of androgyny on the marriage relationship. Many men, being of a more independent spirit, will simply avoid marrying and having children if they face having to give up their independence and engage in "unnatural" nurturing and caretaking roles. And it is not as if they had few alternatives. The old system was largely based on the marital exchange of sex for love. If a man wanted regular sex (other than with prostitutes) he had to marry. Today, more permissive sexual standards and a huge pool of single and divorced women (to say nothing of married women) provide abundant opportunities for sex outside a permanent attachment. This sociocultural reality may help to explain men's current tendency of delaying marriage, and the growing complaint of women that "men will not commit."

Nevertheless, most men eventually do marry and have children, and when they do they receive enormous personal benefits. The real concern, therefore, is not men's delay of marriage (it is largely to the good), but what happens to the marriage afterwards. If the best thing parents can do for their children is to stay together and have a good marriage, one serious problem with the "new-father" alternative, in which dad tries to become mom, is that such a marriage may not prove very enduring. This is an issue seldom discussed by "new father" proponents. Marriages which follow this alternative, especially those in which a "role-reversal" has taken place, have a high likelihood of breakup.

Why should a marriage in which the husband is doing "just what he thought his wife always wanted" be at high risk? We can only speculate about the answer by looking at the nature of modern marriages. Marriages today are based on two factors: (1) companionship, that is, husband and wife are expected to be close friends; and (2) romantic love based on sexual attraction, a biologically rooted phenomenon which expects husband and wife to be each other's exclusive sexual partners.

The joining of these two principles is not without its problems. For a good companion, you want someone with whom you have a great deal in common. But for a sexual partner, people tend to be attracted to the differences in the other. Therein lies a festering tension that must be resolved if modern marriages are to endure—a couple must have enough in common to remain best friends, but be different enough so that sexual attraction is maintained. In strong marital relationships, differences are viewed as complementary;

the relationship is characterized by balanced gender-differentiated behavior and equitable division of labor.

The effect of social androgyny in a marital relationship can best be studied in Northern European countries where modernist trends are most advanced, the social environment is relatively benign, and the pursuit of androgyny has gone further than anywhere else. Despite men's growing reluctance to "commit" to marriage, once the marriage has taken place, it is women who have shown an increased tendency to seek a divorce. One possible new reason for this emerged from a discussion with several prominent Swedish marriage counselors.

Sweden is probably the world's most androgynous society, in the sense that men are more involved in childrearing and other domestic activities than they are anywhere else, and women are freer of domestic duties. I asked the marriage counselors, what has been the biggest change over the past quarter century in the marital problems of their clients.

In the 1960s, it was usually the wife who sought counseling. In time-tested manner, her husband had walked out on her for someone else, and it was she who wanted to hold the marriage together—partly for the sake of the children. Today, the exact opposite situation has arisen. In three quarters of all cases, it is the husband who seeks out the marriage counselor to try to hold the marriage together. Why? Because the wife has walked out on him.

Unlike the errant husband of yore, the wife usually does not have another sexual partner, but has simply become bored with her husband and has lost sexual interest. She wants to try life on her own as a single parent. Most of these cases involve androgynous marriages in which husbands have tried hard to become "moms." Not incidentally, these husbands are very upset about the impending divorce, in part because they have become so involved with their children that they abhor the thought of losing them. (In Sweden, women normally receive child custody.)

This could be a misinterpretation. It could be characteristic of only a handful of middle-class Swedes. And it could be argued that the Swedish wives are only doing what their husbands have been doing for generations. But if true, something is wrong. Is this the way we want the story of modern marriage to end?

A related example of the marital consequences of social androgyny comes from neighboring Norway, via an ethnographic study of marital change over the past few decades among working class couples living in a small town. Anthropologist Marianne Gullestad found that these couples have relentlessly tried to incorporate into marriage and family life the goal of "equality as sameness" that is so prominent in Norwegian economic and public life. "The change in the way of thinking and arguing," she notes, "is one from complementarity to unity, from segregation to jointness, from asymmetry to symmetry." Whereas marriages were once based on an interdependence of tasks and functions, now they are based on endless negotiation, held together exclusively by emotional "loyalty" that stems originally from romantic love.

This aptly describes marital trends in all modern societies. The problem is that such marriages are very fragile and subject to breakup, in Norway as elsewhere. Gullestad identifies one likely reason:

> [T]here is a contradiction between romantic love and the desired equality in the division of tasks, because romantic love implies imagination and mystery, and, therefore, some cultivation of otherness. Romantic love implies imagination, adventure, excitement, that the two genders are able to be a little secretive to each other, and that is doubtless difficult if they strive to define their relationship in terms of being more and more similar, more and more the same.

She concludes that equality as sameness between the spouses within the household makes the expression of gender identity problematical in a culture that stresses sexuality and monogamous romantic love as the basis for marriage. This then seems to create certain tensions between the expectations of love and the gender neutral division of household tasks.

Gullestad may have focused too sharply on the division of tasks. Perhaps the emphasis on "equality of sameness" is more consequential in the expectation that men and women are emotionally the same and have the same talents and abilities. But surely she is on to something. The basis of sexual and emotional attraction between men and women is based not on sameness but on differences. Childrearing couples who have been able to stay together and remain interested in each other for a long period of time (an important area for new research) are not likely to be couples who are relentlessly pursuing the ideal of social androgyny.

There appear to be sound biological and sociological reasons why some gender differentiation of roles within childrearing families is necessary for the good of society. Gender differentiation is important for child development, and probably important for marital stability. While the fully equal participation of both parents in childrearing is essential, fathers are not the same as mothers, nor should they be. Rather than strive for parental androgyny in the home, and be continuously frustrated, we would do much better to acknowledge, accommodate, and appreciate the very different needs, sexual interests, values, and goals of each sex.

Differentiation of roles by gender, however, is now mainly of importance in only one institutional sphere of society—the family—and even there for only the relatively short phase of life when young children are being reared. Gender differentiation no longer applies to life in its entirety, as once was the case. This leaves adults abundant time, in the non-childrearing phases of their lives, for the pursuit of self-fulfillment through social roles of their own choosing.

Understanding the Future of Fatherhood: The "Daddy Hierarchy" and Beyond

Arlie Russell Hochschild

In her memoir, *Sweet Summer*, Bebe Moore Campbell describes a conversation between four African American girls growing up in the urban middle-class in the 1950's. Their fathers had divorced their mothers, but to varying degrees the girls were still "daddies' girls." Comparing their fathers to a prior image of a "good dad" in a "good family" in their time and place, they located them as higher or lower on a "daddy hierarchy." (The children presumed the presence and involvement of their mothers and so declined to arrange a hierarchy of mothers.) At the top of the girls' list was the daddy of a little white girl who spent time with her and built her a beautiful doll's house. In the middle were daddies present but preoccupied with repairing broken cars, or daddies who loved their children but spanked them too hard. At the bottom were "deadbeat dads" who disappeared from their children's lives altogether.[1]

Drawing on the American experience, I trace changes in both the ideals and reality of fatherhood. The older ideal of the father who commands authority and pays the bills has partly given way to the ideal of the nurturant new father who bonds with his child (but still pays some bills.) In addition, this new ideal of the nurturant father is born in a context of multiplying ideals and images of a good father and a good family.

As the ideals of fatherhood have changed and diversified, so too have the realities. Compared to their fathers and grandfathers, modern new fathers are more tightly bonded to their children, while deadbeat dads and anonymous sperm donors in fertility clinics are far less so.

Animating changes in both the ideal and reality of fatherhood, I shall argue, are two clusters of trends. One set of trends presses women to work outside the home, which in turn sets up pressures for men to do more at home. Doing more at home, men often strengthen bonds with their children. The second cluster of trends push in the opposite direction—namely, toward more labile bonds between men and women and weaker bonds between men and children. These two clusters of trends operate in a context of "the capitalization of emotions" and influence the degree to which men emotionally "invest" in their children. Cultural ideals guide the way. The future of fatherhood will reflect the effect of these two clusters of trends on behavior and emotional life. How far these trends go in influencing fatherhood depends, in part, on the degree to which culture shields people against the rationalization of emotional life.[2]

Trends in Ideals: The New Father and the Alternatives

Forty years ago it was socially acceptable for a father to come home after work, pick up the newspaper, wait for dinner, play with his children when he felt like it and enforce his authority as required. That was "good enough." Today, in addition to working for pay, a father is expected to attend the birth of his children, be an interested guide, an engaged friend and warm presence to his children at home (Cutright, 1986). As an ideal in Western culture, the "new father" has come to challenge the older ideal of the breadwinner, even as it sometimes blends with it.[3]

Ideals function differently in postindustrial society than they have in earlier times. In an ever more rationalized capitalist culture, the principles of public life increasingly apply to private life as well. Just as capitalists invest and divest economic capital in more or less profitable enterprises, so fathers invest—and tragically, divest—emotional capital in their children. Ideals guide "investment strategies." I do not mean by this that modern fathers are more detached from their children, only that emotional men increasingly live in a culture of emotion management, a culture that demands a capacity for attachment and detachment.

The modern mix of ideals of fatherhood thus point to a mixture of emotional strategies. Even as men are increasingly encouraged to invest more emotional capital in their children, they are also influenced by a culture of emotional "deregulation" which frees them to "divest" from their children and become deadbeat dads. Much new age ideology of self-discovery may correspond to new notions of "moveable capital" applied to the emotional realm.

Any ideal of fathering corresponds to a certain transfer of resources from father to child. These resources can be seen as various forms of capital—for example, economic, cultural and emotional capital (Bourdieu, 1984). A father may pass on his farm to his son (economic capital). Or he may pass on occupational training (cultural capital). We can think of the "new father" ideal as the promise to transfer a form of emotional cultural capital from father to child.[4]

Some children are "rich" and others "poor" in emotional capital. That is what the "daddy hierarchy" describes. Like material capital, paternal emotional involvement helps reproduce individual class standing. Insofar as social class is correlated to marital stability, and marital stability to fatherly involvement, emotional capital can become a means through which class reproduces itself. Insofar as social class is not related to marital stability or marital stability to a father's emotional investment in his children, emotional capital is its own thing. Whether "emotional capital" is linked to, or independent of social class, the ideal of the involved father is an ideal sustaining the idea of emotional investment in one's progeny.[5]

At the same time, the ideal of the "new father" is a contested ideal. Perhaps most American children who grow up without strong emotional bonds with their fathers, like the girls of *Sweet Summer,* nonetheless accept the ideal of "the new father" (Billingsley, 1993). But in addition, a growing minority have come to question the nuclear family to which the "new father" is the latest adaptation. Other systems of transfer of emotional capital—by those who are not the biological father—are proposed as equally beneficial to a child. Lesbian mothers who conceive children through artificial insemination, heterosexual mothers single by choice or necessity, seek to legitimate alternatives to the "new father." The role and ideal of parent is transferred, as it is in some non-Western societies, to related or unrelated males, or to females. To escape the responsibilities of the new father, Barbara Ehrenreich has argued, some fathers themselves reject the ideal (Ehrenreich, 1983). Thus, on the cultural horizon are those who question both the indispensability of a child's emotional tie to the biological father, and the principle of legitimacy. The daddy hierarchy is no longer a hierarchy with an up and a down, but a set of parallel ideals of parenting. In this view, neither kinship nor gender need guide the emotional investments in children that children need.

Trends in Reality: Diversity of Fatherhoods

Parallel to the growing diversity in ideals of fatherhood is a diversity in the realities of it. The breadwinner/authority father has been marginalized by the new father on the one hand and the deadbeat dad on the other (Gerson, 1993).

Two clusters of trends have fostered this spread of realities. On the one hand, inflation and the globalization of capitalism have reduced male wages, creating a need for women to contribute to the family income. At the same time, a declining birth rate and higher rates of female education and the industrialization of housework have created opportunities for women to work. Today, two out of three American mothers with preschool children work outside the home, and half the mothers of children aged one and under. As ideal and reality, the new father is partly a response to this new reality.

In my research on 50 two-job couples in the San Francisco Bay Area, I found that one out of five working husbands were "new men" in the sense of fully sharing the care of the children and home and fully identifying themselves as men through this sharing (Hochschild, 1989). In Michael Lamb's 1986 review of large-scale quantitative studies on fathering, he distinguishes between engagement (for example, feeding child, playing catch), accessibility (cooking in the kitchen while the child plays in the next room) and responsibility (being the one who makes sure the child gets what he or she needs). When wives go out to work, men become more engaged and accessible but not more responsible for their children (Lamb 1986, pp. 8, 11). In two out of three ways, men are doing more.

At the same time, another cluster of trends points in an opposite direction. The rising rate of divorce and unwed pregnancies is related to weakening bonds between fathers and children. The United States has the highest divorce rate in the world; half of all marriages now end in divorce (in the Netherlands in 1980, one in four marriages ended in divorce). Sixty percent of American divorces involve children. Of divorces involving children, roughly half of the fathers eventually lose touch with their children. In his large-scale study of children of divorce, Frank Furstenberg found that nearly half of the children had virtually no contact with the non-custodial parent (90 percent of whom were fathers) within the last year. One out of six had seen him as regularly as once a week (Furstenberg, 1983).[6] The proportion of children living with two parents has declined from 85 percent in 1970 to 72 percent in 1991.

Fathers who lose touch with their children often retreat into what Judith Wallerstein so poignantly calls phantom relationships with their children, putting a photo of the child on an office desk, and thinking, "My child can call me any time he wants" (Wallerstein, 1989). Such fathers imagine a relationship at one end that a child does not feel at the other.

In addition, based on the National Survey of Children, James Peterson and Nicholas Zill were able to compare the relationship of children (aged 12 to 16) with their parents as this varied according to different types of family situations. Even among children living with both biological or adoptive parents, a scant 55 percent had positive relations with both parents. Of children living only with their mothers, 25 percent had good relations with both parents. Of children living with just their fathers, 36 percent had good relations with both parents perhaps because the mothers stay more involved (Peterson and Zill, 1986).

Parallel to the rise in divorce, is a rise in the rate of out-of-wedlock pregnancies largely (though not exclusively) associated with the growth of poverty. According to Dugger, the percentage of children born to unwed parents increased from 5 percent in 1958 to 18 percent in 1978 to 28 percent in 1988 (Dugger, 1992). The vast majority of unwed mothers know the identity of the father, and many cohabit with them, but the breakup rate is higher for cohabiting than for married couples, and fewer than a fifth of unwed mothers report receiving child support for the year prior to breakup (Furstenberg, 1991).[7] These trends make children more vulnerable to emotional "divestment."

To sum up, one cluster of social trends leads men to become more involved in their children's lives than their fathers or grandfathers were. Another set of trends leads men to become less so while a diminishing number continue the tradition of their fathers as traditional breadwinners. The overall picture is thus one of increasing diversity.

Social Contradictions and Male Identity Making

To be sure, in the past, there were always many ways of being a father. Fathers deserted families under the guise of seeking work, migrating, or under no guise at all. Similarly many breadwinner fathers were very emotionally engaged with their children even when it was not their "role" to be. But the sheer diversity within the realm of values and behavior, and more important, the contradiction between the ideal of the new father and the reality of the disengaged dad are new, at least in scope.[8]

As a consequence, fathers of all sorts have a more active relationship to culture. Fatherhood is increasingly that which one does, and less what one simply is. For example, a married father may come home from work and play ball in the back yard with his son. The boy is there. The yard is there. The ball is there. His father played with him when he was his son's age. The context, the cake of custom, tells him how to be a father. A divorced father, who takes a job in another city, must decide to call his child and may get his ex-wife on the phone instead. He must figure out whether talking to his child is more important to him than avoiding talking to his ex-wife. Will he be the kind of father who waits for his child to call him? Or the kind of father who calls no matter what? The culture provides examples of both kinds. The context, the cake of custom, does not make the key decision for him to quite the degree that it does for the ball-throwing father in an intact marriage. As the kin system weakens its controls on both men and women, fatherhood, like much else in life, becomes more a matter of active choice. As Anthony Giddens notes in *Modernity and Self Identity,* the modern individual does not so much receive an identity as make it (Giddens, 1991). One aspect of this bracing constructionist stance toward culture, for fathers, is the decision about how much emotional capital to invest in one's children. The cultural signals about this are mixed. More investment than ever before is recommended, but less investment than ever before is occurring.

The contradictions facing men are likely to differ according to social class. Here a certain irony unfolds. The cultural ideal of the new active father has changed much faster, especially in the middle classes, than the reality of the new father. In a reversal of Ogburn's theory of "culture lag," we can say that for many middle-class men, there is a reality lag. Many middle-class men will want to be new fathers. They will fall in love with educated women who have or want professional careers, and be able to attract such a wife partly because they offer to be new fathers. But forced into an inhospitable career system, many will live with a contradiction between thinking new father but acting old father.

On the other hand, among working-class men, the reverse may occur. Working-class men (whose less educated wives are more likely to prefer to stay home, and who cannot afford paid help) often cherish a more traditional ideal but nonetheless do a great deal with home and children (Rubin, 1976; Lamphere et al., 1993). In both cases, men are living with ideals that do not fit the reality of their lives.

In the future, we may well see the middle-class ideal of the new active father spreading, as ideals often do, down the social class ladder. This has begun already. At the same time, the working-class reality of economic necessity for two incomes, and the availability of less desirable jobs may well be rising up the class ladder. In the end, for stable couples in both classes, reality and ideal may increasingly point together toward the new active father.

Three Scenarios of the Future

Given present trends, I can visualize three scenarios. In the first one, as more women work, the need for the new father increases as does the appeal of the ideal; the reality and the ideal go together. In the second scenario, family ties continue to loosen, and the ideal of the new father becomes more removed from the reality of life; indeed that ideal increases and develops a powerful life of its own. In the third scenario, family ties loosen, but in doing so, they weaken the ideal of the new father, following the logic: if dads are likely to divorce moms, and if moms go with children, maybe fathers should not get so involved to begin with. I expect the future of fatherhood to reflect the presence of—and a tension between—all three scenarios.

New Father in What Kind of Culture?

The mix of family scenarios will vary depending on economic and political forces. But it will also be shaped by culture, and in particular by ideas about what children need. While the United States is a youth-oriented culture, it is not a child-oriented one. Many parents cherish their children but devalue the work of raising them. They focus on their children, privately, one by one, but largely ignore the social world in which they grow up.

Thus, even if more men become active fathers they may do so in a context of declining state subsidies for children, cuts in public school budgets, shortening of library hours, low wages for daycare teachers, and the absence of family-friendly reforms which would allow parents more time at home. The new father may be moving in the right direction while his society moves in the wrong one.

The new father may have to move against a deeper, more long-term, cultural current as well. I would argue that the terms of understanding about time, appreciation, and honor that had previously been dominant at work are increasingly, if unwittingly, adopted at home. Perhaps this is an instance of what Jurgen Habermas has called the "colonization of the life world." As a colonized sphere, the home has become less able to exert a magnetic draw on men or women, while the workplace has increased its draw. The social and cultural trends pushing both men and women into a highly valued formerly male public work culture—in which work is a major source of self-appreciation, security and enjoyment—are stronger than the trends pushing men or women into active involvement with children at home, a relatively less valued, formerly female realm.

Thus, regardless of gender, in much of the American middle and even working classes, the draw of work seems to be increasing while the draw of family is decreasing and at the same time, for some, work becomes more like home (Hochschild, 1994). More than we have realized, work competes with family as a haven in a heartless world.[9]

Conclusion

Fostering the new father will take both social programs and a more basic shift in culture. Enlightened societies can foster three kinds of support to encourage active fathering. The first could reduce strains on existing families. Work sharing and company family-friendly policies could spread work, give parents more control over work time, more flexible and shorter hours. In the last fifteen years, many large American companies have instituted family-friendly policies. In 1993, President Bill Clinton also signed into law the Family and Medical Leave Act, which permits workers in companies with 50 workers or more twelve weeks of unpaid leave. By reducing the external strain on families, such policies can enable more fathers to live up to the ideals they already hold.

Second, while we cannot, and probably should not, prevent all divorce among parents of small children, we can reduce the negative impact of divorce on children. We could expand mediation programs and counseling services that could help fathers keep contact with children after divorce. Third, we can increase the supply of nurturant men for children who are estranged from their fathers. A number of projects throughout the U.S. are reaching out to men through daycare father outreach program. Children who lack regular contact with their fathers can get contact with fatherly men who volunteer in nursery school or grade school. Thus, we can increase the supply of father substitutes or quasi-substitutes.[10]

Finally, we need to think out the chain of connections between the rationalization of emotional life, the new mobility of emotional capital, and the important needs of small children. As a culture in which capitalism may have made the deepest inroads, the United States may prove to be the handwriting on the wall for countries like Holland, even as the fifties childhood of *Sweet Summer* foretold the future of many children in America. Or it may not. Either way, the future can be one of strengthening the trends that produce daddies children feel lucky to have.

NOTES

1. One purpose of the girls' conversation is to try to agree on the definition of a "good daddy." Is a good daddy one who does not buy a new car for himself before he buys necessities for his child? Is he one who visits and calls often? Is he one that negotiates reasonably with one's mother about one's welfare? Answers to such questions become the basis of their "daddy hierarchy."
2. Trends run on cultural soil. I take the rationalization of emotional life to be a key aspect of that "soil." Drawing from the Frankfurt School of Sociology I focus here on rationalization—a process that applies to four aspects: (a) a mentality (the tendency to "save" time, plan, divide life into means and ends); (b) a

norm (a value on the principle of efficiency); (c) a set of social relations (time and purpose limited, based on the principle of exchange); and (d) a set of wider social circumstances (availability of many services, commodification in advanced capitalism). Thus, the rationalization of emotional life refers to the shaping and management of emotions to adapt individuals to life which has become rationalized in these four ways (e.g. to try to feel what it is useful to feel).

3. As a cultural image, the new father masculinizes a cultural turf formerly seen as the province of women. So in the film *Kramer vs. Kramer* the nurturant dad who kisses his son goodnight is also shown as athletic and undomestic (e.g. eating from cans "the way a man would"). Each cultural notion of a good father thus has differing relations to the image of a manly man.
4. The current research on fatherhood provides material and guideposts for an as yet undeveloped structural theory of fathering. Currently two theories might be mentioned: (a) interest theory, according to which women mainly rear children because it is in men's interest to leave it to women (Polatnick, 1974); (b) psychoanalytic theory, according to which women mainly rear children because, for social reasons, they want to do so more than men do (Chodorow, 1980).
5. We know that the higher the class, the more stable the marriage in general, and the more stable the marriage, the more father–child contact. But we also know that some upper-class marriages are unstable and some children of stable marriages have little real contact with their fathers.
6. In his study of sixty African American mothers and the fathers of their children, Frank Furstenberg found an "unambiguous and universal" norm that biological fathers are obliged to support their children materially and emotionally (Furstenberg, 1991, p. 8). But most of these poor African American young fathers nonetheless gradually lose touch with their children. Men offered different accounts for why this occurred: "It's not my child; Someone else has taken my place; My support isn't going to the child; I don't have the money; She doesn't let me see my child" (1991, pp. 12–14). For their part, the young mothers argue that "men are spoiled," or "selfish, indulges," or "men can't accept the responsibilities of parenthood," and "aren't ready to become fathers" (Ibid. 15). Furstenberg found a high level of mistrust between men and women. At the same time, the father and (though this was less obvious) the mother needed to form what Furstenberg calls an umbrella contract between the pair in order to facilitate the father–child bond.
7. In any society, some sub-groups become a bellwether for the rest of society. In the United States, African Americans of the 1950's and 1960's foretold the future for white Americans in the 1980's and 1990's. The trends that characterized African American families earlier spread to whites later. The tendency for mothers to work outside the home, for men to share the care of the home and childrearing (black men still do more than white men, though still less than black women), and the higher rate of non-marital pregnancy and divorce—increasingly describe white families in American today. Even the bifurcation of fatherhoods into new men and deadbeat dads occurred first for blacks, later for whites. The reason, I believe, has to do with the absence for blacks earlier—and for whites later—of the central breadwinner wage for men. In a sense, then, the new man is not so new.
8. The cultural context of fathering in the future is thus likely to remain plural. This means that each type of father becomes the cultural context for the other. Each type of father is likely to live in a fathering subculture congenial to himself, but different kinds of fathers also know about each other's subcultures. Active fathers see films, hear stories about, have friends who are—and perhaps themselves once were—deadbeat dads. For their part, deadbeat dads see films, hear stories about or have friends who are, or were once themselves, new fathers. They compare themselves to other kinds of fathers. They define themselves as more or less lucky, emotionally richer or poorer in relation to active fathers. Just as unionized companies set the context for non-union companies, and the communist world once set the context for the capitalist world, so different realities of fatherhood set the context for each other. This context-setting is as important as the diversity on which it is based.
9. This is because the family has become more like a workplace; often the tired worker returns to unwashed dishes, unmet human needs, and no time to relax. Ironically, the emotional engineers of late capitalism have made the workplace for some people in certain ways more like home. One can chat and joke with co-workers, get help with problems, and feel appreciated for one's skills (Hochschild, 1994).

10. James Levine, head of the Fatherhood Project at the Families and Work Institute in New York City, notes, "often early childhood communication systems are designed—albeit unintentionally—to promote interaction between female parents and female staff" (Levine, 1993, p. 12). In his activist booklet, *Getting Men Involved: Strategies for Early Childhood Programs,* he describes various initiatives. At the Ounce of Prevention Fund in Chicago, which serves mainly minority families, organizers hire male involvement specialists who recruit fathers and other male relatives to work with the children. Another program combines parenting classes for men estranged from their children with adult education programs, computer training and early childhood certification. Most remarkable of all is a program created by the Texas Migrant Council which organizes daycare vans to follow the migratory pattern of fruit pickers, providing schooling for the children of migrant workers, and opportunities in daycare jobs to men.

REFERENCES

Anderson, E. (1990). *Street Wise: Race, class and change in an urban community.* Chicago/London: University of Chicago Press.

Arendell, T. (1968). *Mothers and divorce.* Berkeley/Los Angeles: University of California Press.

Arendell, T. (1992). After divorce: Investigations into father absence. *Gender and Society 9,* 4, 562–86.

Billingsley, A. (1993). *Climbing Jacob's ladder: The enduring legacy of African American families.* New York: Simon & Schuster.

Bourdieu, P. (1984). *Distinction: A social critique of the judgement of taste.* Cambridge, MA: Harvard University Press.

Campbell, B. M. (1990). *Sweet summer: Growing up with and without my dad.* New York/London: Harper Collins.

Chodorow, N. (1980). *The reproduction of mothering.* Berkeley, CA: University of California Press.

Cutright, P. (1986). Child support and responsible male procreative behavior. *Sociological Focus, 19,* 1, 27–45.

Dugger, C. W. (1992). Establishing paternity earlier to gain child support later. *New York Times* (Jan. 3): A1–B6.

Ehrenreich, B. (1983). *The hearts of men: American dreams and the flight from commitment.* Garden City NY: Anchor Books.

Furstenberg, F., Jr., Nord, C., Peterson, P. & Zill, N. (1983). The life course of children of divorce: Marital disruption and parental contact. *American Sociological Review, 48,* 656–668.

Furstenberg, F. Jr. (1991). *Daddies and fathers: Men who do for their children and men who don't.* Unpublished paper, Sociology Department, University of Pennsylvania.

Furstenberg, F., Jr., & Cherlin, A. (1991). *Divided families: What happens to children when parents part.* Cambridge, MA: Harvard University Press.

Gerson, K. (1993). *No man's land: Men's changing commitments to family.* New York: Basic Books.

Giddens, A. (1991). *Modernity and self-identity: Self and society in the late modern age.* Stanford, CA: Stanford University Press.

Goldscheider, F. K., & Waite, L. J. (1991). *New families, no families? The transformation of the American home.* Berkeley/Los Angeles: University of California Press.

Hareven, T. (1975). Family time and industrial time: Family and work in a planned corporation town, 1900–1924. *Journal of Urban History, 1,* 365–89.

Hertz, R. (1986). *More equal than others: women and men in dual career marriages.* Berkeley/Los Angeles: University of California Press.

Hochschild, A., & Machung, A. (1989) *The second shift: Working parents and the revolution at home.* New York: Avon Books.

Hochschild, A., Lydia Morris, L., & Lyon, S. (Ed.). (1994). Gender relations in public and private: Changing research perspectives. London: Macmillan Publishers (forthcoming).

Lamb, M. E. (Ed.). (1986). *The father's role: Applied perspectives.* New York: John Wiley & Sons.

Lamphere, L., et al. (1993). *Sunbelt working mothers.* Ithaca, NY: Cornell University Press.

Levine, J. A,. et al. (1993.) *Getting men involved: Strategies for early childhood programs.* New York: Scholastic Inc., Early Childhood Division.

Malinowski, B. (1930). *The principle of legitimacy.* In V. F. Calverton & S. D. Schmalhausen (Eds.), *The new generation* (pp. 113–55). New York: Macauley Co.

Peterson, J., & Zill, N. (1986). Marital disruption, parent-child relationships, and behavior problems in children. *Journal of Marriage and the Family, 48,* 295–307.

Pleck, J. H. (1983). Husbands' paid work and family roles: current research issues. In H. Lopata & J. H. Pleck (Eds.), *Research in the interweave of social roles* (Vol. 3). Families and jobs. Greenwich, CT: JAI Press.

Pleck, J. H. (1985). *Working wives, working husbands.* Beverly Hills/London/ New Delhi: Sage.

Poiatnick, M. (1974). Why men don't rear children: A power analysis. *Berkeley Journal of Sociology, 18,* 45–86.

Rodin, N., & Russell, A. (1982). Increased father participation and child development outcomes. In M. E. Lamb (Ed.), Nontraditional families: Parenting and child development (pp. 191–218). Hillsdale, NJ: Erlbaum Publishers.

Rubin, L. B. (1976). *Worlds of pain: Life in the working-class family.* New York: Basic Books.

Sachs, A. (1994). Men, sex and parenthood in an overpopulating world. *World Watch, 7,* 2, 12–19.

Schor, J. (1992). *The overworked American.* New York: Basic Books.

Segal, L. (1990). *Slow motion: Changing masculinities, changing men.* New Brunswick, Rutgers University Press.

Wallerstein, J. S., & Blakeslee, S. (1989). Second changes: Men, women, and children a decade after divorce. New York: Ticknor and Fields.

Weitzman, L. (1985). *The divorce revolution: The unexpected social and economic consequences for women and children in America.* New York: Free Press.

Business and the Facts of Family Life

Fran Sussner Rodgers and Charles Rodgers

Business is a good thing.

Family is also a good thing.

These are simple, self-evident propositions.

Yet the awkward fact is that when we try to combine these two assertions in the new labor force, they stop being safe, compatible, and obvious and become difficult, even antagonistic. Sometimes the most complex and controversial challenges we face have commonsense truths at their roots.

Consider these variations on the same theme:

> Our economy needs the most skilled and productive work force it can possibly find in order to remain competitive.
>
> That same work force must reproduce itself and give adequate care to the children who are the work force of the future.
>
> People with children—women especially—often find themselves at a serious disadvantage in the workplace.
>
> Among Western democracies, the United States ranks number three in dependence on women in the work force, behind only Scandinavia and Canada.

In short, we value both business and family, and they are increasingly at loggerheads.

The Family as a Business Issue

At one time, women provided the support system that enabled male breadwinners to be productive outside the home for at least 40 hours every week. That home-based support system began to recede a generation ago and is now more the exception than the rule. The labor force now includes more than 70% of all women with children between the ages of 6 and 17 and more than half the women with children less than 1 year old. This new reality has had a marked effect on what the family requires of each family member—and on what employers can expect from employees. It is not only a question of who is responsible for very young children. There is no longer anyone home to care for adolescents and the elderly. There is no one around to take in the car for repair or to let the plumber in. Working families are faced with daily dilemmas: Who will take care of a sick child? Who will go to the big soccer game? Who will attend the teacher conference?

Yet employees from families where all adults work are still coping with rules and conditions of work designed, as one observer put it, to the specifications of Ozzie and Harriet. These conditions include rigid adherence to a 40-hour workweek, a concept of career path inconsistent with the life cycle of a person with serious family responsibilities, notions of equity formed in a different era, and performance-evaluation systems that confuse effort with results by equating hours of work with productivity.

Despite the growing mismatch between the rules of the game and the needs of the players, few companies have made much effort to accommodate changing lifestyles. For that matter, how serious can the problem really be? After all, employees still get to work and do their jobs. Somehow the plumber manages to find the key. We know that children and the elderly are somewhere. Why start worrying now? Women's entry into the labor force has been increasing for 20 years, and the system still appears to function.

Nevertheless, we are seeing a rapidly growing corporate interest in work-and-family issues. There are four principal *business* reasons:

First, work force demographics are changing. Most of the increase in the number of working women has coincided with the baby boom. Any associated business fallout—high turnover, lost productivity, absenteeism—occurred in the context of a large labor surplus. Most people were easily replaced, and there was plenty of talent willing to make the traditional sacrifices for success—like travel, overtime, relocation. With the baby boom over and a baby bust upon us, there are now higher costs associated with discouraging entry into the labor force and frustrating talented people who are trying to act responsibly at home as well as at work. In some parts of the country, labor is already so scarce that companies are using progressive family policies as a means of competing for workers.

Second, employee perceptions are changing. Unless we rethink our traditional career paths, the raised aspirations of many women are now clearly on a collision course with their desire to be parents. Before the emergence of the women's movement in the 1960s, many suburban housewives thought their frustrations were uniquely their own. Similarly, for 20 years corporate women who failed to meet their own high expectations considered it a personal failing. But now the invisible barriers to female advancement are being named, and the media take employers to task for their inflexibility.

This shift in women's perceptions greatly changes the climate for employers. Women and men in two-career and single-parent families are much better able to identify policies that will let them act responsibly toward their families and still satisfy their professional ambitions. Companies that don't act as partners in this process may lose talent to companies that do rise to the challenge. No one knows how many women have left large companies because of cultural rigidity. It is even harder to guess at the numbers of talented women who have never even applied for jobs because they assume big companies will require family sacrifices they are unwilling to make.

And it's not just women. In two studies at Du Pont, we found that men's reports of certain family-related problems nearly doubled from 1985 to 1988. (Interestingly, on a few of these items, women's reported problems decreased proportionally, which suggests that one reason women experience such great difficulty with work-and-family issues is that men experience so little.)

In fact, men's desire for a more active role in parenting may be unacceptable to their peers. Numerous reports show that few men take advantage of the formal parental leave available to them in many companies. Yet a recent study shows that many men do indeed take time off from work after the birth of a child, but that they do so by piecing together other forms of leave—vacation, personal leave, sick leave—that they see as more acceptable.[1]

A third reason why more companies are addressing work-and-family issues is increasing evidence that inflexibility has an adverse effect on productivity. In a study at Merck in 1984, employees who perceived their supervisors as unsupportive on family issues reported higher levels of stress, greater absenteeism, and lower job satisfaction.[2] Other studies show that supportive companies attract new employees more easily, get them back on the job more quickly after maternity leave, and benefit generally from higher work-force morale.[3]

Fourth, concern about America's children is growing fast. Childhood poverty is up, single-parent families are on the increase, SAT scores are falling, and childhood literacy, obesity, and suicide rates are all moving in the wrong direction.

So far, the business community has expressed its concern primarily through direct efforts to improve schools. Yet in our studies, one-third to one-half of parents say they do not have the workplace flexibility to attend teacher conferences and important school events. It is certainly possible that adapting work rules to allow this parent–school connection—and trying to influence schools to schedule events with working parents in mind—might have as great a positive effect on education as some direct interventions.

For companies that want to use and fully develop the talents of working parents and others looking for flexibility, the agenda is well defined. There are three broad areas that require attention:

1. Dependent care, including infants, children, adolescents, and the elderly.
2. Greater flexibility in the organization, hours, and location of work, and creation of career paths that allow for family responsibility as well as professional ambition.
3. Validation of family issues as an organizational concern by means of company statements and manager training.

Few companies are active in all three areas. Many are active in none. The costs and difficulties are, after all, considerable, and the burden of change does not fall on only employers. There is plenty for government to do. Individual employees too will have to take on new responsibilities. Corporate dependent-care programs often mean purchasing benefits or programs from outside providers and may entail substantial community involvement. Workplace flexibility demands reexamination of work assumptions by employees as well as employers and often meets with line resistance. A corporate commitment to family takes time to work its way down to the front-line supervisory levels where most of the work force will feel its effects.

Dependent Care

Dependent care is a business issue for the obvious reason that employees cannot come to work unless their dependents are cared for. Study after study shows that most working parents have trouble arranging child care, and that those with the most difficulty also experience the most frequent work disruptions and the greatest absenteeism. Moreover, the lack of child care is still a major barrier to the entry of women into the labor force.

Child-care needs vary greatly in any employee population, and most companies have a limited capacity to address them. But, depending on the company's location, financial resources, the age of its work force, and the competitiveness of its labor market, a corporate child-care program might include some or all of the following:

- Help in finding existing child care and efforts to increase the supply of care in the community, including care for sick children.
- Financial assistance for child care, especially for entry level and lower level employees.
- Involvement with schools, Ys, and other community organizations to promote programs for school-age children whose parents work.
- Support for child-care centers in locations convenient to company employees.
- Efforts to move government policies—local and federal—toward greater investment in children.

Existing child care is often hard to find because so much of the country's care is provided by the woman down the street, who does not advertise and is not usually listed in the yellow pages or anywhere else. Even where lists do exist—as the result, say, of state licensing requirements—they are often out-of-date. (Turnover in family day-care, as this form of child care is called, is estimated at 50% per year.) And lists don't give vacancy information, so parents can spend days making unsuccessful phone calls. Sometimes existing care is invisible because it operates in violation of zoning rules or outside of onerous or inefficient regulatory systems.

In other places—suburban neighborhoods where many women work outside the home or where family income is so high that few need the extra money—there is virtually no child care. Often, too, land prices make centers unaffordable. Infant care is especially scarce because it requires such a high ratio of adults to children. Care for children before and after school and during the many weeks when school is out is in short supply just about everywhere, as is care for "off hour" workers such as shift workers, police officers, and hospital employees.

In addition to the difficulty of finding child care, quality and affordability are always big questions. Cost depends greatly on local standards. In Massachusetts, for example, infant care in centers runs from $150 to more than $200 per week per child due to a combination of high labor costs and strict state licensing standards. Even the highest standards, however, still mean that an infant-care staff member has more to do all day—and more responsibility—than a new parent caring for triplets. In states with lower standards, one staff member may care for as many as eight infants at a time. Up to now, child care in many places has been made affordable by paying very low wages—the national average for child-care staff is $5.35 an hour—and by reducing the standards of quality and safety below what common sense would dictate.[4]

Given all these problems, is it any wonder the companies that want to help feel stymied? While few companies provide significant child-care support today, a very large number are exploring the possibility. We think that number will increase geometrically as the competition for labor grows and more members of the labor force need such support.

One increasingly popular way for companies to address these issues is through resource and referral services. Typically, such services do three things: they help employees find child care suited to their circumstances; they make an effort to promote more care of all types in the communities where employees live; and they try to remove regulatory and zoning barriers to care facilities. Resource and referral services (R&Rs) meet standards of equity by assisting parents regardless of their incomes and their children's ages. And R&Rs work as well for a few workers as for thousands. When the service is delivered through a network of community-based R&Rs, moreover, corporate involvement can also strengthen the community at large.

Although R&R programs can be very helpful, they have limitations. By themselves, they have little effect on affordability, for example, and only an indirect effect on quality, primarily through consumer education and provider training. Also, R&Rs cannot dig up a supply of care where market conditions are highly unfavorable.

A small but growing number of companies provide, subsidize, or contract with outside providers to operate on-site or near-site centers that are available to employees at fees covering at least most of the cost. A North Carolina software company, SAS Institute Inc., provides child care at an on-site center at no cost to employees. The company reports that its turnover rates are less than half the industry average and feels the center's extra expense is justified because it decreases the extremely high cost of training new workers.[5]

Companies that get involved with child-care centers, however, find themselves making difficult trade-offs as a result of the high cost of good care. Many companies won't associate themselves even indirectly with any child care that doesn't meet the highest standards, which means that without a subsidy, only higher income employees can afford the service. But if a company does subsidize child care, it must justify giving this considerable benefit to one group of parents while other parents, who buy child care in some other place or way, get none. One way of avoiding this dilemma is to give child-care subsidies to all lower income employees as an extension of the R&R service, the approach recently announced by NCNB, the banking corporation.

Companies sometimes capitalize centers by donating space or land along with renovation costs or by providing an initial subsidy until the centers are self-supporting. In this way, Du Pont helped a number of community not-for-profit organizations establish and expand existing child-care centers in Delaware. Of course, costs can vary hugely. If a building is already available, renovation and startup costs could be as low as $100,000. In most cases, the bill will run from several hundred thousand to several million dollars.

Businesses are also working more closely with schools to encourage before-school, after-school, and vacation care programs. Such a partnership has been established between the American Bankers Insurance Group and the Dade County, Florida, school system. The school system actually operates a kindergarten and a first- and second-grade school in a building built by the insurance company. In Charlotte, North Carolina, the 19 largest employers have joined forces with the public sector to expand and improve the quality of care.

In any case, employee interest in child care is great, and employees often fix on the issue of on-site care as a solution to the work-and-family conflicts they experience. But helping employees with child care, given the enormity of the problem in the society at large, is a complicated question. More and more companies are taking the kinds of steps described here, but as the pressure grows, business as a whole is likely to focus more attention on public policy.

Of course, dependent care is not just a question of care for children. Studies at Travelers Insurance Company and at IBM show that 20% to 30% of employees have some responsibility for the care of an adult dependent. Traditionally, the wife stayed home and cared for the elderly parents of both spouses, but as women entered the work force, this support system began to disappear. Since the most recent growth in the female work force involves comparatively younger women whose parents are not yet old enough to require daily assistance, the workplace has probably not yet felt the full effects of elder-care problems.

As in the case of child care, studies show that productivity suffers when people try to balance work and the care of parents. Some people quit their jobs entirely. The most immediate need is for information about the needs and problems of the aging and about available resources. Most young people know nothing at all about government programs like Medicare and Medicaid. More often than not, children know very little about their own parents' financial situations and need help simply to open communication.

Unlike child care, elder care is often complicated by distance. In our experience with some 12,000 employees with elderly dependents, more than half lived more than 100 miles from the person they were concerned about. Crises are common. The elderly suffer unexpected hospitalizations, for example, and then come out of the hospital too weak to care for themselves. A service that can help with referrals and arrangements in another city can spare employees time, expense, and anguish. Also, people often need to compare resources in several states where different siblings live in order to make decisions about such things as where parents should live when their health begins to deteriorate.

Conditions of Work

A study at two high-tech companies in New England showed that the average working mother logs in a total workweek of 84 hours between her home and her job, compared with 72 hours for male parents and about 50 hours for married men and women with no children. In other words, employed parents—women in particular—work the equivalent of two full-time jobs.[6] No wonder they've started looking for flexible schedules, part-time employment, and career-path alternatives that allow more than one model of success. For that matter, is it even reasonable to expect people who work two jobs to behave and progress along exactly the same lines as those with no primary outside responsibilities?

Until now, most companies have looked at job flexibility on a case-by-case basis and have offered it sparingly to valued employees as a favor. But increasing competition for the best employees will make such flexibility commonplace. A smaller labor supply means that workers will no longer have to take jobs in the forms that have always been offered. Companies will have to market their own employment practices and adapt their jobs to the demands of the work force.

We all know that the way we did things in the past no longer works for many employees. Our research shows that up to 35% of working men and women with young children have told their bosses they will not take jobs involving shift work, relocation, extensive travel, intense pressure, or lots of overtime. Some parents are turning down promotions that they believe might put a strain on family life. Women report more trade-offs than men, but even the male numbers are significant and appear to be increasing. In our study, nearly 25% of men with young children had told their bosses they would not relocate.

Interestingly enough, few employees seem angry about such trade-offs. They value the rewards of family life, and, by and large, they don't seem to expect parity with those willing to sacrifice their family lives for their careers. Nevertheless, they *are* bothered by what they see as unnecessary barriers to success. Most believe they could make greater contributions and go farther in their own careers—despite family obligations—if it weren't for rigid scheduling, open-ended expectations, and outmoded career definitions. They long for alternative scenarios that would allow them more freedom to determine the conditions of their work and the criteria for judging their contributions.

The question is whether a willingness to sacrifice family life is an appropriate screen for picking candidates for promotions. It would be wrong to suppose that these employees are any less talented or less ambitious than those who don't make the family trade-off. A study we conducted at NCNB showed no evidence of any long-term difference in ambition between people with and without child-care responsibilities. Since fewer and fewer people in our diverse labor force are willing to pay the price for traditional success, to insist on it is only to narrow the funnel of opportunity and, eventually, to lower the quality of the talent pool from which we draw our leaders.

Flexible Schedules In addition to time away from work to care for newborn or newly adopted children, employees with dependent-care responsibilities have two different needs for flexibility. One is the need for working hours that accommodate their children's normal schedules and their predictable special requirements such as doctor's appointments, school conferences, and soccer championships. The other is the need to deal with the emergencies and unanticipated events that are part and parcel of family life—sudden illness, an early school closing due to snow, a breakdown in child-care arrangements.

The most common response to both needs has been flextime. Flextime can be narrowly designed to permit permanent alterations of a basically rigid work schedule by, say, half an hour or an hour, or it can be more broadly defined to allow freewheeling variations from one workday to the next.

Pioneered in this country by Hewlett-Packard, flextime is now used by about 12% of all U.S. workers, while half the country's large employers offer some kind of flextime arrangement. Its effects on lateness, absenteeism, and employee morale have been highly positive.[7] The effects on the family are not as easily measured, but most employees say they find it helpful, and the more scheduling latitude it offers, the more helpful they seem to find it.

A number of companies are considering ways of further expanding the notion of flextime. One alternative, called "weekly balancing," lets employees set their own hours day-to-day as long as the weekly total stays constant. In Europe, some companies offer monthly and yearly balancing. Clearly, this is most difficult to do in situations where production processes require a predictable level of staffing.

In November 1988, Eastman Kodak announced a new work-schedule program that permits four kinds of alternative work arrangements:

1. Permanent changes in regular, scheduled hours.
2. Supervisory flexibility in adjusting daily schedules to accommodate family needs.
3. Temporary and permanent part-time schedules at all levels.
4. Job sharing.

Aetna Life and Casualty too has recently launched an internal marketing effort and training program to help its supervisors adapt to, plan for, and implement unconventional work schedules.

Employees also must assume new roles. In the job-sharing program at Rolscreen Company, for example, employees are responsible for locating compatible partners for a shared job and for ensuring that the arrangement works and that business needs are met.[8] Also, employees are often expected to make themselves available when business emergencies arise. In the best flexible arrangements, employers and employees work as partners.

Part-Time Employment Studies show that a third to half of women with young children want to work less than full time for at least a while, despite the loss of pay and other benefits. Yet we have found in our work with dozens of companies that managers at all levels show firm resistance to part-time work. They seem to regard the 40-hour week as sacred and cannot imagine that anyone working fewer hours could be doing anything useful. Even in companies that accept the need for part-time work, we see managers who refuse to believe it will work in their own departments. Indeed, even the term "part-time" seems to have a negative connotation.

Research on part-time productivity is sometimes hard to interpret, but the studies we've seen indicate that the productivity of part-time workers is, in certain cases, better than their full-time counterparts and, in all cases, no worse. One study comparing part-

Companies That Lead the Way

For years, IBM has steadily increased its efforts to adapt to family needs. It pioneered child-care and elder-care assistance programs. A national resource and referral service network originally put together for IBM in 1984 now serves about 900,000 employees of more than 35 national companies. In 1988, IBM expanded its flextime program to allow employees to adjust their workdays by as much as two hours in either direction and adopted an extended leave-of-absence policy permitting up to a three-year break from full-time employment with part-time work in the second and third years. The company has also been experimenting with work-at-home programs. And earlier this year, it introduced family-issues sensitivity training for more than 25,000 managers and supervisors.

Johnson & Johnson recently announced an extremely broad work-and-family initiative that includes support for elder care and child care, greater work-time flexibility, management training, and a change in its corporate credo.

AT&T recently negotiated a contract with two of its unions that established a dependent-care referral service and provides for leaves of up to one year, with guaranteed reinstatement, for new parents and for workers with seriously ill dependents.

At NCNB, a program called "Select Time" allows employees at all levels, including managers, to reduce their time and job commitments for dependent-care purposes without cutting off current and future advancement opportunities.

Apple Computer operates its own employee-staffed child-care center and gives "baby bonuses" of $500 to new parents. Du Pont has helped to establish child-care centers in Delaware with contributions of money and space. Eastman Kodak has adopted new rules permitting part-time work, job sharing, and informal, situational flextime.

For reasons partly societal and partly strategic, these and scores of other businesses are building work environments that let people give their best to their jobs without giving up the pleasures and responsibilities of family life.

time and full-time social workers found that, hour for hour, the part-time employees carried greater caseloads and serviced them with more attention.[9]

Part-time is not necessarily the same as half-time, as many managers assume. Many parents want 4-day or 30-hour workweeks. Many other assumptions about less than full-time employment are also unwarranted. For example, managers often insist that customers will not work with part-time employees, but few have asked their customers if this is true.

Another axiom is that supervisory and managerial personnel must always be full-time, since it is a manager's role "to be there" for subordinates. This article of faith ignores the fact that managers travel, attend meetings, close their doors, and are otherwise unavailable for a good part of every week.

Career-Path Alternatives It takes a lot of ingenuity and cultural adaptability to devise meaningful part-time work opportunities and to give employees individual control of their working hours. But an even greater challenge is to find ways of fitting these flexible arrangements into long-term career paths. If the price of family responsibility is a label that reads "Not Serious About Career," frustrations will grow. But if adaptability and labor-market competitiveness are the goals, then the usual definition of fast-track career progression needs modification.

The first step, perhaps, is to find ways of acquiring broad business experience that are less disruptive to the family. For example, Mobil Oil has gradually concentrated a wide range of facilities at hub locations, partly in order to allow its employees a greater variety of work experience without relocation.

Another essential step is to reduce the tendency to judge productivity by time spent at work. Nothing is more frustrating to parents than working intensely all day in order to pick up a child on time, only to be judged inferior to a coworker who has to stay late to produce as much. For many hardworking people, hours certainly do translate into increased productivity. Not for all. And dismissing those who spend fewer hours at the workplace as lacking dedication ignores the fact that virtually all employees go through periods when their working hours and efficiency rise or fall, whether the cause is family, health, or fluctuating motivation.

Corporate Mission

Fertility in the United States is below replacement levels. Moreover, the higher a woman's education level, the more likely she is to be employed and the less likely to have children. The choice to have a family is complex, yet one study shows that two-thirds of women under 40 who have reached the upper echelons in our largest companies and institutions are childless, while virtually all men in leadership positions are fathers.[10] If we fail to alter the messages and opportunities we offer young men and women, and if they learn to see a demanding work life as incompatible with a satisfying family life, we could create an economy in which more and more leaders have traded family for career success.

There are four things a company needs to do in order to create an environment where people with dependents can do their best work without sacrificing their families' welfare. It needs to develop a corporate policy that it communicates to all its employees; it needs to train and encourage supervisors to be adaptable and responsible; it needs to give supervisors tools and programs to work with; and it needs to hold all managers accountable for the flexibility and responsiveness of their departments.

The key people in all this are first-line managers and supervisors. All the policies and programs in the world don't mean much to an employee who has to deal with an unsupportive boss, and the boss is often unsupportive because of mixed signals from above.

We have seen companies where the CEO went on record in support of family flexibility but where supervisors were never evaluated in any way for their sensitivity to family issues. In one company, managers were encouraged to provide part-time work opportunities, yet head-count restrictions reckoned all employees as full-time. In another, maternity leave was counted against individual managers when measuring absenteeism, a key element in their performance appraisals. As a general rule, strict absenteeism systems designed to discourage malingerers often inadvertently punish the parents of young children. Yet such systems coexist with corporate admonitions to be flexible. Where messages are mixed and performance measurement has not changed since the days of the "give an inch, they'll take a mile" personnel policy, it is hardly surprising that supervisors and managers greet lofty family-oriented policy statements with some cynicism.

Training is critical. IBM, Johnson & Johnson, Merck, and Warner-Lambert have all established training programs to teach managers to be more sensitive to work-and-family issues. The training lays out the business case for flexibility, reviews corporate programs and policies, and presents case studies that underline the fact that there are often no right answers or rule books to use as guides in the complicated circumstances of real life.

Perhaps the thorniest issue facing businesses and managers is that of equity. Most managers have been trained to treat employees identically and not to adjudicate the comparative merits of different requests for flexibility. But what equity often means in practice is treating everyone as though they had wives at home. On the other hand, it is difficult to set up guidelines for personalized responses, since equity is a touchstone of labor relations and human resource management. Judging requests individually, on the basis of business and personal need, is not likely to lead to identical outcomes.

Seniority systems also need rethinking. Working second or third shift is often the only entry to a well-paying job for nonprofessional employees, but for a parent with a school-age child, this can mean not seeing the child at all from weekend to weekend. Rotating shifts wreak havoc with child-care arrangements and children's schedules. Practices that worked fine when the labor force consisted mostly of men with wives at home now have unintended consequences.

Finally, the message top management sends to all employees is terribly important. In focus groups at various large companies, we hear over and over again a sense that companies pay lip service to the value of family and community but that day-to-day practice is another story altogether. We hear what we can only describe as a yearning for some tangible acknowledgment from top management that family issues are real, complex, and important.

Johnson & Johnson, which sees its 40-year-old corporate credo as central to its culture, recently added the statement, "We must be mindful of ways to help our employees fulfill their family obligations." Du Pont has developed a mission statement that commits it, in part, to "making changes in the workplace and fostering changes in the community that are sensitive to the changing family unit and the increasingly diverse work force."

Throughout Europe, governments have required companies to treat the parenting of babies as a special circumstance of employment and have invested heavily in programs to support the children of working parents. In this country, recent surveys indicate almost universal popular support for parental leave. But our instincts oppose government intervention into internal business practices. We leave decisions about flexibility and the organization of work to individual companies, which means that the decisions of first-line managers in large part create our national family policy.

In this, the United States is unique. But then we are also unique in other ways, including the depth of our commitment to business, to fairness, to equal opportunity, to common sense. Many of our young women now strive to become CEOs. No one intended that the price for business success should be indifference to family or that the price of having a family should be to abandon professional ambition.

REFERENCES

1. Joseph Pleck, "Family-Supportive Employer Policies and Men's Participation," unpublished paper, Wheaton College, 1989.
2. From research conducted by Ellen Galinsky at Merck and Company, Rahway, New Jersey, 1983, 1984, and 1986.
3. Terry Bond, *Employer Supports for Child Care*, report for the National Council of Jewish Women, Center for the Child, New York, August 1988.
4. Marcy Whitebook, Carollee Howes, and Deborah Phillips, "Who Cares: Child Care Teachers and the Quality of Care in America," National Child Care Staffing Study, Child Care Employee Project, Oakland, California, 1989.
5. "On-Site Child Care Results in Low Turnover at Computer Firm," *National Report on Work and Family*, vol. 2, no. 13 (Washington, D.C.: Buraff Publications, June 9, 1989), p. 3.
6. Dianne Burden and Bradley Googins, *Boston University Balancing Job and Homelife Study* (Boston University School of Social Work, 1986).
7. Kathleen Christensen, *A Look at Flexible Staffing and Scheduling in U.S. Corporations* (New York: Conference Board, 1989); and Jon L. Pierce et al., *Alternative Work Schedules* (Newton, Mass.: Allyn and Bacon, 1988).
8. *Work and Family: A Changing Dynamic* (Washington, D.C.: Bureau of National Affairs Special Report, 1986), pp. 78–80.
9. *Part-Time Social Workers in Public Welfare* (New York: Catalyst, 1971), cited in *Alternative Work Schedules*, p. 81.
10. *The Corporate Woman Officer* (Chicago, Ill.: Heidrick and Struggles, Inc., 1986); *Korn/Ferry International's Executive Profile: Corporate Leaders in the Eighties* (New York: Korn/Ferry International, 1986).

CHAPTER 4 Questions for Writing, Reflection, and Debate

READING 15 Men at Work • *Anna Quindlen*

1. "Men at Work" is a humorous treatment of an issue faced by all working parents. What do you think—which is more important, the quality or the quantity of time spent with children? Why?

READING 16 Women, Work, and Family in America • *Suzanne M. Bianchi and Daphne Spain*

2. Bianchi and Spain introduce the idea of the importance of looking at cohorts when examining changes in women's labor force participation. What social events, experienced by your cohort, are likely to shape the balance of work and family roles?
3. The authors argue that increasing involvement in education has contributed to women's greater participation in the labor force. In some areas of study, women are performing better than men. If society allocates jobs to the best qualified people, this should mean women with better grades would get jobs over men with lower grades. Is this likely to happen? What would be the potential effect on families if it did?
4. As this article points out, much of the closing of the pay gap between men and women has been caused by decreases in men's wages. In addition, when married women enter the labor force, they decrease their time spent on domestic tasks, and men contribute more. However, in rebalancing the tasks women do twice as much as men and the total time spent on housework, when you combine both partners' contribution, decreases. Does this imply that role equity will always result in moving to a lower common denominator?

READING 17 Sharing and Reluctance • *Scott Coltrane*

5. In "Sharing and Reluctance," Scott Coltrane investigates the ways Chicano couples negotiate work/family balance. What are the "economics of gratitude" he talks about? How are they used to sway negotiations?
6. Why do you think the responsibility for managing the home remains with the woman, even when men participate in child care and housework? Is this attitude more prevalent in certain ethnic or racial groups? How does this assumption influence the negotiation of roles?

READING 18 Parental Androgyny • *David Popenoe*

7. Popenoe's article reminds us that there are many who do not support role equity for men and women. In fact, this author sees this trend as threatening family life. How prevalent, in our culture, is the idea that egalitarian roles threaten children's development and family stability?

8. What explanations, other than Popenoe's, can you give for the fact that many marriages based on androgynous roles do not last?
9. How do you react to the argument that, because the relationship between fathers and their children has no biological base, culture must foster it?

READING 19

Understanding the Future of Fatherhood: The "Daddy Hierarchy" and Beyond • *Arlie Russell Hochschild*

10. In contrast to the Popenoe article, Hochschild argues that emotional investments in children are no longer bound by gender and/or kinship. However, she cautions that our culture gives fathers mixed signals regarding their role. What are the two stances on fatherhood evident in our society? How are they reinforced by our political, legal, and economic systems?
11. What is your reaction to the author's proposition that America has created a society where work life is more attractive than home life for many people, regardless of gender? If this is true, what are the long-term implications for society? What characteristics of work and home contribute to the possibility of this being true? How do such social forces influence the ways couples negotiate work/family balance?

READING 20

Business and the Facts of Family Life • *Fran Sussner Rodgers and Charles Rodgers*

12. Much of the cultural context that influences the negotiation of work/family roles is not easily altered. However, Rodgers and Rodgers discuss proactive attempts to alter social context through policy initiatives. The core of the argument for such policies is the idea that all of society gains when workers can focus on their jobs and be more productive because they have access to quality programs that make juggling roles easier. The argument against such policies is based on the idea that parenting is a private matter and says that if people choose to work while they have children, it should be their concern how they manage both roles. How do you view the role of employers and the government in setting policies that influence work/family balance?

5

Families in Poverty

How should the U.S. government provide for the welfare of families?

When I was growing up in a large city in the Midwest, the only time I came face to face with poverty was in the city center. As a child, I remember seeing people who were then referred to as "bums." These were almost always men, usually middle-aged or older, and often inebriated. In the last three or four decades, I have come to associate different faces with poverty. The first changes I noticed occurred when the laws regarding institutionalization for mental illness changed. Soon the men on the streets who struggled with alcoholism were joined by people who talked to themselves, acted like children, or sat in a comatose state. At the same time the gender and age composition of people on the street began to change too. Increasingly I saw younger men and women than I had in the past. I remember being troubled that many of these people seemed to be in this situation due to circumstances outside of their control. I was undergoing a transition from viewing poverty as an individual issue to viewing it as a social issue. The readings in this chapter may guide you to that realization too.

Two current issues surrounding poverty stimulated my desire to select readings about this topic. First is the growing number of young families with children in poverty. Just as I saw the faces of poverty change in my lifetime, Marian Wright Edelman draws to our attention the growing number of young families with children swelling the ranks of the financially needy. Among the points made in "Vanishing Dreams of America's Young Families" is the important fact that the impoverishment suffered by children in single-parent families in the past is now reaching two-parent families. In addition, this problem, once associated with young parents of color, is now affecting somewhat older parents across racial groups.

The second current issue that interests me is public reaction to this trend. Someone once said, "You can judge the heart of a country by the way its citizens care for the least able among them." Something in that simple statement appealed to the sociologist in me and stimulated my thinking about questions regarding our responsibilities to help those in need. I hear many of my fellow citizens espouse a capitalist, free-market view that those who have the skills demanded in the labor market will find work, and it's up to those who don't have the qualifications to figure out how to get them. Those who are not resourceful enough to do this deserve to be financially penalized, and the government should not play Robin Hood by taking from those who have been successful in the system to help those who have not.

Lillian Rubin's interviews with families affected by unemployment uncovers the impact of accepting the notion that individuals are responsible for poverty. "When You Get Laid Off, It's Like You Lose a Part of Yourself" reveals the stories of many men whose lives are disrupted or destroyed by the guilt that accompanies being unemployed in a culture that proposes that every person is "the master of his or her own fate." As these stories show, this guilt too often manifests itself in substance abuse and violence that impacts all family members.

On the other end of the continuum from this position are people who believe that poverty is related more to conditions of society than to personal motivation. For example, if public education puts certain groups of people at an educational disadvantage, is it the fault of the individual or the system that this person is unprepared for work? If it is a systemic fault, then why should the individual pay? Another point made to support the argument that poverty is predominantly a structural issue is that a growing number of the families who live in poverty are fully participating members of the labor force. Our free-market capitalist system has taken advantage of the U.S. labor surplus by employing people at wages that cannot support even a moderate standard of living. The women discussed in Kathryn Edin and Laura Lein's "The Choice Between Welfare and Work" provide real-life stories of the struggles of families who try to support themselves through low-paying work. One important message of this piece is that serious problems are caused by attempting to leave public assistance in order to support a family through work. The hidden expenses of employment, transportation, child care, and loss of health care are enough to sabotage even the best intentions.

As we redesign welfare in this country, many more young families will be put in this position. And those who can't find work, and whose assistance is withdrawn, face the distinct possibility of joining the growing numbers of homeless people who seek help from shelters and food pantries run by benevolent groups. Jennifer Glick's research, reported in "Mothers with Children and Mothers Alone: A Comparison of Homeless Families," reminds us that a potential outcome of losing government assistance is the breakup of families. Are children better off separated from their parents than living on the streets or in shelters? Many homeless parents who have the option of leaving their children in someone else's care are forced to make this decision. I, and many others, fear that recent welfare reform will increase the number of parents facing such decisions.

As I write this chapter, I have just returned from a month in Africa. Traveling has shown me that the quality of life for "those who have" suffers significantly when they are outnumbered by "those who have not." As I shared with friends and colleagues my reactions to the poverty I witnessed there, people responded with an eerie sameness. Basically, many Americans see the continuing growth of poverty here as an inevitable aspect of America's future. Part of me hopes they are wrong, part of me fears they are right. The readings in this chapter illuminate the many personal and structural aspects of poverty. I hope that what these authors tell us will inform our response to the question "How should the U.S. government provide for the welfare of families in poverty?" It is my bias that we must do something, because the larger this problem grows the more it affects the well-being of us all.

I keep hoping that if more Americans understand the structural causes of poverty they will be more willing to support what they see as the "Robin Hood" attempts of government to redistribute the wealth of this nation. After all, as William O'Hare shows us in "A New Look at Poverty in America," the citizens of most other Western countries, through their governments, take more responsibility than we do for the well-being of those who can't care for themselves. I believe that many who push for welfare reform and cuts in social needs–based programs do so because they believe many of the myths dispelled by William O'Hare in this reading. Here we see the power of government programs such as Social Security to address poverty, and we find explanations for the failure of others.

This chapter closes with the story of Tyrone Mitchell as told by Studs Terkel. Tyrone's story is a hopeful one. His life illustrates the way many families in poverty seek to use drugs and gangs to move out of their marginalized position in society. Tyrone says his children are the reason he is trying to change his life around. He has marketable skills and entrepreneurial ideas about how he could help himself and his community too. As I read Tyrone's story, I tried to imagine what his life is like now. The story could have many endings. Will he have returned to a life of drugs, gangs, and crimes because he never had a chance to use the skills and ideas he had to better himself? If so, my tax dollars may be supporting him in prison, and they may be supporting his children, too. I tried to imagine different ways my tax dollars could be used to sway the ending of this story into a positive result—one that finds Tyrone contributing to society and his family. I challenge you to do the same as you grapple with the issue of how the U.S. government should provide for families in poverty.

Vanishing Dreams of America's Young Families

Marian Wright Edelman

The future of today's young parents and their small children is now in great jeopardy. Congress and the President must take immediate steps to ensure that every child has a fair start, a healthy start, and a head start.

Americans from all walks of life are profoundly anxious—troubled by what they see around them today and even more by what they see ahead. This anxiety, not only about their own futures but also about the nation's future, is manifested in countless ways: in paralyzing economic insecurity; in an emerging politics of rejection, frustration, and rage; in a growing polarization of our society by race and by class; and in an erosion of the sense of responsibility to help the weakest and poorest among us.

But this anxiety about the future is most vivid when we watch our own children grow up and try to venture out on their own—struggling to get established as adults in a new job, a new marriage, a new home or a new family.

It's true that young families always have faced an uphill struggle starting out in life. But today's young families have been so battered by economic and social changes over the past two decades that the struggle has taken on a more desperate and often futile quality.

And as parents of my generation watch many of their adult children founder—failing to find steady, decent-paying jobs, unable to support families, shut out of the housing market, and often forced to move back home—they know that something has gone terribly wrong. Often they don't know precisely what has happened or why. But they do understand that these young adults and their children may never enjoy the same opportunities or achieve the same standard of living or security that our generation found a couple of decades ago.

Two Generations in Trouble

Young families with children—those headed by persons under the age of thirty—have been devastated since 1973 by a cycle of falling incomes, increasing family disintegration, and rising poverty. In the process, the foundations for America's young families have been so thoroughly undermined that two complete generations of Americans—today's young parents and their small children—are now in great jeopardy. Figure 1 captures the poverty rates of those two jeopardized generations.

Young families are the crucible for America's future and America's dream. Most children spend at least part of their lives—their youngest and most developmentally vulnerable months and years—in young families. How we treat these families therefore goes a long way toward defining what our nation as a whole will be like twenty, fifty, or even seventy-five years from now.

What has happened to America's young families with children in unprecedented and almost unimaginable.

Adjusted for inflation, the median income of young families with children plunged by one-third between 1973 and 1990 (Table 1). This median income includes income from all sources, and the drop occurred despite the fact that many families sent a

FIGURE 1 *Poverty Rates of Families with Children, by Age of Family Head, 1973, 1979, 1982, 1990*

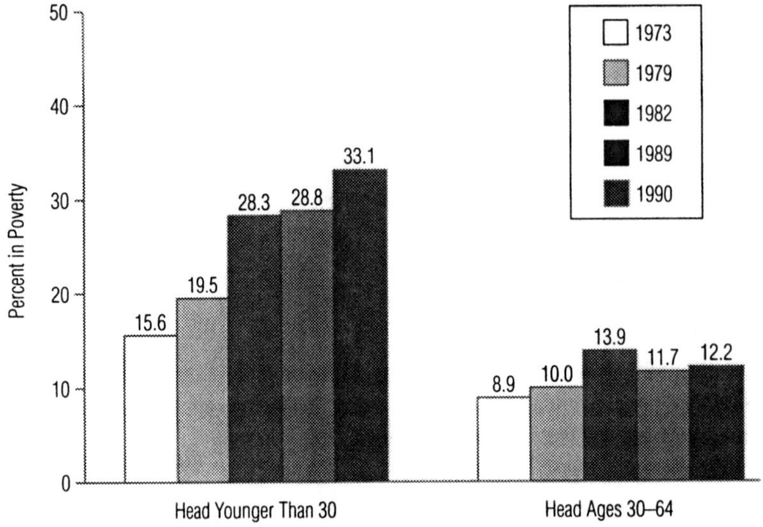

second earner into the workforce. As a result, poverty among these young families more than doubled, and by 1990 a shocking 40 percent or four in ten children in young families were poor.

The past two decades have been difficult for many other Americans as well. But older families with children have lost only a little economic ground since 1973, and families without children have enjoyed substantial income gains. By far the greatest share of the nation's economic pain has been focused on the weakest and most vulnerable among us—young families with children.

This is not a story about the current recession, although the recession surely is having a crushing im-

TABLE 1 *Median Incomes of Families with Children by Age of Family Head, 1973–1990 (in 1990 dollars)*

	1973	1979	1982	1989	1990	Change 1973–1990
All families with children	36,882	36,180	31,819	35,425	34,400	–6.7%
Family head younger than 30	27,765	25,204	20,378	20,665	18,844	–32.1%
Family head age 30–64	41,068	39,978	35,293	39,525	38,451	–6.4%
Young families' median income as a share of older families' income	68%	63%	58%	52%	49%	

Note: The money incomes of families for all years prior to 1990 were converted into 1990 dollars via use of the Consumer Price Index for All Urban Consumers (CPI-U). The U.S. Bureau of Labor Statistics has generated an alternative price index for the years preceding 1983 that conforms to the current method of measuring changes in housing costs. This index is known as the CPI-UXI. Use of this price index would reduce the estimated 1973 real income by about 7 percent, thus lowering the estimated decline in the median income of young families between 1973 and 1990 from 32 percent to approximately 25 percent. None of the comparisons of median income between various groups of families are affected by these changes.

pact on young families. Even comparing 1973 to 1989—two good economic years at the end of sustained periods of growth—the median income of young families with children dropped by one-fourth. Then just the first few months of the recession in 1990 sent young families' incomes plummeting to new depths.

This also is not a story about teenagers. While America's teen pregnancy problem remains tragic and demands an urgent response, only 3 percent of the young families with children we are discussing are headed by teenagers. More than 70 percent are headed by someone aged twenty-five to twenty-nine. The plight of America's young families is overwhelmingly the plight of young adults who are both old enough and eager to assume the responsibilities of parenthood and adulthood, but for whom the road is blocked.

Finally and most importantly, this is not simply a story about someone else's children, about minority children or children in single-parent families or children whose parents dropped out of high school.

All Young Families Affected

Huge income losses have affected virtually every group of young families with children: white, black and Latino; married-couple and single-parent; and those headed by high school graduates as well as dropouts. Only young families with children headed by college graduates experienced slight income gains between 1973 and 1990.

In other words, the tragedy facing young families with children has now reached virtually all of our young families. One in four white children in young families is now poor. One in five children in young married-couple families is now poor. And one in three children in families headed by a young high school graduate is now poor. Nearly three-fourths of the increase in poverty among young families since 1973 has occurred outside the nation's central cities. And poverty has grown most rapidly among young families with only one child (Figure 2).

There is no refuge from the economic and social shifts that have battered young families with children.

FIGURE 2 *Poverty Rates of Children in Young Families, by Characteristics of the Family Head, 1973, 1989, 1990*

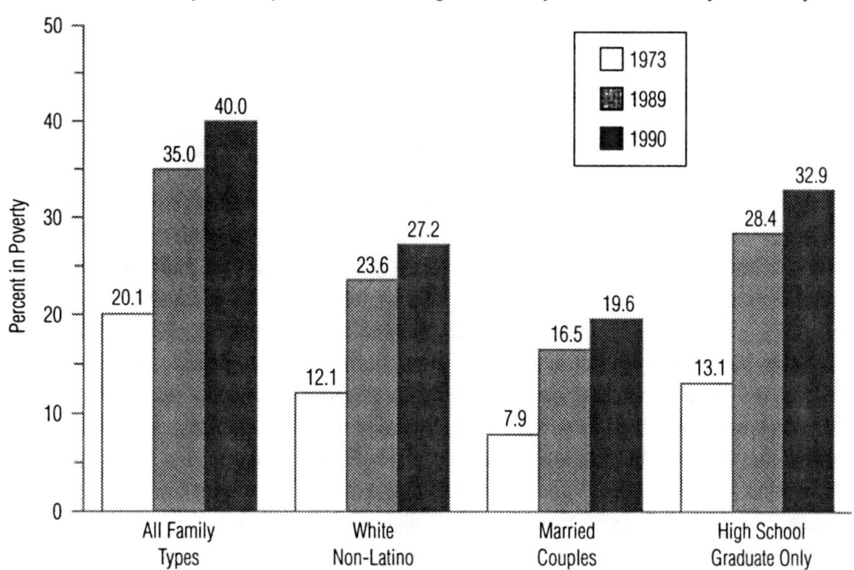

We can pretend that they won't reach our children and our grandchildren. We can pretend that those who play by the rules will be O.K.

We can pretend, but that will not change the reality—the reality that young families have lost a third of their median income, that two in five American children in young families live in poverty, and that these facts have devastating consequences.

Those consequences include more hunger, more homelessness, more low birthweight births, more infant deaths, and more child disability. They also mean more substance abuse, more crime, more violence, more school failure, more teen pregnancy, more racial tension, more envy, more despair, and more cynicism—a long-term economic and social disaster for young families and for the country. In virtually every critical area of child development and healthy maturation, family poverty creates huge roadblocks to individual accomplishment, future economic self-sufficiency, and national progress.

Plummeting incomes and soaring poverty and growing gaps based on age and education and race mean more of all these problems, yet many of our leaders seem not to understand why they are occurring. But there is not really a puzzle, when we recognize that the nation has marginalized and pauperized much of two generations of Americans—young parents and young children.

Young families not only lost income in huge amounts, but as the permanence and quality of their jobs deteriorated, they lost fringe benefits like health insurance as well. In the decade of the 1980s the proportion of employed heads of young families with children whose employers made health insurance available by paying all or part of the cost dropped by one-fifth. And employers cut back on coverage for dependent spouses and children even more than for workers.

Fewer and fewer young pregnant women have been getting adequate prenatal care because they are poorer and less likely to have adequate insurance or any insurance. And our falling vaccination rates and renewed epidemics of measles and other wholly preventable diseases among preschoolers are being driven by plunging incomes in young families, eroding health insurance coverage, and unraveling government programs.

Falling incomes also have devastated young families in an increasingly expensive housing market. One-third fewer young families with children were homeowners in 1991 than in 1980. Young renter families increasingly are paying astronomical shares of their meager incomes for rent. More and more are doubling up or becoming homeless—in some surveys three-fourths of the homeless parents in this country are under age thirty.

Young families are not only suffering from the hunger, housing, health and other problems that their plummeting incomes have caused. They are suffering as well because they are falling further and further behind the rest of the society—imperiling their attachment to the core workforce and to mainstream values and threatening their potential to reacquire the American dream in the decades to come.

In 1973 the income of older families with children was not quite one-and-a-half times that of young families with children. By 1990 it was more than double that of the young families.

Combination of Causes

There is no single cause of young families' plight. Instead, they have been pummeled by a combination of profound changes in the American economy; the government's inadequate response to families in trouble; and changes in the composition of young families themselves.

Much of the increase in their poverty is due to economic shifts and to changing government policies that have made it more difficult for young families to obtain adequate incomes. These changes have hurt all young families with children, regardless of their family structure, race or ethnicity, or educational attainment.

Unlike members of earlier generations, young workers today no longer can be confident of finding stable jobs with decent wages, even if they get a high school diploma or spend a couple of years in college. Since 1973, slower growth in U.S. productivity and declines in blue-collar employment made some drop in inflation adjusted median earnings for young workers inevitable. By last year the average wages of all nonsupervisory workers (of all ages) in the private sector fell to their lowest level since the Eisenhower Administration.

But the losses have been focused disproportionately on young workers. The median annual earnings of heads of young families with children fell a staggering 44 percent from 1973 to 1990. In other words, in the span of less than a generation this nation nearly halved the earnings of young household heads with children (Table 2).

These dramatic earnings losses occurred across-the-board. For example, young white families with children were hit as hard as young Latino families: the median earnings of both groups fell by two-fifths. College graduates as well as high school graduates and dropouts lost big chunks of income. But the drop in median earnings for high school dropouts and for young black family heads has been particularly devastating—in each case more than two-thirds.

The erosion in pay levels (due in part to the declining value of the minimum wage) combined with the growth of temporary or part-time and part-year jobs to put a triple whammy on young workers: far lower annual earnings, less secure employment, and less access to health insurance and other employer-provided benefits.

The huge drop in earnings among America's young workers has not received much attention. In part it has been obscured by the almost Herculean work effort of young parents. Many young married-couple families have tried to compensate for lower wages by sending a second worker into the workforce. These second earners have softened (but not eliminated) the economic blow. But the growing number of young parents working longer hours or coping with two jobs has placed young families with children under tremendous stress and generated new offsetting costs, especially for child care. Many families, moreover, have two jobs that together provide less security and less support and less access to health care than one good job did a generation ago. This two-earner strategy is totally unavailable, moreover, to the growing number of single-parent families.

TABLE 2 *Median Annual Earnings of Heads of Young Families with Children, 1973 and 1990 (in 1990 dollars)*

	1973	1990	% Change 1973–1990
All heads of young families with children	22,981	12,832	−44
Married couple	25,907	17,500	−33
Male-headed	18,547	14,000	−25
Female-headed	2,073	1,878	− 9
White, non-Latino	25,024	15,000	−40
Black, non-Latino	13,860	4,030	−71
Latino	15,924	9,000	−44
Other, non-Latino	17,664	12,000	−32
High school dropout	15,014	4,500	−70
High school graduate	23,817	14,000	−41
Some college	26,496	18,000	−32
College graduate	31,795	25,721	−19

Economic Shifts and Family Changes

Today's young families with children look considerably different from those in the early 1970s. They are more likely to be minority families or single-parent families. Both groups are more likely to be paid low wages and to be poor than other families. So poverty among young families and children as a whole also rises.

The growth in young female-headed families with children is in part a reflection of changing values. But the economic hardships associated with falling earnings and persistent joblessness among young adults have contributed significantly to falling marriage rates and the increasing rates of out-of-wedlock childbearing. And the fastest growth in out-of-wedlock childbearing has occurred among women in their twenties, not among teenagers, a growth driven in significant part by the earnings free-fall for young adults.

The capacity to support a family has a powerful impact on the marriage decisions of young people. More than two centuries ago Benjamin Franklin wrote: "The number of marriages . . . is greater in proportion to the ease and convenience of supporting a family. When families can be easily supported, more persons marry, and earlier in life."

Increases in poverty among young families with children are not the result of young Americans having more children. Indeed, young adults have responded to a tightening economic vise by postponing childbearing and choosing to have fewer children. But these attempts to adapt their behavior have been overwhelmed by the far more rapid pace of economic decline and social disintegration they have encountered.

As a result of these economic and social changes, in 1990 a child in a family headed by a parent under age 30 was

- Twice as likely to be poor as a comparable child in 1973
- If living with both parents, two and a half times as likely to be poor as in 1973
- Nearly three times more likely to have been born out-of-wedlock than his counterpart two decades ago
- One-third less likely to be living in a home owned by her family than just a decade ago
- Three times more likely to see his family pay more than one-half its income for rent

But despite the devastating suffering these numbers suggest, children in young families have been given less and less government help over the last two decades. They were getting less to begin with—government programs are particularly stingy when it comes to helping younger adults and young children. And in the 1970s and especially the 1980s young families saw programs that might help them cut rather than strengthened and reconfigured to adapt to new realities. As a result, government programs were less than half as effective in pulling young families out of poverty in 1990 as in 1979.

Hard-Hit Minority Families

The changes of the last two decades have had a very profound impact on minority young families, especially those that are black. As Table 2 shows, the median earnings of the heads of young black families with children fell 71 percent from 1973 to 1990 (from $13,860 to $4,030 in 1990 dollars). Their total family incomes from all sources fell 48 percent. The median income of these young black families is now below the federal poverty line for a family of three. In 1973 it was nearly double that poverty line. Two out of three children in young black families now are poor.

This crisis for young black families is contributing mightily to the tearing apart of the black community. This society cannot year after year increase the poverty and isolation and hopelessness of black mothers and fathers and children—it can't keep turning the screws tighter and tighter—without appalling consequences. We see those consequences in the emergency rooms and unemployment lines and prisons and homeless shelters and neonatal intensive care wards and morgues of our cities and our suburbs and rural towns. We see it in the omnipresent violence that destroys so many black lives and leaves blacks and whites alike so fearful. More blacks die from firearms each year in this country than died in the century's worth of despicable lynchings that followed the Civil War. More black men die from firearms every six weeks in Detroit than died in the 1967 Detroit "riot." More black and Hispanic men die from firearms in Los Angeles every two weeks than died in the 1965 Watts "riot."

Frankly, though, I would be skeptical that this nation would attack this cataclysm for young black families were it not for the fact that young white families are only a step or two behind in the scope of their economic depression and family disintegration. Perhaps the most important story told in this report is the impact of two decades of this Depression for the young on three types of families we often assume are insulated from hard times:

- From 1973 to 1990 the poverty rate for children living in young white families more than doubled to 27 percent.
- From 1973 to 1990 the poverty rate for children in young married-couple families went up two-and-a-half times—to 20 percent.
- And the child poverty rate in young families headed by high school graduates went up even faster, to 33 percent.

In other words, a generation ago white or married-couple young families or those headed by high school graduates were fairly well insulated from poverty. The damage of the last two decades has cut so broadly and deeply that now one in four white children in young families, one in five children in married-couple young families, and one in three children in families headed by young high school graduates is poor.

Private and Public Response

What response do we see to these problems from private and public leadership? Precious little.

Too much of the business community is wholly untroubled by stripping away from millions of Americans the minimum family-supporting wages, fringe benefits, and job security that could help make our families strong again. The [Bush] Administration has no higher domestic priority than cutting capital gains taxes for the rich. The Congress and the Administration together persist in keeping defense spending above the levels of the average year in the Cold War—impoverishing our society and the world by arming ourselves not only against real external threats but against weapons-justifying fantasies, while letting the internal enemies of poverty, disintegration, violence, and hopelessness rage unabated. The Congress can't mount the political will to get Head Start—a program universally conceded to be effective and cost-effective—to more than one in three eligible children or to pass the refundable children's tax credit that experts from all parts of the political spectrum think is a minimum first step to tax equity and family economic security.

Finally, far too many of the nation's governors and state and national legislators have responded to budget crunches and political turmoil by scapegoating the poor—trying to bolster their political fortunes by pummeling the welfare recipients whose assistance gobbles up a grand total of 2 to 3 percent of state budgets.

In hard times in the past our society usually has had escape valves—an inherent balance that gave to the powerless help from one institution when others turned their backs—from the federal government when the states were at their worst, from the courts when Congress and the executive were unresponsive. Now we seem to be in an awful time when every institution is competing to pander to the powerful and further penalize the poor.

A Fair Start

In response to the economic plight of America's young families, Congress and the President must take three immediate steps in 1992 to ensure that every child has a fair start, a healthy start, and a head start.

A fair start means renewed and sustained economic growth and enough jobs at decent wages to restore the pact our nation used to have with young families—that personal sacrifice and hard work will be rewarded with family-supporting jobs. A fair start also means enactment of a refundable children's tax credit to bolster the incomes of families with children, as proposed in recent months in various forms and amounts by the National Commission on Children and key members of Congress from both parties. Such a credit would reduce federal income taxes for middle- and low-income families and help the lowest-income families that have no tax liability through a tax refund.

While creating no new bureaucracies, a refundable children's credit would target tax relief and economic support precisely to the group—families with children—that has been hardest hit by declining incomes and rising poverty rates since 1973. The [Bush] Administration's alternative proposal—to expand the personal exemption for children—is extremely regressive. It gives $310 to a family with two children if their income exceeds about $100,000; $280 if it is over $50,000; $150 if income is between $15,000 and $50,000; and zero if it is under about $15,000.

Finally, a fair start means creation of a child support insurance system to give all single parents the chance to lift their families out of poverty through work, ensuring that all children who are not living with both parents receive a minimally adequate child support payment from the absent parent or the government when it fails to collect from the absent parent.

What we don't need in this time of great crisis for young families with children is a negative approach rooted in welfare-bashing and welfare cuts that ends up hurting children. Families on welfare are the victims of the recession, not the cause of it. They are victims of budget deficits, not the cause of them. But nearly one-fourth of all young families with children are forced to rely upon Aid for Families with Dependent Children (AFDC) to meet their basic needs, and they are extremely vulnerable to misguided attacks on this essential safety net for children.

Our political leaders know these truths. Yet during an election year too many cannot resist the temptation to direct the public's frustration and anger toward the poorest of poor Americans—those families and children who rely upon welfare for basic income support.

There are ways we can and must improve welfare. For example, we agree with the Administration that welfare parents often don't have enough financial incentive to work because current welfare rules strip them of virtually all of their earnings when they do work. That is why we opposed the repeal of earnings incentives by President Reagan and the Congress in 1981, and why we think they should be restored now for all welfare families, not just those in a few demonstration counties.

But most of the welfare "reforms" now under way in states are little more than crass attempts to slash state budgets without regard to their impact on families with children. Reducing or stopping benefits to newborns when they are the second or third child in a family, as now proposed by several states, is punitive, pointless, and immoral. Only political leaders who are hopelessly out of touch with the realities of poor families' lives could think that an extra $2.50 per day in welfare benefits would cause teen parents to have a second child, or that reducing the added benefit to $1.25/day (as the Governor of Wisconsin and the president now propose for that state) constitutes any serious effort at welfare reform. All they will succeed in doing is taking desperately needed food, clothing, and shelter from infants.

It's time for the president, Congress, and more of our governors to be honest with the American people about the problems facing our economy, our poor families and our children.

The problem is not large numbers of welfare parents trying to "beat the system" by having more children or moving to another state to get higher benefits. The problem is a set of short-sighted, budget-driven welfare rules that make it virtually impossible for parents to work their way gradually off the welfare rolls and a dearth of stable, family-supporting jobs that would allow them to make it on their own.

In many ways, the welfare problem is the same problem facing all young families with children—the result of sharply falling wages, too few job opportunities for those with little education or training, and too little investment in the skills and supports poor parents need to make it in today's economy. And serious solutions begin with a fair start, a healthy start, and a head start for our young families.

A Healthy Start, a Head Start

A healthy start means a national health plan to assure insurance coverage for all Americans. Children and pregnant women need basic health care now, however. As an immediate step, the president and Congress must extend Medicaid coverage to every low-income child and pregnant woman. And to ensure that this insurance provides real access to essential health services, not merely theoretical coverage, children need universal access to vaccines and increased funding for community health centers, and other public health activities.

A head start means full funding of Head Start. A first step in bolstering the productivity of our next generation of workers lies in adequate investments in quality child care and early childhood development. Every dollar invested in good early childhood development programs saves $5.97 in later special education, welfare, crime, and other costs. Yet Head Start still reaches only one in three eligible preschool children.

As recommended by prominent business groups, educators, and a broad range of study commissions that have examined the educational problems of disadvantaged children and youths, the president and Congress should ensure every child a Head Start by

1995 by enacting immediately S. 911, the School Readiness Act, and accelerating the funding increases it provides. A head start also means passing family preservation legislation that will strengthen and preserve families in crisis so that they can better protect, nurture and support their own children. So many of these young parents want to be better parents, and with intensive family preservation services they can get the help they need.

These are essential first steps. To reach them and go beyond them, we're going to have to make the president and Congress come to recognize that child and family poverty and insecurity are a national disaster that requires our addressing them with a pittance of the zeal and shared commitment we now apply to digging out after a devastating hurricane or earthquake or confronting a crisis abroad.

"When You Get Laid Off, It's Like You Lose a Part of Yourself"

Lillian B. Rubin

For Larry Meecham, "downsizing" is more than a trendy word on the pages of the *Wall Street Journal* or the business section of the *New York Times*. "I was with the same company for over twelve years; I had good seniority. Then all of a sudden they laid off almost half the people who worked there, closed down whole departments, including mine," he says, his troubled brown eyes fixed on some distant point as he speaks. "One day you got a job; the next day you're out of work, just like that," he concludes, shaking his head as if he still can't believe it.

Nearly 15 percent of the men in the families I interviewed were jobless when I met them.[1] Another 20 percent had suffered episodic bouts of unemployment—sometimes related to the recession of the early 1990s, sometimes simply because job security is fragile in the blue-collar world, especially among the younger, less experienced workers. With the latest recession, however, age and experience don't count for much; every man feels at risk.[2]

Tenuous as the situation is for white men, it's worse for men of color, especially African-Americans. The last hired, they're likely to be the first fired. And when the axe falls, they have even fewer resources than whites to help them through the tough times. "After kicking around doing shit work for a long time, I finally got a job that paid decent," explains twenty-nine-year-old George Faucett, a black father of two who lost his factory job when the company was restructured—another word that came into vogue during the economic upheaval of the 1990s. "I worked there for two years, but I didn't have seniority, so when they started to lay guys off, I was it. We never really had a chance to catch up on all the bills before it was all over," he concludes dispiritedly.

I speak of men here partly because they're usually the biggest wage earners in intact families. Therefore, when father loses his job, it's likely to be a crushing blow to the family economy. And partly, also, it's because the issues unemployment raises are different for men and for women. For most women, identity is multifaceted, which means that the loss of a job isn't equivalent to the loss of self. No matter how invested a woman may be in her work, no matter how much her sense of self and competence are connected to it, work remains only one part of identity—a central part perhaps, especially for a professional woman, but still only a part. She's mother, wife, friend, daughter, sister—all valued facets of the self, none wholly obscuring the others. For the working-class women in this study, therefore, even those who were divorced or single mothers responsible for the support of young children, the loss of a job may have been met with pain, fear, and anxiety, but it didn't call their identity into question.

For a man, however, work is likely to be connected to the core of self. Going to work isn't just what he does, it's deeply linked to who he is. Obviously, a man is also father, husband, friend, son, brother. But these are likely to be roles he assumes, not without depth and meaning, to be sure, but not self-defining in the same way as he experiences work. Ask a man for a statement of his identity, and he'll almost always respond by telling you first what he does for a living. The same question asked of a woman brings forth a less predictable, more varied response, one that's embedded in the web of relationships that are central to her life.[3] . . .

Men who are denied paid employment as a means of determining their identities may turn to home projects to maintain their self-worth. Mike, a[n unemployed] cement finisher, explains it this way: "If I don't keep busy, I feel like I'll go nuts. It's funny," he says with a caustic, ironic laugh, "before I got laid off my wife was always complaining about me watching the ball games; now she keeps nagging me to watch. What do you make of that, huh? I guess she's trying to make me feel better."

"Why didn't you watch the game last Sunday?" I ask.

"I don't know, maybe I'm kind of scared if I sit down there in front of that TV, I won't want to get up again," he replies, his shoulders hunched, his fingers raking his hair. "Besides, when I was working, I figured I had a right."

His words startled me, and I kept turning them over in my mind long after he spoke them: "When I was working, I figured I had a right." It's a sentence any of the unemployed men I met might have uttered. For it's in getting up and going to work every day that they feel they've earned the right to their manhood, to their place in the world, to the respect of their family, even the right to relax with a sporting event on TV.

It isn't that there are no gratifying moments, that getting laid off has no positive side at all. When unemployment first hits, family members usually gather around to offer support, to buoy a man's spirits and their own. Even in families where conflict is high, people tend to come together, at least at the beginning. "Considering that we weren't getting along so well before, my wife was really good about it when I got laid off," says Joe Phillips, an unemployed black truck driver. "She gave me a lot of support at first, and I appreciate it."

"You said 'at first.' Has that changed?" I ask.

"Hell, yes. It didn't last long. But maybe I can't blame it all on her. I've been no picnic to live with since I got canned."

In families with young children, there may be a period of relief—for the parents, the relief of not having to send small children off to child care every day, of knowing that one of them is there to welcome the children when they come home from school; for the children, the exhilarating novelty of having a parent, especially daddy, at home all day. "The one good thing about him not working is that there's someone home with the kids now," says twenty-five-year-old Gloria Lewis, a black hairdresser whose husband has been unemployed for just a few weeks. "That part's been a godsend. But I don't know what we'll do if he doesn't find work soon. We can't make it this way."

Teenagers, too, sometimes speak about the excitement of having father around at first. "It was great having my dad home when he first got laid off," says Kevin Sollars, a white fourteen-year-old. "We got to do things together after school sometimes. He likes to build ship models—old sailing ships. I don't know why, but he never wanted to teach me how to do it. He didn't even like it when I just wanted to watch; he'd say, 'Haven't you got something else to do?' But when he first got laid off, it was different. When I'd come home from school and he was working on a ship, he'd let me help him."

But the good times usually don't last long. "After a little while, he got really grumpy and mean, jumped on everybody over nothing," Kevin continues. "My mom used to say we had to be patient because he was so worried about money and all that. Boy, was I glad when he went back to work."

Fathers may also tell of the pleasure in getting to spend time with their children, in being a part of their daily life in ways unknown before. "There's a silver lining in every cloud, I guess. I got to know my kids like I never did before," says Kevin's father, who felt the sting of unemployment for seven months before he finally found another job. "It's just that being out of work gets old pretty fast. I ran out of stuff to do around the house; we were running out of money; and there I was sitting on my keister and stewing all day long while my wife was out working. I couldn't even enjoy building my little ships." . . .

. . . [T]he struggles around the division of labor shift somewhat when father loses his job. The man who's home all day while his wife goes off to work can't easily justify maintaining the traditional household gender roles. Therefore, many of the unemployed men pick up tasks that were formerly left to their wives alone. "I figure if she's working and I'm not, I ought to take up some of the slack around here. So I keep the place up, run the kids around if they need it, things like that," says twenty-nine-year-old Jim Andersen, a white unemployed electrician.

As wives feel their household burdens eased, the strains that are almost always a part of life in a two-job family are somewhat relieved. "Maybe it sounds crazy to you, but my life's so much easier since he's out of work, I wish it could stay this way," says Jim's wife, Loreen, a twenty-nine-year-old accounting clerk. "If only I could make enough money, I'd be happy for him to stay home and play Mr. Mom."

But it's only a fantasy—first because she can't make enough money; second, and equally important, because while she likes the relief from household responsibilities, she's also uneasy about such a dramatic shift in family roles. So in the next breath, Loreen says, "I worry about him, though. He doesn't feel so good about himself being unemployed and playing house."

"Is it only him you worry about? Or is there something that's hard for you, too?" I ask.

She's quiet for a moment, then acknowledges that her feelings are complicated: "I'm not sure what I think anymore. I mean, I don't think it's fair that men always have to be the support for the family; it's too hard for them sometimes. And I don't mind working; I really don't. In fact, I like it a lot better than being home with the house and the kids all the time. But I guess deep down I still have that old-fashioned idea that it's a man's job to support his family. So, yeah, then I begin to feel—I don't know how to say it—uncomfortable, right here inside me," she says, pointing to her midsection, "like maybe I won't respect him so much if he can't do that. I mean, it's okay for now," she hastens to reassure me, perhaps herself as well. "But if it goes on for a real long time like with some men, then I think I'll feel different."

Men know their wives feel this way, even when the words are never spoken, which only heightens their own anxieties about being unemployed. "Don't get me wrong; I'm glad she has her job. I don't know what we'd do if she wasn't working," says Jim. "It's just that . . . ," he hesitates, trying to frame his thoughts clearly. "I know this is going to sound pretty male, but it's my job to take care of this family. I mean, it's great that she can help out, but the responsibility is mine, not hers. She won't say so, but I know she feels the same way, and I don't blame her."

It seems, then, that no matter what the family's initial response is, whatever the good moments may be, the economic and psychological strains that attend unemployment soon overwhelm the good intentions on all sides. "It's not just the income; you lose a lot more than that," says Marvin Reed, a forty-year-old white machinist, out of work for nearly eight months. He pauses, reflects on his words, then continues. "When you get laid off, it's like you lose a part of yourself. It's terrible; something goes out of you. Then, on top of that, by staying home and not going to work and associating with people of your own level, you begin to lose the sharpness you developed at work. Everything gets slower; you move slower; your mind works slower. . . .

The men I talked with try to remind themselves that it's not their fault, that the layoffs at the plant have nothing to do with them or their competence, that it's all part of the economic problems of the nation. But it's hard not to doubt themselves, not to wonder whether there's something else they could have done, something they might have foreseen and planned for. "I don't know; I keep thinking I could have done something different," says Lou Coltrane, a black twenty-eight-year-old auto worker, as he looks away to hide his pain. "I know it's crazy; they closed most of the plant. But, you know, you can't help thinking, maybe this, maybe that. It keeps going round and round in my head: Maybe I should have done this; maybe I should have done that. Know what I mean?" . . .

For American men—men who have been nurtured and nourished in the belief that they're masters of their fate—it's almost impossible to bear such feelings of helplessness. So they find themselves in a cruel double bind. If they convince themselves that their situation is beyond their control, there's nothing left but resignation and despair. To fight their way out of the hopelessness that follows, they begin to blame themselves. But this only leaves them, as one man said, "kicking myself around the block"—kicks that, paradoxically, allow them to feel less helpless and out of control, while they also send them deeper into depression, since now it's no one's fault but their own. . . .

For wives and children, it's both disturbing and frightening to watch husband and father sink ever deeper into despair. "Being out of work is real hard on him; it's hard to see him like this, so sad and jumpy all the time," laments Bill's wife, Eunice, a part-time bank teller who's anxiously looking for full-time work. "He's always been a good provider, never out of work hardly a day since we got married. Then all of a sudden this happens. It's like he lost his self-respect when he lost that job."

His self-respect and also the family's medical benefits, since Eunice doesn't qualify for benefits in her part-time job. "The scariest part about Bill being out of a job is we don't have any medical insurance anymore. My daughter got pneumonia real bad last winter and I had to borrow money from my sister for the doctor bill and her medicine. Just the medicine was almost $100. The doctor wanted to put her in the hospital, but we couldn't because we don't have any health insurance."

Her husband recalls his daughter's illness, in a voice clogged with rage and grief. "Do you know what it's like listening to your kid when she can't breathe and you can't send her to the hospital because you lost your benefits when you got laid off?"

In such circumstances, some men just sit, silent, turned inward, enveloped in the gray fog of depression from which they can't rouse themselves. "I leave to go to work in the morning and he's sitting there doing nothing, and when I come home at night, it's the same thing. It's like he didn't move the whole day," worries thirty-four-year-old Deidre Limage, the wife of a black factory worker who has been jobless for over a year.

Other men defend against feeling the pain, fear, and sadness, covering them over with a flurry of activity, with angry, defensive, often irrational outbursts at wife and children—or with some combination of the two. As the financial strain of unemployment becomes crushing, everyone's fears escalate. Wives, unable to keep silent, give voice to their concerns. Their husbands, unable to tolerate what they hear as criticism and blame—spoken or not—lash out. "It seems like the more you try to pull yourself up, the more you get pushed back down," sighs Beverly Coleride, a white twenty-five-year-old cashier with two children, whose husband has worked at a variety of odd jobs in their seven-year marriage. "No matter how hard we try, we can't seem to set everything right. I don't know what we're going to do now; we don't have next month's rent. If Kenny doesn't get something steady real quick, we could be on the street."

"We could be on the street"—a fear that clutches at the hearts and gnaws at the souls of the families in this study, not only those who are unemployed. Nothing exemplifies the change in the twenty years since I last studied working class families than the fear of being "on the street." Then, homelessness was something that happened somewhere else, in India or some other far-off and alien land. Then, we wept when we read about the poor people who lived on the streets in those other places. *What kind of society doesn't provide this most basic of life's needs?* we asked ourselves. Now, the steadily increasing numbers of homeless in our own land have become an ever-present and frightening reminder of just how precarious life in this society can be. Now, they're in our face, on our streets, an accepted category of American social life—"the homeless."

Just how readily accepted they are was brought home to me recently when my husband, who volunteers some time in the San Francisco schools, reported his experience with a sixth-grade class there. He had been invited to talk to the children about career opportunities and, in doing so, talked about his own past as a restaurateur. The students listened, engrossed. But instead of the questions he had expected when he finished, they were preoccupied with how he managed the problem of the homeless. Did he feed homeless people when they asked for food, they wanted to know. He explained that at the time he had restaurants in the Bay Area, there were no homeless on the streets. Undaunted, they wanted to know what he did when he found a homeless person sleeping in the doorway of the restaurant. He reminded them that he had just told them that homelessness wasn't an issue when he was in business. They listened, he said, but they couldn't really grasp the idea of a world without the homeless. How could it be otherwise? At their age, homelessness is so much a part of their daily world that they take it for granted, a phenomenon not of their time but of all times.

As homelessness has increased, even those of us who remember when it was unthinkable have become inured to the sight of the men and women who make their home on the streets. Inured, and also anxious. We recoil as we walk by, trying not to see, unable to meet their eyes, ashamed of our own good fortune, anger and sympathy tugging us in opposite directions. Neither feels good. The anger is a challenge to our belief that we're kind, humane, caring. But the sympathy is even more threatening. To allow ourselves to feel compassion is to open the floodgates of our own vulnerability, of our denied understanding of how delicately our lives and fortunes are balanced.

For Beverly Coleride, as for the other women and men I met, sustaining the denial has become increasingly difficult. No matter how much they want to obliterate the images of the homeless from consciousness, the specter haunts them, a frightening reminder of what's possible if they trip and fall. Perhaps it's because there's so much at stake now, because the unthinkable has become a reality, that anxieties escalate so quickly. So as Beverly contemplates the terror of being "on the street," she begins to blame her husband. "I keep telling myself it's not his fault, but it's real hard not to let it get you down. So then I think, well, maybe he's not trying hard enough, and I get on his case, and he gets mad, and, well, I guess you know the rest," she concludes with a harsh laugh that sounds more like a cry of pain.

She doesn't *want* to hurt her husband, but she can't tolerate feeling so helpless and out of control. If it's his fault rather than the workings of some impersonal force, then he can do something about it. For her husband, it's an impossible bind. "I keep trying, looking for something, but there's nothing out there, leastwise not for me. I don't know what to do anymore; I've tried everything, every place I know," he says disconsolately.

But he, too, can't live easily with such feelings of helplessness. His sense of his manhood, already under threat because he can't support his family, is eroded further by his wife's complaints. So he turns on her in anger: "It's hard enough being out of work, but then my wife gets on my case, yakking all the time about how we're going to be on the street if I don't get off my butt, like it's my fault or something that there's no work out there. When she starts up like that, I swear I want to hit her, anything just to shut her mouth," he says, his shoulders tensed, his fists clenched in an unconscious expression of his rage.

"And do you?" I ask.

The tension breaks; he laughs. "No, not yet. I don't know; I don't want to," he says, his hand brushing across his face. "But I get mad enough so I could. Jesus, doesn't she know I feel bad enough? Does she have to make it worse by getting on me like that? Maybe you could clue her, would you?"

"Maybe you could clue her"—a desperate plea for someone to intervene, to save him from his own rageful impulses. For Kenny Coleride isn't a violent man. But the stress and conflict in families where father loses his job can give rise to the kind of interaction described here, a dynamic that all too frequently ends in physical assaults against women and children.

Some kind of violence—sometimes against children only, more often against both women and children—is the admitted reality of life in about 14 percent of the families in this study.[4] I say "admitted reality" because this remains one of the most closely guarded secrets in family life. So it's reasonable to assume that the proportion of families victimized by violence could be substantially higher.

Sometimes my questions about domestic violence were met with evasion: "I don't really know anything about that."

Sometimes there was outright denial, even when I could see the evidence with my own eyes: "I was visiting my sister the other day, and I tripped and fell down the steps in front of her house."

And sometimes teenage children, anguished about what they see around them, refused to participate in the cover-up. "I bet they didn't tell you that he beats my mother up, did they? Nobody's allowed to talk about it; we're supposed to pretend like it doesn't happen. I hate him; I could kill him when he does that to her. My mom, she says he can't help it; it's because he's so upset since he got fired. But that's just her excuse now. I mean, yeah, maybe it's worse than it was before, but he did it before, too. I don't understand. Why does she let him do it to her?"

"Why does she let him do it to her?" A question the children in these families are not alone in asking, one to which there are few satisfactory answers. But one thing is clear: The depression men suffer and their struggle against it significantly increase the probability of alcohol abuse, which in turn makes these kinds of eruptions more likely to occur.[5]

"My father's really changed since he got laid off," complains Buddy Truelman, the fifteen-year-old son of an unemployed white steel worker. "It's like he's always mad about something, you know, ready to bite your head off over nothing. I mean, he's never been an at-ease guy, but now nothing you do is okay with him; he's always got something to say, like he butts in where it's none of his business, and if you don't jump to, he gets mad as hell, carries on like a crazy man." He pauses, shifts nervously in his chair, then continues angrily, "He and my mom are always fighting, too. It's a real pain. I don't hang around here any more than I have to."

Buddy's mother, Sheila, a thirty-four-year-old telephone operator, echoes her son. "He's so touchy; you can't say anything without him getting mad. I don't mind so much if he takes it out on me, but he's terrible to the kids, especially to my son. That's when I get mad," she explains, passing a hand over her worried brow. "He's got no right to beat up on that kid the way he does."

"Do you mean he actually hits him?" I ask.

She hesitates and looks away, the torment of memory etched on her face. Finally, brushing away the tears that momentarily cloud her vision, she replies, "Yeah, he has. The last time he did it, he really hurt him—twisted his arm so bad it nearly broke—and I told him I'd leave if he ever hit Buddy again. So it's been okay for a while. But who knows? He has a few beers and it's like he goes crazy, like he can't control himself or something."

Many of the unemployed men admit turning to alcohol to relieve the anxiety, loneliness, and fear they experience as they wait day after day, week after week for, as one man put it, "something to happen." "You begin to feel as if you're going nuts, so you drink a few beers to take the edge off," explains thirty-seven-year-old Bill Anstett, a white unemployed construction worker.

It seems so easy. A few beers and he gets a respite from his unwanted feelings—fleeting, perhaps, but effective in affording some relief from the suffering they inflict. But a few beers often turn out to be enough to allow him to throw normal constraints to the wind. For getting drunk can be a way of absenting the conscious self so that it can't be held responsible for actions undertaken. Indeed, this may be as much his unconscious purpose as the need to rid himself of his discomfort. "I admit it, sometimes it's more than a few and I fall over the edge," Bill grants. "My wife, she tells me it's like I turn into somebody else, but I don't know about that because I never remember."

With enough alcohol, inhibitions can be put on hold; conscience can go underground. "It's the liquor talking," we say when we want to exempt someone from responsibility for word or deed. The responsibility for untoward behavior falls to the effects of the alcohol. The self is in the clear, absolved of any wrongdoing. So it is with domestic violence and alcohol. When a man gets drunk, the inner voice that speaks his failure and shame is momentarily stilled. Most men just relax gratefully into the relief of the internal quiet. But the man who becomes violent needs someone to blame, someone onto whom he can project the feelings that cause him such misery. Alcohol helps. It gives him license to find a target. With enough of it, the doubts and recriminations that plague him are no longer his but theirs—his wife's, his children's; "them" out there, whoever they may be. With enough of it, there's nothing to stay his hand when his helpless rage boils over. "I don't know what happens. It's like something I can't control comes over me. Then afterward I feel terrible," Peter DiAngelo, an unemployed thirty-two-year-old truck driver, says remorsefully.

One-fifth of the men in this study have a problem with alcohol, not all of them unemployed. Nor is domestic violence perfectly correlated with either alcohol abuse or unemployment. But the combination is a potentially deadly one that exponentially increases the likelihood that a man will act out his anger on the bodies of his wife and children. "My husband drinks a lot more now; I mean, he always drank some, but not like now," says Inez Reynoso, a twenty-eight-year-old Latina nurse's aide and mother of three children who is disturbed about her husband's mistreatment of their youngest child, a three-year-old boy. "I guess he tries to drink away his troubles, but it only makes more trouble. I tell him, but he doesn't listen. He has a fiery temper, always has. But since he lost his job, it's real bad, and his drinking doesn't help it none.

"I worry about it; he treats my little boy so terrible. He's always had a little trouble with the boy because he's not one of those big, strong kids. He's not like my older kids; he's a timid one, still wakes up scared and crying a lot in the night. Before he got fired, my husband just didn't pay him much attention. But now he's always picking on him; it's like he can't stand having him around. So he makes fun of him something terrible, or he punches him around."

The mother in me recoils at Inez's story. But the psychotherapist understands at least something of what motivates Ramon Reynoso's assault on his young son. For this father, this man who's supposed to be the pillar on which the family rests, who defines himself as a man by his ability to support his family, the sight of this weak and puny little boy is like holding up a mirror to his now powerless self. Unable to tolerate the feelings of self-hatred the image engenders, he projects them outward, onto the child, and rains blows down on him in an effort to distance himself from his own sense of loss and diminishment.

"Does he hit you, too?" I ask Inez.

She squirms in her chair; her fingers pick agitatedly at her jeans. I wait quietly, watching as she shakes her head no. But when she speaks, the words say something else. "He did a couple of times lately, but only when he had too many beers. He didn't mean it. It's just that he's so upset about being out of work, so then when he thinks I protect the boy too much he gets real mad."

When unemployment strikes, sex also becomes an increasingly difficult issue between wives and husbands. A recent study in Great Britain found that the number of couples seeking counseling for sexual problems increased in direct proportion to the rise in the unemployment rate.[6] Anxiety, fear, anger, depression—all emotions that commonly accompany unemployment—are not generators of sexual desire. Sometimes it's the woman whose ardor cools because she's frightened about the future: "I'm so scared all the time, I can't think about sex." Or because she's angry with her husband: "He's supposed to be supporting us and look where we are." More often it's the men who lose their libido along with their jobs—a double whammy for them since male identity rests so heavily in their sexual competence as well as in their work.[7]

This was the one thing the men in this study couldn't talk about. I say "couldn't" because it seemed so clearly more than just "wouldn't." Psychologically, it was nearly impossible for them to formulate the words and say them aloud. They had no trouble complaining about their wives' lack of sexual appetite. But when it was they who lost interest or who become impotent, it was another matter. Then, their tongues were stilled by overwhelming feelings of shame, by the terrible threat their impotence posed to the very foundation of their masculinity.

Their wives, knowing this, are alarmed about their flagging sex lives, trying to understand what happened, wondering what they can do to be helpful. "Sex used to be a big thing for him, but since he's been out of work, he's hardly interested anymore," Dale Meecham, a white thirty-five-year-old waitress says, her anxiety palpable in the room. "Sometimes when we try to do it, he can't, and then he acts like it's the end of the world—depressed and moody, and I can't get near him. It's scary. He won't talk about it, but I can see it's eating at him. So I worry a lot about it. But I don't know what to do, because if I try to, you know, seduce him and it doesn't work, then it only makes things worse."

The financial and emotional turmoil that engulfs families when a man loses his job all too frequently pushes marriages that were already fragile over the brink.[8] Among the families in this study, 10 percent attributed their ruptured marriages directly to the strains that accompanied unemployment. "I don't know, maybe we could have made it if he hadn't lost his job," Maryanne Wallace, a twenty-eight-year-old white welfare mother, says sadly. "I mean, we had problems before, but we were managing. Then he got laid off, and he couldn't find another job, and, I don't know, it was like he went crazy. He was drinking; he hit me; he was mean to the kids. There was no talking to him, so I left, took the kids and went home to my mom's. I thought maybe I'd just give him a scare, you know, be gone for a few days. But when I came back, he was gone, just gone. Nobody's seen him for nearly a year," she says, her voice limping to a halt as if she still can't believe her own story.

Economic issues alone aren't responsible for divorce, of course, as is evident when we look at the 1930s. Then, despite the economic devastation wrought by the Great Depression, the divorce rate didn't rise. Indeed, it was probably the economic privations of that period that helped to keep marriages intact. Since it was so difficult to maintain one household, few people could consider the possibility of having to support two.

But these economic considerations exist today as well, yet recent research shows that when family income drops 25 percent, divorce rises by more than 10 percent.[9] Culture and the institutions of our times make a difference. Then, divorce was a stigma. Now, it's part of the sociology and psychology of the age, an acceptable remedy for the disappointment of our dreams.

Then, too, one-fourth of the work force was unemployed—an economic disaster that engulfed the whole nation. In such cataclysmic moments, the events outside the family tend to overtake and supersede the discontents inside. Now, unemployment is spottier, located largely in the working class, and people feel less like they're in the middle of a social catastrophe than a personal one. Under such circumstances, it's easier to act out their anger against each other.

And finally, the social safety net that came into being after the Great Depression—social security, unemployment benefits, public aid programs targeted specifically to single-parent families—combined with the increasing numbers of women in the work force to make divorce more feasible economically. . . .

Some families find ways to adapt and remain intact. Almost always these are older families with few debts and a short span of unemployment. For many men, returning to work involves a step down to another kind of work, to one of the service jobs that usually pay a fraction of their former earnings—that is, if they're lucky enough to find one. It's never easy in our youth-oriented society for a man past forty to move to another job or another line of work. But it becomes doubly difficult in times of economic distress when the pool of younger workers is so large and so eager. "Either you're overqualified or you're over the hill," Ed Kruetsman, a forty-nine-year-old unemployed white factory worker, observes in a tired voice.

But young or old, when a man is forced into lower-paying, less skilled work, the move comes with heavy costs—both economic and psychological. Economically, it means a drastic reduction in the family's way of life. "Things were going great. We worked hard, but we finally got enough together so we could buy a house that had enough room for all of us," says thirty-six-year-old Nadine Materie, a white data processor in a bank clearing center. "Tina, my oldest girl, even had her own room; she was so happy about it. Then my husband lost his job, and the only thing he could find was one that pays a lot less, *a lot less*. On his salary now we just couldn't make the payments. We had no choice; we had to sell out and move. Now look at this place!" she commands, with a dismissive sweep of her hand. Then, as we survey the dark, cramped quarters into which this family of five is now jammed, she concludes tearfully, "I hate it, every damn inch of it; I hate it."

For Tina Materie, Nadine's fifteen-year-old daughter, her father's lost job has meant more than the loss of her room. The comforts and luxuries of the past are gone, and the way of life she once took for granted seems like a dream. For a teenager whose sense of self and place in the world is so heavily linked to peer group acceptance and to, in Tina's own words, "being like the other kids," the loss is staggering. "We can't afford anything anymore; and I mean *anything*," she announces dramatically. "I don't even go to the mall with the other kids because they've got money to buy things and I don't. I haven't bought a new record since we moved here. Now my mom says I can't get new school clothes this year; I have to wear my cousin's hand-me-downs. How am I going to go to school in those ugly things? It's bad enough being in this new school, but now . . . ," she stops, unable to find the words to express her misery.

Worst of all for the children in the Materie family, the move from house to apartment took them to a new school in a distant neighborhood, far from the friends who had been at the center of their lives. "My brother and me, we hate living here," Tina says, her eyes misting over as she speaks. "Both of us hate the kids who live around here. They're different, not as nice as the kids where we used to live. They're tough, and I'm not used to it. Sometimes I think I'll quit school and get a job and go live where I want," she concludes gloomily....

Companies go bankrupt; they merge; they downsize; they restructure; they move—all reported as part of the economic indicators, the cold statistics that tell us how the economy is doing. But each such move means more loss, more suffering, more families falling victim to the despair that comes when father loses his job, more people shouting in rage and torment: "What the hell's happening to this country?"

NOTES

1. It's not possible to compare the rate of unemployment in these families with those I interviewed two decades ago because the previous sample was made up of men who were employed. But comparing the unemployment rates in 1970 and 1991 is instructive. Among white men with less than four years in high school, 4.5 percent were unemployed in 1970, 10.3 percent in 1991. The figures for high-school graduates are 2.7 percent and 5.4 percent, respectively. For blacks with less than four years in high school, the 1970 unemployment rate stood at 5.2 percent, compared to 14.7 percent in 1991. For black high-school graduates, the rates are 5.2 and 9.9, respectively (*Statistical Abstract,* [U.S. Bureau of the Census, 1992, Table 637, p. 400]). The number of food stamp recipients, which typically rises as the unemployment rate climbs, jumped to an all-time high in 1993, when one in ten Americans were in the food stamp program.
2. Barbara Ehrenreich, *Fear of Falling* (New York: Pantheon Books, 1989), and Katherine S. Newman, *Falling from Grace* (New York: Free Press, 1988), write compellingly about middle-class fears of what Newman calls "falling from grace." But these fears probably are more prevalent among working-class families, and with good reason, since job security is still so much more tenuous there than in the middle class.
3. Cf. Rubin, *Worlds of Pain,* and Lillian B. Rubin, *Women of a Certain Age: The Midlife Search for Self* (New York: Harper Perennial, 1986).
4. A few researchers argue that, since the majority of men who batter their wives are gainfully employed, unemployment is of little value in explaining battering (H. Saville et al., "Sex Roles, Inequality and Spouse Abuse," *Australian and New Zealand Journal of Sociology* 17 [1981]: 83–88; and Martin D. Schwartz, "Work Status, Resource Equality, Injury and Wife Battery," *Creative Sociology* 18 [1990]: 57–61). But the evidence is much stronger in the direction of a relationship between unemployment and family violence; see Frances J. Fitch and Andre Papantonio, "Men Who Batter," *Journal of Nervous and Mental Disease* 171 (1983): 190–191; Richard J. Gelles and Murray A. Straus, "Violence in the American Family," *Journal of Social Issues* 35 (1979): 15–39; New York State Task Force on Domestic Violence, *Domestic Violence: Report to the Governor and Legislature: Families and Change* (New York: Praeger Publishers, 1984); and Suzanne K. Steinmetz, "Violence-Prone Families," *Annals of the New York Academy of Sciences* 347 (1980): 251–265.
5. John A. Byles, "Violence, Alcohol Problems and Other Problems in Disintegrating Families," *Journal of Studies on Alcohol* 39 (1978): 551–553; Ronald W. Fagan, Ola W. Barnett, and John B. Patton, "Reasons for Alcohol Use in Maritally Violent Men," *Journal of Drug and Alcohol Abuse* 14 (1988): 371–392; Fitch and Papantonio, "Men Who Batter"; Kenneth E. Leonard et al., "Patterns of Alcohol Use and Physically Aggressive Behavior in Men," *Journal of Studies on Alcohol* 46 (1985): 279–282; Larry R. Livingston, "Measuring Domestic Violence in an Alcoholic Population," *Journal of Sociology and Social Welfare* 13 (1986): 934–951; Albert R. Roberts, "Substance Abuse Among Men Who Batter Their Mates," *Journal of Substance Abuse Treatment* 5 (1988): 83–87; J. M. Schuerger and N. Reigle, "Personality and Biographic Data That Characterize Men Who Abuse Their Wives," *Journal of Clinical Psychology* 44 (1988): 75–81; and Steinmetz, "Violence-Prone Families."
6. Reported in the *San Francisco Chronicle,* February 14, 1992. The study found that in the same year that unemployment rose from 6.5 to 9.2 percent, there was a 30 percent increase in the number of couples seeking advice from marriage counselors about their waning sex lives.
7. Ethel Spector Person, "Sexuality as the Mainstay of Identity," *Signs* 5 (1980): 605–630.
8. An article in the *San Francisco Chronicle,* October 19, 1992, surveyed several recent studies of divorce, one of which found that when income drops 25 percent, divorce rises by more than 10 percent; another predicted ten thousand divorces for every 1 percent rise in unemployment.
9. Cited in the *San Francisco Chronicle,* October 19, 1992.

The Choice Between Welfare and Work

Kathryn Edin and Laura Lein

This [reading] tells the story of single mothers all over America who face a desperate situation recognized by neither politicians nor the media: neither welfare nor low-wage work provides enough income to cover basic needs.

The federal welfare rules present welfare-reliant mothers with a stark choice: follow the rules—which disallow supplemental income—and subject their families to severe hardship, or break the rules. Virtually all welfare-reliant mothers with whom we spoke during the course of our research chose their family's welfare. In Chicago and Boston, many welfare-reliant mothers coped by taking off-the-books work. Many San Antonio mothers subsisted by purchasing stolen bread, meat, and tennis shoes out of the backs of cars and spent the early morning hours of many weekdays waiting in long lines for government surplus food, used clothing, or assistance in paying an overdue bill. In Charleston, descendants of antebellum Sea Islanders of African heritage did what they had always done to subsist: made and sold handicrafts, worked for cash at fishing docks heading shrimp, and cleaned the houses of rich northern retirees; poor white mothers lived off under-the-table handouts from better-off relatives and friends.

When they could, the welfare-reliant mothers in all these cities persuaded the fathers of their children to circumvent the Child Support Enforcement system, which would have passed through only $50 of the money the father contributed legally, keeping the rest as partial repayment of the family's welfare benefits. Instead, mothers urged fathers to contribute to their children's well-being covertly, enforcing their claims with threats to "turn them in" to child support officials, who could dock their pay, seize their tax returns, revoke their driver's licenses, or, in some cases, throw them in jail if their identities were known.

Some mothers had lost contact with their children's fathers. Other fathers were abusive, violent, addicted, incarcerated, or dead. These fathers rarely contributed in any way, so some mothers turned to other men—boyfriends—whose intermittent earnings could help keep the household afloat. Because of the dire economic needs of their families, some women allowed these men into their homes on a "pay and stay" basis. Sometimes these men became the father of the next child in the family, and sometimes they did not. In either case, when men lost their jobs or stopped bringing home a portion of their paycheck, they were forced back into the households of their own mothers, grandmothers, or sisters or onto the streets. The mercenary tone of these male/female relationships eroded much of whatever trust existed between women and men in poor communities.

For those mothers who did not receive welfare and relied instead on earnings from low-wage work, every dollar they earned meant a decrease in their food stamps and housing subsidies; wage income also threatened their Medicaid eligibility. In addition, working entailed increased costs for child care, transportation, and clothing. Thus, wage-reliant mothers had to generate even more outside income from various survival strategies in order to balance their budgets.

The reality of economic life among unskilled and semiskilled mothers is sharply at odds with the perceptions held by most citizens. Average Americans depend on newspapers, magazines, television, and radio for their information about public policy issues. Journalists and their editors, in turn, publicize those stories that they think will attract the most interest or outrage. Thus, the public has been influenced by stories of welfare queens who used their ill-gotten gain to buy fancy cars and vacation homes or of mothers who exchanged AFDC and food stamps for crack or heroin while their children huddled in filthy unheated hovels.

A lot of Americans also depend on the reports of friends and relatives who have more direct contact with the poor—those whom Michael Lipsky (1980) terms "street level bureaucrats": police officers, social workers, and others whose professional lives place them in close contact with those mothers who are having the most difficulty surviving the harsh world of subsistence living. Finally, ordinary citizens often form opinions based on their own observations in the grocery store, where they inspect the carts of those mothers who pay with food stamps, or when they drive by street corners of poor neighborhoods and see working-aged adults loitering.

Some of the single mothers we interviewed fit these stereotypes in one way or another; some did exchange their food stamps for drugs, and some bought junk food for their kids in the grocery store (though food stamps cannot be used to pay for soda pop, candy bars, frozen dinners, or other nonstaples). The vast majority, however, were managing as well as they could given their resources. Typically, mothers traded food stamps for cash only when they were short on the rent—a common occurrence in cities where rent alone often cost as much as the family received from welfare.

The primary lesson we have taken from their stories is not that the welfare system of the early 1990s engendered psychological dependency or encouraged the formation of a set of deviant behaviors. The real problem with the federal welfare system during these years was a labor-market problem. The mothers we interviewed had made repeated efforts to attain self-sufficiency through work, but the kind of jobs they could get paid too little, offered little security in the short term, and provided few opportunities over time. Meanwhile, mothers who chose to work were even worse off in material terms than their welfare counterparts. To "make it" while working, unskilled single mothers had to be extraordinarily lucky: they had to have a set of special circumstances that artificially lowered the cost of working and they had to be able to employ a set of survival strategies that were consistent with work.

The public tends to see welfare-reliant and wage-reliant single mothers as if they were two distinct populations. In reality, however, a very large proportion of unskilled and semiskilled mothers cycle between the low-wage sector of the economy and the welfare rolls. Welfare- and work-reliant mothers should be seen as two overlapping populations on a single continuum. Roberta Spalter-Roth and her colleagues' (1995) analysis of the Census Bureau's Survey of Income and Program Participation, for example, shows that welfare-reliant mothers have substantial work histories—4.2 years on average. . . . [I]n our sample, 65 percent of the welfare-reliant mothers had worked in a formal sector job during the previous two years, and 84 percent had held such a job during the past five years. We also have shown that about half of all welfare recipients had done some kind of paid work in the past year. We do not know how many low-wage single mothers nationwide have welfare histories, but 60 percent of the work-reliant women in our sample had used welfare recently. In addition, a large fraction of our wage-reliant mothers used federal, state, or community-based assistance while they were working. . . .

What Poor Single Mothers Spend

On average, the 214 *welfare-reliant* mothers we interviewed spent $876 in cash and food stamps each month (in 1991 dollars). They spent 24 percent of their cash on housing, even though half of the sample received housing subsidies and some of the rest were doubled up with relatives. They spent another 30 percent on food, and 39 percent on clothing, transportation, laundry and cleaning supplies, diapers, school supplies, and other necessities. The remaining 7 percent went for nonessentials: a few video rentals or a basic cable subscription, an occasional trip to see relatives or go to an amusement park, a carton of cigarettes or a six-pack of beer, a bingo card or lottery ticket, or an occasional meal at a fast food restaurant. These small expenditures helped mothers survive psychologically and helped them to keep their children in school, off the streets, and out of trouble.

The expenditures of our *wage-reliant* mothers closely matched those of our welfare-reliant mothers with six notable exceptions. First, wage-reliant mothers earned more, so those who lived in a subsidized apartment had to pay more rent. While roughly half of the wage-reliant mothers in our sample received housing subsidies and many more doubled up, the wage-reliant sample as a whole paid about $100 a month more for housing than the welfare-reliant group. This was true even though we recruited our wage-reliant and welfare-reliant mothers from roughly the same neighborhoods.

Wage-reliant mothers also spent more on medical care, child care, transportation, clothing, and nonessentials. Mothers in the wage-reliant group often made substantial copayments for health care or went uninsured. Even though the working mothers and their children were in better health than members of welfare-reliant families, workers still paid three times as much for medical care as the welfare-reliant group. Wage-reliant mothers also spent about a quarter more for clothing, twice as much for transportation, and nearly ten times more on child care than the welfare-reliant mothers. The large difference in child care costs existed even though hardly any of the working mothers paid market rates for child care. Most wage-reliant mothers told us that they could not work if they had to pay market rates. Finally, wage-reliant mothers spent an average of $20 more each month on nonessentials—a rather paltry reward for choosing work over welfare. In other areas, workers' budgets were as bleak as those of the welfare-reliant group.

All told, our wage-reliant group spent $1,243 a month (in 1991 dollars), or roughly $15,000 a year. If these mothers had paid all of their own bills instead of getting housing subsidies, Medicaid, and the like, they would have had to earn at least $16,500 a year before taxes. This figure is a lower-bound estimate of a family's subsistence needs, because the wage-reliant group had fewer and older children, better health, more access to relatives who could watch their children, and received more in-kind help from friends than our welfare-reliant group.

We compared our mothers' expenditures with those of single mothers in the national Consumer Expenditure Survey. Our mothers spent less in almost every category than mothers in the national sample. We concluded that our mothers' expenditures were at the very low end of what constitutes a minimally acceptable living standard in the United States.

City-to-city differences in expenses were largely confined to housing and transportation. Rents were the lowest in San Antonio, which resembled rents in other southern cities including Little Rock, Louisville, Memphis, and New Orleans. Rents in Charleston were higher and typical of rents in other fast-growing cities in the New South. Rents in Chicago were still higher, but typical of rents in other large Midwestern cities. Rents in the Boston area were the highest and representative of many Northeastern and West Coast cities. Our Boston-area mothers spent less on housing than one would expect, however, because most non-subsidized families doubled up with family members or friends.

Transportation expenditures also varied by site. Chicago and Boston had reasonably good public transportation, so central-city residents could ride the bus or subway and seldom needed a car. Suburban mothers in these cities often had to maintain a car, so they paid more for transportation. Charleston provided minimal public transportation in poor neighborhoods, so mothers had to buy automobiles or use neighborhood taxis. In San Antonio, welfare-reliant mothers spent very little on transportation because they pursued survival strategies that did not require them to leave their immediate neighborhood.

Why have so many Americans come to see welfare as "dependency" and work as "self-sufficiency"? This rhetoric persists partly because most discussions of single parents take place without consideration for what welfare provides and what it actually takes to support a family. The official poverty line is of little help since it does not attempt to measure how much mothers need to spend on goods and services.[1] John Schwarz (Schwarz and Volgy 1992, 67–93) has attempted to estimate what the most basic necessities would cost. His budget includes no money for extras or child care. Even so, he determined that a family of four would have needed $20,660 in 1991, or 155 percent of the poverty threshold, to meet its basic needs. Schwarz's budget closely matches our estimates of roughly 150 percent of the poverty line for a working family of three.

Interestingly, both Schwarz's and our estimates closely match public opinion. Since the 1950s, the Gallup poll has asked Americans what they consider to be the minimum amount of income necessary for a family of four to live in their community. During the 1980s, responses ranged from 140 to 160 percent of the poverty line, suggesting that even the American public considers the official poverty measure too low. In 1987, the last time Gallup asked this question, Americans told the pollsters that a family of four would need to spend an amount equaling 160 percent of the poverty line to subsist.

Without an adjustment of the official measure of poverty, policymakers and social scientists will have trouble understanding the true extent and nature of poverty. Currently, both social scientists and policymakers tend to divide the population into two categories—poor and nonpoor—and sometimes use the term "near-poor" to describe those living just above the threshold. It makes little sense, however, to define a group whose income is too low to pay its bills as "near-poor" or to use such unrealistic measures as eligibility criteria for public programs.

Survival Strategies

While making ends meet is far more expensive than previous social science research has indicated, most families are also more resourceful than has been understood. Conventional measures of income miss a lot of the ways in which single mothers make ends meet. In our sample, for instance, welfare-reliant mothers were able to cover only three-fifths of their budgets with welfare, food stamps, and benefits from other means-tested programs in the early 1990s. Wage-reliant mothers could cover about two-thirds of their monthly budgets with wages from their main jobs. Because both welfare- and wage-reliant mothers faced the same fundamental dilemma each month, they relied on similar kinds of survival strategies to generate the additional money they needed to bridge the gap between their income and expenditures.

These survival strategies were dynamic rather than static. They resembled a continuously unraveling patchwork quilt, constructed from a wide variety of welfare- and work-based income; cash and in-kind assistance from family, friends, absent fathers, and boyfriends; and cash and in-kind assistance from agencies. Though welfare- and wage-reliant mothers drew from the same repertoire of strategies, wage-reliant mothers were less likely to rely on supplemental work because they had so little extra time. They were also less likely to get cash help from agencies. For the same reason, they relied much more heavily on their personal networks to meet household expenses. Although maintaining this web of social relations took time, this "work" fit more flexibly into working mothers' schedules. Thus, network-based strategies were much more compatible with full-time work than other strategies.

Mothers living in different cities faced somewhat different constraints and opportunities. For *welfare-reliant* mothers, the feasibility of working an off-the-books job varied by city size and the availability of false IDs. Their ability to get money from family and friends also varied with the strength of local labor markets and that of the local child support system. Robust local labor markets allowed boyfriends, absent fathers, and other network members to contribute more to welfare-reliant mothers' budgets, while a strong child support system helped mothers enforce their claims on fathers. Agency-based differences for welfare-reliant mothers were largely due to the kind of service environment that each site provided.

Wage-reliant mothers' strategies varied by site for much the same reasons that welfare-reliant mothers' strategies did. Supplemental work and network-based opportunities were determined largely by the local labor market. Agency-based strategies differed according to the generosity of local service providers.

Although our data are not longitudinal, we can use them to suggest why so many single mothers in the 1980s and 1990s repeatedly cycled between welfare and work. We have argued that the particular strategies a welfare-reliant mother used to make ends meet either constrained or enhanced her ability to make a permanent transition to work. Ironically, our data suggest that for welfare recipients, work-based strategies did not facilitate permanent departure from the welfare rolls unless they resulted in unusually high and stable earnings. Network-based strategies, however, did facilitate such transitions.

Network-based strategies came in three varieties: contributions from family and friends, contributions from boyfriends, and contributions from absent fathers. Workers and welfare recipients received about the same amount of assistance from their family members, but whites received more than African Americans. Whites received more, but they were more likely to have been married when their child was born and because marriage was associated with a larger network. This may also help explain why disproportionate numbers of African American women were on the federal welfare rolls (Bane and Ellwood 1994). If mothers required strong and generous networks to sustain work without endangering their children, and if minority women had less access to such networks, they would have worked less.

The role of family-based networks was also somewhat paradoxical. On the one hand, the cash-generating strength of mothers' networks enabled mothers to work and cushioned some of the worst economic shocks of working at a subsistence level. On the other hand, even strong networks could not fully protect working mothers from the vagaries of the labor market.

In addition, network support was not free. Because mothers often depended on others only marginally better off than themselves, others in their network expected to receive help if the mothers managed to better their situations. So while our mothers seldom contributed cash to the budgets of other network members, we suspect that better-off mothers (not in our sample) devoted a part of their monthly budgets to assisting others in their network. If obligations are reciprocal in this way, even mothers who manage to escape welfare and the $5-an-hour ghetto might have difficulty getting ahead, and so remain vulnerable to economic shocks even as they move up the income ladder. Furthermore, the mothers we interviewed were expected to invest a good deal of time and energy maintaining relationships with their benefactors—time and energy they could not spend going to school or attending training. These encumbrances also could limit a mother's ability to get ahead.

The second source of network assistance was from boyfriends. Interestingly, the amount of support from this source was not greater for welfare-reliant mothers than for workers. Nor were minority mothers more likely than white women to generate cash from boyfriends. Mothers who had never been married did get a bit more from boyfriends, but part of this difference can be explained by the lack of contributions these mothers received from absent fathers.

Absent fathers also helped mothers cover their expenses. For the welfare-reliant group, we distinguished between contributions received through the formal child support system and those received covertly. In both cases, the existence of strong child support enforcement aided mothers who wanted help from fathers. If mothers and children had few ties to the father, a strong enforcement environment helped them to get more money from the father through formal channels. For mothers who wanted to maintain their children's tie with the father, a strong enforcement environment could still be easily circumvented, and mothers were able to better enforce their claims on covert support because they could threaten to turn the fathers in if they failed to pay. This enabled mothers to keep more of the money and yet preserve the father's relationship with the child. We did not make the formal/covert distinction for working mothers, since participation in the formal system was not a legal requirement for this group. On balance, though, strong child support enforcement did help these mothers get more cash from fathers.[2]

We also found that mothers whose boyfriends lived with them were able to make substantial claims on these men's income. Men who stayed were expected to pay. This finding is sharply at odds with many media stories about inner-city men, which portray them as living off of their girlfriends' welfare checks. There is more economic activity among poor men, both in the formal and informal sectors of the economy, than most assume. Understanding the contribution patterns of unskilled and semiskilled single men as they move between formal, informal, and illegal jobs is crucial. . . .

Material Hardship

Despite the broad range of survival strategies that single mothers employ, most told us they had experienced serious material hardship during the prior twelve months. Not surprisingly, welfare-reliant mothers in high-benefit states experienced less hardship than those in low-benefit states; for workers, mothers with higher wages were better off than those with lower wages. It is more surprising that welfare recipients experienced fewer hardships than workers. In fact, if wage-reliant mothers were to bring their hardship levels down to those of our welfare-reliant mothers, they would have to spend roughly twice as much as the welfare mothers do.

The hardships that single mothers face when they move from welfare to work help to explain why so many unskilled and semiskilled mothers in the last three decades have relied on the federal welfare system, and why so many of those who left welfare for work during that time eventually returned to the government rolls. For these women, while welfare did not work very well, it made more sense than low-wage work, largely because it was so stable.

It is striking that welfare-reliant mothers still experienced substantial material hardship despite the assistance they received from the government. This indicates that even under the old welfare system the safety net was weak and that America's most vulnerable citizens, single mothers and their children, were constantly falling through the loose netting.

There is virtually no social safety net for single mothers who work, even when their wages don't pay enough to make ends meet. Transitional benefits for women who leave state welfare programs for work are undoubtedly necessary, but health benefits and child care assistance end after one or two years of work and the wages of those who have left welfare will seldom rise enough to make up the difference.

Choosing Between Welfare and Work

Unskilled and semiskilled mothers learn vital lessons from their experiences in the low-wage labor market. First, the kinds of jobs these women held in the past (and would get in the future without better skills) did not make them any better off—either financially or emotionally—than they were on welfare. Second, given the unstable nature of the low-wage job market, mothers with whom we spoke believed the transition from welfare to low-wage work might make them worse off and place them and their children at serious risk. Third, no matter how long they stayed at a job and no matter how diligently they worked, few jobs led to advancement. Fourth, past experience made mothers skeptical of the value of job clubs and other work-readiness components of the federal JOBS program, most of which attempted to place mothers in precisely the same types of jobs they held in the past.

Having learned these lessons, many single mothers who had burned out in the low-wage labor market returned to welfare. They did not plan to remain there, however. For these mothers, welfare was often part of a long-term strategy to reenter the labor market more successfully in the future. These plans took two forms. Some planned to stay out of the labor force until the costs of working were lower. This usually meant waiting until their children were in school or they could get a rent or child care subsidy. Others stayed on welfare to get more education and reenter the labor market in a more competitive position.

Most mothers firmly believed that education represented their best hope of breaking out of the $5-an-hour job ghetto. As we have seen, most were cynical about the local JOBS programs. Instead, mothers favored high-quality two- or four-year programs that prepared them for occupations paying a living wage. Pursuing high-quality training required a lot of time—a commodity in very short supply among the single mothers with whom we spoke. Finding time to go to school was even harder for wage earners than for welfare recipients. Furthermore, since they were less likely to qualify for Pell grants or other need-based forms of tuition assistance, going to school was more expensive for workers than for welfare recipients.

Because mothers who relied on wages struggled even harder to make ends meet than those who relied primarily on welfare, most felt that welfare made better economic sense than work did. In a social and psychological sense, however, mothers felt that work held clear merit. They were ambivalent about the toll that work would take on their parenting. On the one hand, mothers feared subjecting their children to the economic and social risks that came with work. On the other hand, they wanted their children to be proud of them and to take them as role models.

Welfare-reliant mothers' estimates of what they would need to leave welfare in the early 1990s reflected the experience of our wage-reliant mothers: they knew they would incur added costs for child care, medical care, transportation, work apparel, and housing. Our welfare-reliant mothers thought they would need $8 to $10 an hour to break even, a rate that closely resembled our estimates based on our wage-reliant mothers' budgets. They also knew they might be able to work for less if they enjoyed a special circumstance that lowered the cost of working or their other expenses.

All else equal, almost all mothers said they would rather work than rely on welfare. They believed work had important psychological benefits and welfare imposed stigma costs. However, mothers who worked knew they must risk their own and their children's well-being to do so. Nevertheless, because most single mothers want to work and most of the public wants them to do so, we advocate work-based solutions to the welfare and poverty problems we describe here. Very simply, then, if we want to make work less costly for unskilled and semiskilled mothers, their earning power must be enhanced. We can raise their wages by making substantial investments in their skill levels, by helping them to get better jobs, or by supplementing the wages and benefits they receive at their current jobs. To make any of these solutions viable, mothers with two children who worked full time and year-round would need to earn at least $16,000 a year—$8 to $10 an hour—in 1991 dollars. This figure assumes very modest child care or health care costs, which, of course, is not necessarily a reasonable assumption to make. Thus, unless affordable, high-quality child care and health care become available, those with expensive health problems or child care needs would have to earn more than we have estimated here or put their children at serious risk. As for the training option, the kind of training necessary to bring unskilled single mothers' earning power up to $8 an hour (1991 dollars) isn't currently available to welfare mothers in most states....

Behavior and Personal Characteristics

To get at some of the most controversial aspects of the welfare debate, we wanted to know whether a mother's personal or social characteristics influenced how much she spent each month, how much hardship her family experienced, and how she bridged the gap between her main income (either welfare or low-wage work) and her expenditures. We looked at whether the receipt of welfare, marital status, family welfare history, neighborhood residence, and race or ethnicity affected the economic behaviors of mothers in our sample. Although our data do not indicate important distinctions between these groups, some effects were surprising and ran counter to conventional theories about the relationship between culture and poverty.

First, mothers who received welfare, mothers who had never married, mothers living in very poor neighborhoods, and minority mothers budgeted as effectively as their more "mainstream" counterparts. In fact, these mothers were more frugal and spent less on nonessentials. We argued that this was because the mothers had more access to the cultural tools necessary to make do with very little. Mothers who had been raised on welfare, however, did not fit this pattern, possibly because they spent some of their money foolishly but more likely because of unmeasured differences in the level of in-kind assistance from kin.

Mothers from the more disadvantaged groups earned as much from both supplemental reported and unreported work as their more advantaged counterparts. There were three exceptions to this rule: Mexican American mothers earned less than whites from reported work; mothers living in very poor neighborhoods earned less from informal work; and mothers who grew up on welfare earned more from underground work. We argued that the first two exceptions were due to local market conditions in San Antonio and the physical isolation of poor neighborhoods, respectively. The last may have resulted from either a greater willingness to engage in crime or a greater access to family members who engaged in criminal trades.

There was only one important group difference in the amount of covert cash help that welfare recipients received from their social networks: Mothers who had grown up on welfare received more. In addition, mothers from more disadvantaged groups had less access to the social resources that might have enabled them to make the transition from welfare to work. This difference might explain why national data show that throughout the 1980s mothers from these groups spent more time on welfare and had more difficulty permanently leaving welfare for work (Edin and Harris forthcoming).

With regard to generating cash help from agencies, mothers from different backgrounds differed little. In fact, mothers from better neighborhoods received more cash from agencies than mothers living in very poor neighborhoods, possibly because agencies use people's address as a proxy for other socially desirable characteristics, like the perceived desire to better oneself.

These findings suggest that mothers' personal and social characteristics cannot explain most of the economic behaviors of the women in our sample. By and large, once mothers from different walks of life found themselves in the same miserable situation, they responded in similar ways. Though some mothers who grew up on welfare did seem to be somewhat more willing to engage in vice to meet their expenses, a majority did not....

Conclusion

In the early 1990s, single mothers chose the harsh world of welfare because that of low-wage work was even more grim. Both our data and national data provide overwhelming evidence for this claim. In the early 1990s, those who managed to survive the low-wage labor market did so because they were unusually fortunate. But, we cannot base policy on the hope that all mothers will be so lucky.

Either substantial wage supplements or high-quality training are essential if the current population of unskilled and semiskilled women is ever to attain self-sufficiency through work. In addition, each of these solutions must include affordable access to reliable child care and health care. While current welfare-to-work programs might move women into jobs in the short term, these mothers and their children will experience much material hardship if they remain trapped in jobs that pay $5, $6, or $7 an hour and offer few benefits.

If, on the other hand, states elect to pay welfare benefits that are even less generous than what [these] women ... received, our data suggest that some mothers will elect to work merely because welfare is even less viable, though the effect will likely be small. Time limits might have a larger effect on family well-being, since mothers will not be able to withdraw from the low-wage labor market to get more schooling or attend to their children's needs. Both of these changes mean that single mothers' children will receive less supervision, spend more time home alone, and become even more vulnerable to the harmful influences of inner-city neighborhoods than is now the case. They will also suffer more material hardship.

It is not unreasonable, therefore, to predict that these children might suffer even higher delinquency, dropout, pregnancy, and incarceration rates than they currently do.[3] These problems, which may not become fully evident for a generation, will certainly prove far more costly in the long run than the "welfare problem" Americans have complained so bitterly about during the 1980s and early 1990s. For those states that choose to invest substantial sums in education and training, child care subsidies, and health care plans for single mothers who work, the future could be much brighter.

NOTES

1. The poverty line is based on the percentage of total income Americans spent for food in the early 1960s (33 percent). At that time, the Department of Agricul-

ture drew up estimates of bare-bones or "minimally nutritious" food budgets for families of various sizes and multiplied it by three. Although this amount is adjusted annually for inflation, the spending patterns of Americans have changed dramatically, and the official measure has not been adjusted to reflect these changes.

Between the mid-1960s and 1980 the proportion of Americans' budgets spent on food decreased to one-fifth. This was because the prices of nonfood items went up, while food costs stayed roughly constant. If we adjusted the poverty line to reflect these changes, the official threshold would need to be raised to 140 percent of the official poverty threshold for a family of four (Ruggles 1990).

2. In sum, a strong enforcement system seemed to be good for mothers' budgets. We are not so sure, however, about the effects of strong enforcement for the well-being of men. In some states with strong enforcement, fathers who fail to pay are jailed. Less dramatically, their driver's licenses can be revoked, their tax returns seized, and so on. These things can occur even when the father loses his job and cannot meet his monthly obligation. Fathers who want to change the award amount they pay must hire a lawyer and go to court, which many cannot afford to do. We suspect that a system of this kind may actually force the least able and willing of fathers to deny paternity or, if paternity is proven, to "go underground"—to take jobs in the informal or underground economy. More research is needed in this area, but policymakers who advocate strong enforcement systems should attempt to learn more about how it affects men's behavior.
3. See McLanahan and Sandefur (1994) for a review of current rates of these social problems among children raised by a single parent.

REFERENCES

Bane, Mary Jo, and David T. Ellwood. 1994. *Welfare Realities: From Rhetoric to Reform.* Cambridge, Mass.: Harvard University Press.

Edin, Kathryn, and Laura Lein. Forthcoming. "Work, Welfare, and Single Mothers' Economic Survival Strategies." *American Sociological Review.*

Lipsky, Michael. 1980. *Street-Level Bureaucracy.* New York: Russell Sage Foundation.

McLanahan, Sara, and Gary Sandefur. 1994. *Growing Up with a Single Parent: What Hurts, What Helps.* Cambridge, Mass.: Harvard University Press.

Ruggles, Patricia. 1990. *Drawing the Line: Alternative Poverty Measures and Their Implications.* Washington, D.C.: Urban Institute Press.

Schwarz, John E., and Thomas J. Volgy. 1992. "Social Support for Self-Reliance: The Politics of Making Work Pay." *American Prospect* (9, Spring): 67–73.

Spalter-Roth, Roberta, Beverly Burr, Heidi Hartmann, and Louise Shaw. 1995. "Welfare That Works: The Working Lives of AFDC Recipients." Report to the Ford Foundation. Washington, D.C.: Institute for Women's Policy Research.

Mothers with Children and Mothers Alone: A Comparison of Homeless Families

Jennifer E. Glick

Homeless families compose the fastest growing segment of the homeless population (Rossi, 1994; Burt, 1992). Most studies on homeless families focus specifically on parents accompanied by minor children in public or private shelters. "When mothers seek emergency shelter alone, they are considered single regardless of their possible contact with children and their own self images as part of a family network" (Brickner et al., 1990: 140). This operational definition of a homeless family excludes those adults who resort to homeless shelters while their children reside elsewhere thus limiting our understanding of the process families go through upon losing stable housing. The present study relies on a sample of homeless mothers both with and without their minor children in a public shelter. Two questions are addressed by this comparison of types of homeless mothers: (1) Are these two groups significantly different from one another? (2) If there are differences between them, do these differences suggest that mothers who resort to the shelter alone are in need of different services than those accompanied by their children?

Studies comparing homeless individuals without children to homeless parents accompanied by children and those comparing housed mothers to homeless mothers suggest that homeless mothers accompanied by children resemble a housed population more so than homeless women without children. For example, homeless women are older and tend to have more psychological problems than homeless mothers with children. Some authors conclude that these two groups are indeed different and in need of different types of services (Johnson and Kreuger, 1989). Comparisons of housed mothers in poverty and homeless mothers find that these groups are similar in age, education, number of children and number of social ties (Goodman, 1991). It appears that homeless mothers are less likely than housed mothers to have the resources to maintain independent housing but are just as likely to have social contacts. While there is some evidence that homeless adults with children are different from those without children, few studies ask about the existence of minor children not in the shelter system (Burt and Cohen, 1989). Consequently, if we rely on the types of studies discussed above, our perception of the extent to which families are affected by homelessness and the needs of such families will be limited to those in which parents and children remain together. By using information on the location of children who are not in a shelter with their mothers, this study can describe the array of strategies used to cope with the loss of stable housing.

Methodology

The data for the study is drawn from a population of homeless families taking shelter in Austin, Texas. The shelter offers a variety of services for those needing assistance and is therefore conducive to collecting a sample of homeless families from a variety of backgrounds and situations. As with many investigations of homelessness, this sample is limited in scope and may not be generalizable to the national population of homeless families. Files of social workers' interviews with homeless women were utilized to create a quota sample of 168 family units with children present or women with dependent children elsewhere. The sample includes women seeking shelter over a recent 18 month period. There were 46 mothers without accompanying children and 81 mothers with at least one child present at the shelter. These groups will be referred to as unaccompanied mothers and accompanied mothers respectively.

Findings

The homeless families in this sample are similar in many ways to the homeless families found in other major studies (Rossi, 1989; Burt and Cohen, 1989). However, the suggestion that homeless families are, for the most part, composed of young mothers leaving their parental home, temporarily homeless and awaiting Public Assistance funds to allow them to set up their own home is not supported (Rossi, 1989). The characteristics of the Austin sample defy this image. The average age and education levels of the mothers in this study, for example, are higher than this image of teen mothers with few skills. There are some differences between the two groups of mothers in this sample, however.

The accompanied homeless mothers are younger (30 vs. 33 years), have younger children (4.1 vs. 8.3 years) and have fewer children (1.9 vs. 2.4) than those unaccompanied by their children. In addition the accompanied mothers are less likely to have completed high school. Nonetheless, half of the accompanied mothers had a high school degree or its equivalent and sixty-six percent of the unaccompanied mothers had at least this much education. The t-tests (not shown) for these demographic characteristics indicate that these are significant differences but neither group fits the stereotype of the young mother leaving her parental home for the first time. A comparison of the work and residential history of these two groups illustrates several similarities between the two groups of mothers.

Few of the accompanied or unaccompanied mothers in this sample are currently employed upon entering the shelter, although many indicate that they have previously worked and gave the reasons for leaving their last place of employment. Table 1 presents the reasons given for leaving employment. Of those specifying a reason, accompanied mothers are more likely to report that they left their jobs due to family responsibilities, pregnancy, child care constraints and illness. The greater likelihood that women accompanied by

TABLE 1 *Employment Status and Reasons for Leaving Last Place of Employment*

	Unaccompanied Mothers		Accompanied Mothers	
	N	Percent	N	Percent
Employed on entering shelter	5	10.9	10	12.4
Reason for leaving last employment				
Move or transportation problem	5	10.9	7	8.6
Job related (i.e. temp job, etc.)	3	6.5	7	8.6
Conflict at work/fired, quit, etc.	5	10.9	5	6.2
Family/Illness, preg., childcare	4	8.7	15	18.5
No reason given/no previous job	24	52.2	37	45.7
Number	46	100.0	81	100.0

Source: Emergency Shelter, Austin, Texas

TABLE 2 *Previous Residences of Unaccompanied and Accompanied Homeless Mothers*

	Unaccompanied Mothers		Accompanied Mothers	
	N	**Percent**	**N**	**Percent**
Own home	9	19.6	19	23.5
Home of spouse/partner	0	0.0	8	9.9
Parents'	4	8.7	10	12.3
Other relatives'	3	6.5	7	8.6
Friends'	13	28.3	18	22.2
Other (tents, shelters, etc.)	5	10.9	6	7.4
Unknown	12	26.1	13	16.0
Number	46	100.0	81	100.0

Source: Emergency Shelter, Austin, Texas

their children lost their jobs due to family responsibilities suggests that the role conflict of breadwinner and care giver may lead to homelessness. These mothers have younger children on average. The difficulties single mothers face in the job market and in finding adequate and affordable child care make this population particularly vulnerable to homelessness (Milburn and D'Ercole, 1989). Since neither group of women is likely to be employed, job programs may be of use to both. Public child care and sick-child care are also needed, especially by those mothers who do not or cannot rely on family and friends for these services and therefore cannot maintain employment.

It is common for families to share living quarters with friends or other family members before resorting to public shelters. As seen in Table 2, accompanied and unaccompanied mothers who came to the Austin shelter are both likely to have come from a family or friend's home. Interestingly none of the unaccompanied mothers came directly from a home shared with a spouse or partner. These mothers are more likely to be formerly married, however, suggesting the availability of a former spouse to house their children when the mothers are unable to do so. The accompanied mothers are more likely to have come directly from a home shared with a spouse. In other words, when marital break-up and homelessness occurred simultaneously, children became homeless with their mothers. Clearly, both unaccompanied and accompanied mothers are likely to have relied upon other social contacts for assistance and housing before resorting to the public shelter.

That social networks are important resources for homeless mothers is also demonstrated by the living arrangements of the children not found at the shelter. Table 3 looks at the living arrangements of all of the children of both groups of mothers. Parents, grandparents and other relatives are important sources of care for these children. Older children are more likely to be in the "other" category, which includes the few children in foster care as well as children who moved in with friends' families.

As the above discussion illustrates, the biggest difference between the accompanied and unaccompanied mothers is the presence or absence of their minor children. Looking only at the 46 mothers who came to the shelter without their minor children, two things are clear. First, the experience of having their children removed from their care by a social service agency is very rare. The reasons mothers separate from their children have much more to do with their financial need rather than abuse or psychological distress. The second observation from these unaccompanied mothers' situations is that family members and

TABLE 3 *Location of the Children of Unaccompanied and Accompanied Mothers*

	Children of Unaccompanied Mothers		Children of Accompanied Mothers	
	N	Percent	N	Percent
Present at shelter	0	0.0	179	88.5
Other parent	30	37.5	8	4.0
Grandparent	28	35.0	8	4.0
Other relative	12	15.0	3	1.5
Other (foster care, self care, etc.)	10	12.5	4	2.0
Number	80	100.0	202	100.0

Source: Emergency Shelter, Austin, Texas

friends are important sources of housing and care for their children. A large minority of all mothers (41.3%) report their children are in the care of their other parent. This means, however, that the majority of the minor children not at the shelter with their mothers are not in the care of either parent but are more likely to be in the care of a grandparent or other relative. Some of the mothers keep their youngest children with them when they move to the shelter but leave older children with others so they can remain in school. For example, one mother came to the shelter with her two toddler children but had left her older son with her parents so that he could stay in school. With only limited resources at their disposal, the mother's parents were doing all they could to help her by housing one of her children.

It is not clear at what point mothers separate from their children, but those who do leave their children in the care of others appear to do so when their social networks become stretched and doubled-up living arrangements become crowded. For example, some unaccompanied mothers report that they did not come directly from the place they left their children but had lived with other family members or moved in with friends for as long as they could. The mothers may move through their social networks before resorting to the public shelter, while using the most secure of their social ties (i.e. their own parents or siblings) to house their children. These situations are frequently intended to be temporary. One mother describes leaving her six year old daughter with her grandparents in another town while looking for permanent housing in Austin. She then intends to be reunited with her daughter once she can afford to care for her. Had this woman been included in other studies of homelessness, she would not have been counted as a homeless family because her child was not with her in the shelter. Nonetheless, the ties between parent and child still exist and the services needed by this woman and those like her are similar to those needed by mothers who bring their children with them to the shelter. Designing programs to support families clearly needs to include families temporarily separated by housing needs.

Summary and Discussion

There are several small differences between those women whose children are housed elsewhere and those who bring at least one of their children to the public shelter in Austin. Unaccompanied mothers are older, have higher educations and [have] older and fewer children. In addition, accompanied mothers are more likely to report that they lost employment due to the need to care for family members. Thus, the precipitating events to homelessness may be different for some of the mothers who bring their children to the shelter than for those unaccompanied.

These two groups of mothers are quite similar in terms of their contact with friends or family suggesting that homeless mothers are not isolated from their social networks. The unaccompanied mothers may have networks with more resources available. For example, the fact that their minor children are not with them reflects the capacity of some mothers' networks to provide instrumental support. In addition, their separation from their children represents choices mothers make in the interest of their children. The fact that some mothers can find friends or relatives to provide care for their children in times of need does not indicate that these mothers have different service needs than those whose friends and family members are unable or unwilling to provide housing.

Including single adults in analyses of family homelessness illuminates the manner in which a lack of stable housing affects all family members, not just those found in emergency shelters. Broadening the analysis of family homelessness to include absent children illustrates the importance of social networks for homeless mothers and the variety of ways families may adapt to a loss of permanent housing. Many homeless mothers have others they can turn to for support and housing for their children and for temporary housing for themselves. The fact that some mothers do not come directly from the same location as their minor children suggests that there are many steps on the route to becoming homeless. Some findings suggest that leaving a stressful, doubled-up living arrangement is the last step before resorting to public shelters (Burt and Cohen, 1989). Apparently another part of this process for some mothers involves establishing secure housing for their children. Such strategies are useful in the face of obstacles for self sufficiency such as low paying jobs and long waiting lists for subsidized housing.

The data of this study call into question the conventional definition of a "homeless family" and suggest that excluding mothers alone in public shelters from services aimed at supporting families without stable housing ignores a significant segment of the population of families affected by homelessness. The findings also have important implications for public policy reforms. For example, the viability of proposals for reform of the welfare system is called into question by the fact that so many mothers indicate that they lost their employment due to family-care needs. Reforms that include job requirements for welfare recipients are untenable without affordable, reliable child care.

The findings of this study also indicate that programs for homeless families are providing services needed by families where only the adults are technically homeless. Many mothers have no one to house them or their children or they must choose which children to leave with relatives. Mothers in the dire economic conditions that lead to homelessness may have friends and family members who wish to help them but cannot afford to do so forcing separations of mothers from their children. A more behaviorally accurate definition of a homeless family would include those adults who lack a fixed permanent address and have minor children dependent upon them, regardless of where their children are located. By including all families in services for homeless families, it may be possible to reunite these children with their parents.

REFERENCES

Bassuk, Ellen L., and John C. Buckner. 1994. "Troubling Families: A Commentary." *American Behavioral Scientist, 37*(3):412–421.

Brickner, Philip W., Linda Scharer, Barbara Conanan, Marianne Savarese and Brian Scanlan (eds.). 1990. *Under the Safety Net: The Health and Social Welfare of the Homeless in the United States.* New York: W.W. Norton and Co.

Burt, Martha R. 1992. *Over the Edge: The Growth of Homelessness in the 1980s.* New York: Russell Sage Foundation.

Burt, Martha R., and Barbara E. Cohen. 1989. "Differences Among Homeless Single Women, Women with Children and Single Men." *Social Problems, 38*(5): 508–524.

Goodman, Lisa A. 1991. "The Relationship Between Social Support and Family Homelessness: A Comparison Study of Homeless and Housed Mothers." *Journal of Community Psychology, 19*(4): 321–332.

Johnson, Alice K., and Larry W. Kreuger. 1989. "Toward a Better Understanding of Homeless Women." *Social Work, 34*:5 37–40.

Milburn, Norweeta, and Ann D'Ercole. 1991. "Homeless Women: Moving Toward a Comprehensive Model." *American Psychologist, 46*(11): 1161–1169.

Rossi, Peter H. 1994. "Troubling Families: Family Homelessness in America." *American Behavioral Scientist, 37*(3): 342–395.

A New Look at Poverty in America

William P. O'Hare

Many of the most contentious social policy issues being debated today—including reform of the nation's welfare, health care, and education systems—involve poverty. Yet the public and policy-makers are often misinformed about the lives and characteristics of America's poor, and even about whether poverty is increasing or decreasing....

Government Assistance for the Poor

The American public is frustrated by the image of soaring social welfare costs and continued high levels of poverty. However, much of this frustration stems from misperceptions about which government programs are driving the increase, a misunderstanding about who receives government assistance, and an exaggerated notion about the amount of assistance that goes to the poor.

Opinion polls indicate that the public is willing to help people in need, but that most people think that the current system does a poor job of alleviating poverty.

The federal government expects to spend over $900 billion to provide benefits directly to individuals in Fiscal Year (FY) 1996, about 58 percent of the total budget. Government programs that provide benefits directly to families or individuals can be divided into two broad categories: (1) social insurance and (2) means-tested benefits. Social insurance benefits are entitlements that go to anyone falling into a specific category, such as the elderly or veterans, regardless of income level.

Social insurance benefits account for nearly three-fourths of all federal money spent on social programs. Social Security, Medicare, and veterans benefits are by far the largest programs. They are expected to consume about $670 billion in FY1996 (see Table 1).

Means-tested programs differ from Social Security and other social insurance programs in that recipients must meet a "means" or income test to qualify. These are welfare programs in common parlance because they are aimed at helping low-income people, primarily families with children. Both federal and state governments contribute toward welfare programs, but the federal government pays the bulk of the costs. In FY1993, the federal government spent twice as much on welfare programs as all the states together.[1] Local governments also contribute to the cost of social programs on a much smaller scale. The federal dollars spent on social programs are being cut back and the states are assuming greater administrative control, but most funds for these programs will continue to flow from the federal government.

Although many people equate welfare with poverty, there is little correspondence between the two terms in government statistics or programs. About one-fourth of the poor do not participate in any major welfare program, and nearly 60 percent of welfare recipients are not officially designated as poor under the current government definition.

Eligibility for welfare programs is based on income, but the income thresholds are not usually tied to the poverty level. Of 70 government programs that assisted low-income people during 1992–1994, only 8

BOX 1 *Common Myths About the Poor*

Public perceptions about the poor, welfare programs, and welfare recipients often are shaped more by myth, anecdote, and misinformation than by research. Some common misperceptions include:

Myth 1. The vast majority of the poor are blacks or Hispanics.
Poverty rates are higher among blacks and Hispanics than among other racial/ethnic groups, but they do not make up the majority of the poor. Non-Hispanic whites are the most numerous racial/ethnic group in the poverty population. They make up 48 percent of the poor, while African Americans make up 27 percent, and Hispanics 22 percent.

Myth 2. People are poor because they do not want to work.
Half of the poor are not in the working ages: About 40 percent are under age 18; another 10 percent are age 65 and older. Many poor people have jobs, but earn below-poverty wages. Nearly 6 million poor adults (ages 18 to 64) worked for 27 or more weeks during 1994—about 30 percent of the working-age poverty population. Many poor individuals report they cannot work because of a serious disability or because they must care for family members.

Myth 3. Poor families are trapped in a cycle of poverty that few escape.
The poverty population is dynamic—people move in and out of poverty every year. Only 12 percent of the poor remain in poverty for five or more consecutive years. [Figure 1]

Myth 4. Welfare programs for the poor are straining the federal budget.
Social assistance programs for low-income families and individuals are expected to cost the federal government about $217 billion in Fiscal Year 1996, which is about 14 percent of projected federal expenditures for that year. A much larger share of the budget (43 percent) goes to other types of social assistance, such as Social Security and Medicare, which mainly go to middle-class Americans, not the poor.

Myth 5. The majority of the poor live in inner-city neighborhoods.
Less than half (42 percent) of the poor live in central city areas, and less than one-quarter live in high-poverty inner-city areas. Over one-third (36 percent) of the poor live in the suburbs, and more than one-fifth (22 percent) live outside metropolitan areas.

Myth 6. The poor live off government welfare.
Welfare accounts for about one-fourth of the income of poor adults. Social Security, which is not welfare because it is not based on need, contributes about 22 percent of the income for the poor. Nearly half of the income received by poor adults comes from wages or other work-related activity. About three-quarters of the poor received some type of welfare benefit in 1994—such as Medicaid, food stamps, or housing assistance—but only about 40 percent received cash welfare payments.

Myth 7. Most of the poor are single mothers and their children.
Female-headed families represent just 38 percent of the poor. About 34 percent of the poor live in married-couple families, 22 percent live alone or with nonrelatives, and the remainder live in male-headed families with no wife present.

Myth 8. Antipoverty programs are designed to reduce poverty.
Most welfare programs are geared to sustain the poor, not pull them out of poverty. Only about 10 percent of the welfare budget goes to education and training programs designed to help people improve their earning potential. About 3 million people were lifted out of poverty by cash welfare assistance in 1994, which lowered the poverty rate only one-half a percentage point.

TABLE 1 *Federal Social Assistance Payments to Individuals in Fiscal Year 1996 by Program or Category*

Category or Program	Estimates in Billions of Dollars	Percent of Total
Social insurance	*$ 671*	*73*
Social Security and railroad retirement	352	38
Medicare	194	21
Federal Employees Retirement and Insurance	84	9
Unemployment assistance	24	3
Veterans health care	17	2
Means-tested programs	*217*	*24*
Medicaid	95	10
Food and Nutritional Assistance (including Food Stamps)	39	4
Supplemental Security Income (SSI)	24	3
Housing assistance	24	3
Earned Income Tax Credit (EITC)	18	2
Aid to Families with Dependent Children (AFDC)	17	2
Other (including student assistance)	*29*	*3*
Total	*917*	*100*

Source: U.S. Office of Management and Budget, Budget of the United States Government: Fiscal Year 1997, Historical Tables: table 11.3.

FIGURE 1 *Length of Poverty Spells, 1968–1987*

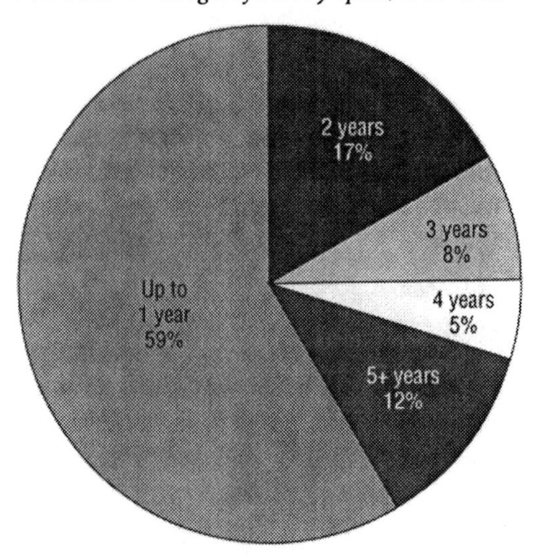

Note: Figures do not add to 100 because of rounding.
Source: Gottschalk, McLanahan, and Sandefur, in *Confronting Poverty: Prescriptions for Change*, eds. Danziger, Sandefur, and Weinburg: 89, Figure 4.1.

limited eligibility to people whose incomes were below 100 percent of the poverty threshold.[2] Families with incomes 130 percent of the poverty level are eligible for food stamps, for example, and the Women, Infants, and Children (WIC) program is available for families at 185 percent of poverty.

Another reason for the lack of correspondence between poverty and welfare statistics is that eligibility for means-tested programs is usually based on monthly income, while poverty status is determined on yearly income. A family's income may slip below poverty level for a month or two, making its members eligible for welfare, but income during the remainder of the year may be enough to keep it above the poverty line.

Those in poverty who do not receive benefits from any programs have various reasons for not participating. Some are discouraged or confused by the bureaucratic procedures involved in applying for programs.[3] Others, especially in rural areas, want to avoid the stigma attached to applying for or receiving welfare. Pride and the desire to remain independent also inhibit some from applying for welfare benefits.

Welfare programs encompass many types of assistance, which can be in the form of cash or noncash benefits. Noncash benefits, such as health care paid by Medicaid, free school lunches, or food stamps, are not counted as income in determining a family's (or individual's) poverty status.

The overwhelming majority of means-tested benefits are designed to sustain people while in poverty rather than help them move out of poverty. Only about 10 percent of means-tested benefits are for programs like education or training that might help welfare recipients become more self-sufficient.

Five programs have accounted for more than two-thirds of federal spending on the poor and disadvan-

TABLE 2 *Receipt of Welfare Benefits by Poverty Status and Family Status, 1994*

Type of Benefit	Total	Poor	Nonpoor	% of Recipients Who Are Poor
		Total population		
Total (1,000s)	261,616	38,059	223,557	
*Percent receiving:**				
Cash (AFDC or other)	10	40	5	56
Food stamps	11	51	4	66
Medicaid	16	56	10	50
Housing subsidies	4	19	2	61
At least one of above	26	73	18	41
		Families		
Total (1,000s)	69,313	8,053	61,260	
Percent receiving:				
Cash (AFDC or other)	9	43	5	54
Food stamps	10	55	4	65
Medicaid	15	60	9	48
Housing subsidies	4	21	2	63
At least one of above	23	77	15	40
		Individuals in nonfamily households		
Total (1,000s)	38,538	8,287	30,251	
Percent receiving:				
Cash (AFDC or other)	8	24	4	62
Food stamps	8	25	3	67
Medicaid	12	32	7	57
Housing subsidies	7	15	5	47
At least one of above	20	46	12	50

*Persons living in households in which someone receives this benefit.
Source: U.S. Bureau of the Census, unpublished tables from the March 1995 Current Population Survey.

taged in recent years. Medicaid, AFDC, Supplemental Security Income (SSI), and Food Stamps each will provide at least $17 billion in assistance in FY1996. In addition, the General Accounting Office has identified a cluster of means-tested housing programs that together provide about $24 billion a year in assistance as a fifth major type of welfare (see Box 2). The federal funds expended on these programs are slated to decrease following the welfare reforms of the mid-1990s. State and local governments may make up some, but not all, of the difference.

More than one-fourth of all Americans received benefits from one of the key welfare programs in 1994, but only 41 percent of the recipients had incomes below poverty after accounting for the income from government assistance programs (see Table 2).

BOX 2 *Government Welfare Assistance Programs—The Big Five*

There are dozens of government welfare programs, but five programs account for most federal dollars spent on antipoverty efforts in recent years. Recent legislation will change many of these programs.

Medicaid: Under Medicaid, the federal government provides matching funds to states to pay part of the medical services for low-income persons who are elderly, blind, or disabled, and for members of low-income families with dependent children. Each state designs and administers its own Medicaid program within federal guidelines, and states have considerable latitude in setting policies regarding eligibility, benefits, and payments to providers of services. In FY1995, Medicaid provided services for about 36 million persons at a cost of about $156 billion ($89 billion in federal funds and $67 billion in state funds).

AFDC—Aid to Families with Dependent Children: A federal-state program established by the Social Security Act of 1935, AFDC provides cash to the families of needy dependent children. Operating within broad federal guidelines, states administer or supervise AFDC programs and set the standards of need and payment levels. The amount states pay varies greatly. In FY1994, typical AFDC payments to a family of three ranged from $120 per month in Mississippi to $923 per month in Alaska. The average for all states was $420. The AFDC program cost about $23 billion in federal and state funds in 1994. Some 5 million families (14 million individuals) received AFDC benefits in FY1994. About two-thirds of the recipients were children. AFDC is being replaced by block grants to states, who will administer their own cash assistance programs.

Supplemental Security Income (SSI): The Supplemental Security Income (SSI) program provides cash directly to elderly, blind, and disabled persons to help bring their incomes up to federally established minimum levels. Most SSI recipients are blind or disabled. About one-third of recipients are elderly. SSI cash payments are administered nationwide by the Social Security Administration, but some states supplement the federal payment and benefit amounts vary by state. In FY1996, about 6.6 million persons received SSI benefits per month, at a cost of about $23 billion in federal and about $3 billion in state funds.

Food Stamps: The Food Stamp program gives vouchers to low-income households to enhance the nutritional value of their diets. State welfare agencies administer the program within broad federal guidelines, but the federal government provides all the funds. In FY1996, nearly 11 million households received food stamps in an average month, at a cost of over $26 billion. The average household received about $180 worth of food stamps monthly in FY1996.

Housing Assistance: Housing Assistance for low-income families is provided through many programs. Some programs target particular areas or groups, such as rural areas or Native Americans. Federal expenditures totaled $24 billion in FY1996.

Section 8 Low-Income Housing is the largest program and accounts for over half of all federal funds on means-tested housing assistance. In FY1992, almost 3 million families received a total of $12 billion in Section 8 housing assistance. Low-rent public housing, which accounts for about one-quarter of all federal housing expenditures, reached about 1.4 million families in FY1992.

Because most welfare programs are targeted at children, people living in families were much more likely to benefit from a government program than people living alone or in a nonfamily household. Over three-quarters of poor families participated in at least one of the five key programs, while less than half of the poor in nonfamily households participated in such programs. None of the programs in Table 2 reach as many as one-third of impoverished individuals in nonfamily households.

Some programs are better than others at reaching the poor. Medicaid and Food Stamps, which have the highest levels of participation among the poor, reach little more than half of all poor individuals. Forty percent of the poor received cash benefits in 1994, primarily from AFDC and SSI, and only 19 percent received housing benefits. The March 1995 Current Population Survey revealed that less than half of all children in poverty lived in families that received cash welfare assistance.

Most poor families receive benefits from two or three programs simultaneously. Participation in one of the main government welfare programs often assures eligibility for other welfare programs. AFDC participants, for example, were automatically eligible for Medicaid and food stamps. With the elimination of the AFDC program and introduction of tighter eligibility criteria in 1996, patterns of welfare use may change.

Welfare's Effect on Poverty

Welfare programs provide basic sustenance, housing, and health care to poor and low-income families, but they have surprisingly little impact on the official poverty rate. Means-tested government cash benefits lifted about 3 million people out of poverty in 1994, lowering the poverty rate by 1 percentage point over what it would have been.[4] Government assistance has a slightly greater effect in reducing poverty among children. However, the United States lags far behind other developed countries in helping poor children (see Box 3).

Ironically, government transfer payments that are not means tested, such as Social Security, are more effective than means-tested programs in reducing poverty, in part because they pay out much more money. The average retired person received nearly $700 monthly from Social Security in 1994. In contrast, recipients of AFDC received less than $150 per person monthly. In addition, AFDC benefits are reduced, dollar for dollar, by any earnings a recipient receives, while Social Security recipients are allowed to earn incomes well above the poverty line before their benefits are reduced.

Cash transfer payments—primarily from Social Security and other entitlement programs—lifted nearly 20 million people out of poverty in 1994. Most beneficiaries were elderly.[5] Without Social Security payments, the poverty rate among the elderly would have been about 50 percent rather than less than 12 percent in 1994.

AFDC benefits are not only much less than Social Security, they vary tremendously by state. Some states offer much more generous AFDC payments than others, but no state provides enough in welfare payments to lift a family out of poverty. Alaska offers the highest monthly AFDC benefit, about $920 for a family of three in FY1994. But even if a family received benefits every month, its annual income would be well short of the $11,821 poverty threshold for a three-person family.

Most states pay a family of three less than $400 monthly in AFDC benefits. Fourteen states, primarily in the South, pay less than $300 monthly to a typical family of three (see Figure 2). Mississippi, the least generous state, pays only $120 per month to such a family.

Average AFDC payments are significantly lower in the states with the highest poverty rates. Differences in living costs among states explain some, but not all, of the relationship between poverty and AFDC payments. Although AFDC payments reduce the percentage of the poor in extreme poverty, they are not sufficient to lower the poverty rate, especially in states where poverty is most prevalent.

Noncash welfare benefits have a greater impact on the well-being of the poor, but they are not included in calculations of poverty. While there is considerable controversy about how to value noncash benefits, especially medical care, the Census Bureau estimates that the poverty rate would be at least 3 percentage

BOX 3 *International Comparisons of Child Poverty*

The family conditions and government programs that affect child poverty in the United States put a large share of U.S. children at a distinct disadvantage relative to children in other developed countries. Many social scientists contend that the lack of investment in our children will put us at a competitive disadvantage in the international marketplace of the 21st century.

A study of child poverty in eight industrial countries published in the late 1980s surprised many when it found that the child poverty rate was higher in the United States than in any of the other countries.[1] A recent update comparing the United States with 16 developed countries suggests things have gotten worse for American children.[2]

Child Poverty in 17 Developed Countries Before and After Government Assistance, Mid-1980s to Early 1990s

Country*	Year	Percent of Children in Poverty: Before Assistance	Percent of Children in Poverty: After Assistance	Percent Children Lifted Out of Poverty by Gov't Programs
United States	1991	25.9	21.5	17
Australia	1989	19.6	14.0	29
Canada	1991	22.5	13.5	40
Ireland	1987	30.2	12.0	60
Israel	1986	23.9	11.1	54
United Kingdom	1986	29.6	9.9	67
Italy	1991	11.5	9.6	17
Germany	1989	9.0	6.8	24
France	1984	25.4	6.5	74
Netherlands	1991	13.7	6.2	55
Norway	1991	12.9	4.6	64
Luxembourg	1985	11.7	4.1	65
Belgium	1992	16.2	3.8	77
Denmark	1992	16.0	3.3	79
Switzerland	1982	5.1	3.3	35
Sweden	1992	19.1	2.7	86
Finland	1991	11.5	2.5	78

*Ranked by post-assistance poverty rate.
Source: Lee Rainwater and Timothy M. Smeeding, "Doing Poorly: The Real Income of American Children in a Comparative Perspective," Working Paper No. 127, Luxembourg Income Study, Maxwell School of Citizenship and Public Affairs (Syracuse, NY: Syracuse University, 1995).

points lower if noncash government benefits were counted as cash income.[6]

Other government programs also help the poor. Benefits may be in the form of services, such as free legal aid from the federally funded Legal Services Corporation, and preschool education for low-income children offered under the Head Start program. These services are not direct payments to in-

The gap between children in the United States and other developed countries reflects differences in private-sector income, but it is accentuated by enormous variations in the role of government. Based on private income alone, the child poverty rate in the United States in 1991 (25.9 percent) was higher than in every other country examined except Ireland and the United Kingdom. However, government assistance programs lifted more than half the impoverished children out of poverty in both of these countries. In the United States, government assistance lifted only 17 percent of poor children out of poverty. After accounting for assistance, the child poverty rate dropped to 12.0 percent in Ireland and to 9.9 percent in the United Kingdom. The U.S. rate, at 21.5 percent, moved to the top of the list (see table).

The post-assistance child poverty rate in the United States was 50 percent higher than the next highest country (Australia at 14 percent) and almost 10 times that of Sweden or Finland. In 11 countries, government assistance reduced the number of children in poverty at least by half. These findings suggest that the public sector in other developed countries does more than the United States to lift poor children out of poverty.

REFERENCES

1. John L. Palmer, Timothy Smeeding, and Barbara Boyle Torrey, *The Vulnerable* (Washington, DC: Urban Institute Press, 1988): 96.
2. Lee Rainwater and Timothy M. Smeeding, "Doing Poorly: The Real Income of American Children in a Comparative Perspective," Working Paper No. 127, Luxembourg Income Study, Maxwell School of Citizenship and Public Affairs (Syracuse, NY: Syracuse University, 1995).

dividuals and are not included in the calculation of poverty status.

The Earned Income Tax Credit (EITC) is another federal program to help low-income families, but it is sometimes overlooked in discussions of welfare programs because it works through the tax code. The EITC provides workers earning less than about $25,000 annually with a tax break that allows low-income families to retain more of their earnings.

The EITC has enjoyed bipartisan support, and has been expanded under Presidents Reagan, Bush, and Clinton. Because of recent program expansion, federal expenditures on the EITC have risen rapidly over the past few years, growing from $4.4 billion in FY1990 to $18.1 billion in FY1996. Seven states have adopted state-level EITC programs.

EITC has several features that appeal to policymakers, especially with the current emphasis on making individuals more responsible for their financial status. First, the EITC rewards work. Only low-income families with earned income are eligible for the benefit. And, up to a point, the more earned income a family has, the more it benefits from EITC. Second, the EITC targets families with children. Third, because it works through the tax code, it does not require additional bureaucracy to administer it.

Because the earnings returned through EITC are technically a tax refund, they are not counted in determining poverty. Poverty is based on pre-tax income. However, if EITC benefits had been treated as additional income in 1994, they would have lowered the poverty rate by 1 percentage point for all Americans, and by 1.5 percentage points for children under age 18.

Dissatisfaction with the Welfare System

There is a widespread belief that the American welfare system does little to help the poor escape poverty. Some critics of the current system claim that it actually perpetuates poverty by encouraging dependency on government handouts. This perspective has been voiced by some researchers, by leaders in both major political parties, and by a large segment of the public. Surveys show that most Americans have a strong negative image of welfare programs in general, although they may support specific programs.[7]

Much of the negative image of the welfare system stems from the segment of welfare recipients who become long-term users. Research shows that about one-third of AFDC recipients stay in the program for six or more years.[8] While relatively few in number, long-term recipients receive a large share of welfare expenditures over time because they receive benefits year after year.

Like the chronically poor, long-term welfare users often lack the education, skills, or work experience to obtain stable employment that pays enough to sustain their families. These long-term users constitute the greatest policy challenge, and are the focus of most current efforts to get people off welfare and into jobs. However, moving people from welfare to work is not easy. Even the most successful workfare programs have met with limited success, and "success" was often determined by the strength of the local economy.[9]

The population that uses welfare is highly dynamic. Many people go on and off welfare over a period of years. More than one-quarter of women who left the AFDC program in the mid-1980s, for example, were back on within a year; two-fifths returned within two years.[10]

While the welfare system may be cumbersome and convoluted, the system has worked fairly well for many who use it. For short-term users, welfare eases the burden of a temporary setback and sustains families until they can improve their situation. However, it does not provide such generous benefits that families are seduced into long-term dependency. About one-third of AFDC recipients stay in the program for less than two years, for example, and another third stay in for two to six years.

The New Welfare System

Welfare reform is currently proceeding along at least three avenues. First, Congress has enacted sweeping changes in welfare programs. Second, some states have substantially changed state-run programs. Michigan, for example, eliminated the General Assistance program that provided help primarily to low-income single adults, and Connecticut has reduced its benefit levels.

Third, individual states are requesting and receiving unprecedented numbers of waivers from federal laws and regulations that allow states to experiment with ways to lower their costs and reduce their welfare loads.

The genesis of current reform efforts goes back at least to the early 1980s. Early in the administration of Ronald Reagan, many social programs were eliminated or cut back severely. In 1984, Charles Murray argued for dismantling the welfare system in his seminal book, *Losing Ground.*[11] Murray's book generated a good deal of support for a wholesale revamping of the system.

In 1988, Congress passed the Family Support Act, which stressed education and job training for welfare recipients, and led to the creation of the Job Opportunities and Basic Skills (JOBS) program.[12] The Act also required that all states expand AFDC eligibility to include impoverished married couples. Previously, some states only allowed single women with children to qualify for AFDC, which many considered a disincentive for low-income mothers to get or stay married. The expanded eligibility, which took effect in 1990, resulted in a surge in the number of married couples receiving AFDC. The caseload for AFDC-UP (unemployed parent), the component of AFDC designed for low-income married couples, nearly doubled between 1989 and 1994.

Although the Family Support Act addressed some of the major concerns about the welfare system, the changes went unnoticed by most of the public, and were deemed insufficient by many analysts. Widespread dissatisfaction with the federal government's management of the welfare system continued. In 1992, President Clinton campaigned for the presidency promising to "end welfare as we know it." In 1996, the Republican majority in the 104th Congress passed legislation designed to completely revamp the welfare system.

In many cases, states have taken the lead in reshaping welfare programs. The movement toward increasing the responsibilities of states and decreasing the voice of the federal government is clearly under way. As a prominent political observer recently stated, "No matter what happens with federal budget negotiations, the Devolution Express—the massive transfer of power from Washington to the 50 states—is unlikely to be derailed."[13]

FIGURE 2 *AFDC Benefits for a Family of Three by State, 1994*

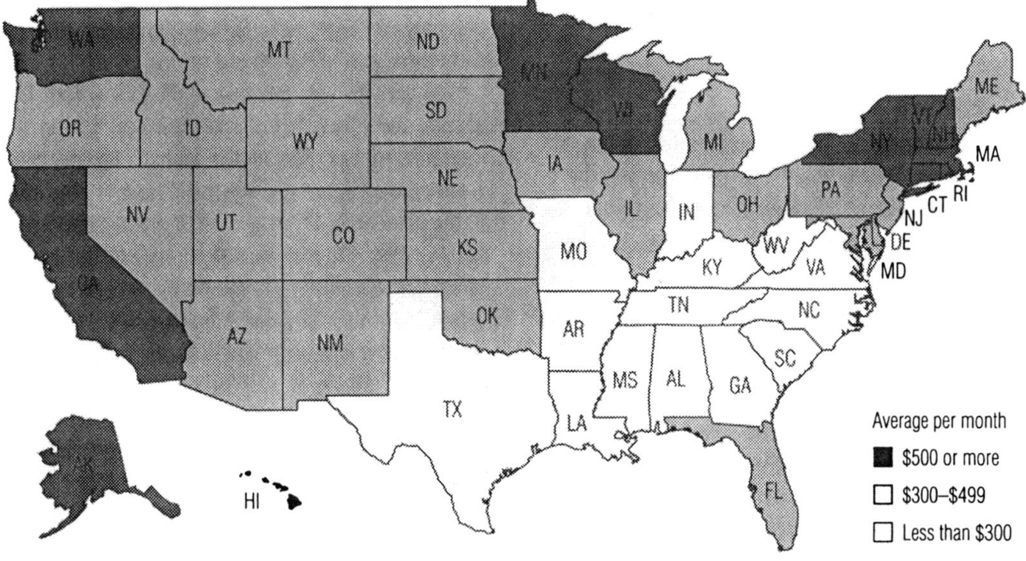

Source: U.S. Department of Health and Human Services.

Most analysts believe that the transfer of power from the federal government to the states will further reduce the total amount of funds allocated for the poor.

Significant welfare reform has already been occurring through the state waiver process. As of summer 1996, about 40 states had received waivers from the federal government that allowed them to experiment with different ways to administer welfare programs. These waivers will supersede changes mandated by recent welfare reforms, at least in the short term. Most of the rule changes sought by states involve ways to help welfare recipients find and keep jobs.

The waiver process requires a rigorous evaluation of the changes implemented, and consequently allows analysts to assess the success of specific state experiments with federal programs. From the standpoint of determining what reforms work best, the waiver process is preferable to the block-grant approach adopted by the welfare reform bill passed by Congress in 1996.

While changes in programs such as Medicaid and Food Stamps may have greater economic consequences for states, the reform of AFDC has been the major focus of public attention.

1996—A Watershed for Welfare Reform

In summer 1996, the U.S. Congress passed the Personal Responsibility and Work Opportunity Reconciliation Act, which fundamentally changes the federal government's relationship with low-income Americans. The legislation had three major elements: (1) cutting welfare spending, (2) giving states more power over program rules, and (3) enacting certain restrictive regulations regarding means-tested assistance (for example, denying benefits to children of unwed teens and many legal immigrants). Among many other features, the new legislation

- Ends the entitlement status of welfare, which guaranteed that any eligible American would receive benefits.
- Replaces the federal AFDC program with block-grant, lump-sum payments to states.
- Imposes a five-year lifetime limit for receiving federal cash assistance for at least 80 percent of each state's recipients.
- Ends cash assistance to able-bodied adults after two years unless they get a job.

- Limits receipt of food stamps by unemployed individuals to three months within a three-year period.
- Denies most types of assistance to legal immigrants who are not citizens.
- Allows states to deny welfare benefits for unwed teenage mothers and their children, unless the mothers attend school and live with an adult.
- Restricts eligibility of disabled children for the SSI program.
- Rewards states with additional cash grants if they reduce births to unmarried women.
- Expands funds for child care.
- Allows individuals who are no longer eligible for welfare to continue receiving Medicaid. Also, individuals who leave welfare to go to work can retain Medicaid coverage for one year.

Many of the provisions will be phased in over time, and the legislation allows states to side-step certain requirements. For example, a state may extend the time limit for federal funds beyond five years for up to 20 percent of its caseload. The law adds about $4 billion for child care to encourage parents on welfare to work, but overall, the reforms will cut federal spending on low-income Americans by an estimated $55 billion over six years.[14]

It is somewhat ironic that welfare reform is occurring at a time when caseloads are already decreasing. The AFDC caseload fell in 42 states between February 1993 and February 1996, yielding a 9 percent decline in the national caseload figure.[15]

Effects of Welfare Reforms

While the passage of welfare reform legislation may signal a new era in the history of government welfare programs, analysts cite several reasons that reform of the AFDC program in particular will have little effect on the official poverty rate. First, only about a fifth of the 23 million adults and half of the 15 million children in poverty are enrolled in the AFDC program. Second, legislation stresses jobs for AFDC recipients but does not insure that adequate money will be allocated for child care, medical care, education, and training programs that the unemployed poor need for the transition to stable jobs. Third, many of the poor who move off welfare through employment will simply become part of the working poor.

Some of the ramifications of the new welfare legislation are clear. The legislation will lessen the amount of money spent on the poor, but it is not likely to have much impact on the official poverty rate. After all the provisions of the law are implemented, the reforms are likely to move an additional 800,000 people below the official poverty threshold, according to an Urban Institute study—a 2 percent rise that would boost the poverty rate by less than one-half a percentage point.[16] However, using a more comprehensive poverty measure (similar to the one proposed by the National Research Council), the Urban Institute calculated that 2.6 million more people, including 1.1 million children, would be poor because of the benefit cuts in the new law.

In the short run, provisions of the law will lower living standards of millions of low-income Americans. Many of the cuts are in noncash benefits, such as Medicaid and food stamps, which are not counted in poverty calculations, and which benefited many people above as well as below the poverty level. The Urban Institute estimated that about 8 million low-income families with children will lose an average of $1,300 each per year under the new law.

The benefit cuts also will erode the incomes and well-being of families who are already poor, pushing them deeper into poverty. The poverty gap will grow by nearly $6 billion by 2001, according to the Urban Institute.

The long-term effects are less clear. Proponents of the new reforms expect them to curb out-of-wedlock births (especially to teens), increase financial self-sufficiency, and produce more stable families. We may never be able to assess the total impact of the legislation because we do not have the baseline data necessary to evaluate the changes undertaken by states in response to the legislation. Data from the Current Population Survey, Survey of Income and Program Participation, Survey of Program Dynamics, and National Health Interview Survey will allow researchers to track changes in the well-being of children and families at the national level. Such data are much more difficult to obtain for states. Researchers will be able to track changes in state welfare caseloads, but may not be able to determine whether states have used their new autonomy to improve the well-being of low-income families. A reduction in welfare caseloads does not necessarily indicate success, because families moving off of welfare may simply be living more austere lives, not moving out of poverty.

Effects on State and Local Areas

States with large immigrant populations will be hardest hit by welfare reforms. In California, which contains 40 percent of the 1.5 million legal immigrants who received federal welfare benefits under the old rules, county governments could be forced to absorb as much as $10.7 billion in medical and social services for noncitizens over the next six years. Counties are legally obligated to maintain a safety net for indigent residents, which includes providing medical care at county hospitals. A Los Angeles County Supervisor noted that, because of the welfare reforms, "All the legal immigrants who are being taken off the welfare rolls are going to end up on our doorstep."[17]

New York is another state with a large immigrant population that stands to pay millions of dollars in additional social welfare because of federal reforms. The state constitution requires the financial support of indigent people, but the state is not likely to be able to pick up the entire cost once federal spending is trimmed back. In New York, as in many other states, public debate is likely to turn to the determination of who is truly needy and what amount of aid is needed to supply "minimal support."[18]

Increased autonomy for states is a key element of the welfare reform bill. However, a number of writers have questioned whether states have the capacity and the will to run welfare programs formerly controlled from Washington,[19] or whether they will pass the responsibility on to local areas that may be even less able to administer such programs.

To make up for cuts in federal funds, many states will need to increase taxes (an unlikely event) or to reallocate their funds. The rapid expansion of Medicaid is already squeezing state budgets, which makes it unlikely that states will divert additional funds to social assistance programs to make up for federal spending cuts.[20]

Some analysts worry that decreased federal involvement will spark "a race to the bottom" in which states compete to cut welfare the most, generated in part by the belief that potential welfare recipients will move to the state that pays the most generous benefits.[21] As states assume more responsibility for social programs, advocates for the poor fear that states will try to reduce welfare as much as or more than other states to keep out welfare recipients and retain corporations and middle-class taxpayers.

Other analysts worry that shifting responsibility for major welfare programs to states will intensify the effect of regional or local economic downturns. During a recession, states will have less revenue to spend on the needy at just the time when more people are likely to be out of work and in need of assistance. Federal funding of social assistance programs provides states with a cushion against economic hard times because costs are spread over all the states.

Summary and Conclusion

The alleviation of poverty is one of the most compelling social issues facing Americans today. The way we choose to deal with poverty has tremendous implications for the fiscal health of federal, state, and local governments, the well-being of our families, and the quality of our future work force. But much of the policy discussion of poverty and welfare is being driven by outdated images of poverty and false assumptions about welfare recipients.

While poverty is lower today than it was in the 1940s and 1950s, the rate has drifted upward since 1973. The poverty rate in 1994 was nearly one-third higher than in 1973, and the number of poor was 15 million greater. Two aspects of the general rise in U.S. poverty are particularly disturbing: the increase in poverty rates among children, and the growing gap between rich and poor.

These two aspects reflect declines in the wages within the low-skill job sector, and demographic changes that have increased the size of the population groups that run a high risk of being poor, such as female-headed families and racial minorities.

The major exception to these trends is the elderly. Poverty rates for the elderly have fallen over the past two decades. Even though the elderly's share of the U.S. population grew from about 10 percent to 13 percent during that time, their share of the poverty population dropped from 15 percent to 10 percent.

While government programs are often blamed for the plight of the poor, a major government program—Social Security—was largely responsible for the dramatic reduction of poverty among the elderly. But, since Social Security consumes about 22 percent of the federal government's budget, it is not a model that is likely to be applied to the rest of the population. Within a climate of reducing government spending, policymakers will look for other ways to deal with poverty among children and families.

Current welfare reform efforts are concerned with getting people into jobs, and off the dole. In the short run, these changes are likely to increase poverty among children. The long-term effects are not clear. Many people favoring the recent welfare reform admit the consequences are unknown, but feel any change is better than the status quo. Conservative spokesman William J. Bennett (former secretary of education) states, "There simply is no other alternative; things have to change if we are to avoid social ruin."[22]

There are three major thrusts to current welfare reforms being discussed by state and federal lawmakers. First, states are taking a more active role in setting policies and administering programs. Because welfare reform involves block grants, which shift more responsibility to the states, reforms are likely to accentuate discrepancies in the amount the poor and near poor receive in different states.

Second, welfare recipients are expected to shoulder more responsibility for their financial situation, and program benefits are more likely to become linked to behavior. Only people who earn income can benefit from the EITC, for example. Mothers who have additional children while on welfare will not receive benefits for that child.

Third, money spent on welfare is likely to be reduced. The movement toward balancing the federal budget, the rapid increase in state Medicaid expenditures, and a clear message from taxpayers to reduce spending will diminish the amount available for other social service programs.

The increased efforts to move welfare recipients into the work force is a move in the right direction, but few policymakers seem to appreciate how difficult this will be. Further, getting a job does not necessarily pull a family out of poverty. Welfare reform is likely to add to the already expanding ranks of the working poor. And, welfare reform is unlikely to stop the growth of female-headed households, erase racial differences in poverty, or slow the geographic concentration of poverty.

But there are reasons for optimism. Even among the groups at highest risk of poverty, getting additional education clearly lowers the likelihood of being poor. The poverty rate for high school dropouts is three times the rate for those who have just a high school diploma. While a high school diploma may not be enough to ward off poverty in the future work force, programs designed to keep teenagers in school and to train and retrain adults will improve poor individuals' chances of getting a job that pays above-poverty wages.

The other optimistic note is that many of the working-age poor already have some attachment to the labor force. Expansion of programs that lower some of the barriers to working by subsidizing child-care and transportation costs or insuring medical coverage for the working poor would enable more of the poor to rejoin the labor force, increase their work effort, and make work more rewarding.

The task is daunting, and the path is littered with obstacles, but with nearly 40 million people below poverty, and another 25 million poised at the brink of poverty, we must continue to search for ways to reduce poverty in America.

REFERENCES

1. General Accounting Office, "Means-Tested Programs: An Overview, Problems, and Issues," T-HEHS-95-76 (Washington, DC: GAO, 1995).

2. National Research Council, *Measuring Poverty: A New Approach* (Washington, DC: National Academy Press, 1995): 321, table 7-2.
3. Margaret Dunkle, "Bottom-Up Look at Welfare Reform: What Happens When Policy-Makers Apply for Assistance from the Programs They Create," *Education Week* 15, no. 13 (Nov. 29, 1995): 21–22.
4. U.S. Bureau of the Census, *Current Population Reports* P60-189: B-28, table B-8.
5. Census Bureau unpublished tabulations from the March 1995 Current Population Survey distributed at press conference, Oct. 6, 1995.
6. U.S. Bureau of the Census, *Current Population Reports,* P60-189: B-28, table B-8.
7. R. Kent Weaver, Robert Shapiro, and Lawrence R. Jacobs, "Trends: Welfare," *Public Opinion Quarterly* 59, no. 5 (Winter 1995): 606–627.
8. Mary Jo Banc and David T. Ellwood, *Welfare Realities: From Rhetoric to Reform* (Cambridge, MA: Harvard University Press, 1994): table 2.3.
9. Daniel Freidlander and Gary Burtless, *Five Years After: The Long-Term Effects of Welfare-to-Work Programs* (New York: Russell Sage Foundation, 1994).
10. Kathleen Mullan Harris, "Life After Welfare: Women, Work, and Repeat Dependency," *American Sociological Review* 61, no. 3 (June 1996): 407–426.
11. Murray, *Losing Ground: American Social Policy 1950–1980* (New York: Basic Books, 1984).
12. Spencer Rich, "Senate Vote Overhaul on Welfare," *The Washington Post,* June 17, 1988: A1.
13. Neal R. Pierce, "A Bigger Bang for the Welfare Buck?" *National Journal,* Feb. 17, 1996: 384.
14. Robert Pear, "Clinton to Sign Welfare Bill That Ends U.S. Aid Guarantee and Gives States Broad Power," *The New York Times,* Aug. 1, 1996: A1, A22.
15. Judith Havemann and Barbara Vobejda, "As Welfare Cases Drop, Politicians Fight for Credit," *The Washington Post,* May 13, 1996: A1, A8.
16. Sheila Zedlewski, Sandra Clark, Eric Meier, and Keith Watson, "Potential Effects of Congressional Welfare Reform Legislation on Family Incomes" (Washington, DC: The Urban Institute Press, 1996).
17. William Claiborne, "California Faces New Welfare Reality: Counties With Large Immigrant Populations Fearful of Financial Hit," *The Washington Post,* Aug. 2, 1996: A9.
18. David Fireston, "New York Area See Huge Costs in Welfare Bill," *The New York Times,* Aug. 1, 1996: A1, A25.
19. Barbara Vobejda, "Few States Ready to Run Welfare," *The Washington Post,* Oct. 25, 1995: A1; and Rochelle L. Stanfield, "Holding the Bag?" *National Journal,* Sept. 9, 1995: 2206.
20. Rebecca Blank, *It Takes a Nation: A New Agenda for Fighting Poverty* (Princeton, NJ: Princeton University Press, forthcoming).
21. Paul E. Peterson, "State Response to Welfare Reform: A Race to the Bottom?" in *Welfare Reform: An Analysis of the Issues,* ed. Isabel V. Sawhill (Washington, DC: Urban Institute 1995): 7–10.
22. William J. Bennett, *The Washington Post,* Aug. 18, 1996: C7.

Tyrone Mitchell

Studs Terkel

He has the build of a professional football player. His resemblance to Bo Jackson is startling. "The kids all call me Bo. A lotta people do." He laughs easily.

He is twenty-five.

A long-time resident of the project who works as a tenants' advocate has known him ever since he was eighteen. "He was ruthless, uncaring, and unloved. His self-esteem was nil. He was disrespectful of his elders and did things in a violent way. He used to stand on the corner with a neck full of heavy gold chains and a baseball bat in his hands. He was a bully, who scared people. Something seems to be happening to him now. Let's cross our fingers."

We hung out on the streets all night, gettin' high, startin' trouble. Rowdy. I was about eighteen, runnin' wild—guns, gangbangin'.

I've been arrested, in and out of county jail. They put me in Division 5, Tier 2B. There I was with robbers, a couple of guys had murder cases, a couple, rape cases. I was supposed to learn from them. I was in for possession. A UUW, unlawful use of a weapon. I been outa jail just two weeks. I was in three and a half months. Aggravated assault, this time.

I should've known better. Not to come back again. But me bein' young, wild, my mind slipped off my work. I done fell off when I come to the projects. The guys said, "Oh, man, come with us." I though it was the thing to do, bein' cool.

Before we come here, I was goin' through a man's stage with a youth body and a youth mind. I had a lot of pressure to take care of my family, to help my mother. My father left a long time ago. To put food in the refrigerator, I dropped out of school in my sophomore year. I was sixteen. By me bein' the intelligent one, I helped my two brothers and a sister with homework. Math and reading. I could figure out things real fast. Even after I dropped out of school.

I met this old guy who had buildings to fix up. He hired me as a helper, doin' carpenter work, hangin' drywall, doin' electrician work. I was learnin' while I was helpin'. I stuck with it.

But when we moved down here, I fell. I was strung out on cocaine. Oh, I had really went a long way with drugs. Now people that is on drugs can come and talk to me, because I reformed myself. I'm not on drugs anymore, and I'm not a Vice Lord anymore. It feels damn good to be back.

I see kids now, the same thing I was doin'. Breakin' in trucks, breakin' in cars. I see myself. Before they get to go into jail, I talk to 'em now, grab 'em. A lotta little kids 'round here know I been in and out of jail. "You wanna end up in and outa jail like I did?" And throwin' rocks. Somebody could put somebody's eyes out.

Most of 'em listen. Then you get some that cuss me out, but I grab 'em, hit 'em a couple of times on their butt. They about six, seven, eight. The older ones, ten, eleven, and twelve, really listen. I can't really get to all of 'em. If I gets this buildin' right here and make it into a center—man! Where they can see, hey, he really tryin' to do somethin'!

I feel like an older man at twenty-five. I started young, so right now I feel mature and wiser. I have three kids, two boys and a girl. My daughter is at my aunt's house and one of my sons is out there playin' with his mother. [*He indicates the lot outside.*] I would like to watch 'em grow up, if nothin' happens to me.

I get up in the mornin' about eight o'clock. I stare out the window about an hour and a half. Just watch. It be quiet in the mornin'. Until the kids come out, about ten and eleven, I be thinkin'. You got a vacant lot here, you got a vacant lot over there. It's been around for a long time. I'd like to put somethin' there, redo the ground. If I could get out there and put somethin' on that ground, the other guys in the gang, seein' me workin' this field, hooked up—I think I can really touch, reach these guys. For one, I'd like to put a laundromat right here. [*He indicates a nearby plot.*] That'd be nice for 'em. Everybody in these buildin's gotta go a long way to the laundromat. To Henry Horner. People from Henry Horner don't like people from another project comin' there. That brings on things.

After hour and a half daydreamin', I come outside. I volunteer for the Salvation Army now. They have free meals at 11 o'clock. Two days ago, a guy wanted to beat up a lady that serve the food. They told him to stand in a line to get a plate and would you take your hat off? He wanted to jump in front of everybody. He gone to cussin' out, callin' out all kind of bad names. He just carried on somethin' terrible.

So I come along and talked to him. "Now hold up, you makin' it bad for everybody they tryin' to feed. Her boss will come here and see that and shut the place down and there be hundreds of people been tryin' to eat here every day. If I let you carry on like this, I be wrong. A lot of these people come a long ways to get this meal." I asked him to apologize to her. I be big, but he was bigger than me. He went to her and apologized.

The guys are still on the corner. They go into bad habits 'cause they don't have nobody to talk to. If I could show 'em what I can do—oh, I just wanna get out there and *do* it.

How come I changed? One day I was puttin' on my clothes, ready to come outside. I had my hat turned to the left, the Vice Lords' way. To be honest, I come out of the bathroom, playin' with one of my sons. He took my hat and straightened it up. I go back to the mirror and turned it back to the left. But then I thought about it. Why did he turn it straight? It was like a sign to me. One of my kids turned my hat straight. He was four then. [*With a touch of awe.*] Had to be a sign, had to be. I look at myself in the mirror, it dawned. That had to be a sign from God. So I straightened my hat out and I come outside.

The gang tried to move my hat to the left. I grabbed their hand, no. They ast me, "What's wrong? You dropped your flag?" I say I'm reformin' myself and I advise you to do the same. Because these guys been gang-bangin' for years and they got nowheres yet. It's time to stand up and be counted for, 'cause nobody ain't countin' no gangbanger unless you dead.

They just hang out, kickin' it, talkin', doin' nothin'. *Nothin'*. Just wastin' time, wastin' their life. Most of 'em is eighteen, nineteen, twenty. They get older, most of my friends is turnin' to drugs, sellin'. That's the road I don't want to go down. They sell it to the neighborhood an' Puerto Ricans, whites.

When they be standin' on the corner, they come out about twelve in the afternoon till four or five the next mornin'. Just hangin' out, drinkin', smokin' reefers. A lot of 'em had dreams. They talk about what they'd like to do, but they don't have that extra get-up-and-go. They have it built up in them, gangbangin', project life, sleepin' with different-type women, the whole nine yards. They tired, just too tired to get out.

I was a good carpenter's helper. If I ever needed anything, short of money, advice, I could always go to this old guy. When I dropped outa school, before I come here, I hung around older guys, fifty-five, sixty years old. I thought I could learn from 'em, because they bein' old, they would not be tellin' me wrong. I learned a lot.

I heard about Martin Luther King from one of the older guys I was workin' for. He showed me the movie of King and the civil-rights movement. Was Martin Luther King's birthday. I got a good idea of what he was really marchin' for. He wanted blacks and whites to come together. Be proud of our brothers and sisters, without fightin' with each other. I really like that. I listened to him tell about Malcolm X. I watched this movie about Medgar Evers about seven, eight times. He was talkin' about peace and freedom. Now we went all the way back, all over again.

I gets along with white people real good. It's now blacks is fightin' blacks, blacks is killin' blacks. Let's see, how can I put this? [*He pauses, closes his eyes.*] If we can really get together—once upon a time we did get together, downhill. Our gang got together with a rival gang and we had a peace treaty. No more gangbangin'. But somebody come out of another neighborhood and the peace treaty broke. There be gang warfare.

It's gotta be more than me, but I think I can start it. It would have to be more than jobs. A lot of 'em need more schoolin', more education. A lotta them dropped out when their friends got killed around here, and go on drugs.

I'm goin' back to school, startin' on the eighteenth of this month. Workin' for my GED at Malcolm X. When I get my GED, I'm goin' for carpentry to get me a certificate. I know a lot about it, but I gotta go back for more. From there, I can see where I'm goin'.

My mother really likes the way I turned myself around. Just this mornin' she told me I look nice. She asked me how long I'm gonna be like this. I say I'm gonna fulfill my dreams. I'm going forward. As a matter of fact, I'm goin' to Brenda Stevenson's mother's house* tomorrow mornin'. I put her kitchen ceiling up already. I'm doin' a new drywall tomorrow. I'm gonna plaster the cracks and paint it. I'll do the roofin' and plumbin'.

When I was young, helpin' in carpentry, I always thought I'm gonna one day get me a pickup truck with my name on it. My name with my *business card*. Pass out my cards, my phone number on it. Call any time. Everybody'll wanna know: How did you get your own business? That would be my office for now. Business pickup, maybe I'd have me a storefront with *my name* on it. It'll be nice. My kids will be there. As they grown older, I will be teachin' 'em. It feels good: Tyrone's back.

---

*A tenants' advocate at the project.

CHAPTER 5 Questions for Writing, Reflection, and Debate

READING 21

Vanishing Dreams of America's Young Families • *Marian Wright Edelman*

1. According Marian Wright Edelman, who are the people affected by poverty in the United States today? Look carefully at the figures she provides. What do they tell us about the relationships among poverty, age, race, and education? What do you think has contributed to the changes in these relationships over time?

READING 22

"When You Get Laid Off, It's Like You Lose a Part of Yourself" • *Lillian B. Rubin*

2. Rubin's work reveals the effects on family life of unemployment. Why is the impact of unemployment felt most keenly by men? Why do so many men feel both guilt and helplessness? How do they play out these emotions in their behavior with their families? What are the relationships among substance abuse, violence, and unemployment?

READING 23

The Choice Between Welfare and Work • *Kathryn Edin and Laura Lein*

3. Studies of poverty in the United States indicate that families move in and out of poverty. Work can help bring some families' income above poverty level, but not always. "The Choice Between Welfare and Work" provides insights into the reasons some families move on and off welfare and into the sacrifices required in both situations. What factors influence these women's decisions to rely on welfare or on work? How do current wage and welfare policies make it difficult to be exclusively one or the other? How do you think recent welfare reforms will alter these women's decisions? What can be done to make the transition from welfare to work more permanent?
4. Why is the poverty line set so low in the United States? Who benefits from this policy, and who loses?
5. How are the survival strategies of families in poverty influenced by geography and social networks?

READING 24

Mothers with Children and Mothers Alone: A Comparison of Homeless Families • *Jennifer E. Glick*

6. One survival strategy used by homeless mothers is to separate from their children. Why would a mother do this? As we create larger holes in the safety net, we may in fact be breaking up families. How can a society justify policies that force parents to make such choices?

READING 25

A New Look at Poverty in America • *William P. O'Hare*

7. Many Americans distinguish between "deserving poor," who are poor because of circumstances outside their control (physical disability, children, etc.), and "undeserving poor," whose poverty is a result of something assumed to be within their control (pregnancy, alcoholism, lack of education). There is less dispute about using public money to assist the "deserving" poor than the "undeserving." How does "A New Look at Poverty in America" dispel the myth that much of the public money spent on assistance programs is used for the undeserving poor?
8. How do you react to this author's statement that welfare is aimed at "sustaining the poor, not pulling them out of poverty?" Many people doubt the potential effectiveness of current welfare reform to pull people out of poverty. What would be required for society to be successful at lowering poverty? Is it ever possible to eliminate poverty? What characteristics of the Social Security benefit, which is not means-tested, have contributed to its success at pulling people out of poverty?
9. What aspects of American history, values, and culture contribute to the fact that the United States is one of the least successful of the developed countries at lifting children out of poverty? What factors account for the fact that many people who are eligible do not receive public assistance?

READING 26

Tyrone Mitchell • *Studs Terkel*

10. Do you think Tyrone Mitchell has been successful at turning his life around? What factors influenced his descent into poverty and his attempts to rise above it? What factors have affected his future?

6

Family Violence

How is domestic violence a social, rather than a personal, problem?

When my students have investigated family violence for papers and presentations, their computer searches turn up a preponderance of literature focused on a clinical approach to understanding victims and perpetrators. Such studies provide interesting reading, but they also encourage us to view family violence as an individual problem rather than as one with social causes and implications. I do not discount any attempts to alleviate the scourge of violence that penetrates many families today, but as a sociologist I believe that people and their actions are influenced by social structures and cultural values. I therefore believe altering the patterns of violence in families requires research that illuminates these connections and policies that aim to alter them.

Family violence is a very personal subject, affecting the lives of millions of people around the world. The book *Paddy Clarke Ha Ha Ha,* by Roddy Doyle, provides a poignant account of a young child's realization that his parents' constant arguing has escalated to physical violence. This story reveals the conflicted feelings that evolve in families when the perpetrator is someone we love and care about. We see the main character attempting to deny what his senses tell him is true.

Young children are not the only ones who struggle with what is and is not abusive behavior. Dawn Currie's work—reported in "Violent Men or Violent Women? Whose Definition Counts?"—reminds us of the complications caused by personal definitions of violence when collecting data on family abuse. Currie shows how individual interpretations of abuse influence the results of studies based on self-reports and interviews.

Another reason some people consider family abuse from an individual perspective is that the potential for abuse cuts across a wide range of relationships within a family. In a three-generation, intact family with two biological parents, there may be wife abuse, husband abuse, child abuse, sibling abuse, or elder abuse. Divorce and repartnering add another layer of potentially violent relationships to families. I have focused these readings on partner abuse, not because I think the other forms mentioned here are less important, but to illustrate that a sociological viewpoint can apply to any form of abuse. Many frameworks for understanding and acting on partner abuse can also be applied to other forms of family violence, and I encourage you to do so. Another reason for focusing on partner abuse is that it allows us to acknowledge that this problem pervades the wide range of ways people form committed relationships. These readings remind us that this problem is faced by legally married couples, cohabitating couples, people who are going out together, gay men, and lesbians.

As you read, you will see that rather than discouraging the application of a sociological framework, the pervasiveness of this problem is precisely the reason to understand it from a cultural context. The Domestic Abuse Intervention Project of Duluth, Minnesota, has produced a visual model that places the personal manifestations of family violence at the core of the problem (see Figure 1). The strength of the model is that it acknowledges that these personal situations are embedded in both an institutional and a cultural context that can work to inhibit or promote such violence.

This model acknowledges that the personal manifestations of this problem are influenced by two sets of factors. The most immediate involves the attitudes and

FIGURE 1 *The Supportive Circles of Social Institutions and Culture*

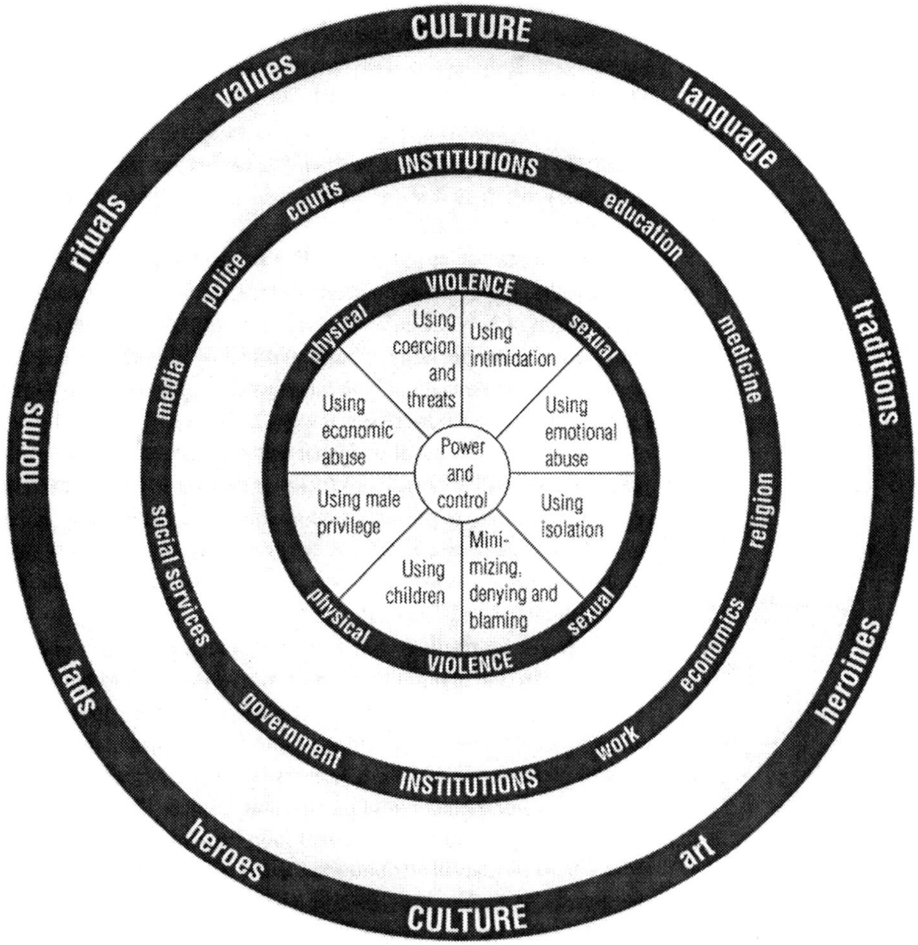

Source: Developed by The Domestic Abuse Intervention Project, 202 East Superior Street, Duluth, MN 55802.

policies found within social institutions such as the police, the legal system, churches, shelters, and support groups. In turn these institutions and the personal manifestations of abuse exist within a more pervasive outer circle of the culture of our society. Elements of culture include social values, gender roles, religious beliefs, media, traditions, and customs.

As you read Steve Friess's "Behind Closed Doors: Domestic Violence," you will find first-hand accounts of ways in which gay and lesbian perpetrators and victims of partner abuse interact with institutions in this middle ring. At the personal level, partner abuse in homosexual couples bears many resemblances to that in heterosexual couples. However, as this reading illustrates, the experience of interacting with representatives of institutions can be very different for homosexual partners.

This outside ring of the Duluth model provides a framework for understanding how culture defines abusive situations. In "Violence Against Women and Girls: The Intolerable Status Quo," Charlotte Bunch explores the influence of culture on partner abuse. This reading illustrates the many ways customs, values, and traditions can be used to sanction partner abuse. After reading this global perspective, challenge yourself to think of ways in which our national culture or your ethnic culture sanctions abuse.

Disciplines, such as psychology, that take an individual, personal approach to family violence provide different theoretical frames from sociology, which seeks explanations of how the larger society impacts individual action. In "Through a Sociological Lens: Social Structural and Family Violence," Richard Gelles offers strengths and weaknesses of both approaches. I agree with Gelles's conclusion that the comprehensive nature of sociological theories is a strength but also a complicating factor. The complication is that although they contribute to understanding, they are less useful in providing a focus for policies and programs. This fact places me in a bind. I believe prevention programs aimed at changing individual behavior, but ignoring institutional and cultural factors that condone family violence, are doomed to failure. Yet this sociological understanding does not provide an easy way to alter this trend.

My search for readings that provide examples of changes in social structures and culture led me first to the historical record provided by Linda Gordon in "The Powers of the Weak: Wife-Beating and Battered Women's Resistance." In this historical account we find a very simple truth: The courageous actions of women who resisted men's dominance in their everyday lives catalyzed change in the institutional and cultural sanctions of partner abuse. Interestingly, in my search for more contemporary examples of change, I found a similar theme. In "The Women of Belize," Irma McClaurin attributes to the collective activism of women much of the decrease in institutional sanctioning of male dominance. Grass-roots women's groups have demonstrated significant potential for social change. As you read this study, consider ways in which such women's groups might stimulate change in other cultures. These readings illustrate that it is possible to decrease family violence by altering social structures and cultural norms.

I have chosen to close this chapter with the selection "Old Problems and New Directions in the Study of Violence Against Women," by Demie Kurz, because I think it is a good reminder of the pervasiveness of family violence into many aspects of women's lives. Of special interest is Kurz's example of how policies aimed at other social problems such as divorce and poverty can deter women from acting to alter their situations. If the lesson we learn from Gordon and McClaurin is that women's collective action is a powerful stimulant for social change, then we must ensure that social policies do not inhibit women's actions.

Like the study of family, the study of family violence lends itself to a multidisciplined approach. In addressing this pervasive problem, social workers, law enforcement officers, counselors, psychologists, and sociologists need to listen to each other and gain from the unique perspective of each discipline. The comprehensive nature of the sociological approach to family violence is both its strength and a potential weakness. These readings document that culture and social structures can be altered to discourage abuse, and that often human agency needs to be fostered and strengthened to promote that change.

Paddy Clarke Ha Ha Ha

Roddy Doyle

The first time I heard it I recognized it but I didn't know what it was. I knew the sound. It came from the kitchen. I was in the hall by myself. I was lying on my stomach. I was charging a Rolls-Royce into the skirting board. There was a chip in the paint and it was getting bigger every time. It made a great thump. My ma and da were talking.

Then I heard the smack. The talking stopped. I grabbed the Rolls-Royce away from the skirting board. The kitchen door whooshed open. Ma came out. She turned quick at the stairs so I didn't have to get out of her way, and went upstairs, going quicker toward the top.

I recognized it now. I knew what the smack had been, and the bedroom door closed.

Da was alone in the kitchen. He didn't come out. Deirdre was crying in the pram; she'd woken up. The back door opened and closed. I heard Da's steps on the path. I heard him going from the back to the front. I saw his shape through the mountainy glass of the front door. The shape broke into just colors before he got to the gate and the colors disappeared. I couldn't tell which way he'd gone. I stayed where I was. Ma would come back down. Deirdre was crying.

He'd hit her. Across the face; smack. I tried to imagine it. It didn't make sense. I'd heard it; he'd hit her. She'd come out of the kitchen, straight up to their bedroom.

Across the face.

*

I watched. I listened. I stayed in. I guarded her.

Nothing happened.

I didn't know what I'd do. If I was there he wouldn't do it again, that was all. I stayed awake. I listened. I went to the bathroom and put cold water on my pyjamas. To keep myself awake. To stop me from getting cozy and warm and slipping asleep. I left the door a bit open. I listened. Nothing happened. I spent ages doing my homework so I could stay up longer. I wrote out pages from my English book and pretended I had to do it. I learnt spellings I hadn't been given. I got her to check me on them, never him.

—S.u.b.m.a.r.i.n.e.

—Good boy. Substandard?

—S.u.b.s.t.a.n.d.a.r.d.

—Good boy. Great. Have you more to do?

—Yes.

—What? Show me.

—Writing out.

She looked at the pages in the book I showed her, two pages with no pictures on them, and at the pages I'd done already.

—Why are you doing all these?

—Handwriting.

—Oh good.

I did it at the kitchen table, then followed her into the living room. When she was putting the girls to bed he was in the room with me, so it was alright. I enjoyed the writing out; I liked doing it.

He smiled at me.

I loved him. He was my da. It didn't make sense. She was my ma.

I went into the kitchen. I was alone. The noises were all upstairs. I slapped the table. Not too loud. I slapped it again. It was the right type of sound. It was duller though, hollow. Maybe it would be different from outside. In the hall where I'd been. Maybe he'd done that, smacked the table. When he was in a temper. That was alright. I did it again. I couldn't make my mind up. I was tempted. I used the side of my hand. She'd come out of the kitchen, straight up to their bedroom. She'd said nothing. She hadn't let me see her face. She'd started going faster before she got to the landing. Not because he'd slapped the table. I did it again. I tried to lose my temper and then do it. Maybe because he'd lost his temper. Maybe that was why she'd gone past me up the stairs, hiding. Maybe.

I didn't know. . . .

Violent Men or Violent Women? Whose Definition Counts?

Dawn H. Currie

In this [reading], I take up the challenge that data about women's use of violence bring to our understanding of violence in intimate heterosexual settings. By now, at least 30 studies show that women use physical violence as frequently as men, if not more. Because most of these data are based on the Conflict Tactics Scale (CTS), survey research has become suspect in many circles. In contrast, this [reading] explores the continued usefulness of survey research on violence in intimate settings. Drawing on qualitative data collected in conjunction with the use of the CTS, I explore the meanings that respondents bring to their accounts of the use of violence. In contrast with those who maintain that survey research overestimates the frequency of male violence against intimate partners, in this [reading] we see that when respondents' interpretations of violence are explored, the opposite, in fact, is more likely the case. In the study discussed below, findings that women employ violence more frequently than men reflect gendered expectations about appropriate behavior. One effect is that whereas men upgrade women's use of violence, women discount or underestimate the violent behavior of male partners.

Although some commentators claim that continued research on violence against women is unwarranted or even a misallocation of funds (see Doob, 1995), this [reading] draws attention to the need for continued research, albeit research "of a different kind." However commonplace the term *male violence* is to feminist researchers and activists today, gendering violence against women has been and remains a protracted struggle. At stake are women's claims for equal protection under criminal law, for social services that extend battered women's options, and for symbolic recognition of women's right to physical autonomy. As reasonable as these types of claims appear within a context of liberal democracy, an entrenched legacy of rape myths and gender stereotypes continues to obstruct women's claims. Until the rise of second-wave feminism, academic criminology in both the United States and Canada was complicit in the perpetuation of this legacy through the dismissal of women as victims of male crime. Similar to other academic fields, it either outright ignored women's experiences or scientifically endorsed misogynist beliefs about violence against women. For example, addressing the issue of sexual violence, Amir (1971) claimed that,

> the victim is the one who is acting out, initiating the interaction between her and the offender, and by her behavior she generates the potentiality for criminal behavior of the offender or triggers this potentiality, if it existed before in him. (p. 259)

Lest this citation be seen as an outmoded or exceptional example, an overview of the literature on wife battery yields claims that the cause of violence can be located in women's low self-esteem, the wife's psychiatric instability, her belittling and tormenting behavior, or her failure to adopt her wifely role (in Ahluwalia, 1987). A sexual double standard that naturalizes men's sexually aggressive behavior and a cultural commitment to the family as a sphere of "intimate" relations foster widespread support for victim precipitation theory as "common sense."

Several decades of feminist research and theorizing have challenged these beliefs about women's responsibility for male violence. In fact, violence against women continues to be one of the fastest growing fields of research and policy (see Straus, 1992). Accompanying this scholarship[1] are the emergence of shelters for battered women, reform of policing practice, and campaigns of public education. Against this trend, however, recent authors argue that women's claims as victims of male violence are based on selective evidence, partial consideration of the issues, or outright misrepresentation (see Fekete, 1994; Sommer, 1997; Straus, 1992). For example, Straus (1993, p.81) raises the possibility that "errors are deliberate distortions intended to discredit scientific findings." Included in this position are findings that women, like men, not only use violence in intimate relations but also initiate violence as frequently as men. Most of these data are based on the Conflict Tactics Scale (CTS), developed as an instrument to capture the dynamics of interpersonal conflict in marital situations. This chapter thus begins with a review of controversies surrounding the CTS as possibly the most widely used scale in survey research on woman abuse. Despite acknowledged limitations, this scale was included in a recent survey of students at my university (University of British Columbia). I provide a brief introduction to this survey in order to discuss findings relevant to physical violence in dating and cohabiting relationships among students. Because these findings raise questions about the validity of abstracted measurements resulting from quantitative surveys, this [reading] also considers the needs of future research on violence in intimate relationships.

Abstracted Violence and Emergence of the "Mutual Combat" Thesis

Developed at the University of New Hampshire, the Conflict Tactics Scale has been widely adopted by researchers in the fields of family violence and woman abuse. Originally designed to measure violence among family members, the CTS asks about means used to resolve "conflicts of interest" (see Straus, Gelles, & Steinmetz, 1980). As a quantitative instrument, the CTS measures three ways of dealing with interpersonal conflict: reasoning, verbal aggression, and physical violence. Discrete incidents characteristic of these three tactics are ordered on a continuum from least to most severe, resulting in 10 items describing nonphysical conflict, followed by 8 items concerning the use of physical violence. To encourage disclosure, questionnaire items are usually introduced with a preamble normalizing conflict in intimate settings. Such conflict is described in terms of disagreements on major decisions, annoyance over something that another person does, or simply as spats or fights that arise when someone is in a bad mood or tired (Straus et. al., 1980). The CTS has been administered in the form of self-completed questionnaires, personal interviews, and telephone interviews. The scale can be used either to self-report abusive behavior or to report about the behavior of other family members. Thus, some researchers calculate rates of woman abuse from the self-reports of male perpetrators (see Brinkerhoff & Lupri, 1988; DeKeseredy, 1988; Kennedy & Dutton, 1989; Lupri, 1990), whereas others use disclosures from female victims (Kelly & DeKeseredy, 1993; M.D. Smith, 1985, 1986, 1987).

Among representative samples, woman abuse researchers estimate that from 11% to 25% of women experience physical abuse by male partners during a typical calendar year. When psychological abuse is included, these estimates are much higher. In their survey of college and university students, DeKeseredy and Kelly (1994), for example, reported an incident rate closer to 80% (men's self-reporting, 74.1%; women's self-reporting, 79.1%). Use of the CTS thus draws attention to the extent of violence against women as a commonplace, rather than exceptional, feature of heterosexual relationships. It is not surprising, therefore, that research on violence against women has became one of the liveliest—if not the most controversial—venues for both academic and public debate (see DeKeseredy & MacLeod, 1997; Sommer, 1997; Straus, 1993). One of the most controversial findings of the 1975 National Family Survey was the high rate of violence by wives (Stets & Straus, 1990).

With at least 30 subsequent studies[2] also finding that women employ violence as frequently as men, women's use of physical violence has captured recent attention. In both Canada (Brinkerhoff & Lupri, 1988) and the United States (Stets & Straus, 1990; Straus & Gelles, 1986; Straus et al., 1980), researchers indicate that women report using violent tactics with the same frequency as their male counterparts. Most of these findings have come from studies on family violence, which treat all family members as equally at risk. At the forefront of debate, Straus and various colleagues challenge the claims of woman abuse researchers and advocates that battered women use violence much less frequently than male partners, that women rarely, if ever, initiate violence, and that violence on the part of women takes the form of self-defense (see Stets & Straus, 1990). Gelles and Straus (1980) cited at least 10 surveys subsequent to the National Family Violence Survey that confirm that women hit and beat their husbands. Stets and Straus (1990) further reported that women initiate violence about as often as men. They noted that sampling may account for some disagreement among researchers. Specifically, research showing gender difference in the use of violence is based on random, nonrepresentative, clinical samples. They suggest four reasons why women might be as violent as men: (a) Battered women may incorporate violence in their own behavioral repertory; (b) women may follow the norm of reciprocating violence; (c) the use of violence in one sphere, such as child care, may carry over into interaction with one's partner; and (d) an implicit norm exists that a woman should use minor violence, such as slapping, on certain occasions.

These types of findings have led to the characterization of intimate heterosexual relations as "mutual combat." At stake is the provision of victim support services for battered women, challenged through an emerging discourse concerning a "battered husband syndrome" (Steinmetz, 1977/1978, 1987). For example, in New Hampshire (not coincidentally the home to the Family Research Laboratory, which developed the CTS), resources were directed away from women's shelters to establish shelters for battered husbands (Rigakos & Bonnycastle, 1997).[3] As well, the claim that husband battering is as prevalent as wife beating was used by officials in Chicago to block funding for a shelter for abused women and their children (DeKeseredy & MacLean, 1990). At the same time, this discourse can negatively affect individual women attempting to escape violent relationships by reducing the plausibility of their "victim status" in the public's mind. For these types of reasons, the CTS has come under severe criticism. Woman abuse researchers have identified major limitations in the use of the CTS as a measure of violent behavior in intimate heterosexual relationships. Given that estimates provided by the CTS are highly reliable, most debate surrounds the interpretation of findings. Three general themes characterize critiques of the CTS.

First, the CTS is criticized because of the rank-ordering of abuse and its failure to provide a context that allows interpretation of measurements, whereas proponents of the CTS are criticized for their failure to interpret findings in a theoretically informed manner. According to Straus et al. (1980), the rank-ordering of violence from least to most aggressive acts is a major advantage of the CTS because it increases the likelihood of disclosure. Straus (1993) himself, however, notes that verbal aggression may be more damaging than physical attacks (also see Vissing, Straus, Gelles, & Harrop, 1991). He justifies distinctions between minor and severe assaults on the grounds that these categories roughly parallel the legal distinction in the United States between "simple assault" and "aggravated assault" (Straus, 1990, p. 58). Woman abuse researchers, in contrast, point out that conventional legal categories have historically contributed to the invisibility of male violence against women. As a consequence, they favor categories that resonate with women's experiences of violence. From the perspective of women rather than researchers or police, psychological humiliation and verbal harassment can be more painful than physical assaults and more devastating in their effects (DeKeseredy & MacLeod, 1997; Walker, 1979).

Second, a related problem concerns what Straus (1990, p. 53) calls the equivalence of violent acts. He notes that kicking a man in the shins, for example, is not the same as kicking him in the groin, whereas both of these incidents are distinct from kicking a pregnant woman in the abdomen. The abstraction and rank-ordering of violent acts thus fail to take into account the degree of injury that is sustained (Brienes & Gordon, 1983; Ferraro & Johnson, 1983). The scope of this problem was highlighted by M. D. Smith (1987). After responding to the CTS, his participants were asked to give an in-depth account of the incident that included its aftermath, such as "whether she sustained an injury, required medical assistance, or needed to call the police" (p. 181). These qualitative accounts necessitated the rerecording of four minor incidents as "severe" acts of violence. They also allowed Smith (1986) to identify some violent acts not included in the CTS: burning, suffocating, squeezing, spanking, scratching, sexual assault, and other forms of psychological abuse. These types of problems draw attention to a third major limitation of the CTS.

Third, related to difficulties in making judgments about the effect of violence is the separation of discrete acts from their situational context. Although nobody denies that women use physical violence against male partners, woman abuse researchers maintain that women rarely initiate attacks against their husbands, lovers, dates, or cohabiting partners. More typically, women's use of violence is proactive or takes the form of self-defense (Berk, Berk, Loseke, & Rauma, 1983; Browne, 1987; DeKeseredy & MacLean, 1997; Dobash & Dobash, 1988; Makepeace, 1986; Saunders, 1986, 1989). At the same time, men's use of violence has been described by battered women as malevolent, a fact not captured when respondents must characterize violence as interpersonal conflict and disagreement (see Straus, 1990, p. 52). In short, the framing of the CTS, together with the extrapolation of incidents of violence out of their everyday context, does not allow the researcher to identify specific motives for behavior. Although experienced and more easily recorded as an episode or event, violence is an extreme expression of one moment in ongoing processes through which heterosexual relationships are "negotiated." One problem is that the CTS assumes—incorrectly—that partners are equal in these negotiations. Against this view, ample research documents ways that cultural standards continue to disempower women by sanctioning male control over many matters of heterosexual negotiation. As Holland, Ramazanoglu, Scott, Sharpe, and Thomson (1996, p. 118) point out in their study of sexual encounters, heterosexual relationships are sites of struggle between the exercise and acceptance of male power and male definitions. By failing to situate findings from the CTS within this framework, quantitative measures of violence have been interpreted, ironically, as evidence of gender equality. Thus, the absence of context parallels another problem associated with findings from the CTS, the absence of theoretical frameworks by many users of he CTS.

In their overview of the "new scholarship" on family violence, Breines and Gordon (1983) characterize the CTS as an "empiricist" approach that can accurately "count" the frequency of abuse but that tells us very little about either the nature or the etiology of domestic violence. As Yllö (1993, p. 47) notes, the latter cannot be adequately understood unless gender and power are taken into account. The problem is that use of the CTS has been largely research- rather than theory-driven. In a review of the emergence of the current dichotomy between researchers emphasizing male violence or violence as gender neutral, Yllö (1993) identifies problems of looking at family violence without a feminist lens. From the very beginning, propositions employed by the family sociologists who pioneered the CTS obscured the importance of gender. Although these sociologists made an important break with the tendency to view family violence in terms of individual psychopathology, systems theory, resource theory, exchange/control theory, and the subculture of violence all ignore gender. Although resource and exchange frameworks, for example, contribute the important insights that power is based on resources and that violence is the ultimate resource for ensuring compliance, by assuming an equal status between players these frameworks ignore the structural limits to women's access to key resources and the cultural ideology of husband dominance.

Even though Gelles and Straus (1988) themselves recognize that power and control are cited by research participants (both men and women) as at the core of events that led to the use of violence, these elements disappear from their interpretations of findings from the CTS. Yllö (1993), in contrast, maintains that domination, and not conflicts of interest, is at the core of family violence. She concludes that a coercive control model of domestic violence is an important theoretical alternative to the conflict tactics model. It identifies violence as a tactic of entitlement and power that is deeply gendered, rather than as a conflict tactic that is personal and gender neutral. Like other proponents of theories that emphasize the patriarchal nature of gender relations (see DeKeseredy & Kelly, 1994; Dobash & Dobash, 1979), she challenges researchers to extend this approach to an understanding of women's use of violence.

It is perhaps interesting that, despite these debates, feminist researchers are among those who continue to employ the CTS. Thus, as Straus (1990) maintains, "for better or for worse, much of the 'knowledge' generated by the large volume of research on 'partner violence' is based on (or critics would say, 'biased by') use of the CTS" (p. 49). From this perspective, the continued use of the CTS, rather than the findings generated by the CTS, has become a subject of intense debate. Recognizing many limitations of the CTS himself, Straus (1990) maintains that one reason for its continued usage is simply the absence of a better alternative. Many of the criticisms discussed above have resulted in a series of revisions to the original scales. Scale items have been expanded, and supplementary questions were added for the National Family Violence Resurvey in 1985. A "severity weighted" scale has also been developed that weighs each CTS item by its relative severity and the frequency with which it occurred.

Acknowledging the limitations of the CTS, DeKeseredy and Kelly (1994) offer one of the first attempts to contextualize female-to-male violence. Authors of the first national survey of violence in dating relationships among Canadian college and university students, these researchers developed a continuum that allows respondents to estimate the extent to which their use of violence occurs as self-defense or fighting back or is initiated by the respondent. Importantly, researchers found that a substantial amount of the total violence used by women was in self-defense (DeKeseredy & MacLean, 1997; DeKeseredy, Saunders, Schwartz, & Alvi, 1997). On this basis, these authors reject the "mutual combat thesis" on empirical, rather than simply ideological, grounds. As shown by their data, violence in intimate heterosexual relations is not symmetrical.

The work of DeKeseredy and his colleagues marks an important advance in debates surrounding the use of violence in intimate settings. For our purposes, it challenges the view held by some feminists that quantitative research necessarily advances patriarchal values and agendas. Proponents of this position base this claim on the grounds that quantitative research has been used historically to support sexist and elitist patriarchal values; it cannot meaningfully address substantive social problems; it mandates an exploitive relationship between researcher and "the researched"; it "objectifies" research participants as "objects" of knowledge; and it cannot capture a "feeling" for the persons under study (see Jayaratne, 1983, p. 142). On these grounds, critics of quantitative research claim that it merely provides an objective "aura" that makes findings convincing and influential.

Although many feminist researchers find merit in these types of critiques, the conclusion that we want to abandon quantitative research altogether is debatable (see McCormack, 1989). Included in its defense is the need for policy-oriented research (see Currie, 1992). One example of such research is a study I conducted[4] while serving on the President's Advisory Committee on Women's Safety on Campus (PACWSOC). The goal of this study was to challenge an administrative discourse on student safety that draws on conventional crime prevention literature. Another goal was to produce data from which to draw reliable and valid generalizations about the frequency and distribution of safety incidents across campus. Although previous efforts by women's groups and organizations were able to contribute to the former, the piecemeal and anecdotal nature of their sources of information about violence on campus ruled out the latter. As a consequence, university administrators depended on reports of violence against women (and men) made available from the local policing authority. The survey conducted by the advisory committee would effectively challenge this monopoly of the production of knowledge about women's safety.

In the next section, I briefly discuss the methodology of the survey, focusing on its employment of the CTS as a measurement of "private," as opposed to "public," safety incidents on campus. By including the notion of private safety, the survey was an attempt to challenge the beliefs, perpetrated by reliance on official crime data, that violence against women on Canadian campuses is infrequent and that the greatest threats to women's safety come from "strangers" who do not belong on campus.[5]

The Current Study

Although several Canadian studies on violence against women appeared throughout the 1980s, at the time of the current project very few data on women's safety in university and campus settings were available. University administrators were forced to rely on either studies conducted on U.S. campuses or crime data generated by official policing agencies. These latter data suggest that the major problem on campus is property crime and that personal crime, especially against women, is a rare event. Against this claim, members of PACWSOC pointed out that violence against women is one of the most underreported crimes in Canada. Furthermore, women's experiences on campus often concern events that are not typically recorded in official crime data. To provide a more accurate assessment of both women's concerns and their risk on campus, a survey of undergraduate and graduate students was undertaken.

A full discussion of the research design and fieldwork is beyond the scope of the current format (see Currie, 1995). Here, I simply note that the study was based on a self-administered questionnaire that covers five topics: (a) demographic variables, (b) use of campus during the previous week and mode of travel around campus, (c) feelings of safety while on campus, (d) experiences of threats to safety while traveling around campus during the previous year, and (e) experiences of threats to personal safety in the intimate situations of heterosexual dating or cohabitation. As well as measuring personal behavior and experiences, attitudinal measures (including beliefs about rape and rape victims) were employed. Items concerning safety while traveling around campus were referred to as measuring "public safety"; those concerning threatening incidents in dating and cohabiting relationships were referred to as measurements of "private safety."

The questionnaire was distributed to a probability sample containing 126 clusters (courses). To identify gendered patterns in safety attitudes and experiences, the questionnaire was distributed to both women and men. As anticipated, although women (and to a lesser extent men) worried most about public travel around campus (especially travel alone to isolated parts of campus after dark), more threats to private than public safety were disclosed on the survey (in well-frequented areas of campus).[6] In this chapter, I limit discussion to disclosures of threatening behavior and use of physical violence that occurred in intimate settings and involved perpetrators known to the respondents.

From the targeted sample, 4,623 questionnaires were distributed. Of these, 2,098 completed questionnaires were returned during the two designated class periods: 1,405 by women and 693 by men, giving a final return rate of 45.9% (women, 67.1% of returned questionnaires; men, 32.9%). Only students with dating or cohabiting heterosexual relationships[7] completed sections of the questionnaire dealing with private safety; data thus underrepresent the extent of interpersonal violence among the university population. In total, 1,611 students completed this section of the questionnaire: 1,103 women and 508 men.

Given its successful deployment in previous surveys on dating violence,[8] a modified version of the CTS was used, asking respondents to report on the frequency of 10 types of physical violence. The CTS was supplemented by open-ended questions that invited respondents to provide detailed accounts of incidents otherwise to be counted as abstracted events. These accounts were to include whether or not incidents were reported to someone in authority, indicating in close-ended format reasons why incidents were not reported. As we shall see, qualitative data were not used simply to provide a context for interpretation of quantitative measurements; in this study, they were used to gain access to respondents' "definitions of the situation." Similar to previous research, this survey was designed to compare the experiences and attitudes of women and men through simple partitioning of data. Reflecting this goal, the questionnaires for women and men were similar, the only difference being that men (but not women) were asked to self-report on their use of psychological and/or physical coercion in sexual relationships with women.[9] Given the limitations of the current format, findings that concern sexual violence will be discussed elsewhere. Below, I discuss the use of physical violence.

Findings: Physical Violence Among Dating Students

Findings from the CTS are summarized in Table 1, which shows the percentages of women and men who disclosed threats or the use of physical violence against them in intimate settings during the previous year. Although we cannot say that physical violence is typical of heterosexual dating relationships, similar to previous research both women and men report incidents of violence. Importantly, the amount of interpersonal (nonsexual) violence perpetrated by partners exceeds that disclosed on the same survey for public threats and/or use of violence, as summarized in Table 2. Measures dealing with public safety yielded 34 reports of being pushed, grabbed, or shoved; 28 reports of having something thrown at the respondent; 35 reports of being physically threatened; and 15 reports of being physically assaulted. Comparable reports of threats and/or use of violence in intimate relations[10] yielded 169 cases in which the respondent was pushed, grabbed, or shoved; 220 reports of having something thrown at the respondent; and 125 reports that constitute assault (59 cases of the respondent being slapped; 51 cases of the respondent being hit, kicked, or punched; 8 cases of the respondent being choked; and 7 cases of the respondent being beaten by his or her partner). In other words, although students claimed they often worry[11] about being accosted by strangers when traveling around campus, they are much more likely to experience physical violence in intimate, dating situations.

For our purposes, it is interesting to note that, in this study, men disclose proportionately more violent incidents against them than women. With the exception of being pushed or shoved, men reported two or three times the rate of incidents than women. Does this mean that women are more violent than men or, alternatively, that men are the victims of battery to the same (or greater) extent as women? To explore these types of interpretations, the context of incidents needs to be examined. In this study, the immediate context of disclosed incidents was explored through open-ended questions that also asked about reporting

TABLE 1 *Incidents of Physical Violence by Heterosexual Partners, by Gender*

	Percentage of Cases Where This Occurred	
	Women (N = 1103)	**Men (N = 508)**
Partner threw, smashed, or kicked something	16.1	11.2
Partner threw something at respondent	2.7	7.1
Partner pushed, grabbed, or shoved respondent	10.3	10.8
Partner slapped respondent	1.5	8.3
Partner kicked, bit, punched respondent	1.1	7.7
Partner hit or tried to hit respondent	1.4	5.5
Partner choked respondent	0.5	0.6
Partner beat respondent	0.3	0.8
Partner threatened respondent with weapon	0.3	0.6
Partner used a weapon	0.1	0.6

Note: Can report more than one incident; Base *N* = respondents with dating or cohabiting relationships.

TABLE 2 *Comparison of Rates of Disclosing Public and Private Safety Threats*

	Public Safety Threats (N = 2,016)	**Private Safety Threats (N = 1,611)**
Grabbed, pushed, shoved	1.7	10.49
Something thrown at respondent	1.4	13.66
Physically threatened	1.7	—
Physically assaulted	0.74	7.8

Note: Rate is incidents per 100 students.

incidents to university "officials."[12] Although not all respondents who disclosed violent incidents on the CTS completed the open-ended questions, the descriptions provided allow us to explore the nature and impact of violence for women and men from respondents' standpoints. No attempt is made here to claim that these descriptions are representative. Given the paucity of qualitative data about incidents measured by the CTS, the discussion that follows is meant to be suggestive and to stimulate further investigation.

Seventeen incidents of physical violence were described by male respondents. The following are descriptions of (what I deemed to be) the five most serious incidents disclosed by men:

> I was teasing my girlfriend and she bit my finger "hard." (24-year-old male, indicated on CTS "partner kicked, bit, or punched the respondent")

> We were verbally fighting about our relationship. She got mad because I wasn't being sensitive to her emotions being experienced at that time, during the fight (argument) she was mad and she hit/pushed me away. (20-year-old male, indicated on CTS "partner pushed, grabbed, or shoved respondent" and that "partner slapped the respondent" in residence room)

> I was angry so I vented my anger on a garbage can. My girlfriend didn't like it so she hit me. (21-year-old male, indicated on CTS "partner slapped the respondent")

> My girlfriend got mad at me for teasing her for revealing a secret. She then threw a glass at me. (19-year-old male, this description accompanies categories on CTS "partner threw, smashed, or kicked something; threw something at you; slapped you; kicked, bit, or punched you; hit or tried to hit you with something")

> My girlfriend is somewhat vain so I tease her, and the incident involved her throwing down her curling iron and throwing a bottle of hairspray at me. (25-year-old male, indicated on CTS "partner threw something at respondent")

Five of the male respondents (or 29% of cases described by men) added the comment that the incident is "not worth mentioning" or that teasing or horseplay was the context:

> We were horsing around and [things] went a little too far. (20-year-old male, indicated on CTS "partner kicked, bit, or punched respondent")

These descriptions stand in contrast with those provided by women. Forty cases of physical violence were described by female respondents. Of these, only two respondents (or 5% of cases) commented that the incident was "not worth mentioning." The following are descriptions of the five most serious cases:

> Physical fights that escalate during end of heterosexual relationship. Starts over some trivial disagreement and ends up with full-scale physical fight, in which both parties are actively involved. Too many to count, too painful to remember just a lot of pain. (24-year-old female, indicated multiple categories of physical violence on CTS, including "partner beat respondent")

> I was married to a very abusive man. I am now receiving counseling at the Women's Centre. Thank you for providing this survey. (21-year-old female, indicated multiple categories of physical violence on CTS)

> An acquaintance from work came to pick me up so we could meet some friends. [He] started verbal sexual innuendoes, then physically came at me. When I told him I was not interested and turned down his advances he became angry. [The] situation got abusive, physically and verbally. (21-year-old female, indicated multiple categories of physical violence on CTS, including "partner beat respondent")

> Twice when we were fighting my husband lost his temper and punched a mirror, breaking it. Once he knocked over a bedside table. Each time, as soon as he had expressed his anger he calmed down and we were able to work through the problem. (20-year-old female, indicated on CTS "partner threw, smashed, or kicked something")

> After an argument I wanted to leave but my boyfriend wouldn't let me and he pushed me into

the door and pinned me against it. (19-year-old female, indicated on CTS "partner pushed, grabbed, or shoved respondent")

From these descriptions, it is not surprising to learn that only women indicated a need for medical attention and that women were much more likely than men to miss school because of the abusive incident.

Not all incidents recorded in Table 1 occurred on campus. One consequence is that very few incidents of physical violence were reported to someone in authority at the university: Only two women and one man reported incidents. Location of the incident, however, was not the only reason for failing to report violence. As shown in Table 3, the most important reason for nonreporting is that the respondent viewed the incident "as trivial." From this perspective, it is important to look at examples given by men and women that reflect their interpretations of incidents as trivial. Although the descriptions suggest that the nature of violence disclosed by men and women differs, they also suggest that men and women differ in the types of incidents they perceive as abusive. For example, three women described incidents in which their partners, but not the respondents, interpreted the context as "play fighting":

> He joined my karate class and after joining began to hit me, lightly and kiddingly in the beginning. After, near the end of the relationship, he hit or kicked me harder. I had to shout at him that I didn't like it and I wasn't going to put up with it and to stop it. He often used karate "play" fighting as an excuse to touch, hit, etc. me. (21-year-old female)

> Joking around. He thought it would be funny to bite my neck—I didn't. (22-year-old female)

Suggested here, but requiring further research, is the way women and men subscribe to differing definitions of what constitutes violent behavior. This difference does not mean that men will discount the violent behaviors of their partners, in the way Adler's (1981) research implies. In this study, it appears that, in terms of disclosing physical abuse by partners on the CTS, men tend to upgrade women's violent behavior. Consider the following descriptions of physical violence reported by men on the CTS:

TABLE 3 *Reasons for Not Reporting Physical Violence by Heterosexual Partners, by Gender*

	Percentage of Cases[1]	
	Women (N = 43)	Men (N = 18)
Respondent saw incident as trivial	53.5	72.2
Would not do any good	20.9	5.6
Incident not at UBC	18.6	16.7
Authorities would see incident as trivial	11.6	22.2
Did not know where to report	9.3	—
Did not want to cause problems	7.0	—
Fear of reprisal	7.0	—
Friends advised against	2.3	5.6
Other reasons	39.5	38.9

Note: Base *N* = all incidents of physical abuse not reported to someone in authority.

[1]Could give more than one reason.

> I do not remember the incident. My wife and I had an argument over something, she got really angry and threatened to throw a teddy bear at me. (23-year-old male who indicated on CTS "partner threw something at respondent")

> My girlfriend thought I wasn't paying enough attention to her and threw something to get my attention (a very heavy book). She also broke my TV remote control because I was playing with it too much (flipping channels). All very trivial. (25-year-old male who indicated on CTS "partner threw something at respondent")

Although men indeed may find women's violence amusing or, at most, annoying in the way Adler (1981) suggests, this does not mean it will not appear in surveys that employ the CTS. On the contrary, qualitative data from this survey suggest that men upgrade women's behaviors. They also suggest, in contrast, that women may downgrade the significance of men's behaviors. Women in this survey who disclosed violence by male partners often went on to excuse their partners' behaviors as understandable:

> It was principally out of anger at me, himself and the situation we were in. I don't think he necessarily meant to hurt me—that is, I don't think he rationalized and then decided to do it. His temper got the best of him and that was almost the scariest part. (21-year-old female who indicated on CTS "partner choked respondent")

> We were alone in the apartment. He has a tendency to argue later in the evenings. Got so angry that he kicked the wall, causing a dent in the wall—just above me where I was lying in bed. I was frightened, but it seemed out of character so I didn't attach much importance to it. (23-year-old female who indicated on CTS "partner smashed or kicked something")

> I know my partner would never hurt me physically; punching his own fist into something is *self* destructive (it hurts me more mentally to see him so upset). (19-year-old female who indicated on CTS "partner smashed something")

As well as providing a glimpse of how respondents define violent behavior, qualitative data also give a sense of how the respondent views the dynamics of interpersonal violence. In 7 of the 40 descriptions given by women, the women described their own behavior as contributing to the dynamics of men's violence. One result is that female respondents often assumed responsibility for the violent incident:

> I allowed the situation.... Things happened a lot faster than I had anticipated/expected. I didn't trust him—the situation may have been avoidable. I shouldn't have been alone with someone I did not trust. (22-year-old female)

> I was punching him first because he was with another girl after we'd gone out for a year.... I was the violent one. (19-year-old female)

If women described men's violence as understandable, they often claimed that it was better to "not aggravate matters" by reporting the incident even if the violence was severe. In short, self-blame, the tendency to excuse men's behavior, and the worry that the respondent's reaction to the incident could contribute to further violence or conflict all contribute to women's failure to report incidents of violence to someone in authority:

> It only happened once and he apologized. (19-year-old female, indicated on CTS "partner hit respondent")

> I excused it because of the alcohol. (20-year-old female, indicated on CTS "partner beat respondent")

> I did not report it because I was not hurt and I thought it would be best to leave this matter alone. (19-year-old female who indicated on CTS "partner tried to hit or hit respondent")

As seen above, men also often described their role in violent confrontations. The most common account by men was to describe women's violence as a response to teasing on the part of the respondent. We have also seen that men tend to upgrade women's behaviors on the CTS. In the final analysis, the qualitative data tell us that differing definitions underlie women's and men's assessments that the incident was "not worth mentioning" or "unimportant." When we look at open-ended descriptions rather than close-ended responses on the CTS, it appears that women tend to downplay men's violence by seeing it as excusable or understandable, whereas men tend to upgrade women's behavior when disclosing incidents of physical abuse. From this perspective, the assessment that an incident is "too trivial to report" carries a different meaning for men than women. Further, we see in Table 3 that women give a broader range of reasons for nonreporting; even though many of the described incidents could be classified as involving severe violence, women were likely to indicate that "it would not do any good" to report the matter or that they did not want "to cause further problems." In contrast, although almost 75% of men indicated that the incident was "too trivial" to report, almost 25% of men indicated that "authorities would see the incident as trivial."

In summary, the extent of physical violence disclosed in intimate settings exceeds that disclosed for public safety. Very little of this violence comes to the attention of university authorities because both women and men, overall, claim that the incidents were too trivial to report. When we look at open-ended descriptions, however, we see that assessment of what constitutes "trivial" behavior differs for women and men. Although the finding that men report more physical abuse than women on the CTS concurs with previous research, here I suggest that this finding reflects respondents' definitions of the situation. Specifically, if respondents accept gender expectations in heterosexual relations, their judgments about the behavior of partners can resonate with sex role stereotypes. In this study, on the one hand, women normalize male violence as "understandable" or "excusable," a response that conforms to the notion of men as typically aggressive. On the other hand, because women are traditionally typified as passive, violent behavior on their part is notable or remarkable although not a serious threat to men. In short, the way gender plays itself out in heterosexual violence is far more complex than can be captured through research designed to "count" violent incidents. Does this mean that we can never know true estimates of the extent of heterosexual violence or understand its nature? Although I do believe that gains can and have been made in understanding the dynamics of intimate heterosexual relations, I would never attempt to answer these types of questions from simply the research presented here. The concluding section thus moves away from interpretations of the current study to reconsider debate surrounding the measurement of violence as the singular basis from which to argue for or against feminist interventions.

Beyond Counting: The Politics of Research

Similar to previous research, the use of the CTS in this study generated findings that, if taken at face value, suggest women are (at least) as violent as men. Unlike previous research, however, this study enables us to interpret these findings by drawing on respondents' rather than researchers' definitions of the situation. These definitions provide empirical grounds to question the validity of the CTS as an accurate measure of either the extent or the nature of violence in heterosexual relations. In short, answers to many questions that continue to place woman abuse and family violence researchers in opposite camps depend on whose definition counts. From this perspective, answers to the question "who is violent" are not empirical, but rather political, because they are linked to larger issues surrounding the production of knowledge. As seen at the opening of this chapter, conventional criminology is characterized by male-centered definitions of male violence: In this chapter, I have suggested that although appearing to be gender-neutral, use of the CTS is simply an extension of this legacy.

It is interesting that researchers on both sides agree that men underreport their use of severe forms of violence when the CTS is used for self-report (see DeKeseredy & MacLean, 1990; Straus, 1990). Straus (1990, pp. 156, 157) himself maintains that because men understate their use of severe violence, "data on violence by men obtained from men needs to be treated with skepticism." He therefore disregards these data obtained from men, basing his analysis on reports of "minor only" violence and women's reports about both their own and men's use of violence. Although there may be, at best, logical grounds for maintaining the position that men mis-estimate only their own use of severe forms of violence, there are, I believe, stronger grounds for questioning the possibility that gender-specific definitions underlie all responses to the CTS.

Fekete (1994), along with other critics "without data,"[13] claims that if woman abuse researchers would ask respondents to describe their interpretations of their own experiences, the incidence and prevalence estimates of violence against women would be shown to be much lower than currently claimed (see DeKeseredy & MacLeod, 1997). It is perhaps ironic that, in doing so, this current study suggests that the opposite is actually the case. The general pattern revealed here is one in which women as well as men mis-estimate violence as measured by the CTS. However, error is not limited to simply the problem that men tend to normalize their own behavior. Although further research is required, in this [reading] I have suggested that both men and women make judgments about their partners' behaviors that resonate with dominant gender expectations. In this study, one consequence is the overestimation by men on the CTS of women's use of violence. One further consequence is a tendency for women to downplay the effect of violence used against them. This minimization takes the form of viewing men's behavior as "understandable" and also of self-blame. Although no similar tendency appears in men's open-ended accounts, this absence may reflect simply the method of data collection. In summary, this [reading] supports previous claims that both women and men will underestimate men's use of violence on the CTS and further suggests that men also overestimate women's use of violence.[14]

Only future research can settle issues raised by the study discussed in this [reading]. As illustrated by this [reading], however, answers to the questions whether women are as violent as men and whether women employ violence in similar ways as men cannot be answered through large-scale quantitative research alone. This is not to say that survey research does not have its place in the study of violence against women. In the current study, quantitative data were useful in establishing, for example, patterns of campus use that graphically characterize the university as gendered space (see Currie, 1994). Moreover, quantitative research is necessary to assess the impact of intervention; even strategies promoted by feminists can have unintended effects (see Currie, 1990; also see Buzawa & Buzawa, 1993). The point here is that the partitioning of quantitative data has encouraged researchers on both sides to treat sex as an explanatory rather than a descriptive variable (see Eichler, 1980). Although partitioning data according to the gender of respondents is useful in identifying patterns in responses, it implies not only equivalence of violent acts but also equivalent definitions of the situation on the part of respondents. It is interesting that so many social scientists have been unwilling to move beyond descriptive data in public debate over heterosexual violence. In my opinion, this reluctance testifies to male privilege, not simply in everyday heterosexual encounters, but also in the production of knowledge. Few academics, it would appear, are willing to deal with the question of whose definition counts when it comes to their engagement in public discussion of everyday social problems.

In conclusion, it is time to move beyond simply "counting" the violence in everyday life. Although there are good reasons why improvements in counting are always on the social scientific agenda, there are equally compelling reasons to improve the definitions we bring to everyday social problems. This is not the same as saying that we simply adopt respondents' definitions of the situation. In this study, for example, a significant minority of female students framed their understanding of personal safety in terms of dominant patriarchal definitions; that is, many women expressed concern over strangers on the university campus, and almost one third of women, like men, disagreed with the statement that "men, not women, are responsible for rape." Thus, a significant proportion of young women agreed that women's behavior (e.g., drinking, mode of travel) is responsible for sexual assault. Although researchers may thus want to begin from participants' definitions of the situation to better understand the everyday dynamics of heterosexual encounters, these definitions themselves require investigation. Clearly, understanding the social world in this way is a much more complex undertaking than can be discussed here. I direct the reader to a vast literature that treats "women's everyday world as problematic" (see D. Smith, 1974, 1980, 1981, 1987, 1990). I continue to be surprised that criminological research generally remains uninformed by this literature. One consequence, explored here, is overreliance on research methodologies that favor verification over discoveries that could lead to a reformulation of the questions that motivate criminological investigation. For example, as noted in this [reading], whose definition of violence in everyday encounters is verified by the continued uncritical use of the CTS?

NOTES

1. This comment is not meant to imply a causal relationship between academic research and the emergence of community interventions. Although many academics are engaged in practical as well as discursive struggles against violence against women, the development of

shelters for women is a direct consequence of community-based feminism.

2. See Straus (1993).
3. These shelters have since closed because of the complete absence of clientele.
4. My thanks to the Educational Measurement Research Group, who drew the sample and administered the questionnaire. The survey was funded by the Provost.
5. University administrators emphasized the presence of vagrants on campus and, as a result, advanced some legalistic solutions to campus safety.
6. The Student Union Building and university residences, for example, were the typical location for private safety threats.
7. The decision to include only heterosexual violence was based on two considerations: (a) the nature of campus politics, which are frequently homophobic; and (b) the belief that an entirely different approach and research methodology is required to study interpersonal violence in same-sex relationships. This decision was not meant to deny the documented fact of violence in lesbian relations (see Hart, 1986).
8. Previous research includes Deal and Smith Wampler (1986); Arias, Samios, and O'Leary (1987); and Makepeace (1986). Importantly, the first National Survey of Woman Abuse in Dating Relations was being conducted concurrently by Katharine Kelly and Walter DeKeseredy. Because initial intentions were to produce data that would be comparable to this survey, several measurements employed by the national survey were replicated in my study.
9. Unlike most previous research, men were also asked to disclose unwanted sexual experiences in heterosexual encounters.
10. The base N for the dating subsample is smaller than that for the sample reporting on public safety.
11. In this sample, 57.1% of women indicated that they worry "a fair amount" or "a lot" about their personal safety while on campus, compared with 12.6% of men.
12. The options provided in close-ended format were local policing authority, campus security, an administrator, student counseling, an instructor, and the Women Students' Office.
13. I find this phrase, coined by DeKeseredy, insightful because it distinguishes criticisms informed by empirical investigation from those based on ideological commitment to the status quo.
14. I cannot comment on women's self-reports of their use of violence, as self-reporting was not included in the current survey.

REFERENCES

Adler, E. S. (1981). The underside of married life: Power, influence, and violence. In L. H. Bowker (Ed.), *Women and crime in America*. New York: Macmillan.

Ahluwalia, S. (1987). *A history of domestic violence: Implications for medical intervention in Saskatchewan*. Unpublished master's thesis, University of Saskatchewan.

Amir, M. (1971). *Patterns in forcible rape*. Chicago: University of Chicago Press.

Arias, I., Samios, M., & O'Leary, K. D. (1987). Prevalence and correlates of physical aggression during courtship. *Journal of Interpersonal Violence, 2,* 82–90.

Berk, R. A., Berk, S. F., Loseke, D. R., & Rauma, D. (1983). Mutual combat and other family violence myths. In D. Finkelhor, R. J. Gelles, G. T. Hotaling, & M. A. Straus (Eds.), *The dark side of families: Current family violence research*. Beverly Hills, CA: Sage.

Breines, W., & Gordon, L. (1983). Review essay: The new scholarship on family violence. *Signs: Journal of Women in Culture and Society, 8,* 490–531.

Brinkerhoff, M. B., & Lupri, E. (1988). Interpersonal violence. *Canadian Journal of Sociology, 13,* 407–434.

Browne, A. (1987). *When battered women kill*. New York: Free Press.

Buzawa, E. S., & Buzawa, C. G. (1993). The scientific evidence is not conclusive: Arrest is no panacea. In R. J. Gelles & D. R. Loseke (Eds.), *Current controversies on family violence*. Newbury Park, CA: Sage.

Currie, D. H. (1990). Battered women and the state: From the failure of theory to a theory of failure. *Journal of Human Justice, 1*(2), 77–96.

Currie, D. H. (1992). Feminism and realism in the Canadian context. In J. Lowman & B. D. MacLean (Eds.), *Realist criminology: Crime control and policing in the 1990s*. Toronto: University of Toronto Press.

Currie, D. H. (1994). Women's safety on campus: Challenging the university as gendered space. *Humanity and Society, 18*(3), 24–48.

Currie, D. H. (1995, June). *Student safety at the University of British Columbia: Preliminary findings of a study of student safety* (Report submitted to the Provost).

Deal, J. E., & Smith Wampler, K. (1986). Dating violence: The primacy of previous experience. *Journal of Social and Personal Relationships, 3,* 457–471.

DeKeseredy, W. (1988). *Woman abuse in dating relationships: The role of male peer support.* Toronto: Canadian Scholars' Press.

DeKeseredy, W., & Kelly, K. (1994). Woman abuse in university and college dating relationships: The contribution of the ideology of familial patriarchy. In D. H. Currie & B. D. MacLean (Eds.), *Social inequality; Social justice.* Vancouver, BC: Collective Press.

DeKeseredy, W., & MacLean, B. D. (1990, May). Researching woman abuse in Canada: A realist critique of the Conflict Tactics Scale. *Canadian Review of Social Policy, 25,* 19–27.

DeKeseredy, W., & MacLean, B. D. (1997). "But women do it too": The contexts and nature of female-to-male violence in Canadian, heterosexual dating relationships. In G. S. Rigakos & K. D. Bonnycastle (Eds.), *Unsettling truths.* Vancouver, BC: Collective Press.

DeKeseredy, W., & MacLeod, L. (1997). *Woman abuse: A sociological story.* Orlando, FL: Harcourt Brace Jovanovich.

DeKeseredy, W., Saunders, D. G., Schwartz, M. D., & Alvi, S. (1997). The meanings and motives for women's use of violence in Canadian college dating relationships: Results from a national survey. *Sociological Spectrum, 17,* 199–222.

Dobash, R. E., & Dobash, R. P. (1979). *Violence against wives.* New York: Free Press.

Dobash, R. E., & Dobash, R. P. (1988). Research as social action: The struggle for battered women. In K. Yllö & M. Bograd (Eds.), *Feminist perspectives on wife abuse.* Newbury Park, CA: Sage.

Doob, A. N. (1995). Understanding the attacks on Statistics Canada's Violence Against Women Survey. In M. Valverde, L. MacLeod, & K. Johnson (Eds.), *Wife assault and the Canadian criminal justice system.* Toronto: University of Toronto Press.

Eichler, M. (1980). *The double standard: A feminist critique of feminist social science.* London: Croom Helm.

Fekete, J. (1994). *Moral panic: Biopolitics rising.* Montreal: Robert Davies.

Ferraro, K. J., & Johnson, J. M. (1983). How women experience battering: The process of victimization. *Social Problems, 30,* 325–338.

Gelles, R. J., & Straus, M. A. (1980). *Intimate violence.* New York: Simon & Schuster.

Gelles, R. J., & Straus, M. A. (1988). *Intimate violence: The causes and consequences of abuse in the American family* (2nd ed.). New York: Simon & Schuster.

Hart, B. (1986). Preface. In K. Lobel (Ed.), *Naming the violence: Speaking out about lesbian battering.* Seattle, WA: Seattle Press.

Holland, J., Ramazanoglu, C., Scott, S., Sharpe, S. & Thomson, R. (1996). "Don't die of ignorance," I nearly died of embarrassment: Condoms in context. In S. Jackson & S. Scott (Eds.), *Feminism and sexuality: A reader.* New York: Columbia University Press.

Jayaratne, T. E. (1983). The value of quantitative methodology for feminist research. In G. Bowles & R. D. Klein (Eds.), *Theories of women's studies* (pp. 140–161). Boston: Routledge & Kegan Paul.

Kelly, K., & DeKeseredy, W. (1993). The incidence and prevalence of woman abuse in Canadian university and college dating relationships. *Canadian Journal of Sociology/Cahiers canadiens de sociologie, 18*(2), 137–159.

Kennedy, L. W., & Dutton, D. (1989). The incidence of wife assault in Alberta. *Canadian Journal of Behavioral Science, 21,* 40–54.

Lupri, E. (1990). Hidden in the home: Wife abuse in Canada—Selected findings from a 1987 national survey. In C. McKie & K. Thompson (Eds.), *Canadian social trends.* Toronto: Thompson.

Makepeace, J. M. (1986). Gender differences in courtship violence victimization. *Family Relations, 35*(3), 383–388.

McCormack, T. (1989). Feminism and the new crisis in methodology. In W. Tomm (Ed.), *The effects of feminist approaches on research methodologies.* Calgary: Calgary Institute for the Humanities.

Rigakos, G. S., & Bonnycastle, K. D. (1997). *Unsettling truths.* Vancouver, BC: Collective Press.

Saunders, D. (1986). When battered women use violence: Husband-abuse or self-defense? *Violence and Victims, 1,* 47–60.

Saunders, D. (1989). *Who hits first and who hurts most? Evidence for the greater victimization of women in intimate relationships.* Paper presented at the Annual Meetings of the American Society of Criminology, Reno, NV.

Smith, D. (1974). The ideological practice of sociology. *Catalyst, 8,* 39–54.

Smith, D. (1980). An analysis of ideological structures and how women are excluded: Considerations for academic women. In P. Grayson (Ed.), *Class, state, ideology, and change: Marxist perspectives.* Toronto: Holt, Rinehart & Winston of Canada.

Smith, D. (1981, January 20). *The experienced world as problematic: A feminist method.* The Twelfth Annual Sorokin Lecture, University of Saskatchewan, Saskatoon.

Smith, D. (1987). *The everyday world as problematic: A feminist sociology.* Boston: Northeastern University Press.

Smith, D. (1990). *The conceptual practices of ruling: A feminist sociology of knowledge.* Toronto: University of Toronto Press.

Smith, M. D. (1985). *Woman abuse: The case for surveys by telephone* (LaMarsh Research Programme on Violence and Conflict Resolution, Report No. 12). Toronto: York University.

Smith, M. D. (1986). Effects of question format on the reporting of woman abuse: A telephone survey experiment. *Victimology, 11,* 430–438.

Smith, M. D. (1987). The incidence and prevalence of woman abuse in Toronto. *Violence and Victims, 2,* 173–187.

Sommer, R. (1997). Beyond a one-dimensional view: The politics of family violence in Canada. In G. S. Rigakos & K. D. Bonnycastle (Eds.), *Unsettling truths.* Vancouver, BC: Collective Press.

Steinmetz, S. K. (1977/78). The battered husband syndrome. *Victimology, 2,* 499–509.

Steinmetz, S. K. (1987). Family violence: Past, present, and future. In M. B. Sussman & S. K. Steinmetz (Eds.), *Handbook of marriage and the family.* New York: Plenum.

Stets, J. E., & Straus, M. A. (1990). Gender differences in reporting marital violence and its medical and psychological consequences. In M. A. Straus & R. J. Gelles (Eds.), *Physical violence in American families: Risk factors and adaptations to violence in 8,145 families.* New Brunswick, NJ: Transaction Books.

Straus, M. A. (1990). The Conflict Tactics Scales and its critics: An evaluation and new data on validity and reliability. In M. A. Straus & R. J. Gelles (Eds.), *Physical violence in American families: Risk factors and adaptations to violence in 8,145 families.* New Brunswick, NJ: Transaction Books.

Straus, M. A. (1992). Sociological research and social policy: The case of family violence. *Sociological Forum, 7,* 211–237.

Straus, M. A. (1993). Physical assaults by wives: A major social problem. In R. J. Gelles & D. R. Loseke (Eds.), *Current controversies on family violence.* Newbury Park, CA: Sage.

Straus, M. A., & Gelles, R. J. (1986). Societal change and change in family violence from 1975 to 1985 as revealed by two national surveys. *Journal of Marriage and the Family, 48,* 465–479.

Straus, M. A., Gelles, R. J., & Steinmetz, S. K. (1980). *Behind closed doors: Violence in the American family.* Garden City, NY: Anchor.

Vissing, Y. M., Straus, M. A., Gelles, R. J., & Harrop, J. W. (1991). Verbal aggression by parents and psychosocial problems of children. *Child Abuse & Neglect, 15,* 223–238.

Walker, L. E. (1979). *The battered woman.* New York: Harper & Row.

Yllö, K. A. (1993). Through a feminist lens: Gender, power, and violence. In R. J. Gelles & D. R. Loseke (Eds.), *Current controversies on family violence.* Newbury Park, CA: Sage.

Behind Closed Doors: Domestic Violence

Steve Friess

Dana walked into the only battered women's shelter in her Midwestern city with a bloody nose, bruises across her chest, and a couple of fingers as mangled as her spirit.

A caseworker raced to her, first calling in a doctor to tend to her wounds and then leading her to a room where she could rest. "You're in a safe place now," the caseworker comforted. "You can relax."

Dana believed her and sank into a soft cot, falling asleep without even slipping under the covers. For three days she let her paranoia subside, letting down her guard to tell her caseworker of the daily abuse she suffered at the hands of her lesbian partner.

On the fourth day Dana ran out of the shelter screaming. She had walked into a common area to find, sitting casually on a couch, the woman who for two years slammed Dana's face against the kitchen counter whenever Dana came home a few minutes late. "I just got the hell out of there, got into my car, and drove 600 miles to my mother's in Chicago," says Dana, who doesn't want her last name used. "I later found out that [her partner] told them she was abused, and because she's a woman they just checked her in too."

At least Dana had three days of peace. When Curt Rogers of Boston fled the torment of a lover who had restrained him for three hours and threatened his life, there was no place to turn. "A gay man cannot get into a shelter, period," says Rogers, who found hideouts with the help of coworkers in the weeks after he ran away. "A lesbian, depending on the shelter or her willingness to hide her sexuality, can go somewhere. The gay man is left hanging."

Neither gay men nor lesbians have good options, and gay groups around the nation don't seem eager to touch the issue, according to the second annual "Report on Lesbian, Gay, Bisexual and Transgender Domestic Violence," published in October by the National Coalition of Anti-Violence Programs. The report's release gave the issue its greatest surge of publicity yet, prompting a spate of stories in the mainstream media that advocates hope will lead to a broader recognition that domestic violence doesn't happen just to straight women. "We are about 20 years behind the battered women's movement in terms of information and certainly in terms of the amount of resources available," says Susan Holt, project coordinator for domestic-violence programming at the Los Angeles Gay and Lesbian Center.

No shelters exist for men, though in San Francisco, Boston, and a few other cities, battered males can obtain hotel vouchers from domestic-violence agencies. None of the nation's 1,500 battered women's shelters are devoted to lesbians, although some have caseworkers who focus on lesbian clients. And most crisis hot-line operators answer the phone expecting traditional battered-women situations. "I called one of those once," Dana says, "but I hung up when the woman on the phone asked me if 'he' was still there or if I expected 'him' to come home soon."

If few services are available to the gay abuse survivor, even fewer exist for the batterer. In Boston a group for lesbian and bisexual abusers was formed this year by Emerge, the nation's oldest agency treating batterers, and the organization hopes to start one for gay male abusers in the next year. Emerge's clientele consists almost entirely of straight men court-ordered into therapy, something that judges rarely demand of gay or lesbian abusers. "They'll more often be self-referred or urged to do this by a therapist or partner," says clinical director Susan Cayouette.

Hundreds of male batterer groups meet across the nation, but Cayouette and others don't allow gay abusers to attend because they believe it creates a volatile situation among men already prone to violence. "What we found is that straight men have an added problem—that there's racism and homophobia there as well as the sexism that makes them abusive to women," says John Hokanson, community liaison and chief educator at End Violence Now, an Atlanta-based group offering support services for victims and education for perpetrators of same-sex domestic violence.

Little reliable data are available to measure how pervasive gay domestic violence is, but activists frequently refer to several unscientific surveys over the past decade that claim that 25% of gay and lesbian partners are battered. The study by the antiviolence coalition cited 2,352 cases of abuse in 1996 in 12 U.S. cities, a vast undercounting but still the most the coalition has ever been able to document, says report coauthor Greg Merrill.

If the 25% figure is correct, that's the same percentage estimated for women in heterosexual relationships. Yet activists meet with widespread resistance when they push the issue among gay men and lesbians. "Many people have expressed to me a great deal of embarrassment that we've revealed this report to the mainstream media because we shouldn't be promoting negative information about us," says Merrill, director of client services at San Francisco's Community United Against Violence. "People just don't want to talk about it. Gay people feel immune to domestic violence the same way straight people in the beginning of the AIDS epidemic felt immune to HIV."

Instead gay men and lesbians focus on combating hate crimes or winning various legal rights. "You tend to say, 'OK, I'm going to go home to the person who knows I'm gay, and I want to believe it's a safe place,'" says Hokanson. "There are a lot of hate crimes, but then to come back to our own relationships and be bossed around or abused is something we don't want to talk about."

Unlike the way gays worked early on to involve heterosexuals in the AIDS crisis, heterosexuals aren't rushing to insist that domestic violence is also a gay concern. Rather, gay domestic-abuse activists actually spend time trying to convince society that gay domestic violence exists. They're also working to debunk the notion that people in same-sex relationships ought to be able to defend themselves. "The person who is contacted for help will often assume that this is a mutual battering situation, which is a myth," says Lynn Frost, a lesbian abuse survivor on staff with Little Rock, Ark.'s Women's Project, one of the reporting agencies for the study. "Because both persons are of the same sex, the counselor or volunteer assumes there is not a power issue involved."

Police officers often make the same assumptions, a key reason gay men and lesbians rarely file domestic abuse reports. "I would never call the police, because police officers are notoriously not safe," says Connie Burk, executive director of Advocates for Abused and Battered Lesbians, based in Seattle. "Very often officers can't figure out who is the abuser, so survivors are arrested instead because they're bigger or more butch."

It's Sgt. Norman Hill's job to fix that in Boston. Hill, the police department's liaison to the gay community, gives recruits a six-hour training program on gay issues, an hour of which is devoted to handling same-sex disputes. "I think the workshops are working, because I have had instances where people have walked up to me on the street to say, 'I had a problem with my significant other, and we had to call the police, and they were excellent.'" Such success heartens activists but remains rare, with many officers denying that ignorance of gay issues interferes with their work. Sgt. Ernest Whitten of the Little Rock Police Department's domestic-violence division bristled at the suggestion that his officers might need training on how to handle gay domestic abuse, insisting, "We deal with them all the same." And abuse survivor Rogers, who started Boston's Gay Men's Domestic Violence Project after leaving his abusive relationship, says some officers in other parts of Massachusetts dismiss his message: "Try going into a room with 16 police officers in uniform who don't want to hear your story. Then you have to explain your gay relationship to them and tell them how it went bad, knowing you're confirming their inner thoughts about how it was doomed."

Progress on any front has been slow, primarily because few of the millions of dollars spent annually by state and federal agencies are earmarked for the gay-related component of the problem. "It burns me up that a gay man who is a victim of domestic violence is being denied services simply because he is a gay man," Rogers says. "We're talking about life-and-death services, protection from a batterer. Things will eventually change, but people are going to have to get angry, get noisy."

First, though, the community needs to be educated. Unlike the battered women's movement—which received a huge boost from the publicity surrounding the murder of O. J. Simpson's ex-wife and allegations that Simpson had abused her—gay abuse survivors have few famous examples or national talk shows to mirror their plight. Olympic diving star Greg Louganis revealed intimate details of domestic abuse in his autobiography *Breaking the Surface* and on the speaking circuit, but advocates say even his efforts weren't enough to jar a reticent community into alarm.

Indeed, many gay men and lesbians prefer not to believe a problem exists, says Dana, the abuse survivor. In the year after she left her lover, she tried to describe her experiences to her new lesbian friends but found they didn't want to listen. . . . "If that kind of thinking exists in our community," she muses, "how can we expect anyone else to care?"

Violence Against Women and Girls: The Intolerable Status Quo

Charlotte Bunch

Violence against women and girls is the most pervasive violation of human rights in the world today.... It is so deeply embedded in cultures around the world that it is almost invisible. Their sufferings are compounded by systematic discrimination and humiliation in the home and workplace, in classrooms and courtrooms, at worship and at play.

Yet once recognized for what it is—a construct of power and a means of maintaining the status quo—it can be dismantled.

Yet it is rarely acknowledged that violence against women and girls—more than half of humanity, many of whom are brutalized from cradle to grave simply because of their gender—is the most pervasive human rights violation worldwide.

Gender violence is also a major health and development issue, with powerful implications for coming generations as well as society in general. Eliminating the violence is essential for peace: Peace at home and peace at large. Without it, the notion of human progress is merely a fantasy. A few facts:

- Roughly 60 million women who should be alive today are "missing" because of gender discrimination.
- In the United States a woman is physically abused by her intimate partner every nine seconds.
- In India, more than 5,000 women are killed each year because their dowries are inadequate according to their husbands.
- In some countries of the Middle East and Latin America, husbands are often exonerated from killing an unfaithful or disobedient wife.
- Rape as a weapon of war has been documented in seven countries recently, though its use has been widespread for centuries.
- Throwing acid to disfigure a woman's face is so common in Bangladesh that it warrants its own section of the penal code.
- About 2 million girls each year (6,000 every day) are genitally mutilated—the female equivalent of amputation of all or part of the male penis.
- More than 1 million children, overwhelmingly female, are forced into prostitution every year, the majority in Asia.

It is clear that these crimes are, in the main, vastly underreported. As social scientists are now discovering, the sheer scope and universality of violent acts against women and girls defy even the most educated perceptions. The fact is that most gender violence not only goes unpunished but is tolerated in silence—the silence of society as well as that of its victims. Fear of reprisal, censorship of sexual issues, the shame and blame of those violated, unquestioning acceptance of tradition and the stranglehold of male dominion all play their part. In many countries, so does the active or passive complicity of the state and other institutions of moral authority.

Gender violence, in all of its varied manifestations, is not random and it is not about sex. It serves a deliberate social function: asserting control over women's lives and keeping them second-class citizens.

Constant vigilance is needed to protect the fragile gains made thus far, to continue along the road to equality—and to bring an end to the torrent of daily violence that degrades not only women but humankind in its entirety.

The Intimate Enemy

For tens of millions of women today, home is a locus of terror. It is not the assault of strangers that women need fear the most, but everyday brutality at the hands of their husbands and lovers. Battering at home constitutes by far the most universal form of violence against women and is a significant cause of injury for women of reproductive age.

Domestic violence occurs across education, class, income and ethnic boundaries. An analysis of 35 recent studies from industrialized and developing countries shows that one quarter to one half of all women have suffered physical abuse by an intimate partner. The prevalence and pattern of domestic violence are remarkably consistent from one culture to the next. In most countries today, domestic abuse is officially regarded as a private family matter.

Statistics on rape from industrialized and developing countries show strikingly similar patterns: Between one in five to seven women will be victims of rape in their lifetime. While sexual and physical assault are broadly accepted as crimes outside the home, the law in most countries is mute when it comes to attacks within the family.

Laws that stop at the doorstep of the family are . . . moral hypocrisy. There are compelling reasons why the issue cries out for urgent public attention. . . . Children of violent fathers are often physically abused alongside their mothers. In addition, studies show that children of violent fathers repeat that behavior with their own wives and commit violent acts in the larger society. Next, there are clear parallels between behavior within and outside the home. If the systematic oppression of women and girls is tolerated widely at the family level, society at large will be shaped accordingly. . . . Finally, it is a matter of public health. Violence debilitates women and girls physically, psychologically and socially, sometimes with lifelong results.

Family violence affects the healthy development and productivity of society at large. Protecting women's rights and raising their status are essential to endeavors of economic development ranging from family planning to food production. Women's aspirations and achievements are powerfully inhibited by [the] threat of male violence.

Harmful Traditions

The most enduring enemies of a woman's dignity and security are cultural forces aimed at preserving male dominance and female subjugation often defended in the name of venerable tradition. In industrialized societies like the U.S. . . . rap music insults women as "whores"; popular men's magazines celebrate gang rape . . . sexual harassment of women trying to integrate into the armed forces. . . .

In developing countries, violent practices against women are often recognized and defended as cultural traditions. Wife-beating, for example, is considered part of the natural order in many countries—a masculine prerogative celebrated in songs, proverbs and wedding ceremonies.

At their most extreme, expressions of gender violence include "honor" killings, female genital mutilation and dowry deaths, as well as deepseated, even murderous, preference for male children. In courts of law, the "honor defense" is institutionalized in some Middle Eastern and Latin American countries, allowing fathers or husbands to walk away from murder. In 12 Latin American countries, a rapist can be exonerated if he offers to marry the victim and she accepts. The family of the victim frequently pressures her to marry the rapist, which they believe restores the family's honor.

The concept of male honor—and fear of female empowerment—also underlies the practice of female genital mutilation (FGM). . . . Traditions also feed the practice of "dowry death." . . .

As war becomes less a battle between countries and more a struggle for supremacy between ethnic groups, women and girls increasingly face rape and forced pregnancy in times of conflict. Well over 20,000

Muslim women were known to be raped in Bosnia and Herzegovina during the Balkan war, and more than 15,000 women were raped in one year in Rwanda. Just in recent years, mass rape has also been reported as a weapon of war in Cambodia, Liberia, Peru, Somalia and Uganda....

There is nothing immutable about the violent oppression of women and girls. It is a construct of power, as was apartheid, and one that can be changed.

Through a Sociological Lens: Social Structure and Family Violence

Richard J. Gelles

The core of the sociological perspective is the assumption that social structures affect people and their behavior. The major social structural influences on social behavior in general, and family violence in particular, are age, sex, position in the socioeconomic structure, and race and ethnicity. In addition, the structure of social institutions also influences social behavior. In the case of family violence, the structure of the modern family as a social institution has a strong overarching influence on the occurrence of family violence.

Social Facts and Social Influences

Age

Violence in intimate relationships follows the same general patterns with regard to age as does violence between nonintimates. The rates of violence (both victimization and offending) are highest for those between the ages of 18 and 30 years (Gelles & Straus, 1988; Straus, Gelles, & Steinmetz, 1980; U.S. Department of Justice, 1991; Wolfner & Gelles, 1993). Family violence, with the exception of the victimization of the elderly, is a phenomenon of youth, thus explanations for family violence need to consider issues such as life-span development, stage in the family life cycle, and human development if explanatory models are to reflect accurately the relationship between age and violence.

Sex

Interpersonal violence outside of intimate relationships takes place primarily between male offenders and male victims. The data on sex and family violence are somewhat different and often controversial. Much of the research on child maltreatment indicates that mothers are as, or more, likely to maltreat their children as are fathers (Burgdorf, 1980; National Center on Child Abuse and Neglect, 1988; Straus et al., 1980; Wolfner & Gelles, 1993). The sex difference, however, is not as clear as it might appear. First, the social construction of child maltreatment, especially the process of designating a perpetrator in official reports of child maltreatment, leaves females and mothers vulnerable to being identified as abusers and neglecters even if they are not directly responsible for the harm their children experience. Mothers are nearly always cited as offenders in cases of child neglect, not because they are the ones who directly caused harm to their children, but because cultural and societal views hold mothers responsible for the welfare of their children. Similarly, mothers are sometimes cited as maltreaters in official reports of child sexual abuse even when the perpetrator was the male partner or some other male, because child protective workers often assume that mothers have the responsibility for protecting their children from sexual abuse.

The data on physical abuse also indicate that females are nearly as, or more, likely than males to assault and abuse their children physically. However, as

Margolin (1992) explains, these data fail to consider the different levels of responsibility males and females have for child care. When the level of responsibility for child care is controlled—for instance, comparing abuse committed by male and female baby-sitters (Margolin, 1991) or comparing abuse by single parents (Gelles, 1989)— males are actually more likely to be physical abusers than are females.

The data on physical violence and abuse between spouses are even more controversial than the data on child abuse (see Kurz, 1993; Straus 1993). Some students of family violence, especially those who use a feminist perspective (see Kurz, 1993; Yllö, 1993), argue that females are vastly disproportional victims of adult intimate violence. Their point of view is supported by data on wife abuse derived from shelters and other helping agencies (see, for example, Dobash, Dobash, Wilson, & Daly, 1992). On the other hand, Murray Straus, among others, argues that there are far more women using violence toward men than the shelter data indicate. Although I cannot resolve this issue in this [reading], the data do suggest that males are the more likely offenders and females the more likely victims of family violence, consistent with a gender pattern of interpersonal violence found in other settings and groups.

Sex is also a factor in abuse of the elderly. Data indicate that women are the most likely victims of elder abuse. Data on offenders are somewhat more controversial. Steinmetz's (1993) conceptualization of elder abuse supports the claim that middle-aged females who are under stress from their caretaking obligations are the most likely abusers of the elderly. Pillemer's (1993) conceptualization that abuse is a result of the dependency of the offender is more neutral on which sex would be the most likely offender.

Position in the Social Structure

Wife abuse, child abuse, elder abuse, and other forms of family violence tend to occur in all social and economic groups. Violence and abuse can be found among truck drivers and physicians, laborers and lawyers, the employed and the unemployed, the rich and the poor. The fact that violence can be found in all types of homes leads some people to conclude that social factors, especially income and employment, are not relevant in explaining family violence. But although family violence does indeed cut across social and economic groups, it does not do so evenly. The risk of child abuse, wife abuse, and elder abuse is greater among those who are poor, who are unemployed, and who hold low-prestige job (Gelles & Straus, 1988; Pelton, 1978; Straus et al., 1980; Wolfner & Gelles, 1993). One of the mechanisms that explains why family violence is more likely to be found among those who are poor and unemployed or holding low-prestige jobs is social stress. The more stressful experiences individuals and families have to deal with, the greater the likelihood of the occurrence of some form of family violence (Milner & Chilamkurti, 1991; Starr, 1988; Straus, 1980a, 1990; Straus et al., 1980).

Race and Ethnicity

The data on family violence and race and ethnicity are somewhat contradictory. If one looks at official report data on child abuse, blacks and other minority racial groups are vastly overrepresented among those reported for child maltreatment (see, for example, American Association for Protecting Children, 1988; Gil, 1970). On the other hand, two national surveys of recognized and reported child maltreatment found that blacks were not overrepresented among those recognized for child maltreatment (Burgdorf, 1980; National Center on Child Abuse and Neglect, 1988). Other studies have found that blacks have lower rates of child maltreatment than do whites (Billingsley, 1969). Survey data indicate that blacks are more likely to use violence and abusive violence toward their children (Hampton & Gelles, 1991; Straus et al., 1980). This higher rate is the result of blacks having lower incomes and higher rates of unemployment than whites (Cazenave & Straus, 1979).

Official report data and survey data both agree that the rate of violence toward women is higher among blacks than among whites (Goetting, 1989; Hampton, Gelles, & Harrop, 1989).

The Second National Family Violence Survey, conducted in 1985, included an oversample of Hispanic families. The rates of husband-to-wife violence and parent-to-child violence among Hispanic respondents were significantly higher than those among non-Hispanic whites (Straus & Smith, 1990). As with blacks, the higher rate of violence in Hispanic homes is largely a function of the strong links among family violence, low income, urbanization, and youthfulness. Hispanic families are likely to have lower incomes than are white non-Hispanic families, are more likely to live in urban areas, and are younger than non-Hispanic whites.

I should point out that some official records, particularly official reports of child abuse and data from criminal justice agencies on wife abuse, reflect both the reality of the greater risk of abuse and violence in these groups *and* the fact that abuse and violence in these groups are overreported to official agencies. Newberger, Reed, Daniel, Hyde, and Kotelchuck (1977) and Hampton and Newberger (1985) found that poor and minority children are more likely to be correctly *and incorrectly* reported for child abuse, whereas white and middle- and upper-class families are much less likely to be correctly and incorrectly reported for abuse. Similarly, wife abuse and elder abuse in lower-income and minority families is much more likely to come to the attention of the police and courts than is violence in more affluent homes.

Structure of the Family as a Social Institution

The psychological perspective, because it looks for the causes of violence within the individual perpetrator, ignores the special and unique structure of the family as a social institution. The feminist perspective focuses only on the influence of gender and gender-structured relations on the institution of the family and the violence and abuse therein. The family, with the exception of the military in times of war and the police, is society's most violent social institution (Straus et al., 1980). The likelihood of being a victim of violence at the hands of a stranger or on the streets is measured in terms of risk per 100,000 people, but the risk of family violence is measured in terms of a rate per 100 individuals (Gelles & Straus, 1988). Thus a comprehensive perspective that explains family violence must consider the attributes of the family as a social institution that create such a high risk for violence.

In work published in 1979, Murray Straus and I identified the unique characteristics of the family as a social group that contribute to making the family a violence-prone institution (Gelles & Straus, 1979). Later, Straus, with his colleague Gerald Hotaling, noted the irony that these same characteristics we saw as making the family violence-prone also serve to make the family a warm, supportive, and intimate environment (Straus & Hotaling, 1980). Briefly, these factors are as follows:

1. *Time at risk:* The ratio of time spent interacting with family members far exceeds the ratio of time spent interacting with others, although the ratio varies depending on the stage in the family life cycle.
2. *Range of activities and interests:* Not only do family members spend a great deal of time with one another, the interaction ranges over a much wider spectrum of activities than does nonfamilial interaction.
3. *Intensity of involvement*: The quality of family interaction is also unique. The degree of commitment to family interaction is greater. A cutting remark made by a family member is likely to have a much larger impact than the same remark in another setting.
4. *Impinging activities:* Many interactions in the family are inherently conflict structured and have a "zero-sum" aspect. Whether a disagreement involves a decision about what television show to watch or what car to buy, there will be both winners and losers in family relations.
5. *Right of influence:* Belonging to a family carries with it the implicit right to influence the values, attitudes, and behaviors of other family members.
6. *Age and sex differences:* The family is unique in that it is made up of different ages and sexes. Thus there is the potential for battles between generations *and* between sexes.

7. *Ascribed roles:* In addition to the problem of age and sex differences is the fact that the family is perhaps the only social institution that assigns roles and responsibilities based on age and sex rather than interest or competence.
8. *Privacy:* The modern family is a private institution, insulated from the eyes, ears, and often rules of the wider society. Where privacy is high, the degree of social control will be low.
9. *Involuntary membership:* Families are exclusive organizations. Birth relationships are involuntary and cannot be terminated. There can be ex-wives and ex-husbands, but there are no ex-children or ex-parents. Being in a family involves personal, social, material, and legal commitment and entrapment. When conflict arises it is not easy to break off the conflict by fleeing the scene or resigning from the group.
10. *Stress:* Families are prone to stress. This is due in part to the theoretical notion that dyadic relationships are unstable (Simmel, 1950). Moreover, families are constantly undergoing changes and transitions. The birth of children, maturation of children, aging, retirement, and death are all changes recognized by family scholars. Moreover, stress felt by one family member (such as unemployment, illness, bad grades at school) is transmitted to other family members.
11. *Extensive knowledge of social biographies:* The intimacy and emotional involvement of family relations reveals a full range of identities to members of a family. Strengths and vulnerabilities, likes and dislikes, loves and fears are all known to family members. Although this knowledge can help support a relationship, the information can also be used to attack intimates and can lead to conflict.

Sociological Theories of Family Violence

Position in the social structure is clearly and strongly related to family violence. In order to illustrate how the sociological perspective applies and uses the empirical data on proximate correlates of family violence and the unique features of the family as a social institution, this section presents summaries of four primarily sociological theories of family violence: general systems theory, resource theory, exchange/social control theory, and subculture of violence theory.[1]

General Systems Theory

Murray Straus (1973) and Jean Giles-Sims (1983) developed and applied a social system approach to explain family violence. Here, violence is viewed as a system product rather than as the result of individual pathology. The family system operations can maintain, escalate, or reduce levels of violence in families. General systems theory describes the processes that characterize the use of violence in family interactions and explains how violence is managed and stabilized. Straus (1973) presents eight propositions to illustrate how general systems theory relates to family violence:

1. Violence between family members has many causes and roots. Normative structures, personality traits, frustrations, and conflicts are only some.
2. More family violence occurs than is reported.
3. Most family violence is either denied or ignored.
4. Stereotyped family violence imagery is learned in early childhood from parents, siblings, and other children.
5. The family violence stereotypes are continually reaffirmed for adults and children through ordinary social interactions and the mass media.
6. Violent acts by violent persons may generate positive feedback; that is, these acts may produce desired results.
7. Use of violence, when contrary to family norms, creates additional conflicts over ordinary violence.
8. Persons who are labeled violent may be encouraged to play out a violent role, either to live up to the expectations of others or to fulfill their own self-concepts of being violent or dangerous.

Giles-Sims (1983) elaborates Straus's basic model and identifies six temporal stages that lead to wife battering:

1. establishing the family system
2. the first incident of violence
3. stabilization of violence
4. the choice point
5. leaving the system
6. resolution or more of the same

Resource Theory

The resource theory of family violence assumes that all social systems (including the family) rest to some degree on force or the threat of force. The more resources—social, personal, and economic—a person can command, the more force he or she can muster. However, according to William Goode (1971), the more resources a person actually has, the less he or she will actually use force in an open manner. Thus a husband who wants to be the dominant person in the family but has little education, has a job low in prestige and income, and lacks interpersonal skills may choose to use violence to maintain the dominant position. In addition, family members (including children) may use violence to redress grievances when they have few alternative resources available.

Exchange/Social Control Theory

In earlier work I have elaborated on the basic propositions of an exchange theory of aggression and developed an exchange/social control model of family violence that proposes that wife abuse and child abuse are governed by the principle of costs and rewards (Gelles, 1983). Drawing from exchange theory, I have noted that violence and abuse are used when the rewards are higher than the costs. Drawing from social control theories of delinquency, I have proposed that the private nature of the family, the reluctance of social institutions and agencies to intervene—in spite of mandatory child abuse reporting laws—and the low risk of other interventions reduce the costs of abuse and violence. The cultural approval of violence as both expressive and, in the case of disciplining children, instrumental behavior raises the potential rewards for violence.

Subculture of Violence Theory

The subculture of violence theory is perhaps the most fully developed and widely applied sociocultural explanation of violence (see Wolfgang & Ferracuti, 1967, 1982). This theory asserts that social values and norms provide meaning and direction to violent acts, and thus facilitate or bring about violence in situations specified by these norms and values. Subculture of violence theory explains why some sectors, or subcultures, of society or different societies are more violent than others, especially when they have cultural rules that legitimate or require violence.

The Attractiveness of Psychological Explanations

The initial discussions of child abuse and wife abuse tended to overlook or downplay the relevance of social factors in explaining or helping to understand family violence. By and large, this was a consequence of the medical, or psychiatric, model that was applied by those who first discussed child abuse in the professional literature (see, for example, Kempe, Silverman, Steele, Droegemueller, & Silver, 1962; Steele & Pollock, 1968). As Barbara Nelson (1984) points out, the first people to identify a problem often shape how others will perceive it (p. 13).

The early writings on family violence discounted social factors as playing *any* causal role in the etiology of abuse. As Steele and Pollock (1968) put it, "If all the people we studied were gathered together, they would not seem much different than a group of people picked by stopping the first several dozen people one would meet on a downtown street" (p. 92). They went on:

> Social, economic, and demographic factors ... are somewhat irrelevant to the actual act of child abuse. Unquestionably, social and economic difficulties and disasters put added stress in people's lives and contribute to behavior which might otherwise remain dormant. But such factors must be considered incidental enhancers rather than necessary and sufficient causes. (p. 94)

For Steele and Pollock and other early students of child abuse, the explanation for abuse was that abusers suffered from significant psychopathology.

Leroy Schultz's (1960) examination of 4 cases of wife assault from a caseload of 14 spouse assaulters focused on mother–child dynamics as a means of explaining wife assault. Schultz noted that each assaulter was characterized by a domineering-rejecting mother relationship in which the child experienced primary rejection. The result was a passive-submissive individual who avoided conflict at all costs. Schultz noted that a uniformly poor mother–child relationship makes for a frustrated dependency in which the child's emotional needs are never met. He went on to explain that children who cannot permit aggressive impulses to break through during youth have difficulty as adults in entering into interpersonal relationships that do not duplicate their original dependency as children. These individuals seek to recreate dependent relationships with their spouses, but when their dependency needs are frustrated, the men tend to attack the objects of their frustration—their wives.

Current psychological explanations of child abuse, wife abuse, and family violence are considerably more sophisticated than the earlier notions of psychopathology or frustration-aggression arising out of disturbed patterns of mother–child relationships. Psychological theories of family violence also draw heavily on social learning as an explanation for child abuse, spouse abuse, elder abuse, and other forms of family violence (see O'Leary, 1988; . . .). However, psychological explanations of violence continue to overlook and minimize the contributions of social and structural factors to the occurrence and persistence of violence and abuse in intimate relationships.

The notion that social factors are not relevant, or not as relevant as psychological factors, in explaining family violence is often manifested in assertions and statements such as "Family violence can be found in all social groups and in all income levels." Anecdotal examples of violence and abuse in wealthy families, or among physicians or lawyers, are also offered as proof that social factors play only a minimal causal role in family violence.

There continues to be a heavy psychological bias in most theoretical conceptualizations about the causes and explanations of child abuse, wife abuse, elder abuse, and other forms of family violence. The enduring stereotype of family violence is that the abuser is mentally disturbed or truly psychotic and that the victim is a defenseless innocent. The typical reaction to a description of a case of domestic violence or a photo of an abused woman or child is that "only a sick person" would do such a thing. The stereotype is so strong that unless the offender fits the profile of the mentally disturbed, psychotic alien and the victim is portrayed as innocent and defenseless, there is a tendency not to view the event as "abuse." The stereotype is so strong that some women who have been abused fail to define their experiences as abuse because the violence was not as severe as that depicted in such popular media accounts as the television movies *The Burning Bed* and *A Cry for Help: The Tracey Thurman Story.* Thus considerable public attention is focused on the most sensational cases of intimate violence. Horrible torture of women and children, sexual abuse in day-care centers, and the killing of babies and the elderly make news, not only because such cases are somewhat unusual (although less unusual than the public thinks), but because they fit the stereotype of what really is "family abuse."

We want to believe that the family is a safe, nurturant environment. We also do not want to see our own behavior and the behavior of our friends, neighbors, and relatives as improper. Thus most people want to envision "family violence" as terrible acts committed by horrible or bizarre people against innocents. This allow us to construct a problem that is perpetrated by "people other than us."

The theory that abusers are sick is often supported by a circular argument. Those who use the psychological level of analysis sometimes note that one of the character disorders that distinguishes child abusers is an "inability to control aggression." This seems a simple enough diagnosis. However, it is circular. How do we know that these people cannot control their aggression? Because they have abused their children. The abuse is thus the behavior to be explained *and* the means of explaining the behavior. When clinicians try to assess individuals without knowing whether or not they have abused their offspring or spouses, they find that they cannot accurately determine whether someone abused a family member based only on a psychological profile. In fact, only about 10% of abusive incidents are caused by mental illness. The remaining 90% are not amenable to purely psychological explanations (Steele, 1978; Straus, 1980b).

The Attractiveness of Feminist Theory

Feminist theory is becoming the dominant model for explaining violence toward women. There are significant strengths in the feminist explanation of wife abuse, as well as some important weaknesses. One major strength of the feminist approach is its "praxis" or "advocacy" approach. Feminist theory is about women's victimization as a social problem and the need to do something about the patterned, continuing, and harmful use of psychological and physical coercion to control and dominate women. To say that feminist theory is "politically correct" is to damn the theory and the theorists with faint praise. Feminist theory provides the explanation *and* the formulation to both explain and end violence toward women.

A second strength of feminist theory is the diverse, yet consistent, empirical support for the proposition that gender inequality explains violence toward women. A number of recent studies by different researchers who used different methodological approaches on different populations have all found that gender inequality explains variations in the incidence and rates of violence toward women. Rebecca Morley (in press) used both in-person interviews and mailed questionnaires to study wife abuse in Papua New Guinea. Her findings cast significant doubt on the traditional hypothesis that modernization and the resulting social disorganization of modernization produce increased risk of wife beating. Although modernization does produce new pressures, expectations, and changes in women's support systems, the underlying explanation for the abuse of women is the husband's perceived right to control his wife and a social structure that "allows" husbands to assert this right. Murray Straus (in press) analyzed data from the Second National Violence Survey as well as aggregate-level data to examine patterns of wife assault in the 50 U.S. states. Straus's findings parallel Morley's New Guinea data. Social disorganization does not entirely explain variations in the rates of violence toward women in the United States. The greater the inequality between men and women and the greater the degree of social disorganization, the higher the rate of assault on wives.

Two additional researchers have employed data from newspapers to examine the abuse of women. Devi Prasad (in press) conducted a formal content analysis of newspaper articles on dowry-related violence in India, and Ko-Lin Chin (in press) used a less formal analysis of newspaper reports on violence toward "out-of-town brides" in the Chinese American community. The anecdotal data presented by Prasad and Chin add further weight to a gender inequality model of wife assault. The structurally inequitable positions of out-of-town Chinese brides and Indian women increase their risk of victimization.

The recent studies cited above support the earlier work of feminist scholars and sociologists who found that structured gender inequality is strongly associated with violence toward women. In addition, Straus and Morley both compare the explanatory power of competing theoretical models (social disorganization versus gender inequality) and find stronger support for the gender inequality model. Finally, the results come from a range of scholars who examined wife abuse using different methodologies and different theoretical approaches.

A final strength of the feminist perspective is that many feminist scholars, such as Yllö, Kurz, Dobash and Dobash, and Pagelow, are sociologists. They apply the sociological imagination, social facts, and sociological frames of reference to explaining violence toward women. Thus their approach is not entirely different from the theoretical approach used by sociologists, or those Yllö has labeled "family violence researchers."

The limitation of feminist theory is the other side of the coin of the theory's strength. Although the "gendered lens" provides a clear focus on violence toward women, the lens is a telephoto lens, not a wide-angle lens. The telephoto focus on violence toward women examines factors such as patriarchy, dominance, and control, and excludes from the vision other salient and important aspects of social structures and social institutions. The main problem with the feminist perspective is that it uses a single variable, patriarchy, to explain the existence of wife abuse. Moreover, the theory fails to account for the lack of variance of this single variable across time and cultures. Although the feminist perspective provides a politically attractive theory that is amenable to broad social action, it does not provide a useful theory to explain the complex nature of family violence. Feminist theory offers a single-variable analysis, albeit a powerful one, in a multivariable world. Moreover, feminist theory is an analysis of only one type of violence or victimization. The gendered lens does not, and apparently cannot, account for a wide range of objective phenomena that fall under the general label of "family violence." Neither Yllö (1993) nor other feminist scholars and theorists have been able to apply the feminist perspective to child abuse, sibling abuse, violence by women, or abuse of the elderly.

Summary

The sociological perspective provides the widest and most inclusive perspective from which to understand and explain family violence. A sociological perspective neither excludes nor diminishes the contributions of psychological or social psychological variables; rather, it places these variables within a wider explanatory framework that considers the impact of social institutions and social structures on social behavior. Similarly, sociological theory offers a more complex formulation for the varied phenomena of violence and abuse between intimates and is applicable to a wider range of victimization than is feminist theory.

Yet the sociological perspective has a major drawback. Because the sociological perspective *does not* focus on a single characteristic of social life (e.g., personality or gender inequality), sociological theories are by definition complex. The sociological theories reviewed in this chapter are complicated, and such theories do not lead to simple solutions, either in clinical or practice settings or in terms of social policy. One cannot easily use a sociological theory to inform clinical practice. Nor can one use it to develop a simple legislative package for a state or federal legislative body. Those who seek simple answers and simple solutions will find little of value in the sociological perspective.

NOTE

1. Two theoretical frameworks that have been applied to violence, symbolic interaction and conflict, are not reviewed here because they have not been widely applied to the study of family violence. Theories that are not primarily sociological, such as social learning theory (O'Leary, 1988), ecological theories (see Belsky, 1980; Garbarino, 1977), and sociobiological or evolutionary biology theory (see Burgess, 1979; Burgess & Draper, 1989; Daly & Wilson, 1988), are also not reviewed. In addition, this section does not review patriarchy or feminist theory, as this perspective is examined . . . by Kersti Yllö (1993).

REFERENCES

American Association for Protecting Children. (1988). *Highlights of official child neglect and abuse reporting, 1986.* Denver: American Humane Association.

Belsky, J. (1980). Child maltreatment: An ecological integration. *American Psychologist, 35,* 320–335.

Billingsley, A. (1969). Family functioning in the low-income black community. *Casework, 50,* 563–572.

Burgdorf, K. (1980). *Recognition and reporting of child maltreatment.* Rockville, MD: Westat.

Burgess, R. L. (1979). *Family violence: Some implications from evolutionary biology.* Paper presented at the annual meetings of the American Society of Criminology, Philadelphia.

Burgess, R. L., & Draper, P. (1989). The explanation of family violence: The role of biological, behavioral, and cultural selection. In L. Ohlin & M. Tonry (Eds.), *Family violence* (pp. 59–116). Chicago: University of Chicago Press.

Cazenave, N., & Straus, M. A. (1979). Race, class, network embeddedness, and family violence: A search for potent support systems. *Journal of Comparative Family Studies, 10,* 280–299.

Chin, K.-L. (in press). Out-of-town brides: International marriage and wife abuse among Chinese immigrants. In R. J. Gelles (Ed.), Family violence [Special issue]. *Journal of Comparative Family Studies.*

Daly, M., & Wilson, M. (1988). *Homicide.* New York: Aldine DeGruyter.

Dobash, R. P., Dobash, R. E., Wilson, M., & Daly, M. (1992). The myth of sexual symmetry in marital violence. *Social Problems, 39,* 71–91.

Garbarino, J. (1977). The human ecology of child maltreatment. *Journal of Marriage and the Family, 39,* 721–735.

Gelles, R. J. (1983). An exchange/social control theory. In D. Finkelhor, R. J. Gelles, G. T. Hotaling, & M. A. Straus (Eds.), *The dark side of families: Current family violence research* (pp. 151–165). Beverly Hills, CA: Sage.

Gelles, R. J. (1989). Child abuse and violence in single parent families: Parent absence and economic deprivation. *American Journal of Orthopsychiatry, 59,* 492–501.

Gelles, R. J., & Straus, M. A. (1979). Determinants of violence in the family: Toward a theoretical integration. In W. R. Burr, R. Hill, F. I. Nye, & I. L. Reiss (Eds.), *Contemporary theories about the family* (Vol. 1, pp. 549–581). New York: Free Press.

Gelles, R. J., & Straus, M. A. (1988). *Intimate violence: The causes and consequences of abuse in the American family.* New York: Simon & Schuster.

Gil, D. (1970). *Violence against children: Physical child abuse in the United States.* Cambridge, MA: Harvard University Press.

Giles-Sims, J. (1983). *Wife-beating: A systems theory approach.* New York: Guilford.

Goetting, A. (1989). Patterns of marital homicide: A comparison of husbands and wives. *Journal of Comparative Family Studies, 20,* 341–354.

Goode, W. (1971). Force and violence in the family. *Journal of Marriage and the Family, 33,* 624–636.

Hampton, R. L., & Gelles, R. J. (1991). A profile of violence toward black children. In R. L. Hampton (Ed.), *Black family violence: Current research and theory* (pp. 21–34). Lexington, MA: Lexington.

Hampton, R. L., Gelles, R. J., & Harrop, J. (1989). Is violence in black families increasing? A comparison of 1975 and 1985 national survey rates. *Journal of Marriage and the Family, 51,* 969–980.

Hampton, R. L., & Newberger, E. H. (1985). Child abuse incidence and reporting by hospitals: The significance of severity, class, and race. *American Journal of Public Health, 75,* 56–60.

Kempe, C. H., Silverman, F. N., Steele, B. F., Droegemueller, W., & Silver, H. K. (1962). The battered-child syndrome. *Journal of the American Medical Association, 181,* 17–24.

Kurz, D. (1993). Physical assaults by husbands: A major social problem. In R. J. Gelles & D. R. Loseke (Eds.), *Current controversies on family violence* (pp. 88–102). Thousand Oaks, CA: Sage

Margolin, L. (1991). Abuse and neglect in nonparental child care: A risk assessment. *Journal of Marriage and the Family, 53,* 694–704.

Margolin, L. (1992). Beyond maternal blame: Physical child abuse as a phenomenon of gender. *Journal of Family Issues, 13,* 410–423.

Milner, J. S., & Chilamkurti, C. (1991). Physical child abuse perpetrator characteristics: A review of the literature. *Journal of Interpersonal Violence, 6,* 345–366.

Morley, R. (in press). Wife-beating and modernization: The case of Papua New Guinea. In R. J. Gelles (Ed.), Family violence [Special issue]. *Journal of Comparative Family Studies.*

National Center on Child Abuse and Neglect. (1988). *Study findings: Study of national incidence and prevalence of child abuse and neglect: 1988.* Washington, DC: U.S. Department of Health and Human Services.

Nelson, B. J. (1984). *Making an issue of child abuse: Political agenda setting for social problems.* Chicago: University of Chicago Press.

Newberger, E. H., Reed, R. B., Daniel, J. H., Hyde, J. N., Jr., & Kotelchuck, M. (1977). Pediatric social illness: Toward an etiologic classification. *Pediatrics, 60,* 178–185.

O'Leary, K. D. (1988). Physical aggression between spouses: A social learning perspective. In V. B. Van Hasselt, R. L. Morrison, A. S. Bellack, & M. Hersen (Eds.), *Handbook of family violence* (pp. 31–56). New York: Plenum.

Pelton, L. (1978). Child abuse and neglect: The myth of classlessness. *American Journal of Orthopsychiatry, 48,* 608–617.

Pillemer, K. (1993). The abused offspring are dependent: Abuse is caused by the deviance and dependence of abusive caregivers. In R. J. Gelles & D. R. Loseke (Eds.),

Current controversies on family violence (pp. 237–250). Thousand Oaks, CA: Sage.

Prasad, D. (in press). Dowry-related violence: A content analysis of news in selected newspapers. In R. J. Gelles (Ed.), Family violence [Special issue]. *Journal of Comparative Family Studies.*

Schultz, L. G. (1960). The wife assaulter. *Journal of Social Therapy, 6,* 103–111.

Simmel, G. (1950). *The sociology of Georg Simmel* (K. Wolf, Ed.). New York: Free Press.

Starr, R. H., Jr. (1988). Physical abuse of children. In V. B. Van Hasselt, R. L. Morrison, A. S. Bellack, & M. Hersen (Eds.), *Handbook of family violence* (pp. 119–156). New York: Plenum.

Steele, B. (1978). The child abuser. In I. Kutash, S. B. Kutash, L. B. Schlesinger, and Associates (Eds.), *Violence: Perspectives on murder and aggression* (pp. 285–300). San Francisco: Jossey-Bass.

Steele, B., & Pollock, C. (1968). A psychiatric study of parents who abuse infants and small children. In R. E. Helfer & C. H. Kempe (Eds.), *The battered child* (pp. 103–147). Chicago: University of Chicago Press.

Steinmetz, S. K. (1993). The abused elderly are dependent: Abuse is caused by the perception of stress associated with providing care. In R. J. Gelles & D. R. Loseke (Eds.), *Current controversies on family violence* (pp. 222–236). Thousand Oaks, CA: Sage.

Straus, M. A. (1973). A general systems theory approach to a theory of violence between family members. *Social Science Information, 12,* 105–125.

Straus, M. A. (1980a). Social stress and child abuse. In C. H. Kempe & R. E. Helfer (Eds.), *The battered child* (3rd ed., pp. 86–102). Chicago: University of Chicago Press.

Straus, M. A. (1980b). A sociological perspective on the causes of family violence. In M. R. Green (Ed.), *Violence and the family* (pp. 7–31). Boulder, CO: Westview.

Straus, M. A. (1990). Social stress and marital violence in a national sample of American families. In M. A. Straus & R. J. Gelles (Eds.), *Physical violence in American families: Risk factors and adaptations to violence in 8,145 families* (pp. 181–201). New Brunswick, NJ: Transaction.

Straus, M. A. (1993). Physical assaults by wives: A major social problem. In R. J. Gelles & D. R. Loseke (Eds.), *Current controversies on family violence* (pp. 67–87). Thousand Oaks, CA: Sage.

Straus, M. A. (in press). State-to-state differences in social inequality and social bonds in relation to assaults on wives in the United States. In R. J. Gelles (Ed.), Family violence [Special issue]. *Journal of Comparative Family Studies.*

Straus, M. A., Gelles, R. J., & Steinmetz, S. K. (1980). *Behind closed doors: Violence in the American family.* Garden City, NY: Anchor/Doubleday.

Straus, M. A., & Hotaling, G. T. (Eds.). (1980). *The social causes of husband–wife violence.* Minneapolis: University of Minnesota Press.

Straus, M. A., & Smith, C. (1990). Violence in Hispanic families in the United States: Incidence rates and structural interpretations. In M. A. Straus & R. J. Gelles (Eds.), *Physical violence in American families: Risk factors and adaptations to violence in 8,145 families* (pp. 341–367). New Brunswick, NJ: Transaction.

U.S. Department of Justice. (1991). *Criminal victimization in the United States.* Washington, DC: Government Printing Office.

Wolfgang, M., & Ferracuti, F. (1967). *The subculture of violence.* London: Tavistock.

Wolfgang, M., & Ferracuti, F. (1982). *The subculture of violence* (2nd ed.). London: Tavistock.

Wolfner, G., & Gelles, R. J. (1993). A profile of violence toward children. *Child Abuse and Neglect, 17,* 197–212.

Yllö, K. A. (1993). Through a feminist lens: Gender, power, and violence. In R. J. Gelles & D. R. Loseke (Eds.), *Current controversies on family violence* (pp. 47–62). Thousand Oaks, CA: Sage.

The Powers of the Weak: Wife-Beating and Battered Women's Resistance

Linda Gordon

The basis of wife-beating is male dominance—not superior physical strength or violent temperament (both of which may well have been effects rather than causes of male dominance), but social, economic, political, and psychological power. It is less useful to call male dominance the cause of wife-beating, because we usually mean something more specific when we speak of cause; after all, most men, including many very powerful and sexist men, do not beat women.[1] But it is male dominance that makes wife-beating a social rather than a personal problem. Wife-beating is not comparable to a drunken barroom assault or the hysterical attack of a jealous lover, which may be isolated incidents. Wife-beating is the chronic battering of a person of inferior power who for that reason cannot effectively resist.

Defining wife-beating as a social problem, not merely a phenomenon of particular violent individuals or relationships, was one of the great achievements of feminism. Women always resisted battering, but in the last hundred years they began to resist it politically and ideologically, with considerable success. While that success is far from complete, it is important to recognize the gains, and to give credit where it is due. Wife-beating is now not only illegal but also, to a majority of Americans, shameful. The contemporary alarm about wife-beating is an emblem of this achievement. The fact that many find it unacceptable that wife-beating continues at all is a sign of the greater respect that women have won, in large part as a result of 150 years of feminist consciousness-raising. Moreover, women have gained substantially, if unevenly, in the economic and psychological strengths needed to escape abusive men.

If the achievements of feminism in countering wife-beating have been inadequately recognized, those of battered women themselves have been practically invisible. It is not a denial of their victimization to notice also their bravery, resilience, and ingenuity, often with very limited resources, in trying to protect and nurture themselves and their children. Elizabeth Janeway has eloquently called such gifts the "powers of the weak."[2] This [reading] argues that in the process of protecting themselves, battered women helped to formulate and promulgate the view that women have a right not to be beaten.

This [reading] also examines *how* male dominance is enforced by, and produces, violence against women. Wife-beating usually arises out of specific domestic conflicts, in which women were by no means always passive, angelically patient, and self-sacrificing. To analyze these conflicts, and women's role in them, does not mean blaming the victim, a common distortion in the literature on wife-beating. That women are assertive in domestic power struggles is not a bad thing; women's suppression of their own needs and opinions is by far the greater danger. Victorian longings for women without egos or aggression should be understood as misogynist myths. Examining the construction of specific marital violence in historical context may contribute to understanding how male supremacy worked and is resisted....

The condemnation of wife-beating and child-beating had similar roots, and had made substantial progress by the late nineteenth century. Contrary to some common misconceptions, wife-beating was not generally accepted as a head-of-household's right at this time, but was considered a disreputable, seamy practice, and was effectively illegal in most states of the United States by 1870.[3]

Although wife-beating was not widely considered legitimate, neither was public discussion of it. . . . Feminists as well as more conservative moralists preferred it to remain a hidden or at least whispered subject. [So] it is not surprising that [the women whose case reports to child-protection agencies were used as data for this study also reveal] an indirect approach. In the strategies adopted by these battered women, we can see the outlines of a veritable history of the changing meanings of wife-beating among the immigrant working class. Many of the pre-industrial communities from which these clients, largely immigrants, had come tolerated a male privilege to hit ("punish") wives. However, one should not suppose that prior to modern feminism women never objected to or resisted beating. A better if rough paradigm with which to understand "tolerance" of wife-beating is as a tense compromise between men's and women's, patrilineal and matrilineal interests. Unlimited family violence was never tolerated, and there were always standards as to what constituted excessive violence. Recently such notions as the "rule of thumb"—that a man might not use a stick thicker than his thumb to beat his wife—have been cited as evidence of the extremes of women's humiliation and powerlessness. On the contrary, such regulation was evidence of a degree of women's power, albeit enforceable mainly through the willingness of others to defend it. But women often did have allies within the patriarchal community. If that much abused word "patriarchy" is to have any usefulness, it must be used to describe a system larger than any individual family, a system which required regulation even of its privileged members. While patriarchal fathers could control their households, they in turn were subject to sanctions—social control—by the community, whose power brokers included not only fellow patriarchs but also women, particularly senior women. The agency clients were accustomed to appealing to fathers as well as mothers, brothers as well as sisters and friends, for support against abusive husbands.

Nevertheless, in the nineteenth and early twentieth centuries, many women clients did not seem to believe they had a "right" to freedom from physical violence. When social workers expressed disgust at the way they were treated, the clients sometimes considered that reaction naive. They spoke of the inevitability of male violence. Their refusal to condemn marital violence in moral terms must be interpreted carefully. It did not mean that these women were passive or accepted beatings. They often resisted assault in many ways: fighting back, running away, attempting to embarrass the men before others, calling the police. And they did express moral outrage if their men crossed some border of tolerability. There is no contradiction here. The language of absolute "rights" is only one legitimate approach to self-defense. In a patriarchal system there were neither institutions nor concepts defending absolute rights, but rather custom and bargaining. Because the client women did not conduct a head-on challenge to their husbands' prerogatives does not mean that they liked being hit or believed that their virtue required accepting it. (Failure to make this distinction is the result of flat and ahistorical conceptions of what patriarchy and female subordination have been like. There was no society in which women so "internalized" their inferiority, to use a modern way of explanation, that they did not struggle to improve their situation.)

What was new in the nineteenth-century middle-class reform sensibility was the notion that wife-beating was entirely intolerable. Family reformers proposed, like abolitionists toward slavery and prohibitionists toward drink, to do away with physical violence in marriage altogether. This differed from their attitude toward child abuse, because they did not propose to do away with spanking. By contrast, many poor battered women had a more complex view of the problem than their benefactors: welcoming all the help they could get in their individual struggles against assault, they also needed economic help in order to provide a decent family life for children. Given a choice, they might have preferred economic aid to prosecution of wife-beaters.

Feminist reformers also avoided women's violence toward men, whether offensive or defensive. The Victorian sensibility made them feel they should offer charity only to "true women," peaceful and long-suffering. There were political advantages to their myopia: they kept the focus on battered women and declined to redefine the problem as mutual marital violence; they knew that it was a whole system of male power, not just physical violence, that made women battered.[4] On the other hand, their view of women's proper role ruled out the possibility that women could create independent lives and reject violent husbands. To these nineteenth-century child-savers, women's victimization meant virtue more than weakness; women who submitted to abuse were more praised than those who left their husbands. For example, in the random sample of this study, battered women frequently left or kicked out their husbands, then repeatedly reconciled or reunited with them.[5] In the 1960s such a record would probably have made a social worker question a woman's sincerity and doubt the point of continuing to offer help. In the nineteenth century these women's ambivalence was interpreted as evidence of their commitment to fulfilling wifely duties.

A "Right" Not to Be Beaten

... Surprisingly, women's complaints about wife-beating escalated just as feminism was at its nadir. The 1930s were the divide in this study, after which the majority of women clients complained directly rather than indirectly about wife-beating. In 1934, for example, a young mother of three, married through a matchmaker at sixteen to an Italian-born man, repeatedly made assault-and-battery complaints against him. He was also a non-supporter, but her logic differed from that of earlier clients, and it was the beating that appeared actionable to her. It should not be surprising that this was an American-born woman much younger than her immigrant husband, a woman who may have had higher or perhaps less conventional aspirations than was the average among family-violence clients. Her husband's probation officer described her as a "high-type Italian," and the caseworker thought she expected "people to do things for her."[6] Women ... allege[d] child abuse in order to get agency help, but in the investigations they tended to protest about their own abuse more strongly. One MSPCC [Massachusetts Society for the Prevention of Cruelty to Children] agent complained in 1940 that the mother was not really very interested in her son's problem but only wanted to talk about herself.[7]

In other cases in that year, women rationalized their battering in new ways: not as an inevitable part of the female condition, as a result of the male nature, but as something they individually deserved. One woman said, "This is my punishment for marrying against my mo.'s wishes."[8] Even in blaming themselves women expressed a new sensibility that wife-beating should not be the general lot of women.

Wife-beating accusations stood out even more because of the virtual disappearance of non-support complaints. This striking inverse correlation between non-support and wife-beating complaints stimulates an economistic hypothesis: economic dependence prevented women's formulation of a sense of entitlement to protection against marital violence, but it also gave them a sense of entitlement to support; by contrast, the growth of a wage labor economy, bringing unemployment, transience, and dispersal of kinfolk, lessened women's sense of entitlement to support from their husbands, but allowed them to insist on their physical integrity. It is a reasonable hypothesis that the Depression, by the leveling impact of its widespread unemployment, actually encouraged women regarding the possibilities of independence.

An oblique kind of supporting evidence for this process of consciousness change is provided by wife-beaters' defenses. Men did not often initiate complaints to agencies, but they frequently responded with counter-complaints when they were questioned. Their grievances were usually defensive, self-pitying, and opportunistic. They remain, however, important evidence of a consensus among men about the services they expected from wives—or about what complaints might be effective with social workers. Men accused of wife-beating usually countered that their wives were poor housekeepers and neglectful mothers, making themselves the aggrieved parties. The men's counter-accusations were, of course, a means of seeking to reimpose a threatened domination. Yet they simultaneously expressed a sense of an injustice, the violation of a traditional and/or contractual agreement, and their dismay at the historical changes that made women less able or willing to meet these expectations.

Often male and female expectations of marital responsibilities were consonant. Women as well as men professed allegiance to male-supremacist understanding of what relations between the sexes should be like. These shared assumptions, however, by no means prevented conflict. Women's assumptions of male dominance did not mean that they quit trying to improve their situations. Husbands expected dominance but also expected women's resistance to it. Clients of both sexes expected marriage and family life to be conflict-ridden—they did not share the bourgeois denial of family disharmony—and demonstrated no shyness in exposing their family hostilities to social workers. Female clients often both "accepted" that men were violent—that is, they did not approve but expected it—and also tried to stop it.

By emphasizing mutual conflict as the origin of wife-beating, I do not mean to suggest an equality in battle. Marital violence almost always resulted in the defeat of women and served to enforce women's subordination. Nor did every act of marital violence emerge from an argument. Contestation could be chronic, structured into the relationship. Male violence often became a pattern, virtually normal, appearing regularly or erratically, without relation to any particular interaction. One man who eventually murdered his wife beat her because their children "had no shoe lacings."[9] Some men simply came home drunk and angry enough to hit anyone in the way. But their drinking . . . was often an assertion of privilege, as was their violence an assertion of dominance. . . .

Victims' Resistance

While the first-wave women's movement had asserted women's rights to personal freedom even in marriage, it had not provided any organized, institutional means for poor women to secure and defend that right, a power which was necessary for women really to believe in their own entitlement. Until the revival of feminism and the establishment of battered-women's shelters in the 1970s, wife-beating victims had three resources: their own individual strategies of resistance; the help of relatives, friends, and neighbors; and the intervention of child-welfare agencies. None was adequate to the task. The first two were easily outweighed by the superior power of husbands and the sanctity of marriage itself, and the last did not well represent the interests of the women themselves. Still, on some occasions victims were able to use these inadequate resources to construct definite improvements, if not permanent solutions.

Women in abusive relationships with men still face great difficulties in extricating themselves. These difficulties in turn weaken their ability to insist that the men's behavior change, since the woman's threat to leave is often her most powerful lever and his only incentive to change. Such difficulties were greater fifty or one hundred years ago, and greater for the poor and uneducated women who dominated in these cases. Their difficulties were essentially those faced by single mothers. The biggest obstacle for most women facing abusive men was that they did not wish to lose their children; indeed, their motherhood was for most of them (including, it must be emphasized, many who were categorized as abusive or neglectful parents) their greatest source of pleasure, self-esteem, and social status. In escaping they had to find a way simultaneously to earn and raise children in an economy of limited jobs for women, little child care, and little or no reliable aid to single mothers. They had to do this with the often low confidence characteristic of women trying to take unconventional action. Moreover, these women of the past had the added burden of defying a social norm condemning marital separation and encouraging submission as a womanly virtue. . . .

[One] response to beatings was fighting. For differing reasons, both feminists and sexists have been reluctant to recognize or acknowledge women's physical aggression. Yet fighting was common and accepted among poor women of the past, more so than among "respectable" women and contemporary women.[10] Fourteen percent of the marital violence cases contained some female violence—8 percent mutual violence and 8 percent husband-beating.

Most of the women's violence was responsive or reactive. This distinguished it from men's violence, which grew out of mutual conflict, to be sure, but was more often a regular tactic in an ongoing power struggle. Some examples may help to illuminate this distinction. Women's violence toward husbands in these records fell into three typical patterns. The most common was mutual violence. Consider the 1934 case of an Irish Catholic woman married to a Danish fisherman. He was gone at sea all but thirty days a year, and there was violence whenever he returned. One particular target of his rage was Catholicism: he beat his sons, she claimed, to prevent them going to church with her and loudly cursed the Irish and the Catholics—he was an atheist. The neighbors took her side, and would hide her three sons when their father was in a rage. The downstairs tenants took his side. They reported that she swore, yelled, hit him, and chased him with a butcher knife; that she threw herself down some stairs to make it look as if he had beaten her. Amid these conflicting charges it was certain, however, that she wanted to leave her husband, but he refused to let her have custody of the children; after a year of attempted mediation, the MSPCC ultimately lent its support for a separation.[11] In this case the woman responded with violence to a situation that she was eager to leave, while he used violence to hold her in the marriage. Her violence, as well as her maintenance of neighborhood support, worked relatively effectively to give her some leverage and ultimately to get her out of the situation. An analogous pattern with the sexes reversed could not be found—indeed, probably could not occur. Women's violence in these situations was a matter of holding their own and/or hurting a hated partner whom they were not free to leave. The case records contain many plaintive letters from wife-beaters begging for their wives' return: "The suspense is awful at times, especially at night, when I arrive Home, I call it Home yet, when I do not hear those gentle voices and innocent souls whisper and speak my name."[12]

A second pattern consisted of extremely frightened, usually fatalistic wives who occasionally defended themselves with a weapon. In 1960, for example, the MSPCC took on a case of such a woman, underweight and malnourished, very frightened of her profane, abusive, alcoholic, and possibly insane husband. One day she struck him on the head so hard he had to be hospitalized.[13] This is the pattern that most commonly led, and leads, to murder. Female murderers much more commonly kill husbands or lovers than men do; the overwhelming majority (93 percent) claim to have been victims of abuse.[14]

In a third pattern, the least common, women were the primary aggressors. One 1932 mother, obese, ill, described as slovenly, kicked and slammed her six children around, locked them out of the house, knocked them down the stairs, and scratched them, as well as beating her husband and forcing him and an oldest daughter to do all the housework. His employer described him as "weak and spineless, but very good-hearted." Ultimately this woman was committed to a state mental hospital at her own request on the basis of a diagnosis of an unspecified psychosis.[15]

Of the three patterns of female violence, the latter two usually involved extremely distressed, depressed, even disoriented women. The fighting women in mutual violence cases were not depressed, and may have been better off than more peaceful ones. Over time there appeared to be a decline in mutual violence and women's aggression.[16] The apparent decline in women's violence was offset by an increase in women's leaving marriages. A likely hypothesis is that there is a trade-off between women's physical violence and their ability to get separations or divorces.

Although women usually lost in fights, the decline in women's violence was not a clear gain for women and their families.[17] Condemnation of female violence went along with the romanticization of female passivity which contributed to women's participation in their own victimization. Historian Nancy Tomes found that a decline in women's violence in England between 1850 and 1890 corresponded to an increase in women's sense of shame about wife-beating, and reluctance to report or discuss it.[18] In this area feminism's impact on women in violent families was mixed. The delegitimization of wife-beating increased battered women's guilt about their inability to escape; they increasingly thought themselves exceptional, adding to their shame. First-wave feminism, expressing its relatively elite class base, helped construct a femininity that was oppressive to battered women: by emphasizing the superiority of women's peacefulness, feminist influence made women loathe and attempt to suppress their own aggressiveness and anger. . . .

Wife-Beating, Gender, and Society

This [reading] has been an extended demonstration that wife-beating is a social problem. It has been sanctioned and controlled through culture—religious belief, law, and, most importantly, the norms of friendship, kinship, and neighborhood groups. One assault does not make a battered woman; she becomes that because of her socially determined inability to resist or escape: her lack of economic independence, law enforcement services, and, quite likely, self-confidence. Battering behavior is also socially determined, by a man's expectations of what a woman should do for him and his acculturation to violence. Wife-beating arose not just from subordination but also from contesting it. Had women consistently accepted their subordinate status, and had men never felt their superior status challenged, there might have been less marital violence. To focus on women's "provocations," and to examine men's grievances against their wives, is not to blame women but, often, to praise them. It is to uncover the evidence of women's resistance.

To some extent the female gender itself has been influenced by millennia of violence, and a socialization toward passivity. But the relationship between battering and feminity is more complex. Women have been as aggressive, irrational, and self-destructive as men in marital conflict. But by and large, because women had the most to lose in relationships structured by coercion, women developed greater cooperative, socially manipulative skills. Their much-reputed wicked tongues were evident in these case records. Indeed, women's verbal skills were often honed to sharpness precisely to do battle against men's superior power, including violence. Their verbal style was a better tool for creating familial and community cohesion than was violence. This superiority was not, however, a result of moral superiority, as the nineteenth-century reformers believed; rather, it was a collective characteristic developed as a result of the structural position of the gender.

Wife-beaters' behavior was also highly gendered. Accustomed to supremacy, acculturated to expect service and deference from women, and integrating these expectations into the ego itself, men were understandably disoriented to encounter resistance and unskilled at negotiating compromises. Within this context, some men have a smaller range of responses to anger, less constructive responses to stress and frustration than others. Wife-beaters are by no means commonly crazy or even temporarily disoriented, but they may indeed have more self-destructive behaviors than less violent men....

The batterers I have described were not ideologues defending the dominance of their sex. Neither were they necessarily insecure. They were using violence to increase their control over particular women, defending real, material benefits. Beatings kept women from leaving, kept them providing sexual, housework, and child care services (or were intended to do so).[19] Wife-beating was not usually a mere emotional expression of annoyance, or a symbolic display of power. It did not result from an individual man's "need" to demonstrate masculinity; if masculinity was threatened, that threat arose in a struggle with another person. Some beatings resulted from demands for deference or from conflicts about status, apparently symbolic issues. But in these relationships the symbolism of power functioned to organize and reinforce real power relationships, which in turn provided real benefits and privileges to the "boss."

Sociologist William Goode's "resource" theory has some useful explanatory elements regarding wife-beating. It identifies the sources of power, or resources, with which people try to win benefits. Using a market metaphor, it considers violence a more "costly" resource which will not be spent unless "cheaper" resources, such as love, approval, or money, are absent or depleted. This theory would explain why poor and low-status men, who lack other resources, may use violence more readily than rich and prestigious men. (The fact being explained, however, has been challenged: as with all family violence, it is difficult to distinguish reported incidents from actual incidence.) A problem with this theory is the assumption that violence is somehow more "expensive," a tactic of last resort; it is evident that many, mainly men, turn to violence quickly, long before they have exhausted other tactics.[20] Nevertheless, the metaphor of resources reminds us that violence is a tactic in a struggle for an end, not merely a ritual behavior which could be altered by retraining.

Batterers were not necessarily conscious of their goals. Often they felt so wounded by women's behavior, and so desperately longed for a wife's services, that they experienced their violence as uncontrollable; they felt they had no recourse. Their sense of entitlement was so strong it was experienced as a need. Their wives did not feel so entitled. And when, stimulated often by responsibility for children, they gave up trying to wheedle and pacify, and tried to escape, they found what they had always suspected: a set of obstacles, any one of which might have been definitive—poverty, motherhood, isolation, and the hostility or indifference of social control agencies. When the context is supplied, many seemingly ineffective responses to wife-beating, including resignation, pandering, and changes of mind, are revealed to be rational, trial-and-error, even experienced and skilled survival and escape tactics.

Battered women's defeats are losses for everyone. Wife-beating molds not only individual relationships but also the overall social definitions of heterosexual relations. Wife-beating sends "messages" to all who know about it or suspect it; it encourages timidity, fatalism, manipulativeness in women. Men's violence against some women (extra- as well as intrafamilial) reinforces all women's subordination and all men's dominance. This does not mean that wife-beaters got what they wanted. On the contrary, wife-beating, even more than nonfamily violence against women, is often dysfunctional even for the assailants. In many of the cases reviewed here, men longed for the impossible, for sycophantic service and selfless devotion, which they would have hated had they gotten it, and their violence brought them no gain. On the contrary, in most marriages, even in extremely patriarchal societies, men's and women's interests have been complementary as well as adversary, especially because their economic futures were joined. This contradictory nature of marriage and the family—requiring cooperation among unequals—helps explain why wife-beating is not universal. Men benefited more from camaraderie, mutual respect, and friendship. Cooperation, especially in work, promoted men's as well as women's values: prosperity, health, calm, leisure.

If battered women's failures were costly to all, their successes were beneficial to all. The victims' own struggles were hard to see until the last two decades, when battered women organized themselves as part of a feminist movement. In fact, battered women's self-image, interpretation of their problem, and strategies of resistance had always been influenced by organized feminism. In turn, they also influenced social and legal policy, particularly through their interactions with social workers and other authorities: at worst they kept the issue from being completely forgotten, at best they provided a pressure for such solutions as we have today—liberalized divorce, AFDC, prosecution. Even women who have never been struck have benefited from the "disestablishment" of marriage that is now taking place, the process of transforming it from a coercive institution, inescapable and necessary for survival, to a relationship that is chosen.

NOTES

1. No studies to date have identified characteristics which distinguish wife-beaters from other men. The reason may be the either-or approach, some scholars looking for psychological disorders and others for social-stress factors. E.g., Frank A. Elliott, "The Neurology of Explosive Rage: The Dyscontrol Syndrome"; John R. Lion, "Clinical Aspects of Wifebattering," M. Faulk, "Men Who Assault Their Wives," and Natalie Shainess, "Psychological Aspects of Wifebattering," all in Maria Roy, ed., *Battered Women. A Psychosociological Study of Domestic Violence* (New York: Van Nostrand, 1977); Evan Stark and Anne H. Flitcraft, "Violence Among Intimates: An Epidemiological Review," forthcoming in *Handbook of Family Violence*, eds. Vincent B. Van Hasselt et al. Some might hypothesize that batterers are men who think they can get away with it. This study did not provide the evidence to study violent men. In these case records, women were virtually the only adults interviewed. This tilt was overdetermined: the caseworkers considered women the responsible parents and consulted with them primarily; the women were voluble in complaining about marital violence—and all personal problems—while the men were not; as aggressors, the men felt it in their interest to make themselves scarce.
2. Elizabeth Janeway's *Powers of the Weak* (New York: Knopf, 1980) is an important and underrated contribution to feminist theory. In the tradition of Eugene

Genovese's and Herbert Gutman's understanding of slave resistance, and of many labor historians' accounts of workers' resistance, it is followed now by James Scott's *Weapons of the Weak: Everyday Forms of Peasant Resistance* (New Haven: Yale University Press, 1986), an analysis of peasant resistance.

3. Elizabeth Pleck, "Wife Beating in Nineteenth-Century America, *Victimology* 4, 1 (1979), pp. 60–74. This did not mean that courts reliably found against wife-beaters. The nineteen most often cited precedents from 1823 to 1876 defy a generalization that there was steady motion toward women's rights to physical protection from their husbands: People v. Winters, N.Y. 1823; Bradley v. State, Miss. 1824; Perry v. Perry, N.Y. 1831; Poor v. Poor, N.H. 1836; The State v. Buckley, Del. 1838; People v. Mercein, N.Y. 1842; Commonwealth v. Fox, Mass. 1856; Richards v. Richards, Penn. 1856; Gholston v. Gholston, Ga. 1860; Joyner v. Joyner, N.C. 1862; State v. Black, Ala. 1864; Commonwealth v. Wood, Mass. 1867; Adams v. Adams, Mass. 1868; State v. Rhodes, N.C. 1868; State v. Mabrey, N.C. 1870; Fulgham v. State, Ala. 1871; Knight v. Knight, Iowa 1871; Commonwealth v. McAfee, Mass. 1871; Shackett v. Shackett, Vt. 1876. (Research by Kathy Brown.) The uneven development was partly because there is such variation among the states, and partly because the relevant decisions were responding also to custody contests. But opinions were common that specifically denied that men had *any* right to physical chastisement of wives. For example, in Commonwealth v. McAfee, Mass. 1871: "A man has no right to beat or strike his wife even if she is drunk or insolent." Or, in Fulgham v. State, Ala. 1871: "The husband can not commit a battery upon his wife, by way of inflicting upon her 'moderate correction' in order to enforce obedience to his just commands."
4. The few cases of beaten husbands were not used, as they might have been later, to call attention to women's culpability, but the men were portrayed as "of low type," drunken, immoral. #2008, 2561. [Numbers are case codes created by the researcher as she went through the agency files.—Ed.]
5. For just a few examples, #0315A, 0813A, 2003, 2008, 2054A, 2058A.
6. #4007A
7. #4584
8. #4284
9. #1825A
10. These case records contain, for example, instances of fights among women, particularly among neighbors, but also among family members.
11. #4060. See also, e.g., #2008, 2561, 3541, 3546, 5085.
12. #2024.
13. #6042. See also, e.g., #3363, 5543.
14. Jane Totman, *The Murderess. A Psychosocial Study of Criminal Homicide* (San Francisco: R. and E. Research Associates, 1978), pp. 3, 48.
15. #3024. See also #4261, 4501, 6086. I cannot resist the only partly humorous observation that if there is a pattern of "masochism" in violent marriages, it describes male better than female behavior, since it is mainly the men who appear to want to continue the violent relationships.
16. Particularly noticeable was the disappearance of women attacking other women. In the first decades of this study the random sample turned up several cases like that of a 1910 Irish-American woman who had "drinking parties" with other women, not infrequently ending in name-calling and fights; she and her daughter fought physically in front of an MSPCC agent; and her daughter was arrested for a fight with another girl. (See #2047.) As previously, we do not have enough data on women's violence to support this impression statistically. Several experts on contemporary marital violence have found, contrary to my impression, continuing high rates of mutual violence and women's violence—e.g., according to Murray Straus, as much as 49.5 percent, of couples reporting any violence, although women remain the more severely victimized. See Murray Straus, "Victims and Aggressors in Marital Violence," *American Behavioral Scientist* 23, 5 (May–June 1980), pp. 681–704. The studies reporting female-to-male violence have been sharply criticized for producing misleading data; see, for example, Elizabeth Pleck, Joseph H. Pleck, Marlyn Grossman, and Pauline Bart, "The Battered Data Syndrome: A Reply to Steinmetz," *Victimology* II, 3–4, 1977–78, pp. 680–83.
17. It is possible that there was no decline in women's violence but only in the reporting of it.
18. Tomes, "A Torrent of Abuse."
19. Two works that make this argument are Susan Schechter's *Women and Male Violence,* chapter 9, and Emerson R. and Russell Dobash's *Violence Against Wives: A Case Against the Patriarchy* (New York: The Free Press, 1979).
20. For a fuller discussion of resource theory and references, see Breines and Gordon, pp. 514 ff.

Women of Belize

Irma McClaurin

Domestic violence is one of the most pressing issues for women in Belize. No ethnic group seems to be immune to the practices of wife/partner beating, verbal abuse, and emotional abuse. Further, occupation and educational levels are not barriers to domestic violence—it touches women of every ethnicity, occupation, and social status, and is generally perpetrated by their partners or spouse. According to the literature disseminated by Women Against Violence, "In every District of Belize women report being physically assaulted by their husbands, common-law partners, boyfriends, ex-partners and relatives. Approximately 90% of violent crimes against women are committed by someone close to the woman. Few assaults are committed by strangers."[1] Given this, it seems reasonable to suggest that one of the underlying tenets of Belize's culture of gender is an acceptance of domestic violence. Zola presents a scenario that depicts how domestic violence, especially marital rape, is often overlooked in Belize:

Zola: That is what you call rape—you has damaged that woman, you has been so cruel to her and yet want to use her as your partner or your wife, which as you know it is your duty. But from the way he treated you and abused you, him lets you become his enemy. So it's just like someone walks in and wants to abuse you and wants to have a sexual relationship with you in the motion of a rape. It's something just like that. And yet, in Belize they claims to say that a husband cannot rape a wife! These are the things we are trying to protect women against. These are the things that the mens in Belize are not educated enough to realize. . . .

But we wish that we could do it to let the people know, let mens know that the Women Commission is not fighting to let the women has the power over the men. It is not that. It is just that we want them to realize that we are a part of them. We need them and they need us, but treat us like human beings.

Like some men they have a wealth, they are wealthy. They live with their wife until death. And there he dies and the wife is left with nothing. Sometimes the mother inherits it; sometimes the sister inherits it; sometimes a kid in the home inherits it [knocks over microphone while gesturing]. Things like this.

Irma: So if someone dies and there is no will you don't inherit anything? It is divided among other relatives?

Zola: You can't have inherited anything. This all comes from the men, from the law to show that rarely woman in Belize, very few can seek out a right as a wife.

Another thing is as I said up in the Women's Commission at the first time they elected me, and this is why they elected me. Not [that I am] proud or boasting of who I am. Because some women came in and we really relate our affairs, what we went through in the past and let us be one as family. The women, I let them know straight out and clean that I used to be in their shoes, where when I make my [shopping] list my husband scratch what I doesn't need. You know? It came to a stage that he'd give his mom money to buy me clothing and I had to tolerated it. I was dumb to life.

In making the simple statement, "I was dumb to life," Zola both acknowledges and critiques her earlier actions and attitudes. Today she is overtly political and aggressive, but understands her past life as one of emotional and psychological dependency. It is a life emblematic of women in the role of minors—in Zola's case this translates into a perception that she is not responsible enough or adult enough to determine when and if she should buy her own clothes. I did not encounter . . . any other examples of men creating dependency in this manner, but such personal matters are not openly discussed by men or women. What Zola's example does show is that the means by which men and society elicit women's compliance to subordination are wide and varied.

Change: Its Catalysts and Obstacles

As is often the case in an individual's changing consciousness, a single event may alter material conditions in a way that forces the person to reflect on her current circumstances. In Zola's case, the event was her partner's illness; his sudden departure to the United States for medical treatment disrupted the relative economic security that she relied upon.[2] Suddenly she found herself alone. In coming to terms with this insecure status, which left her without any financial support, Zola was forced both to confront the nature of her economic and personal relationship to her partner and to assess her own future—one in which she might find herself alone. She comments on the process.

Zola: He got sick and had to get a heart surgery. We didn't had anything. Then I have to be the man. I don't know where I had that strength from and I came out and I said to myself, "No, I can't sit back. I have four kids in high school and I have one at home [knocks over the microphone], and I have him to take care of and I'll get out." I get out and then I learned that yes I didn't have to depend on you [the husband]. What was I depending on you all of these years for?

He used to mean treat me, he used to beat me. Then I started to think back and see if I had done it. I had to sustain him for such a while until he could have gained his health. I got my kids out of high school. I said, well, fine, we have fifty-fifty now. If I could have done it there are many women who could have done it.

I was involved in the Women's Bureau [Department of Women's Affairs] women's group. [I was] going out and leaving my home for the first time, really getting involved. He said to me, what is this all about? He had been hearing some rumors about the organization. I said, I don't know what this is all about but when I come back I will let you know what it is all about. I went out and I let myself be clear and known out there. They said, hey, if she could have done it, we can do it. It wasn't no hassle, which she tolerated—he got weak and she got strong and she realized where she was needed.

In this situation illness is the catalyst for a number of changes, especially in roles, as Zola's husband becomes economically dependent on Zola. Moreover, Zola must create strategies of survival that look beyond this particular event. In the process she discovers and creates a greater sense of self, which gets fostered by her participation in a women's group. In joining the latter she does not directly challenge the state, but she is certainly challenging state-formed relations (like marriage) that require women to be subordinate to men.

Zola: This is the weird [thing], the men have to realize that you are needed or that you are worthwhile; it doesn't make sense. Some of them don't survive to know that you are needed. And then you left with that fear to go on living. [You think] well, okay I was in a rented home and he died and he didn't leave anything for me so I leave with the children. So I need somebody to take care of me. There you go and live the same life, these children [are] left [and] abandoned, because maybe this man doesn't want these kids.

Here Zola describes women's acceptance of the economic-sexual cycle that structures many women's lives.

Zola: That is what I said; we are trying to become a strong Women's Commission. That we can show the women outside we are building from ourselves and we are trying to help them with some of the things we have gone through to promote themselves, to promote women in Belize, but we cannot promote them without a law enforcement.

We need the police, we need the doctors, we need the ministers to do these things. Also, by ourselves, we have no law, we are not allowed to go into a home and say, hey, I heard you been beating you wife last night, you know, things like this. We need cooperation from the law and from different organizations. We need the law.

Zola points out that some change can only come about if men accept that women are equal, but she is fully aware that legislation (institutional support for change) is needed as well.

Irma: So at this point there is no inheritance law that will protect women. Are there any laws that are in effect now that need to be changed, that would help women more?

At this point I followed up on Zola's emphasis on the importance of laws in the process of change. The question, however, also reflected a standard academic focus—I wanted to know about inheritance law, a question that momentarily diverted me away from Zola's story of transformation. This shift on my part is a typical example of how researchers sometimes don't listen. I was interested in learning about the structural elements that shaped Zola's life while her story was about personal things.

In response to my academic digression, Zola ignored me and remained focused on her agenda, going on to talk about the way violence is perpetuated not just in the home but also in the school. As a result of her own experience, she is well aware that abuse creates dysfunctional families and sets up cycles that can persist for generations. She uses the analogy of an unattended infection to describe how abusive behavior can be reproduced.

Zola: It's something like you have a little sore and you doesn't take care of it and it really hurts. You go through it [domestic violence], the kids go through it and they see what is going on and if you have nine boys and they see what is going on you have nine men growing up that will carry on what their father was doing. So that in particular, we need a lot of law in that, as good as in the school—the teachers cannot beat the kids. It should be in the home that you cannot beat your wife. In this the family court, it is there but it is protecting the children. It is not fully enforced to protect women.

Zola's observations about the effects of abuse on women are echoed in the literature produced by Women Against Violence. In the excerpts below, the group describes the severity of domestic violence in Belize and the fear and anxiety it generates.

> Many women in our country are not safe in their own homes. Women have reported their partners for beating them with such weapons as knives, crowbars, machetes, electric wire, pint bottles, mop sticks, rocks, boards and rope. Some women are threatened with guns on a regular basis.
>
> Few women press charges against their husbands for fear of going to court and of further beatings. Rather than lay charges some women decide to apply for a legal separation and maintenance. Others feel they cannot leave or report the violence because they have several children to support, they do not have money or a place to live, they still love their partner and hope he will change, they want the children to have a father, they don't want their partner to go to jail or they are afraid they can't cope on their own.[3]

These statements articulate the personal, economic, and physical constraints that confront many women in Belize.

Institutionalizing Change

After this intense discussion of domestic violence, I shifted the interview focus in order to learn more about how women become agents of change.

Irma: Can you just explain how you were elected to the position? Were you nominated by someone in the ministry or was it a vote in town?

Zola's answer to this question is a testament to her strong sense of the interrelationship between her personal life and political processes. The control she exerted over her narrative throughout the interview, the commitment she exhibited in shaping her story—making herself through a narrative mode—is analogous to the control she exerted over the circumstances of her life.

Zola: [I tell my daughter] if I had the education you all had I wouldn't be in that house cooking now, washing, and making baby. I gave it to you and you didn't use it. I didn't had it; I doesn't even have a primary certificate but I had more than what you have. I have something—I was born with common sense. And I use it. And [with] that I make myself who I am today.

Moreover, Zola is well aware that she has been transformed.

Zola: Really, well, I am an active person and being an active person in politics, in my community and different things. Dorla [President of the National Women's Commission] asked someone here to recommend me. And there were several of us who still went up, and as I said everyone had a point of view themselves what they have been going through, see if they are strong enough to be in this Women's Commission.[4] I wasn't ashamed to say what I went through because it has made me who I am today.

Here Zola makes a clear, indisputable connection between her process of shifting from an abused woman with a minor's status to the fully responsible person she is today. More significant, it is not her knowledge of formal politics or bureaucracy that makes her a valuable representative, but the specific details of her life, which serve as pedagogical tools. She has had to circumvent the hidden bureaucracy and structures of the gender system in order to be who and where she is at this point.

Zola: I doesn't have that fear anymore [where I have] to come out and say can I go to this meeting or can I go into town? I had to do that. I doesn't have to do that anymore because my husband understands. From the mere fact that I had to pick up the burden of him and five kids, that means I am responsible enough to have that right. I don't know if it is part of fear that he is sick now or whatever, but he gained understanding.

Zola says her partner has adjusted to her newfound autonomy and sense of empowerment, but she recognizes that their role reversal has spawned a fear. Such fear I found common among many Belizean men, who believe that if women become involved outside of the home, if they exhibit behaviors or attitudes that express autonomy, independence, or self-confidence, they will come back home thinking "dey the boss." Based on this male attitude toward women and some women's acceptance of it, controlling a woman's movements—restricting her ability to see friends, go to social events, and the like—is an important weapon in the cycle of domestic violence.

Again, the Belize Women Against Violence movement has been most successful in identifying this form of control as part of a pattern. Beyond the most common, and thereby easily recognized, forms of physical and sexual abuse that men use to exert "*power* and *control* over their partners," the group describes the more subtle forms that men use to force women's compliance. The WAV brochure states:

> How do men abuse their partners? . . .
> Emotionally, by:
> continually insulting her
> controlling what she can do and who she can see
> controlling all the money
> treating her like a servant
> making her feel bad about herself
> having extra-marital affairs
> keeping her from getting a job

. . . [I]t is within the boundaries of the institutional and ideological dimensions of gender that the dynamics of power and subordination among Belizean men and women are very often set up and executed through physical and mental abuse.

Although men exercise real economic power over women, they still exert their authority and control through physical battering as well. Up until 1992, when the Domestic Violence Bill was . . . [written], this behavior was viewed by the police and much of Belizean society as a "private ting." Domestic disputes were regarded as something to be worked out between the individuals, regardless of how severe the battering, in the privacy of their home. Violent acts thus go unacknowledged by the community, unreported to the police, and remain rarely discussed among friends and families. Domestic violence is so acutely invisible that it did not warrant any mention in one of the most important official documents produced by the police, the "Report of the Belize Crimes Commission."[5] Yet the local papers run stories several times a year of women mutilated, burned, and murdered by their husbands or partners in domestic disputes.

. . . Zola provide[s] evidence of the way domestic violence manifests itself in individual women's lives. As [her story indicates], patterns of physical and mental abuse appear early in a woman's life cycle. These are often established when teenage girls become involved with an older boyfriend or older man—the latter representing a de facto father authority figure. One way men exercise their gender power is by controlling women's movements. A retired schoolteacher describes what this pattern looked like in her life: "That is it! I was a prisoner! That is the word. I was a prisoner. He was afraid that other men would talk to me. So I was a prisoner. So you see they afraid. To me, most of the time he would talk, I would sort out that he felt like inferior. Inferior. That's the way some men feel, you know." From a man's perspective, women who freely travel about the town and city alone are rebellious and need to be controlled and contained. When a couple begins to live together, males expect that a woman's mobility will be limited by the requirement that she attend to her home duties. In this traditional way of life, a woman's path is generally confined to the market, school, church, and home. Dances, bars, and other activities that place a woman in a public space and in the scrutiny of other men are discouraged and in effect deemed taboo.

Another way men exercise their gender power is through social isolation. Some women are denied the privilege of having friends and visits in their home, a restriction most commonly explained by saying that women gossip when they get together and men condemn this behavior.[6] A woman's friendships then are limited to those relationships established prior to her marriage with school friends, women kin, or those neighbors who travel the same circumscribed route from kitchen to market and back again. Some women told me that although they had lived in their communities for years, they had little more than a "good morning" relationship with most of their female neighbors.

One of the most powerful holds that men have over women is the threat of abandonment. To avoid abandonment, women in Belize practice endurance. Since the intent of the threat is to gain compliance, women's failure to do anything that provokes such a threat can be read as acquiescing to male needs and authority. Bullard describes women's behavior under these circumstances: "A Belizean woman will accept much abuse and tolerate a great deal of offensive behavior from her mate in return for the economic security of his wages, so great is her dread of being left alone. This is a fairly common fate, and almost one-third of the women classified as 'mothers' in the 1960 census had never been, or were not currently, living with a mate."[7] Erica Jong, an American writer, once commented that the worst slave was the one who beat herself—making the point that women participate in their own subordination. They seldom challenge male behaviors or expectations for fear of being cast in a negative role. . . .

In a study of battered women in Iowa, Julie L. Stiles and Douglas Caulkins describe how women become "prisoners in their own homes,"[8] Drawing on Lewis Okun's theory of coerced control, they conclude: "Our interpretation reinforces the notion that abusive isolation is a critical component of domestic violence as practiced in Iowa or elsewhere." They also state:

"Instead of implying that women necessarily invite their abuse, our interpretation shows that abuse creates an aberrant social structure from which it is difficult to escape."[9] Their study shows that women's participation in women's groups is of crucial value in breaking this cycle of abuse. They argue: "One of the effective paths for women escaping abusive relationships is ... [through] ... collectivist support organizations."[10] Given this consequence of collective participation, men's hostile response to women's groups, especially their antagonism toward WAV, becomes understandable. Women's collective action is interpreted as a definite threat to men's efforts to control. Men rationalize their antagonism by asserting that women gossip and that no good can result when women get together. Their dismissal of the worth or value of women's group, I surmise, derives from their fear of precisely what happened to Zola—she changed and was no longer tolerant of the limitations her partner tried to impose on her.

Zola: [Now] whenever there is a meeting, unless I can't [because] I have a family meeting, but whenever there is somewhere I have to go, I go. Now I am asked to go, like campaigning for a town board election; they asked me to go and campaign with them. I can campaign. To go out with them to meetings in San Ignacio [in the west of Belize], it was no problem to come out and work on the election day; it was no problem.

Social scientists are constantly grappling with the question of how to determine that change has occurred.[11] Zola gives us some useful criteria. For her, one indicator of change is the ability to control her own mobility. As we have seen, women in Belize are often kept socially isolated and their movements are severely restricted by men as a way of controlling them. In her new direction in life, not only does Zola recognize the importance of being asked to attend political functions, which indirectly affirms her power in the community arena, but she is well aware that her ability to go to these functions without having to seek permission is a visible sign that she is in control of her own life. Another indicator is the economic autonomy and sense of security she has achieved through the ownership of land.

The source of these changes Zola sees as primarily internal. Women must find their inner resourcefulness first, before they rely upon external forces in the form of organizations or the law. According to Zola's philosophy, each of us possesses the capacity to change the circumstances of our lives.

Zola: Sometimes we woman has to learn and let *something grow into us that we are woman*, we are no more a kid, and we have to show that we can't depend on the law for everything.[12] We has to learn to depend on ourselves. Let ourselves grow and this is what we are trying to do. If we let ourself grow strong, a strong Women Commission, and stand up and keep what we have together, we can make a law and we can change a law. We need that strength, we need that cooperation, we need that understanding. I didn't know who was elected at Belmopan. We went for four days workshop and then I got back and had a letter from the minister pointing out I had been elected.

These words form a powerful statement that reflects Zola's current consciousness. Moreover, she is aware that the process of change is multifaceted and occurs at several levels. At the individual or personal level, she suggests that it is incumbent on women to come to some self-awareness that they are not children, not minors. Such change, she posits, must be inner directed. That is, women cannot depend upon the law to give them this inner resourcefulness or knowledge of their situation. At the same time, she also asserts that through collective action, women have the ability to transform the external circumstances of their lives; thus she attributes tremendous significance to unity and collective action, especially in the form of women's groups. Zola's "analysis" reflects a fully developed awareness of the personal, cultural, and structural basis of women's subordination. In her eyes, change on one level cannot occur without change at the other levels. In her cosmology, personal change is interrelated with social change.

This last aspect is particularly evident in Zola's narration of her involvement in women's groups, which also becomes a narration of her life. I interpret this to mean that women's groups often provide an arena for particular women's performances of empowerment. As such, women are actors within the arena of women's groups. In Zola's case, the specific circumstances of her life had already created the conditions for change, and she immediately took advantage of them. Yet, as she tells us, changing her life does not lead to community acceptance of such change and can result in alienation. Women's groups provide an alternative environment of support and nurturing, reinforcing the path of change women may be on.

What is most instructive about Zola's development is her recognition of the close relationship between personal change and collective action. She demonstrates for us that individuals do not operate in isolation; also that we are constantly in search of affirmation that the choices we make are correct, not only in terms of our individual assessment of them, but in terms of the community's assessment. Women's groups act as critical sources of affirmation. They become communal sounding boards against which women can test new ideas, interpretations, and strategies for change, all within a safe environment. This suggests that if the choices women make challenge the traditional rules and behaviors of their communities, then they must form alternative sources of support so that they may continue to function within the ever-present social constraints or boundaries. Even as women change, they are negotiating within an environment that may appear static or unassailable. As they confront the problems such a disjuncture creates, they do so with a new strength that is a combination of their own personal desires and the support of a specialized community of like-minded women.

Because of Zola's awareness and her sense of empowerment, I thought it important to determine the degree to which her reading of "the woman's problem" in Belize corresponded to other readings I had elicited from survey data and interviews. I asked her: "Are there other problems that you see that Belizean women face or women in this particular community face? [Are there] other problems that they have to deal with?" She responded:

Zola: In our area, I mean this is something from inheritance, but as I said you women must open your eyes and see where your advantages stands and where your advantage is taken. I am not an educated person but I am an active person, outspoken person, and this is where I get around. I think that if you fight yourself out of it, you can get out of it. As I said education is a very great thing. It is something like a bank but common sense is as good and that is what I think I use.

From way back especially in Lemongrass, the mens are the men. The men is the man of the home, whether you are the one who maintains the home they stands being the men. Okay? Basically it is very rare when you hear a woman saying, "That is my piece of land; I am living on it, or those cattle out there are mine." It is something very scarce to know. These are the things I feel we must look upon. My mother had twenty-two of us, and there was one chosen son that my father left everything for. So it is something from way beyond, way, way back. Well yes, he was the chosen one. He doesn't drink, he doesn't smoke, he is a respectful brother. If we had problems we could always go to him. My mom is basically take care of by him. But how many children are there? How many boys are there that can find a wife like that?

At this point, Zola not only gives us insight into women's exclusion from acquiring property through marriage, but also suggests that wives can limit the extent to which a son will provide for his mother's needs. In her own family, Zola's sister-in-law did not have much control and so her brother took care of their mother. But because she was "loaned" out as a young child, Zola has few details about many of the dynamics of family interpersonal relations.[13] Nor did I pursue this topic as thoroughly as I might have.

Zola's concern about inheritance practices as a major issue for women illustrates the extent to which cultural traditions and life experience structure women's activism.... In general, the specific problems women confront in their lives (domestic violence, inheritance) or their concern about specific issues (nutrition, breast feeding, family planning) are the motivation for their activism. These variables explain to some degree their decision to participate in particular women's groups....

NOTES

On March 1, 1993, two years after my interview with Zola, the Domestic Violence Bill was signed into law. It focused on providing immediate intervention in order to ensure the safety of the party at risk, who may be a spouse or a child. The bill provides a protection order to restrain the abuser from "being on the premises," "engaging in conduct of an offensive or harassing nature," "speaking or sending unwelcome messages," "taking possession of specified personal property," and other acts intended to intimidate or otherwise disrupt the life of the at-risk person (Domestic Violence Act, 1992, 10–11). This goal is affirmed in the bill's opening statement: "An Act to afford protection in cases involving domestic violence by the granting of a protection order; to provide the police with powers of arrest where a domestic offense occurs and for matters connected therewith or incidental thereto" (Gazette, June 20, 1992, 1). Of central concern to women's groups like BOWAND and WAV is the hope that this bill, which clearly delineates the role of the police in domestic violence, will move law enforcement agencies away from their current role of complicity toward a new one of prevention and intervention.

1. Belize Women Against Violence, "Domestic Violence, Woman Abuse, Wife Assault, Spouse Abuse, Battering," information brochure, Belize City, n.d.
2. Zola did not reveal how her husband financed his medical trip to the United States, but relatives often take up a collection for such emergencies. Many Belizeans also have relatives in the States who may be able to assist them while they seek treatment.
3. Belize Women Against Violence, "Domestic Violence, Woman Abuse."
4. The National Women's Commission's "origins lie in the Women's Bureau, established in August 1981, 'to upgrade the situation of Belizean women.' The Women's Bureau recommended in 1982 that a Cabinet-appointed National Women's Commission on the Status of Women in Belize be established" (National Women's Commission, "Objectives and Structure," March 1991). In November 1982, the fourteen-member group was appointed. All members are volunteers who represent the country's regional and ethnic/cultural diversity. They are appointed by the Cabinet and serve five-year terms. Between the formation of the commission and 1993, four women served as president: Jane Usher, Kathy Esquivel, Martha Marin, and Dorla Bowman.
5. Published in 1993.
6. Ruth Borker has argued that gossip is one example of women's power, which they use to regulate male behavior or to shape community attitudes. Certainly men fear being the topic of women's gossip and characterize women's talking as trivial and unproductive. She states: "Researchers have been especially interested in women's gossip because it is seen as an important source of social power for women in these communities. Researchers in southern Europe have seen women's gossip as an important mechanism for social control and for asserting social values, supporting the general anthropological view of the functions of gossip." Ruth Borker, "Anthropology: Social and Cultural Perspectives," in *Women and Language in Literature and Society*, ed. Sally McConnell-Ginet, Ruth Borker, and Nelly Furman (New York: Praeger, 1980), 34.
7. M. Kenyon Bullard, "Hide and Secrete. Women's Sexual Magic in Belize," *Journal of Sex Research* 10, 4 (1974): 261.
8. Julie L. Stiles and Douglas Caulkins, "Prisoners in Their Own Home: A Structural and Cognitive Interpretation of the Problems of Battered Women," paper presented to the Iowa Academy of Sciences (1989).
9. Ibid., 8. Cf. Lewis Okun, *Woman Abuse: Facts Replacing Myths* (Albany: State University of New York Press, 1986).
10. Stiles and Caulkins, "Prisoners in Their Own Home," 8.
11. Cf. Ann Bookman and Sandra Morgan, eds., *Women and the Politics of Empowerment* (Philadelphia: Temple University Press, 1988); Teresa Pires de Rio Caldeira, "Women, Daily Life, and Politics," in *Women and Social Change in Latin America*, ed. Elizabeth Jelin (London: Zed Books, 1990); and Jelin, ed., *Women and Social Change in Latin America*.
12. Emphasis added.
13. Zola indicated that she was "adopted out" at the age of five.

Old Problems and New Directions in the Study of Violence Against Women

Demie Kurz

Until recently, scholars rarely acknowledged the topic of violence against women. Fortunately, in the past two decades some researchers have focused their attention on this serious problem. These researchers have made significant progress in understanding this violence, its origins, and its consequences (Bart & Moran, 1993; DeKeseredy, 1995; Dobash & Dobash, 1992; Kurz, 1997; Stark & Flitcraft, 1996; Yllö, 1993). Their findings not only have contributed to our knowledge of this topic but also have been used by activists and reformers to raise awareness of this issue and to bring about changes in public policy (Dobash & Dobash, 1992; Yllö, 1993).

Despite this progress, however, violence against women all too often remains invisible. Two factors converge to prevent us from seeing how extensively this violence permeates our society and how it can affect all aspects of women's lives. The first is the current conceptualization of this issue in gender-neutral terms. When discussing violence against women, many social science researchers, policymakers, and those in the media use terms such as *family violence, intimate violence,* or *domestic violence* (Jones, 1994; Lamb, 1991). As they are currently used, these constructs convey the impression that violence is directed by all family members against all other family members, and they mask the facts of who is being violent to whom (Dobash, Dobash, Wilson, & Daly, 1992; Lamb, 1991, 1995). I argue in this [reading] that although a variety of types of violence occur in the family, the use of gender-neutral terms such as *family violence* can conceal the serious problem of violence against women.

The second major factor that obscures our knowledge of the prevalence of violence against women and its impact on their lives is the compartmentalization of this issue in research and social policy; that is, although more researchers and policymakers are aware of domestic violence as a general problem and refer to it in their work, they all too often fail to investigate how violence may be an important part of the particular problem they study (Kurz, 1995). Thus, researchers study the family *or* violence, or work *or* violence, *or* divorce or violence, rather than understand that violence against women is an integral part of women's experiences in all of these social arenas (Kurz, 1996).

In this [reading], I examine how these two practices—the framing of violence against women as a problem of family violence and the compartmentalization of this issue in current research—obscure the prevalence and impact of this important problem. It is critical that we obtain accurate data about the nature of violence against women and its impact on all areas of social life, both for research purposes and to create social policies that help women and do not put them at greater risk. This task is particularly urgent, given the prevalence of physical violence against women. It is estimated that 10% to 20% of women are beaten by a male intimate in a given year, and one quarter to one half will be beaten by a male intimate at least once in their life (Straus & Gelles, 1986). Witnessing the physical abuse of their mothers can also have a negative impact on children (Jaffe, Wolfe, & Wilson, 1990; "Silent Victims," 1993).

Dominant Conceptualizations of Violence Against Women

As noted, the dominant conceptualization of violence against women as a problem of family violence is widespread. Unfortunately, many social science researchers view physical violence against women as a gender-neutral problem of "family violence," with women being as violent as men (Brinkerhoff & Lupri, 1988; Gelles, 1993; Gelles & Cornell, 1985; Gelles & Straus, 1988; McNeely & Mann, 1990; McNeely & Robinson-Simpson, 1987; Shupe, Stacey, & Hazelwood, 1987; Steinmetz & Lucca, 1988; Stets, 1990; Straus, 1993; Straus & Gelles, 1990). According to this view, the problem of violence in the family is a problem of "spousal abuse," "assaultive partners," and "violent spouses" (Arias, Samios, & O'Leary, 1987), in which husbands and wives resort to physical force to solve their conflicts.

Social science data play an important role in legitimating and responding to social problems in our society. Thus, through its use of terms such as *family violence* or *domestic violence,* social science contributes to the marginalization of the problem of violence against women in our culture. Some have been critical of this terminology (Dobash et al., 1992; Jones, 1994; Kurz, 1997; Lamb, 1991, 1995; Wardell, Gillespie, & Leffler, 1983; Yllö, 1993), but many social scientists continue to use it, believing that men and women are equally violent toward each other. Because a widespread knowledge of the gendered structure of power is absent in our society, few discourses provide an alternative view of the gender-based nature of violence against women.[1]

Why do social scientists continue to use gender-neutral paradigms when so much evidence indicates that it is principally women who are battered? I argue that this happens when researchers adopt models of family behavior and instruments for measuring violence that are based on misleading assumptions about the family. Family violence researchers typically interpret their data within gender-neutral frameworks such as systems theory (Gelles, 1993). In this theoretical perspective, the "family system" operates to "maintain, escalate, or reduce levels of violence in families" (Gelles, 1993, p. 36). These researchers argue that all family members are part of the family system, contribute to family patterns and events, and bear responsibility for what happens in a family. Although all family members, like all individuals, are responsible for their actions, this paradigm overlooks the fact that gender norms and gendered relations of power enter into and partially constitute all social relations and activities and pervade the entire social context in which a person lives. In the case of marriage, norms promoting male dominance and males' right to use force in heterosexual relationships set the parameters for behavior in marriage.

Family violence researchers also claim that family norms are affected by norms of the wider society. They believe that a "culture of violence" in the United States and the stressful nature of modern life contribute to the creation of family violence and that violent people are also influenced by different "subcultures of violence" (Gelles, 1993; Steinmetz, 1987). These researchers cite evidence of a widespread cultural acceptance of violence in television programming, folklore, and fairy tales and in surveys showing widespread public acceptance of violence in U.S. culture (Straus, Gelles, & Steinmetz, 1980). They believe that husbands and wives come to accept these norms, which condone violence as a means of solving conflict. The problem with this perspective, however, is that the culture of violence framework fails to explain why, overwhelmingly, women are the targets of male violence.

Finally, those who believe that stress is a critical factor in causing violence believe that violence in the contemporary U.S. family is caused by a variety of social-structural factors, including stresses from difficult working conditions, unemployment, financial insecurity, and health problems (Gelles & Cornell, 1985; Gelles & Straus, 1988). Once again, however, this explanation does not account for why women are the primary targets of the violence. Many women, as well as men, experience stress but do not use violence.

Researchers who use family violence frameworks base their conclusions on data produced by research instruments called the Conflict Tactics Scales (CTS; Straus, 1979). Authors of surveys based on these scales conclude that "women are about as violent within the family as men" and that women as well as men are perpetrators and victims of physical violence. I argue that these scales produce misleading data. The CTS require respondents to identify those conflict resolution tactics, listed in the scales, that they have used in the previous year. These range from nonviolent tactics (calm discussion) to the most violent tactics (use of a knife or gun). Using these scales, researchers (Straus & Gelles, 1986) find similar percentages of husbands and wives using violent tactics. The CTS have been used in many surveys here and abroad, and Straus's findings have been replicated in many surveys (Brinkerhoff & Lupri, 1988; Nisonoff & Bitman, 1979; Stets, 1990), including studies of dating violence (Arias et al., 1987; DeMaris, 1987; Lane & Gwartney-Gibbs, 1985).

Findings from the 1985 National Family Violence Survey (Straus, 1993), based on women's responses to the CTS, show that both wife and husband were violent in 48.6% of cases, the husband only was violent in 25.9% of cases, and the wife only was violent in 25.5% of cases. Straus (1993, p. 94) concluded from these data that "regardless of whether the analysis is based on all assaults, or is focused on dangerous assaults, about as many women as men attacked a spouse who had not hit them during the one-year referent period." In earlier studies based on the CTS, Straus et al. (1980) found that 12.8% of husbands used violent tactics in conflicts with their wives and that 11.7% of wives directed these tactics against their husbands. Straus et al. (1980, p. 36) concluded that whereas "traditionally men have been considered more aggressive and violent than women," looking at the couples in which the husband was the only one to use violence and those in which both used violence, "the most common situation was that in which both used violence."

Researchers who use a violence-against-women framework (Berk, Berk, Loseke, & Rauma, 1983; Dobash & Dobash, 1979; Dobash et al., 1992; Kurz, 1993; Pleck, Pleck, Grossman, & Bart, 1977–78; Saunders, 1988; Stark & Flitcraft, 1985; Yllö, 1993) argue that the data claiming an equivalence of male–female violence, particularly data based on the CTS, are flawed and that the CTS fail to provide reliable data. They believe that the validity of the CTS is undermined for several reasons. First, the continuum of violence in the scales is so broad that it fails to discriminate among very different kinds of violence (Dobash & Dobash, 1979; Dobash et al., 1992; Stark & Flitcraft, 1985). Further, the scales do not ask what acts were done in self-defense, who initiated the violence, or who was injured. If these questions were asked, they believe, the picture would be clear: Overwhelmingly, men abuse women and women use violence primarily for self-defense (Breines & Gordon, 1983; Brush, 1990; Dobash & Dobash, 1979; Dobash et al., 1992; Pleck et al., 1977–78; Saunders, 1988, 1989).

Critics of the CTS argue that data from interview studies show that when women use violence, it is typically in self-defense. Saunders (1989) found that, in the vast majority of cases, women attributed the violence to self-defense and fighting back. Emery, Lloyd, and Castleton (1989), in an interview study based on a small sample of women victims of dating violence, found that most women spoke of self-defense. They also found that women spoke of using violence in frustration and anger at being dominated by their partners and in retaliation for their partners' violent behavior. Second, there may be significant male–female differences in self-reporting. Some researchers argue that men are more likely than women to underreport the extent of their violent acts (Okun, 1986).

Violence-against-women researchers also point out that the CTS focus narrowly on counting acts of violence (Dobash et al., 1992; Yllö, 1993). Such a focus overlooks related patterns of abuse in relationships, including psychological abuse and sexual abuse, and does not address other means of nonviolent intimidation and domination, including verbal abuse, use of suicide threats, or use of violence against property, pets, or children or other relatives (Yllö, 1993). Although it is important to focus on acts of violence, and although stopping all violent acts is a very important goal, it is important to remember that it would hardly stop abuse. Decreasing the levels of abuse would require reducing the inequality in the distribution of power and control in relationships.

Unfortunately, numbers that claim violence between men and women is "equal" and "mutual" have also been used against women to cut funding for battered women's shelters (Lewin, 1992; Loseke, 1992). Many fear that claims about mutual violence can be used to deny violence against women in the creation of custody and visitation agreements (Fineman, 1995) and to absolve social institutions of responsibility for failing to respond to violence against women (Kurz, 1993). Still others fear that the family violence perspective will reinforce the individualist bias in the field of counseling—that counselors will focus on clients' individual and personal problems without identifying the inequality between men and women, which is the context for battering (Adams, 1988). They disagree with those family violence proponents who argue that violence is caused primarily by frustration, poor social skills, or inability to control anger.

By taking the family as the basis of their analysis, family violence researchers who use the frameworks and methods described here obscure the issue of violence against women. Although it is valuable for some purposes to take the family as a unit of analysis, it is misleading to assume that the family exists predominantly above and beyond the interests of its members. Men, women, and children do frequently act together as a unit in social and family activities; however, they also have different and potentially conflicting interests. The family is highly gendered. The actions of members are influenced at every point by gender norms and opportunities. When researchers take gender into account, it becomes clear that physical violence against women should be compared, not with elder abuse and child abuse, but with related types of violence against women, such as rape, marital rape, sexual harassment, and incest, all of which also result from male dominance (Wardell et al., 1983).

Compartmentalization of the Study of Violence Against Women

Another factor that prevents researchers from presenting the full extent of violence against women in social life is the compartmentalization of the study of this violence; that is, they view it and analyze it as distinct and separate from other issues. As noted earlier, researchers study women in the family, or women in the workplace, or violence against women in separate categories and fail to see the impact of violence on many aspects of women's lives. In this section, I demonstrate the impact of violence against women on two areas of social life that we do not usually associate with this issue: (a) women's experiences of divorce and (b) the lives of poor women. For women in divorce situations and for poor women, violence can seriously threaten their physical and economic well-being.

Divorce

Divorce is widespread in contemporary society and has a significant impact on family life and on social life more generally (Furstenberg & Cherlin, 1991). Thus, it is particularly unfortunate that the study of divorce typically fails to consider violence in the family or violence against women (Kurz, 1995). In my study (Kurz, 1995), divorced mothers reported experiencing high rates of violence at the hands of their husbands during their marriages. They also stated that this violence had a significant impact on their experience of the entire divorce process. The study and the interviews focused on how women viewed the ending of their marriages, how they managed on their reduced incomes, how they negotiated for resources from ex-husbands and the state, and what custody and visitation agreements they made. For these women, violence played a role in all aspects of the divorce process.

My study of divorce is based on interviews with a random sample of 129 mothers of diverse backgrounds.[2] These women reported that they experienced high rates of violence during marriage: 50% of women experienced violence at least two or three times in their marriages; 16% experienced violence once; and 4% experienced violence after the separation. Of the 50% who experienced violence at least two or three times, 37% experienced serious and frequent violence (defined as more than three incidents of violence, or one very serious incident of violence).[3] These rates are higher than those found among married couples, cited earlier in this chapter, and they are comparable with those found in other studies of divorcing women.[4] Interestingly, however, most studies of divorce do not analyze the impact of violence on the divorce process; they cite violence as another "factor" in divorce, and usually not an important one (Ahrons & Rodgers, 1987; Emery, 1988; Price & McKenny, 1988).

Women in my study cited reasons for the violence that were similar to those reported in other studies (Dobash & Dobash, 1992; Yllö, 1993). Women most commonly reported that violence occurred when they attempted to act independently. For example, women reported violence when they "started to change" or when their ex-husbands found that things did not go their way, when dinner was late, and when certain types of food were not available. These women also spoke of generally controlling behavior on the part of their ex-husbands, who sometimes didn't allow them to have friends, to go to work, or to go back to school. Some said they were not allowed to be in the company of other men because their husbands thought this would lead to sexual relationships. The following quote from a woman illustrates this point: "He was violent when I would go out and do things on my own. He didn't like that. For example, I went out and found myself a job. He didn't want that. He wanted me to always be home. My father gave me a car and I would take the kids places. He didn't like that."

These divorced women reported that violence affected their lives in several ways. First, they reported violence as one major reason that they left their marriages: 19% of women reported domestic violence as the reason for their divorce; most women reported leaving because they believed that witnessing the violence had begun to have a negative impact on the children or because they experienced a particularly serious incident of abuse. As one woman said: "All the violence was hard on my son. He saw me injured when he was 2 years old. He saw a lot. It's affected my son. He's mixed up. I left because I was afraid of what this was doing for my son." The women who left because of violence were among those who experienced the most repeated incidents of violence. Some of these women reported staying in violent relationships longer than they wanted to because they had nowhere to go; others did not leave sooner because they could not find work and feared they would not be able to support themselves and their children. Several women spoke of having stayed in relationships for extended periods of time because they wanted to try to make their marriages work.

Second, some women reported that violence played a significant role in their negotiations for child support, custody, and visitation. These women stated that their experience of violence during the marriage or separation made them fearful during divorce negotiations. In fact, 30% of women stated they were fearful during their negotiations for child support, and 38% reported fear during negotiations for custody. In both cases, these fears were related to women's experience of violence during marriage. A statistically significant relationship was found between women's experience of violence during marriage and separation and their fear during negotiations for child support and for custody. The more serious or frequent the violence these women experienced, the more fearful they were during these negotiations.

The 38% of women who said they were fearful during negotiations for custody spoke of their ex-husbands' harassing behavior. Some women reported that their ex-husbands threatened to take the children. One woman said: "I was fearful the whole time [I was negotiating for child support]. He was always threatening. By this time, I had been to abuse court. He was under orders not to come near the house, the kids, or me. The teachers were notified . . . I was terrified he would take the kids." Several other women stated that their husbands had fought for and won custody of the children even though the mothers believed that the fathers didn't really want custody. Arendell (1995) reports that substantial numbers of fathers undertake serious legal challenges to obtain custody of their children, their goal being not actually to obtain custody of their children but to try to "balance out the power of their former wives by prohibiting maternal custody, which was the prime example of men's losses and women's disproportionate authority in divorce" (p. 81). Two women reported that their husbands had kidnapped their children. Some women remained fearful of additional events of violence while they were negotiating for custody, and a few had court orders of protection mandating that their husbands stay away from the family homes. As is described below, their fears led some of these women to compromise on their demands for resources during divorce negotiations because they were afraid that if they did not give in on resources, then they could lose custody of their children.

Even for those mothers who resolved custody issues, however, this was not necessarily the end of their conflicts with their ex-husbands over children. Many of the women who faced harassment during negotiations for custody also experienced harassment during negotiations for agreements about when the fathers would visit the children. In fact, 29% of mothers who stated that visitation took place described conflict with their ex-husbands over visitation,[5] a rate of conflict similar to that found in other studies (Maccoby & Mnookin, 1992). These mothers believed that the fathers were using visitation as a way to check up on and control them, and some were afraid that their ex-husbands would be violent again. These mothers, many of whom had experienced violence in their marriages, reported that they wanted less, not more, visitation.

In addition to problems with custody and visitation, indications are strong that their fears caused some of these women to reduce their requests for child support. A statistically significant relationship was found between women's fears during negotiations for child support and for custody and their receipt of child support. Only 34% of women who reported being fearful during negotiations for custody, for example, received regular child support, in contrast with 60% of those who did not report fear during negotiations for child support. The reports of women who received no child support because they were fearful confirm this view. They spoke of being afraid that their ex-husbands would become physically violent if they tried to get child support, and a few mentioned being afraid that their ex-husbands would kidnap the children. Others reported that, because of their fear, they compromised and accepted a child support award that was lower than what they were entitled to. Still others gave up trying to obtain increases in their child support awards at a later time, when they were entitled to additional funds because their expenses for their children had increased.

The role of violence in divorce has serious ramifications for our divorce policies (Cahn, 1991). In the case of child custody after a divorce, the current recommended policy is automatic joint legal custody, which gives the noncustodial parent, typically the father, the right to participate in important decisions about his child's life, such as education and religious training, even if he does not have physical custody of the child (Fineman, 1995; Maccoby & Mnookin, 1992). The high level of violence toward women in marriages and the ongoing serious conflict between a significant minority of couples after marriage, however, make it imperative that guarantee of mandatory joint residential custody not be automatic. Further, the law must be ready to limit or deny access to fathers who use visitation for harassment or who engage in other threatening or dangerous behavior. Courts should consider evidence of abuse at every point in negotiations for custody and visitation. Further, because passing laws does not guarantee that judges will apply or enforce them, judges should be trained about domestic violence and its legal, sociological, and psychological implications (Arendell, 1995). Similarly, because some state legislators continue to promote mandatory joint physical custody, ignoring how much harm this could inflict on women and children, they too should be trained about violence against women.

As for child support policies, to increase fathers' financial participation, states have begun to require mothers to locate and identify the fathers of their children in order to obtain child support, a procedure called *paternity establishment* (Roberts, 1994). Requiring mothers to identify fathers in order to get child support is part of a general trend to try to make fathers take more financial responsibility for their children. Unfortunately, laws such as these can put an undue burden on mothers who have serious conflicts with their former partners, especially those women who have suffered physical abuse. Recent research has demonstrated that many women remain in danger of experiencing violence after a separation (Kurz, 1996).

In conclusion, we see that violence against women can play an important role in the divorce process; however, as noted earlier, this fact is not addressed in most of the divorce literature. The violence literature, in contrast, does document how women leave marriages and intimate relationships with male partners because of violence. This literature has not been integrated into the literature on divorce, however, or into that of many other fields. Although the rate of male violence toward women is high, and therefore it is logical to think that we would find a high rate of violence reported by divorced women, we are accustomed to thinking of violence as a distinct issue, separate from other important areas of study. We must look for evidence of violence against women in all aspects of our research and our social policies.

Women's Poverty

In the course of my research on mothers and divorce, I found a second arena where violence had a serious impact on the lives of women but where it had not been addressed—the study of poverty. As in the case of violence and divorce, violence and poverty are not usually studied together, and thus violence is invisible in the lives of poor women. Researchers and activists have emphasized the fact that women across a range of backgrounds and income levels have experienced male violence.[6] As is described in this section, however, increasing evidence indicates that poor women experience even higher rates of violence than other women.

In my study, poorer women experienced the most violence, with the women on welfare, the poorest women in the sample, experiencing more violence than any other group: 71% of these women reported experiencing violence at least two or three times during their marriages or separations, in contrast with the 50% average for the sample. In addition to experiencing the most violence, welfare women experienced more serious violence (defined as more than three incidents of violence, or one very serious incident of violence) than women of any other group: 58% of these women experienced serious violence, whereas 37% of the sample as a whole experienced serious violence.

Large-scale survey data also show that the poorer women are, the more likely they are to experience violence at the hands of their intimate partners. According to the U.S. Department of Justice's National Crime Victimization Survey (1995, p. 4), "women with an annual family income under $10,000 were more likely to report having experienced violence by an intimate than those with an income of $10,000 or more." Further, rates of violence at the hands of intimates decrease with each increase in income category. Thus, the rates of "violent victimizations" per 1,000 females age 12 or older for those with an annual family income of $9,999 or less was 19.9%; for those with annual family incomes of $10,000 to $14,999, 13.3%; $15,000 to $19,999, 10.9%; $20,000 to $29,999, 9.5%; $30,000 to $49,999, 5.4%; and $50,000 or more, 4.5%. In their nationwide survey of the use of violence by married couples, Straus and Gelles (1986) found that the lower a woman's education, income, and occupational levels, the more likely she is to be battered. Other evidence on the rates of violence experienced by poor women comes from recently publicized data on the high levels of violence experienced by women who have been on Aid to Families With Dependent Children (AFDC), or welfare. Several studies have reported that roughly two thirds of women on welfare have experienced violence in their lifetime (Davis & Kraham, 1995, p. 1145; McCormack Institute & Center for Survey Research, 1997; Raphael, 1996). In a study of victims of violence in Massachusetts, researchers found that 20% of current welfare recipients had been abused by former or current boyfriends or husbands within the past 12 months (McCormack Institute & Center for Survey Research, 1997).

What is the reason for the higher levels of violence reported by low-income women? Is it the case that poorer women are more forthcoming about the amount of violence they experience, or that more of them report the violence to the police because they have less access to other kinds of legal assistance? This is always a possibility, but at this point no data are available on this question. It is also possible that something about the circumstances of those living in poverty contributes to the higher rate of male violence. For example, men from lower-income groups may have a stronger belief in the legitimacy of violence than other men because they typically hold more traditional gender ideologies than other men. It is not clear, however, that lower-income men actually behave in more gendered ways than other men (Hochschild, 1989).

Another explanation for the higher rates of violence reported by poorer women could be that lower-income men have fewer ways of controlling their partners than other men do. The higher men's social class, the more ability they have to control their female partners through their greater economic resources. Many women in all social classes in my sample volunteered that their husbands tried to control decisions about family life and family finances. Interviewers did not specifically ask about this kind of behavior, but many women volunteered that they thought their husbands were controlling and that they did not like this. So, although according to women's volunteered statements, men of all different backgrounds were controlling, perhaps higher-income men, who control greater amounts of family income and property than poorer men do, believed that they did not need to resort to violence to control their female partners and to control family decisions.

Whatever the reason for the high rates of violence they experience, when poor women leave violent relationships, they often have few resources and experience particular hardships. Davis and Kraham (1995) point out that batterers commonly isolate their partners from financial resources. Many battered women do not have access to cash, checking accounts, or charge accounts (Lerman, 1984). When poor women do leave violent relationships, they have difficulty gaining access to secure employment, the most important ticket to economic well-being. When women leave relationships and seek job training, their batterers may continue to harass and abuse them. A study conducted by the Taylor Institute in Chicago demonstrated that violence can inhibit or prevent poor women from participating in welfare-to-work programs and from obtaining employment (Raphael, 1995, 1996). When women do gain employment, abusers may continue to harass them on the job (Zorza, 1991). Given these problems, it is not surprising that the rate of unemployment among battered women is higher than that among other women (Strube & Barbour, 1983). Homelessness is a particular problem for battered women, as evidenced by the fact that a significant portion of the homeless population is battered women (Zorza, 1991).

In my sample, abused women who were poor had a particular need for resources. Some had lost their homes after the divorce because they had no money to pay for mortgages, and in a few cases their ex-husbands kept the homes and the women were forced to leave. Some women had not been allowed to work while they lived with their batterers and so had no job skills or experience. As one woman said: "I was not allowed to go out. I wasn't really allowed to talk on the phone . . . . I wasn't allowed to have a job. I wasn't allowed to have friends." Almost none of these women received child support from their batterers. For many, welfare assistance was critical as they pulled their lives back together, got more education and training, and looked for work.

The data presented here have implications for both research and policy. They demonstrate that when we study any issue related to poverty, we must look for evidence of violence against women. These data also indicate serious problems with "welfare reform" as it has been recently enacted (Fineman, 1996). Elimination of the entitlement to welfare has made it much more difficult for women to live outside marriage. We must change this situation and provide all mothers, especially poor mothers, with more social welfare measures: day care, flexible working hours, health care, jobs with a decent minimum wage, health, pension and other benefits, child allowances, and basic income guarantees. We must also, of course, pass stricter enforcement measures to protect women against abusers.

The new federal welfare bill did stipulate that states may consider abused women to be exempt from some of the bill's requirements, such as the provision that women may receive assistance for only 5 years. Unfortunately, however, the federal government made this option only voluntary. Because states can exempt only 20% of eligible people from welfare, and because other people also deserve exemptions, many abused women will not be granted a reprieve from the requirements of the new welfare system (Vobejda, 1997). This is a particularly serious situation because, as noted above, so many women who qualify for welfare have been abused by husbands and boyfriends and may be at further risk of abuse.

In concluding, I raise one note of caution about focusing attention on poor women and violence. Drawing attention to the high levels of violence that poor women experience raises the possibility that some people will use this fact to further stigmatize poor women. The consequences of violence for poor women are so great, however, that we must focus our attention on this problem while being clear that levels of violence are unacceptably high for all women.

Conclusion

The data presented in this chapter have serious implications for both research and policy. Researchers must develop new paradigms that are not based on gender-neutral concepts (e.g., family violence) but that take violence against women as their frame of reference, as well as other types of violence against women, including rape and sexual harassment. Similarly, researchers must develop new instruments and measures that specify exactly who initiated the violence in intimate relationships, how serious the acts of violence were, and what the consequences were.

Researchers must also integrate the study of violence into other areas of research. As demonstrated in this chapter, despite greater recognition of issues of violence against women, many researchers still compartmentalize this issue and still think of violence as an activity or problem distinct from other issues that affect women. The result is that we fail to see how significant a role violence can play in women's lives. The examples given here of the role of violence in the divorce process and in the lives of poor women show how important it is to look for violence in all areas of social life. Data presented here demonstrate that male violence can deprive divorced women and poor women of resources and can make their lives even more difficult. Policymakers must understand the extent to which violence affects women in all areas of social life and develop more wide-reaching policies to reduce the incidence of this pervasive problem.

NOTES

1. For further debates on the family violence perspective and the violence against women perspective, see Straus (1993) and Kurz (1993).
2. The sample included 61% white women, 35% black women, and 3% Hispanic women. Interviews lasted from 1½ to 3 hours and included both open-ended and fixed-choice questions. Women were also asked questions about violence. A further discussion of the methodology is available in Kurz (1995).
3. In addition, 4% of women experienced violence during the separation only, and an additional 13% experienced violence during both the marriage and the separation. The rate of violence that women experienced during the separation was probably even higher, but one question on the survey was worded in such a way that an accurate determination of the number of these women could not be made. The levels of violence that women experienced were determined by using a modified version of the Conflict Tactics Scales (Straus, 1979).
4. In a study of 362 separating husbands and wives, Ellis and Stuckless (1993) reported that over 40% of separating wives and 17% of separating husbands stated they were injured by their partners at some time during the relationship. Studies by Fields (1978) and by Parker and Schumacher (1977) found that between 50% and 70% of divorcing wives reported being assaulted by their husbands at least once during their marriages. According to Schulman (1979), two thirds of divorced women in a Harris poll reported violence in their former relationships.
5. Altogether, 25% of women in the sample reported conflict over visitation. The 25% figure includes women who reported no visitation at all. For those women who reported visitation, 29% reported conflict.
6. For example, some of the early and most widely read texts on battering and family violence never mention social class issues. See Dobash and Dobash (1979), Pagelow (1984), and Straus et al. (1980).

REFERENCES

Adams, D. (1988). Treatment models of men who batter. In K. Yllö & M. Bograd (Eds.), *Feminist perspectives on wife abuse* (pp. 176–199). Newbury Park, CA: Sage.

Ahrons, C., & Rodgers, R. (1987). *Divorced families: A multidisciplinary developmental view.* New York: Norton.

Arendell, T. (1995). *Fathers and divorce.* Newbury Park, CA: Sage.

Arias, I., Samios, M., & O'Leary, K. D. (1987). Prevalence and correlates of physical aggression during courtship. *Journal of Interpersonal Violence, 2,* 82–90.

Bart, P., & Moran, E. (Eds.). (1993). *Violence against women.* Newbury Park, CA: Sage.

Berk, R., Berk, S. F., Loseke, D., & Rauma, D. (1983). Mutual combat and other family violence myths. In D. Finkelhor, R. Gelles, Hotaling, & M. Straus (Eds.), *The dark side of families: Current family violence research* (pp. 197–212). Beverly Hills, CA: Sage.

Breines, W., & Gordon, L. (1983). The new scholarship on family violence. Signs: *Journal of Women in Culture and Society, 8,* 490–531.

Brinkerhoff, M., & Lupri, E. (1988). Interspousal violence. *Canadian Journal of Sociology, 13,* 407–434.

Brush, L. D. (1990). Violent acts and injurious outcomes in married couples: Methodological issues in the National Survey of Families and Households. *Gender & Society, 4,* 56–67.

Cahn, N. (1991). Civil images of battered women: The impact of domestic violence on child custody decisions. *Vanderbilt Law Review, 44,* 1041–1097.

Davis, M. F., & Kraham, S. J. (1995). Protecting women's welfare in the face of violence. *Fordham Urban Law Journal, 22*(4), 1141–1157.

DeKeseredy, W. K. (1995). Enhancing the quality of survey data on woman abuse: Examples from a Canadian study. *Violence Against Women, 1*(2), 158–173.

DeMaris, A. (1987). The efficacy of a spousal abuse model in accounting for courtship violence. *Journal of Family Issues, 8,* 291–305.

Dobash, R., Dobash, R. E., Wilson, M., & Daly, M. (1992). The myth of sexual symmetry in marital violence. *Social Problems, 39*(1), 71–91.

Dobash, R. E., & Dobash, R. (1979). *Violence against wives.* New York: Free Press.

Dobash, R. E., & Dobash, R. (1992). *Women, violence, and social change.* London: Routledge & Kegan Paul.

Ellis, D., & Stuckless, N. (1993). *Hitting and splitting: Predatory preseparation abuse among separating spouses* (Mediation Pilot Project Report #7). Submitted to the Attorney General of Ontario.

Emery, R. (1988). *Marriage, divorce, and children's adjustment.* Newbury Park, CA: Sage.

Emery, R., Lloyd, S., & Castleton, A. (1989). *Why women hit: A feminist perspective.* Paper presented at the Annual Conference of the National Conference on Family Relations, New Orleans, LA.

Fields, M. D. (1978). Wife-beating: Facts and figures. *Victimology, 2*(3–4), 643–647.

Fineman, M. (1995). *The neutered mother, the sexual family, and other 20th-century tragedies.* New York: Routledge.

Fineman, M. (1996). The nature of dependencies and welfare "reform." *Santa Clara Law Review, 36,* 1401–1425.

Furstenberg, F. F., Jr., & Cherlin, A. (1991). *Divided families.* Cambridge MA: Harvard University Press.

Gelles, R. (1993). Through a sociological lens: Social structure and family violence. In R. Gelles & D. Loseke (Eds.), *Current controversies on family violence.* Newbury Park, CA: Sage.

Gelles, R., & Cornell, C. (1985). *Intimate violence in families.* Beverly Hills, CA: Sage.

Gelles, R., & Straus, M. (1988). *Intimate violence.* New York: Simon & Schuster.

Hochschild, A. (1989). *The second shift.* New York: Viking.

Jaffe, P., Wolfe, D., & Wilson, S. K. (1990). *Children of battered women* (Developmental Clinical Psychology and Psychiatry, Vol. 21). Newbury Park, CA: Sage.

Jones, A. (1994). *Next time, she'll be dead: Battering and how to stop it.* Boston: Beacon.

Kurz, D. (1993). Physical assaults by husbands: A major social problem. In R. Gelles & D. Loseke (Eds.), *Current controversies on family violence* (pp. 88–103). Newbury Park, CA: Sage.

Kurz, D. (1995). *For richer, for poorer: Mothers confront divorce.* New York: Routledge.

Kurz, D. (1996). Separation, divorce, and woman abuse. *Violence Against Women, 2*(1), 63–81.

Kurz, D. (1997). Violence against women or family violence? Current debates and future directions. In L. L. O'Toole & J. R. Schiffman (Eds.), *Gender violence: Interdisciplinary perspectives.* New York: New York University Press.

Lamb, S. (1991). Acts without agents: An analysis of linguistic avoidance in journal articles on men who batter women. *American Journal of Orthopsychiatry, 61*(2), 250–257.

Lamb, S. (1995). Blaming the perpetrator: Language that distorts reality in newspaper articles on men battering women. *Psychology of Women Quarterly, 19,* 209–220.

Lane, K. E., & Gwartney-Gibbs, P. A. (1985). Violence in the context of dating and sex. *Journal of Family Issues, 6,* 45–59.

Lerman, L. G. (1984). Model state act: Remedies for domestic abuse. *Harvard Journal on Legislation, 21,* 61, 90.

Lewin, T. (1992, April 20). Battered men sounding equal-rights battle cry. *New York Times,* p. 12.

Loseke, D. (1992). *The battered woman and shelters: The social construction of wife abuse.* New York: State University of New York Press.

Maccoby, E. E., & Mnookin, R. H. (1992). *Dividing the child: Social and legal dilemmas of custody.* Cambridge, MA: Harvard University Press.

McCormack Institute & Center for Survey Research. (1997). *In harm's way? Domestic violence, AFDC receipt, and welfare reform in Massachusetts.* Boston: University of Massachusetts.

McNeely, R. L., & Mann, C. (1990). Domestic violence is a human issue. *Journal of Interpersonal Violence, 5,* 129–132.

McNeely, R. L., & Robinson-Simpson, G. (1987). The truth about domestic violence: A falsely framed issue. *Social Work, 32,* 485–490.

Nisonoff, L., & Bitman, I. (1979). Spousal abuse: Incidence and relationship to selected demographic variables. *Victimology, 4,* 131–140.

Okun, L. (1986). *Woman abuse.* Albany: State University of New York Press.

Pagelow, M. D. (1984). *Family violence.* New York: Praeger.

Parker, B., & Schumacher, D. (1977). The battered wife syndrome and violence in the nuclear family of origin: A controlled pilot study. *American Journal of Public Health, 67*(8), 760–761.

Pleck, E., Pleck, J. H., Grossman, M., & Bart, P. (1977–78). The battered data syndrome: A comment on Steinmetz' article. *Victimology, 2,* 680–684.

Price, S. J., & McKenny, P. (1988). *Divorce.* Newbury Park, CA: Sage.

Raphael, J. (1995). *Domestic violence: Telling the untold welfare-to-work story.* Chicago: Taylor Institute.

Raphael, J. (1996). *Prisoners of abuse: Domestic violence and welfare receipt.* Chicago: Taylor Institute.

Roberts, P. G. (1994). Child support orders: Problems with enforcement. *The future of children: Children and divorce, 4*(1), 101–120.

Saunders, D. (1988). Wife abuse, husband abuse, or mutual combat? In K. Yllö & M. Bograd (Eds.), *Feminist perspectives on wife abuse* (pp. 90–113). Newbury Park, CA: Sage.

Saunders, D. (1989). *Who hits first, and who hurts most? Evidence for the greater victimization of women in intimate relationships.* Paper presented at the American Society of Criminology, Reno.

Schulman, M. (1979). *A survey of spousal violence against women in Kentucky* (Study #792701 for the Kentucky Commission on Women). Washington, DC: U.S. Department of Justice.

Shupe, A., Stacey, W., & Hazelwood, R. (1987). *Violent men, violent couples: The dynamics of domestic violence.* Lexington, MA: Lexington Books.

Silent victims: Children who witness violence. (1993). *Journal of the American Medical Association, 269*(2), 262–264.

Stark, E., & Flitcraft, A. (1985). Woman battering, child abuse, and social heredity: What is the relationship? In N. Johnson (Ed.), *Marital violence* (pp. 147–171). Boston: Routledge & Kegan Paul.

Stark, E., & Flitcraft, A. (1996). *Women at risk: Domestic violence and women's health.* Thousand Oaks, CA: Sage.

Steinmetz, S. (1987). Family violence: Past, present, and future. In M. Sussman & S. Steinmetz (Eds.), *Handbook of marriage and the family* (pp. 725–765). New York: Plenum.

Steinmetz, S., & Lucca, J. (1988). Husband battering. In V. Van Hasselt, R. Morrison, A. Bellack, & M. Hersen (Eds.), *Handbook of family violence* (pp. 233–246). New York: Plenum.

Stets, J. (1990). Verbal and physical aggression in marriage. *Journal of Marriage and the Family, 52,* 501–514.

Straus, M. (1979). Measuring intrafamily conflict and violence: The Conflict Tactics (CT) Scales. *Journal of Marriage and the Family, 41,* 75–88.

Straus, M. (1993). Physical assaults by wives: A major social problem. In R. Gelles & D. Loseke (Eds.), *Current controversies on family violence* (pp. 67–87). Newbury Park, CA: Sage.

Straus, M., & Gelles, R. (1986). Societal change and change in family violence from 1975 to 1985 as revealed by two national surveys. *Journal of Marriage and the Family, 48,* 465–479.

Straus, M., & Gelles, R. (1990). How violent are American families: Estimates for the National Family Violence Resurvey and other studies. In M. Straus & R. Gelles (Eds.), *Physical violence in American families* (pp. 95–112). New Brunswick, NJ: Transaction Press.

Straus, M., Gelles, R., & Steinmetz, S. (1980). *Behind closed doors: Violence in the American family.* Garden City, NY: Doubleday.

Strube, M. J., & Barbour, L. S. (1983). The decision to leave an abusive relationship: Economic and psychological commitment. *Journal of Marriage and the Family, 45,* 785–793.

U.S. Department of Justice, Bureau of Justice Statistics. (1995, August). *Special report, National Crime Victimization Survey, violence against women: Estimates from the redesigned survey.* Washington, DC: Author.

Vobejda, B. (1997, April 12). Welfare waiver for abused women may cost states. *Washington Post,* p. A10.

Wardell, L., Gillespie, C., & Leffler, A. (1983). Science and violence against women. In D. Finkelhor, R. Gelles, H. Hotaling, & M. Straus. (Eds.), *The dark side of families: Current family violence research* (pp. 69–84). Beverly Hills, CA: Sage.

Yllö, K. (1993). Through a feminist lens: Gender, power, and violence. In R. Gelles & D. Loseke (Eds.), *Current controversies on family violence.* Newbury Park, CA: Sage.

Zorza, J. (1991). Woman battering: A major cause of homelessness. *Clearinghouse Review, 25,* 421.

CHAPTER 6

Questions for Writing, Reflection, and Debate

READING 27

Paddy Clarke Ha, Ha, Ha • *Roddy Doyle*

1. In *Paddy Clarke Ha Ha Ha,* we participate in a young child's realization that his father has physically abused his mother. What are the boy's emotions revealed in this story? Do you think this is a typical reaction for a child? How are the father and mother acting? Do you think their behavior is typical?

READING 28

Violent Men or Violent Women? Whose Definition Counts? • *Dawn H. Currie*

2. Recent research has made much of the fact that partner abuse is a problem for both sexes. Many of these studies are based on self-report questionnaires. What examples does Dawn Currie offer to suggest that these statistics are misleading?
3. In what other ways might research on family violence be affected by how people define violence? Can you think of any ways to eliminate this methodological problem?

READING 29

Behind Closed Doors: Domestic Violence • *Steve Friess*

4. "Behind Closed Doors" reminds us that partner abuse is not limited to heterosexual couples. What similarities and differences in partner abuse exist for heterosexual, lesbian, and gay male couples?
5. This reading provides some firsthand accounts of interactions of both perpetrators and victims with social institutions such as the police, shelters, and support groups. What problems do homosexuals confront when interacting with elements of the social structure that are more used to dealing with violence in heterosexual couples?
6. Why is there some resistance within the homosexual community to "going public" with the problem of partner abuse?

READING 30

Violence Against Women and Girls: The Intolerable Status Quo • *Charlotte Bunch*

7. Why do so many cultures condone violence against women?
8. How do elements of culture such as art, rituals, values, traditions, and myths encourage or discourage partner abuse?
9. How does society lose out when women are kept in submission to men? How does it gain?

READING 31

Through a Sociological Lens: Social Structure and Family Violence • *Richard J. Gelles*

10. How do psychological and sociological explanations of family violence differ? Does the fact that partner abuse is prevalent in a range of age, class, and racial groups negate the value of sociological explanations?

11. Which of the sociological theories presented here seems the most useful in understanding partner abuse?
12. How can culture work against treatment programs based on a psychological model?

READING 32

The Powers of the Weak: Wife-Beating and Battered Women's Resistance • *Linda Gordon*

13. "The Powers of the Weak" offers historical evidence of the role of individual action in altering the structural context in which abuse takes place. Why is it important that women claim a right not to be beaten?
14. How does the description of victims' resistance in this reading influence your interpretation of the data regarding women's violence in Reading 28, "Violent Men or Violent Women?"

READING 33

The Women of Belize • *Irma McClaurin*

15. How does the issue of partner abuse in developing nations differ from that in industrialized nations?
16. How does this reading help you understand the importance of women's collective action in promoting change? What can society do to facilitate such actions?

READING 34

Old Problems and New Directions in the Study of Violence Against Women • *Demie Kurz*

17. Kurz offers examples of how policy in one area of family life, such as divorce, can spill over into other areas of family life. How does this idea relate to systems theory as described in Reading 31, "Through a Sociological Lens"? Is there any way to prevent this spillover effect?

7

Divorce and Remarriage

What can be done to reduce the financial, emotional, and social costs of divorce?

In the 1950s, I was a kid. Those of you who spend nights watching reruns of old TV situation comedies will know what I mean when I say it was the time of *Father Knows Best, My Three Sons, The Donna Reed Show, Ozzie and Harriet,* and countless other examples of media families that represented the "All-American Ideal" in family life. There is much debate today over how much various aspects of the lifestyle portrayed, in these shows, represented reality. TV has never been known for its accurate representation of life; it usually capitalizes on ideals. As I watched these shows, even my immature mind understood that the situations and solutions these shows conveyed bore little resemblance to real family life. But I never questioned the reality of the family structure on which these series were based. As I looked around the outer ring, urban track, neighborhood in which my family lived, and the families of the children with whom I went to school, there were few, if any, families affected by divorce. There were no children in my school, neighborhood, or family that disappeared on the weekend or in the summer to spend time with a noncustodial parent. But by the time I started teaching in the 1970s, things were very different. In a few decades divorce had gone from a rare experience to something that touched the lives of many of my neighbors, students, and family. And this trend was as pervasive in the large major urban area I'd left as in the small rural community where I was hired.

In the "Family Life" courses I taught at this time, we focused on the good aspects of the changes in divorce laws that had improved the accessibility of divorce, allowing couples who had lived in unsatisfactory relationships for years to move on to a more fulfilling life. The phenomenon was so new that researchers hadn't had time to investigate long-term consequences of divorce. The work of Diane Vaughan relayed here in "Uncoupling: The Breakdown of the Cover-Up" provides great insight into the process many couples go through in deciding to dissolve marriages under the new more liberal divorce laws. These are very personal accounts that reveal the patterns of interactions in a disintegrating relationship. As you can see from these accounts, the initiator's experience is very different from that of the partner.

Soon researchers began to question the effects of divorce on individual well-being. It was too early to investigate long-term implications, and many of these studies focused on negative emotional impact on children's lives. Research on the effects of divorce on children's emotional development still offers mixed findings. Improvements in well-being seem most common in children whose intact families were violent. Some studies indicate that early effects of divorce disappear over time. These studies are complicated by considering the degree of problems children had before the divorce, and the impact of the custodial parent's repartnering. Researchers have not put aside the question of the impact of divorce on children, but they have added several new concerns, which I have used to focus the readings in this chapter.

These new studies ask more sociological questions regarding the financial consequences of divorce and the role of policy in alleviating the poverty that affects women and children after divorce. The reading by Frank Furstenberg and Andrew Cherlin, "The Economic Consequences of Divorce," reveals the stark facts of the unequal financial impact of divorce on men and women and children. This reading also gives us insight into the role of divorce, alimony, and child support laws in contributing to these inequities.

The work of Terry Arendell, discussed in "After Divorce: Investigations into Father Absence," reveals the anger and need for control that most noncustodial fathers experience after divorce. As you read this piece, note that fathers withhold money as a way to exert power over ex-wives. In addition, they justify abdicating financial and emotional responsibilities to children on the basis of feeling that they have been marginalized in their children's lives.

Some researchers and policymakers are promoting legislation aimed at redistributing the financial burden of divorce. Current initiatives range from restricting access to divorce to publishing the pictures and identification of parents who have reneged on paying child support. Some recent policy initiatives are aimed at returning to the times when divorce was more difficult to obtain, with the intent of forcing people to work harder at making marriage work. This position rests on the assumption that families are better off struggling with their circumstances than parting from them. These advocates promote using public policy for a version of "social engineering," which raises two interesting questions. In the short term, can society alter the financial impact of divorce, and in the long term, can we institute laws that will reverse the trend of divorce?

In "Divorcing Reality," Stephanie Coontz proposes that such reactionary policies are motivated by "pop psychology" reported in the media rather than sound research. Her article takes a strong position against retracting more liberalized divorce laws, because women who have the fewest resources or are in abusive relationships would suffer most from such a change. She finds more value in attempts to coerce noncustodial parents into meeting financial obligations.

It is common for custodial parents to cohabitate or remarry, and biological parents who fade out of children's lives seem to do so expecting that someone else has taken on their emotional and financial commitment to their children. Lynn White offers several theoretical explanations for why someone else is not likely to have done so, in her article "Stepfamilies over the Life Course: Social Support." White suggests that the bonds in stepfamilies are not usually as strong as those in biological families, but this should not be interpreted as meaning that stepfamilies have no benefit for the well-being of children whose biological parents have divorced. Mary Ann Mason and Jane Mauldon's research, relayed in "The New Stepfamily Requires a New Public Policy," shows that there are significant financial benefits for children in stepfamilies. In addition, they note that stepparents provide financial support even in the absence of legislation that requires them to do so, whereas biological parents required to provide financial support often do not. This raises interesting questions regarding the power of policy to ensure that parents meet their obligations to children. These authors suggest granting stepparents a stronger legal position in recognition of, and in hopes of promoting, their involvement in their stepchildren's financial and emotional well-being.

When I return to that area of the Midwest where I grew up, I find that divorce has touched the lives of many family, friends, and neighbors. At a personal level the emotional and financial toll of divorce is evident, but I also see that, in time, many of these families reestablish themselves with new partners and new beginnings. I am not an advocate of social engineering through family policy, but I do think it's time to reconsider the financial and emotional impact of current trends in divorce and remarriage legislation. This chapter challenges you to understand the social implications of these trends and to consider creative solutions for the future.

Uncoupling: The Breakdown of the Cover-Up

Diane Vaughan

The collaborative cover-up [of problems in a relationship] eventually breaks down as a result of interaction between the two partners which is so subtle, so complex, so volatile, so dynamic that using words to describe it imposes an order contradicted by reality.[1] The initiator [of the breakup] controls this moment, for the partner will not admit the relationship is in serious trouble until the initiator displays discontent with sufficient clarity and force so that the partner can no longer avoid this conclusion. To get the partner's attention, the initiator increases the frequency, intensity, and visibility of signals previously unnoticed—or else interjects new ones sure to do the job.

Direct Confrontation

One new signal is a direct confrontation in which the initiator reveals secrets in full, clear, and painful detail.[2]

> The violent part was all during a two week period. Actually, it had been the month before that we stopped sleeping together. He never really came out and said he no longer loved me till the violence broke out. He would scream out, "I can't stand you. You bore me." [FLORAL DESIGNER, AGE 28, SEPARATED AFTER LIVING TOGETHER 4 YEARS]

> My then-husband came home for dinner. He had been acting very quiet for weeks and I had been asking him what was wrong. He had a very responsible job and was under a lot of pressure and I assumed that it was because of his work that he was just so quiet, and he kept telling me it was just work. Well, he came home for dinner with some kind of downtrodden look on his face and I said to him, "What is wrong?" We were all assembled at the dinner table and he asked me to come upstairs, said that he just couldn't stand it any longer. So I left the children and went up into the bedroom and he closed the door and he told me he loved somebody else. Just whammo. So he was very confused. What was bothering him was that he really thought that he loved me, too. He just didn't know what to make of it. So I was trembling all over. He said then he wanted to tell me and he wanted to go to her right then. [TEACHER, AGE 35, DIVORCED AFTER 11 YEARS]

...

The initiator directly confronts the partner not only with negative feelings, but with the wish to end the relationship. Taken together, these two messages are so powerful that the partner is forced to alter the frame of reference that she or he has held on to so dearly—though by attempting to break the news gently, the initiator sometimes still clouds the issue by giving mixed signals even now.... ("So he was very confused. What was bothering him was that he really thought he loved this other person, but he thought he loved me, too"). The initiator who follows up this speech by simply and spontaneously walking out the door or by insisting that the partner leave gives a powerful signal, however.

Initiators directly confront the partner only when absolutely certain about their feelings....

Many speak of a precise moment when they "knew the relationship was over," when "everything went dead inside," when "I walked through this house and felt none of it was mine," when "I realized I didn't belong there anymore."[3] The partnership is clearly incompatible with the initiator's own sense of self, life, and values. Being in the relationship now unambiguously *detracts from and stands in opposition to* the person the initiator has become. Initiators experiencing this moment of certainty sometimes say that it is accompanied by the feeling of being a guest or a stranger in familiar surroundings. Others remember it as a moment when it was not themselves, but the partner who seemed to be the stranger....

The initiator's certainty results not only from the recognition that he or she no longer belongs in the relationship, but also from a sense of belonging elsewhere. Self-validation is now coming from other sources. The relationship has been replaced with something else that confirms the initiator's sense of self, and this resource now provides stability—the sense of being "at home." The initiator finds the differences between worlds so clear and so great that not only is no remedy possible, but continuing in the relationship is out of the question.[4] Although the initiator has experienced this conflict between individual identity and the identity bestowed by the relationship all along, these feelings are now sharply defined and keenly felt. More important, the initiator identifies and articulates this conflict as the source of unhappiness.

Often the initiator's moment of certainty occurs after the initiator has had an intense experience elsewhere, then rejoins the world shared with the partner. Perhaps the initiator returns from a reunion with family or old friends, a business meeting in another city, a stolen weekend with a lover, or from a satisfying time alone. The juxtaposition of the two worlds creates a "re-entry shock," as the initiator experiences the home environment with the other experience freshly in mind....

Once convinced of feelings about the partnership, initiators no longer dwell on the question of whether or not to go. Instead, they begin to think about *how* to go.[5] Uncoupling has many costs. Many people remain in unhappy relationships because of their unwillingness to suffer the economic, emotional, and social costs of leaving: loneliness, disruption, a decreased standard of living, the loss of other relationships, the misery of the partner, the astonishment and anger of parents or in-laws, the sorrow of children, the condemnation of the Church, the holier-than-thou attitude of friends who "have had trouble too, but we stuck it out." Initiators who immediately confront the partner with their wish to terminate the relationship do so because they believe they can manage these other costs of the transition.

Planning

Not all who experience certainty about the partner act immediately, however. Unable or unwilling to endure the costs of leaving, they plan instead to confront the partner when some future event occurs that decreases these costs: "after Tommy starts school"; "when I get the raise"; "when Dad dies"; "when I finish the degree"; "when the kids are gone"; "when I get a job"; "when his health improves." I am reminded of an old story about a couple in their eighties who filed for divorce. The judge questioned the partners extensively about their mutual wish to end their relationship so late in life. Finally convinced that they both were deeply unhappy and had been so for decades, he asked, "Why have you taken so long to do this?" They responded, "We were waiting for the children to die."

For some initiators, the plan to end the relationship in the future may remain a fantasy, a rehearsal for something that never happens. This fantasy may even become the alternative that allows them to continue in the relationship. Fantasies about leavetaking are not restricted to those who are uncoupling or wishing to uncouple. The happily coupled also indulge. A miserable experience with the other person may trigger images of being somewhere else with someone else—or even alone. But it remains a fleeting notion, unacted on. For those whose discontent is longstanding however, the fantasy is more than a passing thought. It is regularly incorporated into the initiator's life, and evokes thoughts about leaving that assume an organized, means–ends character, eventually leading to action. Planning becomes serious business and initiators become very efficient about it.[6] They consider the financial implications of a separation.[7]

> I had started to think about living someplace else on my own. The thought came into my head—financially, how do I do it. I would find myself going to work.... I did my thinking while driving, and adding up column A and column B and seeing if I could make it, you know? And if I couldn't have financially done it then, I probably would have stayed on until I could. But I had made up my mind, yeah, prior to leaving, and when I did that was one of my thoughts. [POTTER, AGE 32, SEPARATED AFTER LIVING TOGETHER 9 YEARS]

...

Some initiators' plans include preparing the partner to live alone. Initiators can decrease the cost of leaving by decreasing the partner's social, emotional, and financial dependence on the relationship. Consequently, initiators sometimes encourage partners toward alternative resources of their own. Earlier in the transition, initiators may have urged partners in these pursuits to make them more attractive and thus improve the relationship. Now the urgings have a different thrust....

> When we first become lovers, we stopped doing the bar scene. It is such a threat. Everyone is looking and it's almost like a game there to try to take a lover from someone else, so we just didn't go. People are so desperate for meaningful relationships, you know. After I started seeing Paul, I wanted to go again because Scotty is a great dancer and he always attracts a lot of attention, so like I thought if we started going to the bar again, Scotty would meet someone and then I could be with Paul. [ACCOUNTANT, AGE 29, SEPARATED AFTER LIVING TOGETHER 3 YEARS]

...

As the planned date of confronting the partner approaches, initiators may turn their attention to tying up the loose ends of living with the other person. They may announce their departure after completing some commitment the partners have made together, after they have the house in good repair, or even after a thorough cleaning. Not only do they get the house in order, but also themselves and their belongings: having shoes or car repaired, visiting doctors or dentists for check-ups, getting braces on the kids, sorting through clothes, documents, or keepsakes. They suggest to a friend that they may suddenly need to move in for a few days. Some look for an apartment, or even rent one.

The initiator's planning often culminates in a precise moment when, according to schedule, resources at hand, and with a well-rehearsed speech, the initiator confronts the partner about wanting to end the relationship....

> She had a good friend coming to stay with us for the weekend, so I thought that would be a good time to tell her. I told my sister and made arrangements to stay with her until I could find a place of my own. I made dinner that night and was so nervous I was nauseous. The minutes dragged. It was awful. I kept thinking, "This is it. If I don't do it now, the timing will probably never be so right again." Also, I had told my sister, so that meant I really had to go through with it. We had dinner. We talked about our days. We cleaned up. After dinner, it was her habit to have coffee and read the paper. I sat down beside her. I said, "You know that I haven't been truly happy for some time now." She put her coffee down and looked at me. I went on. I got through it. [SECRETARY, AGE 26, SEPARATED AFTER LIVING TOGETHER 5 YEARS]

Changing Plans

Life, however, usually does not allow us to be quite this orderly. The initiator's carefully thought-out plan is often changed because of some unforeseen event that shifts the costs and benefits of leaving or staying. The leavetaking can be delayed again and again, or instead occur precipitously. The initiator may alter the scheduling of the announcement because, unexpectedly, the cost of staying in the relationship dramatically increases. The initiator loses an important resource and in its absence, tension between the partners escalates. Perhaps a close friend moves away, a project is completed, a love affair dissolves, or a parent dies. With the resource gone, the relationship seems unbearably bleak and, in fact, intolerable. Then, disregarding inadequate preparation, the initiator directly confronts the partner before the planned date.[8]

Usually, the initiator not only drew self-validation from the now-absent resource, but also gave it time and energy. As a consequence, the initiator spent less time with the partner. With the loss of the alternative (unless the initiator quickly finds a substitute), interaction between initiator and partner increases, exposing incompatibility and generating conflict. A change that increases the initiator's leisure time can contribute to the decision to confront in another way. Social disruption is one of the costs of uncoupling. The initiator may have delayed bringing matters to a head before because work or other commitments did not leave time for the predicted disorganization that would follow. The alternative commitment absorbed time, and the initiator now has time to spare.

Sometimes it's not the real loss of some resource but the *threat* of loss and the *potential* of increased interaction with the partner that causes the initiator to confront earlier than planned....

> We had been seeing each other for some time, when all of a sudden this man moved in next door to her. They started in right away being friends, and I thought nothing of it. Our plan was to marry when my kids were grown, at that time it was three years away, then I would divorce Madge. And she seemed comfortable with that idea. But then the shock came. She started talking about how her kids needed a father and one night I was over there and she said Bert (that's his name) had offered to come across with a ring right then. I moved fast. I went out and got a ring of my own and took it over there, then told Madge I wanted out. [DIVISION MANAGER, AGE 54, DIVORCED AFTER 19 YEARS]

The cost of staying also may increase because of events affecting the partner's life. Perhaps the partner begins some transition that promises to change the couple's lifestyle in a way that is not in the initiator's best interest.[9] The partner makes a choice—or contemplates making a choice—and the initiator acts....

Perhaps the partner experiences some loss: becomes seriously ill, loses a job or a close friend. As a consequence, the partner's dependence on the initiator increases. Initiator and partner spend more time together with the result that the couple's incompatibilities become more obvious, causing the initiator to act before the scheduled confrontation.

The initiator may confront the partner earlier than planned because an unexpected event makes life apart from the partner seem more appealing and manageable. Consequently, the benefits of leaving increase. Although accumulating resources all along, the initiator discovers that some critical gap is suddenly filled, providing the catalyst for departure. Some find a transitional person. Some have a new opportunity in their job or a sudden financial windfall. Others describe a brief romance that did not offer a promise of anything permanent, but gave them the courage to confront because of increased self-confidence, or the suggestion that something better might be out there waiting for them.

Many initiators are prepared to cope with all the costs of leaving except one: the social consequences of being responsible for the break-up. How will the partner respond? Threats, crying and pleading, violence, suicide? If there are children, how will they react? Will they be so hurt and angry that the initiator can only end the relationship at the cost of losing the children's affection? And what about friends and relatives? While the potential negative repercussions can never be completely avoided, initiators often try to reduce them....

Indirect Methods: Shifting the Burden

The partner responds to the increasingly visible negative signals by redefining the relationship as troubled. A confrontation results, but not because the initiator takes responsibility for it and directly reveals wishes and feelings. It happens because the initiator displays discontent so that the responsibility for the confrontation is shifted to the partner. Confrontation has the potential to change the course of the relationship in unpredictable ways. By transferring the blame for it, the initiator also transfers potential negative consequences to the other person. These indirect methods are not new, by the way. They have been part and parcel of the initiator's expression of unhappiness all along. Though they do not necessarily occur separately, I will distinguish them.

Fatal Mistake

The partner responds to the initiator's escalating display of discontent by committing some grievous error. The initiator seizes on it, confronts the partner with the wish to terminate, and points to the partner's behavior as the reason for all the ills in the relationship. The initiator confronts, but the partner's failure is the reason the relationship has come to this crisis. The partner admits to this failing and concedes its consequences. Not only does the partner redefine the relationship as troubled, but the partner assumes the responsibility for the problems—and the confrontation. The partner's fatal mistake can occur several ways. A partner may respond to the initiator's discontent with unprecedented reactions that reinforce the initiator's negative opinions:[10] creating angry scenes, becoming demanding and interrogative, withdrawing from the initiator, bestowing suffocating attention, becoming physically abusive, refusing or insisting on sex, exhibiting hysteria, destroying furniture or belongings, or taking a lover. With this evidence of discrediting behavior at hand, the initiator confronts the partner: how could the initiator possibly continue living with someone who behaves in this manner?

The partner can make another kind of fatal mistake. He or she fails by virtue of some longstanding characteristic. Previously a source of minor complaints, the irritating trait now becomes a major stumbling block, for as tensions increase between the couple, the objectionable characteristic has a tendency to occur despite the partner's best efforts to control it. Not only does it occur, but it occurs more often and is more exaggerated. Suppose the initiator has always objected to the partner's drinking. The partner tries to control it, but in response to mounting tension and personal unhappiness goes on a binge. This episode becomes the last straw. Or suppose the initiator is disturbed by the partner's emotional highs and lows. In a discussion in which the initiator lists all the ways in which the partner fails, the partner seethes with rage, then becomes hysterical. The initiator scores the point. Or suppose the initiator complains that the partner is an unsatisfactory sex partner. Under pressure to perform, the partner becomes frigid or impotent. The initiator points out the flaw. The partner makes the fatal mistake of displaying the characteristic one more time. The initiator then confronts the partner, justifying the wish to terminate on the basis of the partner's obviously unremediable failing—in essence playing a trump card held all along.

The partner's fatal mistake may be an error of omission, rather than of commission. The initiator asks something of the partner that the initiator knows in advance the partner cannot do. The idea of making someone an offer he or she can't refuse is reversed: the initiator makes the partner an offer she or he can't accept. For example, in one relationship where the partner valued monogamy, the initiator suggested that the couple's relationship be expanded to include other sex partners.[11] The partner found this solution impossible and could not agree to the arrangement. In another, the partner was a woman who could not bear children. The initiator complained that her infertility (formerly viewed as beneficial) was a source of unhappiness. The partner conceded her inability to meet the other person's needs and the initiator made his point.

Decreased Interaction

A second indirect method initiators use to force partners to redefine the relationship as seriously troubled is to gradually decrease time spent together. Initiators are likely both to withdraw psychologically while in the home and to disappear more and more into the outside world. They've been doing this all along, of course, but as their unhappiness increases, their absences become more prolonged. Initiators may do this in order to provide time out from the relationship and its consequences (conflict, boredom, displeasure, fear) or to pursue more pleasurable activities elsewhere.[12]

Initiators may even use short separations as a test of their own ability to operate independently. They may try a weekend away or a vacation by themselves. They may visit relatives and friends: one week increases to two, and two increases to six. They may take a job, work late, or extend their other activities so that the couple's time together is reduced. If presented with the option, initiators may increase the amount of traveling they do on the job, stay overnight with a friend in the city occasionally, or the more affluent may even take an apartment in town. Sometimes the separation is of even longer duration. The initiator may take a job-training course in another city, sign up for the armed services, or engage in a lengthy period of education at a distant school.

If you want to see what a man loves, Oscar Wilde notes, observe how he spends his time.[13] Whatever the initiator gives as the rationale ("big push at the office," "my friend is having a crisis"), the fact is that the couple is spending less and less time together. Their interaction diminishes to the point where the partner eventually gets the message that something is amiss. The partner accepts the unwanted solitude for the negative signal it is, and confronts the initiator about the status of the relationship.

Rule Violation

A third indirect method involves violating the rules of the relationship.[14] The initiator breaks some rule (either spoken or unspoken) about proper conduct toward the partner and the initiator's behavior is so contrary to expectations that the partner is shocked to attention. A person who cares would not behave this way. The transgression is a breach of trust so great that the partner's self-concept is put in jeopardy; personal dignity is challenged to the extent that the partner cannot continue in the relationship without losing face.[15] The costs of ignoring the initiator's display of discontent now overwhelm the benefits. Compelled to acknowledge that the relationship is in serious trouble, the partner confronts the initiator, demanding an explanation....

The partner's threshold for tolerating these rule violations, these assaults on dignity and identity, sometimes depends on how much the violations become public. Although we may lose face before the person we live with, the indignities we suffer may still be tolerable because no one else knows (or we *think* no one else knows). Should the initiator's rule violations become visible to others, the partner will usually act to avoid social embarrassment.[16]

The violations that are most public, and thus most likely to move the partner to confront the initiator, are those pertaining to sex with other people.[17] By their very nature, they are public at the most fundamental level.

> He didn't tell me. That's what made me angry. Up to then we had a fairly open relationship. We were seeing each other, sleeping together, living together, but if we really wanted to see somebody else, as long as it was, "I'm going to do this, I won't be in, and do you like this person," it was OK. I have this philosophy. I do not own anybody's body, I maybe want your soul, which is what I wanted, but I don't own your body, and as long as you—I told him this, we told each other this—as long as you come back to me, and as long as I'm first, do what you want. Then he started an affair with a woman named Darlene but he didn't tell me about it. He started sneaking around like we were married to one another and he was some 40 year-old man going through his second childhood. It was very insulting and it made me very angry and very upset. I thought about killing myself a couple of times, like what am I doing wrong, what could I be doing that he could do such a rotten thing to me? Because it was a pretty damn rotten thing to do, not having the consideration to at least tell me, which I think is what hurt most of all. [STUDENT, AGE 21, SEPARATED AFTER LIVING TOGETHER 2 YEARS]

...

Partner as Detective

The partner picks up on the signals and confronts, but the initiator does not open up to the partner. Thwarted in the attempt to get to the bottom of things, the partner finds the signals of trouble too strong to ignore. Propelled by a suspected infidelity, the partner assumes the role of detective, searching for evidence to confirm the status of the relationship—whatever it may be.[18]

> One day, I was emptying the trash. I think this was probably in the bedroom where the waste basket had a piece of paper torn up into a dozen little pieces and it intrigued me, so I put them together and it turned out to be a receipt from a florist for some flowers that had been sent to a girl in care of a dentist's office. So I called the dentist's office and asked for the girl and she got on the phone and I asked her if she knew Ronnie Sacco, and she said, "He's a patient here," and I said, "Well, this is his wife and I was wondering why he would send you flowers." [EXECUTIVE SECRETARY/SUPERVISOR, AGE 34, DIVORCED AFTER 9 YEARS]

...

Sometimes the initiator's rule violation comes to light not through the partner's own detective work, but because someone else sees the initiator in some setting or engaging in some activity that is suspicious and uncharacteristic. Perhaps a friend observes an intimate lunch in an out-of-the-way restaurant, or a neighbor catches a glimpse of a quick embrace at a group outing. Those who witness this evidence bear a heavy burden. What to do? The heaviest burden often falls to those living in the same household who have loyalties to both partners: the child, who is taken on an outing as camouflage for one parent's Saturday lunch with a lover; the child, who witnesses a frequent visitor to the house when one parent is away; the child, who picks up an extension phone and hears a whispered conversation not meant for other ears.

Witnesses often remain silent, unburdening themselves only after the couple has separated. But occasionally, moved by the obvious breach of conduct and a connection to one of the partners or to the couple, a witness steps forward, presenting evidence. The partner may deal with this news of the initiator's transgression as with previous negative signals: by ignoring, misperceiving, or denying the information. Indeed, the partner may discredit the witness and cut ties with the bearer of bad tidings. Sometimes the news triggers the partner's search for evidence of confirmation or contradiction. When the information comes from a third party, the partner is often moved to act because what perhaps had been only a vague doubt gains legitimacy due to the observations of another person. Moreover, the partner's actions now have a witness. Someone else has evidence of the initiator's rule violation, and what the partner previously chose to ignore can no longer be overlooked. To save face, the partner must do something.

When partners engage in a search for evidence, the search becomes all-consuming. Their preoccupation is total, as they devote large portions of their thought and energy to confirming the status of the relationship. The initiator's denials only serve to spur the partner on....

When the detective work leads to evidence that is incontrovertible, the victory is scarcely sweet. The partner is forced to acknowledge the worst. In a direct confrontation, the partner reveals the initiator's secrets, exposing the subterfuge—and the impoverished state of the relationship.

For initiators to use decreased interaction, the fatal mistake, and/or rule violation rather than assume the responsibility for confrontation themselves seems not only to be the coward's way out but, if intentional, also unspeakably cruel. Consider, for a moment, the question of intent. An initiator is a person caught in an unhappy situation. Wanting to make the break and unable to directly confront the partner or to leave, the initiator's unhappiness finds expression in other ways: irritability, drunkenness, r udeness, displays of temper, absences, silence, withholding love.[19] Negative signals increasingly appear or—perhaps more accurately—the little symbols of civility and affection we reserve for our beloveds disappear. These signals are, in part, a reflection of the tension and stress initiators are experiencing.[20] They are weighing whether to stay or to go, and showing the agitation that normally accompanies making any major decision. In addition, concern about how to approach the partner also produces stress and changes in behavior.[21]

Further tension results because initiators are engaged, to varying degrees, in two different lifestyles and the strains eventually begin to show. As initiators become increasingly involved elsewhere, the signs of their transition—changed habits, conversation topics, new interests, physical changes—become more numerous. Controlling all the signals becomes more difficult. In addition, initiators are no longer as committed to maintaining the cover-up as they once were.[22] All these factors combine to exacerbate the display of discontent. As a result, initiators challenge the partner's definition of self and of the relationship. And while initiators may not deliberately set out to create such dissonance for the partner, they nonetheless realize their actions will provoke it.[23]

> I could never have left Julie. She was so vulnerable, I just could not do it even though I didn't love her anymore. But I started disappointing her in lots of ways. I realized I was becoming someone that I knew she couldn't like. [HIGH SCHOOL TERACHER, AGE 26, SEPARATED AFTER LIVING TOGETHER 4 YEARS]

...

Although many factors contribute to the indirect methods described, the question of the initiator's intent remains a difficult issue to sort out. Sometimes, initiators retrospectively question their actions. Did they intentionally act to get their partners to assume the responsibility for changing the course of the relationship? "I didn't think so at the time, but now I wonder if I took my lover out publicly so my wife would catch me at it." On the other hand, some initiators admit they knew their actions were intentional. Initiators deliberately resort to the fatal mistake, rule violation, and decreased interaction when they not only no longer love the other person, but more commonly no longer like the partner. Interest in their own well-being overrides consideration for that of the partner. The security initiators derive from their alternatives and their desperation to be out of the relationship combine to create a situation in which the end justifies the use of any means to attain it.

In sum, when initiators shift the responsibility for the confrontation to the partner through these indirect methods, we cannot say that intent is always present, or that it is always absent. If we want the answer to this question, we must look carefully at the details of the individual case. Even then, we may not resolve the question, for many initiators do not know the answer themselves.

NOTES

1. For a discussion of the importance of studying action, reaction, and interaction between partners, see Gerald R. Miller and Malcolm R. Parks, "Communication in Dissolving Relationships," in Steve Duck, ed., *Personal Relationships. 4: Dissolving Personal Relationships* (London: Academic Press, 1982), p. 146.
2. Davis describes various conversation forms that can occur in a confrontation. See Murray A. Davis, *Intimate Relations* (New York: Free Press, 1973), pp. 217–235.
3. See John Lofland and Rodney Stark's discussion of "turning points" in "Becoming a World-Saver: A Theory of Conversion to a Deviant Perspective," in Howard Robboy, Sidney L. Greenblatt, and Candace Clark, *Social Interaction, Introductory Readings in Sociology* (New York: St. Martin's, 1979), p. 465.
4. J. W. Thibault and H. H. Kelley, *The Social Psychology of Groups* (New York: John Wiley, 1959).
5. Barney G. Glaser and Anselm L. Strauss, *Status Passage* (London: Routledge and Kegan Paul, 1971), pp. 83–84.
6. For a discussion of identity as a determinant of agendas, see George J. McCall and J. L. Simmons, *Identities and Interactions* (New York: Free Press, 1966), pp. 244–249. Goffman calls such plans "strategic secrets": intentions and capacities that are concealed to prevent others from adapting effectively to an intended goal. Erving Goffman, *The Presentation of Self in Everyday Life* (New York: Anchor, 1959), p. 142.
7. Lucia H. Bequaert, *Single Women: Alone and Together* (Boston: Beacon, 1976), p. 27.
8. Lofland and Stark, p. 465.
9. Glaser and Strauss, pp. 17, 21–24.
10. See also Judith S. Wallerstein and Joan Berlin Kelly, *Surviving the Breakup: How Children and Parents Cope with Divorce* (New York: Basic Books, 1980), p. 136. See also Carmen DeMonteflores and Stephen J. Schultz, "Coming Out: Similarities and Differences for Lesbians and Gay Men," *Journal of Social Issues* 34(1978): 62.

11. See also Robert S. Weiss, *Marital Separation* (New York: Basic Books, 1975), p. 22.
12. Peter M. Blau, *Power and Exchange in Social Life* (New York: John Wiley, 1964), p. 84.
13. Oscar Wilde, "The Picture of Dorian Gray," in Richard Aldington, *The Indispensable Oscar Wilde* (New York: The Book Society, 1950).
14. Rule violation is related to the distribution of power in a relationship.... See ... Norman K. Denzin, "Rules of Conduct and the Study of Deviant Behavior: Some Notes on the Social Relationship," in George J. McCall, Michal M. McCall, Norman K. Denzin, Gerald D. Suttles, and Susan B. Kurth, *Social Relationships* (Chicago: Aldine, 1970), pp. 62–94.
15. Davis, pp. 265–267.
16. Erving Goffman, *Interaction Ritual: Essays on Face-to-Face Behavior* (Garden City, N.Y.: Anchor Books, 1967), pp. 5–45, 97–112.
17. See also Weiss, p. 30. He notes that the extraordinary power of violation of the fidelity norm comes from its symbolic meaning. Such a violation not only questions the initiator's commitment, but suggests the partner's sexual inadequacy; thus it is potentially damaging to both the public esteem and the self-esteem of the partner.
18. Erving Goffman, *Frame Analysis: An Essay on the Organization of Experience,* (New York: Harper & Row, 1974), p. 463.
19. The tendency for initiators to prefer avoidance styles rather than confrontational ones is also noted in R. E. Kaplan, "Maintaining Interpersonal Relationships: A Bipolar Theory," *Interpersonal Development* 6 (1976): pp. 106–119, and Steve Duck, "The Personal Context: Intimate Relationships," in P. Feldman and J. Orford, eds., *Psychological Problems: The Social Context* (London: John Wiley, 1980), pp. 73–96.
20. Paul Ekman, *Telling Lies: Clues to Deceit in the Marketplace, Politics, and Marriage* (New York: Norton, 1985).
21. Steve Duck, "A Topography of Relationship Disengagement and Dissolution," in Duck, ed., p. 22.
22. Erving Goffman, *Interaction Ritual* , pp. 126–129.
23. Erving Goffman, *The Presentation of Self in Everyday Life,* p. 210.

The Economic Consequences of Divorce

Frank F. Furstenberg, Jr., and Andrew J. Cherlin

Academics and clinicians debate how much divorce affects the personality, values, and behavior of children. But the economic effects are hardly ever disputed. Divorce often results in a sharp drop in the standard of living of children and their custodial parents. It carries many families into poverty and locks others into economic disadvantage for lengthy spells.

Here are the bleak facts. In 1988, 16 percent of all families with children under age 18 were poor—by itself, a shocking figure. But among female-headed families with children under 18, 45 percent were poor. In contrast, 7 percent of families with children headed by a married couple were living in poverty. In other words, families with children are more than six times as likely to be poor if they are headed by a mother alone than if they are headed by two parents.[1] Researchers who have followed families over time have found that these spells of poverty often begin just after the disruption of a marriage. According to one study, about 10 percent of white children and 14 percent of black children whose parents separated fell into poverty the following year.[2] Moreover, the study showed that families with a single parent were much less likely to escape from poverty. Overall, white children growing up in single-parent families spent an average of 3.2 years in poverty during their childhood, whereas those who lived continuously in two-parent families spent just 0.5 years in poverty.[3]

This [reading] explains why a lowered standard of living is a likely consequence of marital dissolution for children. In exploring the reasons why, we inevitably confront one of the most pressing policy issues of our day: what, if anything, can be done to break the link between family change and economic disadvantage? . . . [T]he evidence suggests that our peculiarly American view of marital and family responsibilities is strongly implicated in the economic plight of children. In America we leave it up to individual couples to sort out their personal affairs when their marriages break up. The state only becomes an active party when they cannot resolve their differences. But this strong commitment to letting divorcing couples reach their own private agreements often works to the disadvantage of mothers and children.

How the System Works

Just as there are no guidelines for couples on how to accomplish an emotional divorce, there are no guidelines for working out an economic settlement either. In fact, getting an economic divorce raises many of the same issues that couples face in sorting out marital and childrearing responsibilities. Once again, the gender-based division of labor in the family is at the heart of the problem. The family has always functioned as an economic distribution system, channeling resources from productive members to dependents such as children and the elderly. During the early years of industrialization, many older children worked for wages which they contributed to the family. Most wives contributed largely through household tasks, childrearing, or taking in boarders and lodgers. Then, in the early part of this century, children's earnings became less central to the household economy, and men, to a growing extent, became the primary wage earners. More recently, working wives have shared this role, although their earnings rarely equal their husbands'.[4]

The fact that men and women contribute unequal amounts to the total family income poses no problem so long as the family continues to operate as a common unit. Men and women trade labor inside and outside the home to the benefit of all family members. But divorce disrupts the family distribution system. Most fathers still rely on their former wives to care for the children, and most mothers still expect financial support from their former husbands—even though they don't live together. Dependent children must be guaranteed domestic stability and economic security even though parents no longer pool resources. Unfortunately, there is no formula for converting a two-parent household economy into two separate units.

Lacking such a formula, it is up to the conflicting parties to arrive at a fair settlement for dividing economic assets and obligations. And if the husband and wife are unable to agree, they have recourse to the legal system. In this sense, they bargain, as two legal scholars have put it, in the "shadow of the law."[5] These private negotiations are supposed to lead to an equitable economic settlement which also protects the children's interests. Unfortunately, the evidence suggests otherwise. Generally, what happens is that the economic pie is sliced into unequal parts, and men are served a larger helping at the expense of the rest of the family.

A good way to illustrate the problem is to see how Herb and Helen managed their economic divorce. At the time of their separation, Herb was the assistant manager of a fast-food franchise. He was earning about $24,000 per year. Helen had recently returned to work as a part-time secretary. She made $8,000 in the year before the marriage broke up. They had moved into a $60,000 condominium garden apartment only a year before the separation occurred. Although both Herb's and Helen's parents had helped them make the down payment, the furnishings and monthly payments had depleted their modest savings, leaving almost no cash reserves at the time of their separation.

The custody of the children was never an issue between them. Herb was deeply attached to his five-year-old son but assumed that the kids would stay with Helen. Her concern for the children led Herb to make what he thought was a generous settlement. He gave up his modest equity—about $10,000—in the condominium and took only a few of the furnishings for his small apartment. He got to keep the car, which permitted him to visit the children. In return, Helen agreed not to request any alimony and to accept a monthly child-support payment of $300, 15 percent of Herb's gross income.

Herb and Helen soon faced the harsh economic realities of divorce. It is far more expensive to maintain separate households. Even though Helen began to work full time, her financial situation worsened. She was now fully responsible for the mortgage and for the remaining payments on the furnishings. In addition her child-care costs doubled. She had counseling bills for her son, who was having trouble at school; and she badly needed a car to manage her complicated work and child-care schedule. She was starting to have trouble meeting the monthly payments. A move to a smaller place would have been less expensive, but she would have saved only about $100 a month and would have had to move out of the neighborhood, which was within walking distance of her parents' home.

Helen soon asked Herb to increase his monthly payments, but he resisted. After meeting his apartment expenses, the car payments, and his child support, he had little discretionary income. Some months after his final separation Herb started dating Alice, whose husband had stopped paying child support a few years after he moved in with a new partner. Herb realized that if he and his new partner got serious, he would have to cut down on the monthly payments to Helen in order to help support this new household. Naturally, he had not yet broached the subject of reducing the monthly payments with Helen, who was still complaining about living on her tight budget.

Herb and Helen epitomize the normal American way of reallocating economic responsibilities following divorce. If anything, they managed the situation more equitably than do most couples. But it does not take an accountant to recognize that in the post-marital bargaining, Helen and the two kids lost out economically when Herb moved out. She assumed full responsibility for supporting the children on a family income of $17,600 ($14,000 salary plus $3,600 in child support). Herb may feel disadvantaged relative to his standard of living before the separation, but in fact he is making out pretty well. Herb has discharged his economic responsibilities by providing $3,600 a year, retaining $20,400 for his own support.

When marriages dissolve, the shift in family responsibilities and family resources assumes a characteristic form. Women get the children and, accordingly, assume most of the economic responsibility for their support. Men become nonresidential parents and relinquish the principal responsibility for their children's support. Men, like Herb, may believe that they are making economic sacrifices equivalent to their former wives. But the facts don't generally support their contention.

Economic Status Before and After Divorce

Herb and Helen are by no means among the most disadvantaged of divorcing couples. And Helen, despite her very real economic troubles, is undoubtedly better off than many divorcing women. Together, she and Herb were earning close to the median family income, and they had some property to divide. Forty percent of all couples who divorce make no property settlement because there is nothing of value to divide.[6] (If we were to add to that figure the couples who separate but never divorce, the proportion of those without property would be even higher.)

In addition, Herb agreed to let Helen keep the condominium for the sake of the children, forgoing his share of the equity. Had he not done so, Helen would have been forced to sell the property immediately. Most divorcing couples with some tangible assets are in Herb and Helen's situation. Their home is the only item of real value. So in order to divide the property evenly, it is often necessary to sell the house. In Lenore Weitzman's study of divorce settlements in California, about a third of court-ordered divorces required the sale of the family home.[7] Other studies indicate the situation in California is not unique. The proportion of couples that liquidate the family home after divorce seems to be rising.[8] The loss of the family home is one of the most disruptive consequences of divorce. As we shall soon see, moving away from a familiar environment often starts a chain of social transitions resulting in profound, and sometimes damaging, psychological changes.

Furthermore, Herb is dutifully paying close to 15 percent of his annual income in child support. Not nearly enough, you may be saying, but Herb's perception from talking to the guys is that he is paying more than most. He is correct. The government publishes figures on the amount of child support that women with children under 21 receive from fathers living outside the children's home.[9] Among women who had been married (as opposed to those who had given birth out of wedlock), about one fourth had no child-support agreement in 1987. And among those who had an agreement, one fourth reported that no support payments were made in that year. Thus, roughly two fifths of all previously married fathers contributed nothing at all to their children's support. Among the other three fifths, the amount of child support paid averaged about $235 a month. Child-support payments amount to only about 10 percent of the income of separated and single mothers and 13 percent of the income of divorced mothers. By this standard, Herb's modest contribution of $3,600, which amounts to 20 percent of Helen's income, is relatively generous.

In noting that Herb is paying more than most men, we are not implying that he is paying a fair share. Helen and her two children clearly are not making out very well, and his limited contribution is part of the reason. The cost of raising a preschool child in 1985 at a modest standard of living was about $4,500 a year, excluding child-care expenses.[10] It is difficult to estimate the additional cost of a second child, but let's assume it is about 50 percent of the cost of the first. At best, then, Herb is only covering about two fifths of his children's real expenses—and much less if child care is included. Furthermore, the costs of raising children rise as they get older. But Herb's agreement with Helen, like most, makes no allowance for their changing needs or even for inflation. Consequently, although Herb is making more than half again as much as Helen, he is paying far less than she is for their children's support, and the level of support he is providing will undoubtedly shrink in purchasing power over time.

According to one national study that followed families throughout the 1970s, women suffer about a 30 percent decline, on average in their income in the year following a separation, whereas men experience a 15 percent increase.[11] The divergence can be explained partly by men's greater earning capacity. On average women earn only about two thirds of what men earn. As long as families pool their incomes, this disparity is of no special economic consequence to women and children. But when marriages dissolve, women must support themselves and their children with their lower incomes and with child-support payments, if any, from their former husbands. It is debatable whether divorce ought to change the relative economic standing of men and women. Some economists would argue that women who seek divorce are trading off a better standard of living for improved psychological well-being. But this reasoning takes no account of the children.

How much difference would it make to children if ex-husbands like Herb paid a fairer share? It is not easy to arrive at a precise formula, but one widely accepted standard was devised by Irwin Garfinkel, a leading authority on child support.[12] Taking into account both children's needs and the capacity of an absent parent to pay, Garfinkel developed a scheme that is now being applied in Wisconsin; variations on it have been adopted by about half of the other states. Noncustodial fathers are required to pay a fixed proportion of their gross income that varies according to the number of children they have, ranging in Wisconsin from 17 percent per year for one child to 34 percent per year for five or more.

Garfinkel's standard would greatly alter the agreement Helen and Herb reached on their own. Instead of paying $3,600, Herb would have to contribute $6,000—an increase of two thirds. Their standard of living would be more equal: Herb would end up with a gross income of $18,000 and Helen and her children would have $20,000. In families of modest means, like Herb and Helen's, raising the level of child support to a fairer standard makes a dramatic difference in the family income of a female household head and her children. If implemented nationwide, the Garfinkel formula would increase the total amount of child support due by about 69 percent—just by applying a common standard. And if money were collected from all absent fathers, the current level of child support provided to women would be tripled.[13] At 1985 levels, the average amount of child support would rise to more than $6,500. This figure, however, assumes perfect compliance, which no one thinks is possible....

Some skeptical observers have questioned how well this policy of redistributing resources from men to women and children will actually work. It is based on the assumption that most males can afford to pay a larger share of their earnings than they do now. At higher income levels, especially for formerly married fathers, such an assumption may be warranted. Even for men of more modest means, like Herb, there is probably substantial room for redistribution. But at lower income levels, it may not be so easy or profitable to collect from formerly married men, not to mention never married fathers. Thus, we should not hold out high hopes that a better system of child support will solve the problem of poverty arising from marital dissolution among low-income families. Estimates indicate a modest reduction in poverty can be achieved, and it will be most evident for white women and their children, whose ex-husbands are better able to pay adequate support levels. Among black men, who more often have limited earning capacities, increasing child-support payments will be difficult to accomplish and may do little more than rearrange which families are poor.[14]

What Economic Decline Means to Children

Even if elevating the level of child support would not solve all economic problems, it would be likely to diminish the extreme poverty of some female-headed families and to cushion the economic fall of many others. The national data show that currently divorced women and their children do not regain their predivorce standard of living until five years after the breakup, on average.[15] And these aggregate figures conceal sharp differences according to women's circumstances. Women who were married to men with high earnings typically experience a sharper decline in living standard than women who were already poor, and they recover more slowly. Like Helen, most women cannot immediately regain their economic losses by working more. If they have been out of the job market or only working part time, it takes a long while before they acquire skills and experience needed to earn enough to support a family. One study shows that, in time, many improve their situation by career advancement.[16] In the short term, though, the surest means to a quick economic recovery is through remarriage. This is one of the main reasons why black women—who generally have poorer prospects for remarriage—are less likely than whites to recover economically from a divorce.

So a woman's economic position remains tied to her marital status even after she divorces. Many separated and divorced women will remain near or below the poverty line for several years unless they remarry. What this means for women and children is that every year a million families experience a drop in income that is similar to what families went through during the Great Depression.[17] This income loss is usually accompanied by a host of related life changes. To see this more clearly, let's return to our case history and watch what happened to Helen and her children after her divorce became final.

Remember that Helen was wondering whether she could afford to meet the payments on her new condominium. After six months of falling further and further behind, she decided that she had to sell it. She and the children temporarily took refuge with her parents while she paid off some of her debts. This gave her time to look around for suitable housing, but it was difficult for her son Mickey, who did not want to move away from the kids on his block. Unfortunately, the move also required that the children switch schools. Mickey, who had experienced some difficulties in kindergarten before the move, began to have academic problems in the first grade, which were aggravated by the adjustment to a new school. The teacher advised Helen to have Mickey repeat the first grade.

Helen's parents were supportive and helpful, but she did not feel comfortable staying with them indefinitely. Her younger brother was still at home, and he often woke the kids up when he came in late. Sally was sharing a room with Helen, an arrangement that would not work indefinitely. Helen did not want to put the two children in the same room. So she found an apartment near her work. It was in a less desirable section of the city, farther away from her parents, but it provided the necessary space at an affordable price.

A few weeks before Helen and her kids moved into the apartment, Herb told Helen that he could not afford to keep up the payments at the existing level. He had begun living with Alice, the woman he had been dating, and her two children. Already anticipating marriage, he expected to help support his new family. Herb's visits with Mickey and Sally were becoming more irregular. Although still attached to his son, Herb sometimes felt that Mickey was a stranger to him. And he felt as though he hardly knew his daughter. Herb began to ask himself why he should pay for kids whom he seldom saw. He was rapidly becoming more attached to Alice's children, feeling that they were just as much his responsibility as his own children. Besides, Helen was working full time. She had cleared some money from the condominium sale and was now living in less expensive housing.

Herb's threats to reduce his support payments and his decreased contact with the kids were making Helen feel desperate. She was furious about Herb's relationship with Alice, and her rage sometimes spilled over to Mickey, especially when he wanted to talk about his visits with his dad and Alice and her family. Helen did not know whether to see a therapist or a lawyer. In the second year after her divorce, she ended up seeing both, further draining her limited economic reserves. Even with her lower rent, Helen was barely making it; so she reluctantly accepted overtime work at the office to help meet the expenses. This gave her even less time with the children, who she knew needed more of her attention.

Helen's experiences are typical of many women in the aftermath of a divorce.[18] So, too, are her children's experiences. Katherine Newman describes the economic trajectory of most divorced women as a process of "falling from grace."[19] Helen, like many others after divorce, fell from the middle-class and took her children with her. She moved twice in less than two years, trading home ownership for cheaper housing in a less desirable neighborhood. She increased her working hours and decreased the time she was able to spend with her children.

Sara McLanahan, in a study of the economic consequences of divorce, discovered that almost two fifths of divorced mothers move in the first year after divorce, a rate far higher than the occurrence for stably married families during the same interval.[20] Even after the first year, divorced women continue to move at a rate of about 20 percent a year, about one third more often than women in intact marriages. More of the moves reported by divorced women resulted from necessity than choice, especially in the immediate aftermath of divorce. During the first year after divorce, 15 percent of the divorced women were forced to move—seven times the rate of forced moves among stably married women.

Many women, like Helen, take temporary shelter in their parents' households. About 7 percent of formerly married women with children were living with their parents at the time of the 1980 census. Moreover, this snapshot picture greatly underestimates the proportion who ever move back home after a divorce. Among separated and divorced women with younger children—who are probably nearer to the time of the separation—17 percent were living with their parents in 1980.[21] Doubling up has some economic and psychological benefits for divorced mothers and their children. However, it is not generally the preferred arrangement of most divorced women or even of their parents, who often find the arrangement stressful. One study reported a sharp rise in family conflict after members began sharing a household.[22] A national study of grandparents revealed that parents frequently provided substantial assistance to recently divorced women and their children but that intergenerational relations sometimes became intrusive. Divorced women complained that grandparents all too frequently interfered with their parental authority or intruded in their family affairs.[23] And another investigation concluded that heavy reliance on parents sometimes hinders a divorced mother's ability to adjust to her new status.[24]

Many women also share Helen's experience of longer hours at work and less time at home. One study found that the proportion of women working 1,000 hours a year or more rose from 51 to 73 percent after a separation, and their incomes increased proportionately.[25] Women who remain unmarried are especially likely to increase their working hours, often taking new jobs in order to do so. Another study discovered higher rates of both quitting and being laid off in the year after divorce.[26]

These residential and occupational changes are part of a familiar pattern that divorced women exhibit. As Helen's case illustrates, economic and emotional losses are often confounded. Even among women who welcomed the dissolution of their marriages, the confluence of events that accompany a sharp drop in income and living standard is bound to create great stress. And this stress often reverberates in ways that affect the children. Helen had less time to spend with Mickey and Sally. They were subjected to several unexpected and unwanted moves. For Mickey, the timing was especially unfortunate, for it exacerbated his school problems. Helen's anger over Herb's economic threats (and his new relationship) complicated her dealings with her son. In ways such as these, the children of divorce are both directly and indirectly affected by a sudden economic drop. In this respect, they share many of the emotional problems experienced by the children who went through the Great Depression.[27]

We are happy to report that things eventually settled down for Helen and her children. Herb backed down from his threat to reduce child support when he was notified by Helen's lawyer that he would face court action. In addition, Herb realized that he was not making out badly at all. The child-support payments had not gone up in the two years since his divorce, whereas his salary had risen to $28,500, an increase of nearly 20 percent. Herb's change in attitude improved his relationship with Helen. His visits with the children, however, were becoming even more irregular since Helen stopped prodding him to see them.

For her part, Helen learned to make do with less. She made friends in the apartment building and joined a baby-sitting co-op. But Mickey still is not doing well in school. The second move created more academic problems, but at least he is getting along better with his schoolmates. Helen is feeling less depressed and is managing her life much better. She is calmer with the kids and has learned from her therapy just to listen sympathetically to Mickey when he talks about his dad. But despite Helen's recovery, she and her children are not living as well—materially at least—as they were three years earlier. Moreover, they are not nearly as well off as Herb and his new family. Herb is combining his income of $28,500 (effectively $24,900 when child-support payments are deducted) with Alice's earnings of $15,000. Together, their family income of $43,500 places them both in a higher income bracket than either was in during their first marriages. Compared with Helen, Mickey, and Sally, Herb has little to complain about. In contrast, Herb's former wife and children have been subjected to a series of stressful events partly or wholly because they became poorer.

Why Don't Fathers Pay More?

Government statistics on alimony and child support were not compiled before the early 1970s. Between 1970 and 1978, no significant changes occurred. For example, only 10 percent of women actually received alimony in 1978 (of the 14.3 percent women awarded alimony, 70 percent collected some payments). By 1985 this figure had risen ever so slightly to 10.7 percent (73 percent of the 14.6 percent awarded alimony). Weitzman marshals some evidence to show that recent divorce settlements provide less well for women than those occurring a decade or more ago. But her argument holds, at best, for middle-class women leaving long-standing marriages. This group is probably significantly worse off today than they were in the past. We doubt that other formerly married women were better off a few decades ago; rather, we think they were not provided for then and they are still not provided for today.

We suspect that most men, divorced fathers included, strongly adhere to the belief that fathers should be obliged to pay their fair share of their children's support.[28] If this is true, how can we then account for the seemingly callous behavior of many males? One reason, mentioned earlier, is that some men do not have the money to provide much assistance. Unemployed, underemployed, and low-wage-earning fathers may have little to share with their children. Yet this explanation hardly seems sufficient to explain why two fifths of all formerly married fathers pay nothing at all. Evidence from a variety of sources indicates that almost all formerly married fathers could pay something, and most are capable of paying far more than they do.[29] And men who may not be able to pay much at the time of divorce often improve their economic circumstances in time.

The Census Bureau conducted a survey on child support in 1987 in which women without a child-support agreement were asked why they had failed to get one.[30] A small proportion had settled for a property settlement instead of child support. The rest were evenly divided between women who said they had not asked for child support and those who had asked for child support but could not obtain an award. Why should any mother with children not insist on a child-support award? There is no single answer to this question. Some divorcing women refuse a settlement as a way of distancing themselves from their former spouses. Some bargain away child support for exclusive custody rights. In their view, it is not worth a few thousand dollars to have to deal with their former spouse or to have him involved with their children. Women who have remarried or reentered relationships may be especially inclined to dismiss their former spouses from economic obligations. Many women, one study reports, came to expect so little economic support from their ex-husbands while they were married that they were relieved at not having him "take money out of the till."[31]

For one reason or another, then, a substantial number of women are willing to pay, in effect, to stop their former husbands from interfering in their affairs. Many more women, however, would like to collect support but are unable to get a binding agreement. Some fathers simply disappear when the marriage ends—they don't stay around long enough to negotiate a settlement. In the Census Bureau survey, the most common reason that women report for why they were not able to reach a child-support agreement was that they were unable to locate the father. And even when fathers remain in the picture, it is often difficult to get them to pay. In Terry Arendell's study of middle-class divorced mothers, several women told her that they had given up on trying to obtain child support because they lacked the resources, energy, or persistence to obtain a binding agreement. Arendell concludes that women often are intimidated by the legal system, which appears to them as unresponsive, if not unsympathetic, to their children's needs.[32]

Her conclusions echo a theme sounded by others who have examined how the judicial system functions. Weitzman argues that, in child-support disputes, the courts are biased in favor of fathers. In response to a series of hypothetical cases, American judges were much less sensitive to children's economic needs and more sensitive to men's economic needs than their British counterparts. She concludes that our system seems to put "fathers first," ahead of mothers and their children. We agree that the legal system must receive some of the blame for the low levels of child support in this country.[33]

But there are other, more subtle, factors at work that are linked to gender differences in the way that men and women deal with economic matters. Researchers such as Weitzman and Arendell point out that men (and their attorneys) are more likely than women to view the care and custody of children as part of the settlement package which includes property and child support. Women, on the other hand, usually view custody as a separate issue of the highest importance. Therefore, in order to retain custody, women tend to compromise more easily on the economic matters.[34]

Many of the same factors that prevent women from obtaining a child-support award in the first place also prevent them from enforcing awards that are in effect. In 1985, over 75 percent of divorced women received child-support awards, but only three quarters of these women were collecting any payments. Just about half received the full amount of support due. Consistent with the pattern of visitation described in the previous chapter, support payments generally dwindle as time goes on. Just as Herb was tempted to do, many men fail to make good on their original agreement—even though most child-support payments shrink in purchasing power because they are not adjusted for cost of living. And again, the legal system is often not much help:

> We've got a court order for child support. Last year, when I knew my ex-husband had a job and where, I tried to go through my attorney to get enforcement. But all that happened was that I got stuck with court costs and attorney's fees, and I didn't get any money in return. I did have a lien on his paycheck once. I got like two hundred dollars that way, and he promptly got fired. That's the only thing I ever got from him. It just didn't pay off.[35]

There are many reasons, including economic ones, why men fail to pay the amount agreed upon. Remarriage by either partner often erodes a man's commitment to child support. When they remarry, men like Herb often feel entitled to reduce their support payments because they have a second family to support. When custodial mothers remarry, fathers frequently see this as an opportunity for reducing support payments, reasoning that their economic assistance is needed less.[36] Father's rights groups contend that many men fail to live up to their support agreements because their former wives deny them access to their children. But as far as we can tell, fathers who must resort to economic retaliation because they are locked out of their children's homes are a tiny minority. These incidents tend to receive more attention than the much more common instances of families in which men withdraw economic support before or concurrently with their diminished involvement with their children.[37]

The weight of the admittedly incomplete evidence suggests to us that most men pay little or no child support because they can get away with it. The gender-based division of labor in the family leads many men to see their children as women's responsibility. In the abstract, most men will say that a father's obligation to his children is inviolable. In practice, many fathers cut loose from their children soon after their marriages end.

Yet just as often, men find themselves in Herb's situation. They begin their postdivorce lives with a strong commitment to support their children. Over time, their resolution weakens as relations with their children become emotionally less rewarding or they acquire a new set of family commitments. In effect these men trade in old obligations for new ones. From their point of view, they are not callously disregarding their family responsibilities but rather redefining them as they move from one marriage to the next. This pattern points to an unresolved policy issue. How do we reconcile the economic obligations of parents to their different sets of offspring and to their biological and sociological dependents? In contrast with many other Western nations, the United States has until recently favored a voluntary approach to child support. In effect, fathers have been allowed to put their money where their heart is. And many fathers have opted to invest most if not all of their resources in their current household rather than their former one. If all women reentered marriage soon after divorce, this arrangement would work well enough. But fewer women than men remarry, and our society has begun to realize the high costs to children of letting fathers decide whether their first loyalties are to their present or past families.

Where Do We Go from Here?

To the extent that fathers retain an emotional stake in their children, we think this will increase their sense of economic obligation. But the process of cultural change in the role of fathers is likely to be gradual. In the meantime, we must look to other means of securing economic support from fathers.

Concern about the meager level of child support and about the dependency of many female-headed families on public assistance led to the passage of federal child-support legislation in 1984 and 1988. These laws mandate state guidelines for minimum support levels and better mechanisms, including wage garnishing, for collecting support payments. A decade earlier, such measures would have been unimaginable in the United States. Evidently, we are slowly abandoning our voluntary, do-it-yourself approach to child support. It remains to be seen whether this nationalized approach to child-support collection will work as well in the United States as it appears to in some European nations....

During the 1980s, we witnessed a growing wave of interest in fatherhood. Indeed, if one takes seriously the popular and professional outpouring of materials, it would be only a slight exaggeration to conclude that American males are currently in the midst of a cult of fatherhood. More fathers are participating in the delivery process, taking time off from work to care for their children, and sharing more of the care of their preschoolers. Yet it is hard to determine just how many men have been enlisted in this movement; we suspect that there is still more talk than action. Nonetheless, the trend toward greater involvement of fathers is real and could have consequences for fathers' behavior after divorce. As commitment to fatherhood rises, a shrinking fraction of men will sever the bonds with their children simply because they no longer reside with them.

NOTES

1. This difference in family structure goes a long way toward accounting for the enormous racial disparity in poverty rates. To be sure, even within family types, black families are still poorer than white families; but the racial gap in poverty shrinks considerably when the marital status of the household head is taken into account. U.S. Bureau of the Census, Current Population Reports, series P-60, no. 166, *Money Income and Poverty Status in the United States: 1988* (Advance Data from the March 1989 Current Population Survey) (Washington, DC: U.S. Government Printing Office, 1989).
2. Greg J. Duncan and Saul D. Hoffman, "Economic Consequences of Marital Instability," in M. David and T. Smeeding, eds., *Horizontal Equity, Uncertainty, and Economic Well-Being* (Chicago: University of Chicago Press, 1985), pp. 427–470.
3. Greg J. Duncan and Willard L. Rodger, "Longitudinal Aspects of Childhood Poverty," *Journal of Marriage and the Family* 50 (November 1988): 1007–1021.
4. Susan Bianchi and D. Spain, *American Women in Transition* (New York: Russell Sage Foundation, 1986).
5. R. H. Mnookin and L. Kornhauser, "Bargaining in the Shadow of the Law: The Case of Divorce," *Yale Law Journal* 88 (1979): 950–997.
6. Judith Seltzer and Irwin Garfinkel, "Inequality in Divorce Settlements: An Investigation of Property Settlements and Child Support Awards," *Social Science Research* 19 (1990): 82–111.

7. Lenore Weitzman, *The Divorce Revolution: The Unexpected Social and Economic Consequences for Women and Children in America* (New York: The Free Press, 1985).
8. James B. McLindon, "Separate but Unequal: The Economic Disaster of Divorce for Women and Children," *Family Law Quarterly* 3 (Fall 1987): 351–409; Charles E. Welch III and Sharon Price-Bonham, "A Decade of No-Fault Divorce Revisited: California, Georgia, and Washington," *Journal of Marriage and the Family* 45 (May 1983): 411–418.
9. U.S. Bureau of the Census, Current Population Reports, series P-23, no. 167, *Child Support and Alimony: 1987.* (Washington, DC: U.S. Government Printing Office, 1990).
10. C. S. Edwards, "Updated Estimates of the Cost of Raising a Child," *Family Economics Review* 4 (1985): 26.
11. Greg J. Duncan and Saul D. Hoffman, "A Reconsideration of the Economic Consequences of Marital Dissolution," *Demography* 22 (November 1985).
12. Irwin Garfinkel, "The Role of Child Support Insurance in Anti-Poverty Policy," *Annals*, AAPS, 479 (May 1985): 119–131; Irwin Garfinkel, "A New Approach to Child Support," *Public Interest* 75 (Spring 1984): 111–122.
13. Ann Nichols-Casebolt, "Economic Impact of Child Support Reform on the Poverty Status of Custodial and Noncustodial Families," *Journal of Marriage and the Family* 48 (November 1986): 875–880.
14. Nichols-Casebolt, "The Economic Impact of Child Support Reform."
15. Duncan and Hoffman, "Reconsideration of the Economic Consequences of Marital Dissolution"; Duncan and Hoffman, "Economic Consequences of Marital Instability."
16. Richard R. Peterson, *Women, Work, and Divorce* (Albany: State University of New York Press, 1989).
17. Richard V. Burkhauser and Greg J. Duncan, "Economic Risks of Gender Roles: Income Loss and Life Events over the Life Course," *Social Science Quarterly* 70 (March 1989): 3–23.
18. Ruth A. Brandwein, Carol A. Brown, and Elizabeth M. Fox, "Women and Children Last: The Social Situation of Divorced Mothers and Their Families," *Journal of Marriage and the Family* 36 (August 1974): 498–514; Robert S. Weiss, *Going It Alone* (New York: Basic Books, 1979); Lenore Weitzman, *The Marriage Contract* (New York: The Free Press, 1981); Sara McLanahan, "Family Structure and Dependency: Early Transitions to Female Household Headship," *Demography* 25 (1988): 1–16; Terry Arendell, *Mothers and Divorce: Legal, Economic, and Social Dilemmas* (Berkeley, CA: University of California Press, 1986).
19. K. Newman, *Falling from Grace: The Experience of Downward Mobility in the American Middle Class* (New York: The Free Press, 1988).
20. Sara S. McLanahan, "Family Structure and Stress: A Longitudinal Comparison of Two-Parent and Female-Headed Families," *Journal of Marriage and the Family* (May 1984): 347–357.
21. James A. Sweet and Larry L. Bumpass, *American Families and Households* (New York: Russell Sage Foundation, 1988).
22. Weiss, *Going It Alone.*
23. Andrew J. Cherlin and Frank F. Furstenberg, Jr., *The New American Grandparent: A Place in the Family, A Life Apart* (New York: Basic Books, 1986).
24. Leigh A. Leslie and Katherine Grady, "Changes in Mothers' Social Networks and Social Support Following Divorce," *Journal of Marriage and the Family* 47 (August 1985): 663–673.
25. Duncan and Hoffman, "Economic Consequences of Marital Dissolution."
26. McLanahan, "Family Structure and Stress."
27. Glen H. Elder, Jr., *Children of the Great Depression* (Chicago: University of Chicago Press, 1974).
28. There are few surveys that have tapped men's attitudes toward child support; but in those that have, men strongly endorse child support. Weitzman, *Divorce Revolution;* Ron Haskins, "Child Support: A Father's View," in Alfred J. Kahn and Sheila B. Kamerman, eds., *Child Support: From Debt Collection to Social Policy* (Newbury Park, CA: Sage Publications, 1988), pp. 306–327.
29. Robert I. Lerman, "Child-Support Policies," in Phoebe H. Cottingham and David T. Ellwood, eds., *Welfare Policy for the 1990s* (Cambridge, MA: Harvard University Press, 1989).
30. U.S. Bureau of the Census, *Child Support and Alimony: 1987.*
31. Janet A. Kohen, Carol A. Brown, and Roslyn Feldberg, "Divorced Mothers: The Costs and Benefits of Female Family Control," in George Levinger and Oliver C. Moles, eds., *Divorce and Separation: Context, Causes,*

and Consequences (New York: Basic Books, 1979), pp. 228–245.

32. Arendell, *Mothers and Divorce.*
33. Weitzman, *Divorce Revolution.* See also Alfred J. Kahn and Sheila B. Kamerman, eds., *Child Support: From Debt Collection to Social Policy* (Beverly Hills, CA: Sage Publications, 1988).
34. Arendell, *Mothers and Divorce;* Weitzman, *Divorce Revolution.*
35. Arendell, *Mothers and Divorce,* p. 21.
36. Frank F. Furstenberg, Jr., and Graham B. Spanier, *Recycling the Family: Remarriage after Divorce* (Beverly Hills, CA: Sage Publications, 1984).
37. Frank F. Furstenberg, Jr., "Marital Disruption, Child Custody, and Visitation," in Kahn and Kamerman, *Child Support,* pp. 277–305.

After Divorce: Investigations into Father Absence

Terry Arendell

On the basis of in-depth interviews with 75 divorced New York fathers, the phenomenon of postdivorce paternal absence is investigated. The accounts provided by the interviewees suggest that father absence is more than a literal practice: it is also a perceived option and a standard of comparison. Father absence is a strategy of action, the objective of which is to control situations of conflict and tension and emotional states. That the majority of the fathers in the study shared common explanations with regard to father absence is an indication of their participation in a "masculinist discourse of divorce." A primary theme in the discourse is the rhetoric of rights....

Absence as a Strategy of Action

A strategy of action is a persistent way of ordering action through time (Swidler 1986). The objective of postdivorce absence as a strategy of action is the avoidance or circumvention of further conflicts, tensions, and certain emotional states and uncertainties. It involves three interrelated and overlapping components: actual practice, a perceived or optional line of action, and emotion management.

A Practice and Optional Line of Action

The boundary between absence as a practice and absence as an optional line of action is ambiguous and problematic: the point at which the option became practice for the fathers in the study was not always discrete and identifiable since absence often came about as a result of omission rather than commission. That is, absence was not always a deliberate decision taken at a particular point but often became a condition over time as the gaps between contacts or visits with the children were allowed to lengthen. Further, certain fathers perceived their relations with their children to be so tenuous as to make the possibility of becoming a fully absent parent realistic.

Postdivorce absence as a practice and an option was, specifically, a strategy of action aimed at controlling particular situations. Primarily to be controlled was interpersonal conflict with the former wife and, occasionally, with the children of the ended marriage.[1] With the exception of the small number of fathers characterized as androgynous, the vast majority were engaged, or had been until their withdrawal from their children's lives, in highly antagonistic relations with their former wives. These conflicts, for the most part, were a continuation of the character of the marital relationship, at least as it unraveled. Dissension between the former spouses often involved issues of contention carried over from the marriage and the specifics of the divorce property settlement. But most conflict was centered on their children who constituted the primary postdivorce link between them. Conflicts over children were complex: not only did specific issues evoke disagreement but the former partners often viewed child-related matters from distinctive positions and in divergent ways. One noncustodial father, for example, told of a continuing quarrel with his former wife.

> I gave my son a shotgun for his 12th birthday and she had a fit. She was afraid he was suicidal, he was so depressed over the divorce, his failing grades in school, and our fighting. She was sure I was giving him the message it was okay for him to shoot himself. I viewed the gun as a symbol to him that I trusted him, that I had confidence in him to pull through all of this. I was right too: he never did hurt himself and he seems normal now [at sixteen]: he likes to hang out with his buddies and drink beer and he likes girls. I understand what it is to be a boy growing up; she doesn't. It's important for him to learn to be a man and he can't do that without my input. But she'd like to see me completely out of their lives.

Concerns and disagreements about children's well-being were often based on some form or another of evidence and were sometimes grave. Parents intent on using whatever was available to them in their ongoing interpersonal struggles, however, could deliberately exaggerate or fabricate concerns regarding their children who were both readily available and relatively powerless pawns in the postdivorce hostilities. The specifics of disagreements over children varied, centering generally on issues related to their living arrangements, care, treatment, and economic support. Mothers and fathers accused each other of neglecting their children including, for example, leaving them alone without supervision or sending them to school or to the other parent's home in dirty and torn clothing. In a unique twist, one custodial father accused his former wife of neglect for leaving their child with him rather than assuming primary custody as he believed mothers should. Quarrels over children's educational experiences and moral and religious guidance were common. Allegations of physical and sexual abuse were exchanged between the former spouses and, according to the fathers, were disproportionately attributed to them. Ten fathers had been formally investigated on the basis of such allegations.[2]

Postdivorce conflicts, although not universal, were common experiences among the men participating in the study regardless of their level of discontent with the divorce outcome. Contributing to conflicts was the defining of family relationships within the context of rights. As a constitutive element of the shared masculinist discourse of divorce, rights had a distinctive connotation: generally, *rights* was used synonymously with that which was expected, desired, and believed to be deserved as a *man.* The rhetoric of rights, widely available in the culture at large and appropriated from political and legal theory and practice, encompassed beliefs about individualism, autonomy, choice, control, and authority, beliefs that are also central to the cultural conventions and norms of masculinity (Jaggar 1983; Pateman 1990). Rights as used by these divorced men with regard to their family situations and experiences then was a euphemism for male privilege within the stratified gender system. Especially at issue were the privileges of position and roles of husband and father as held in the family prior to separation and divorce. Thus framed in the language of rights were matters of control and authority in the postdivorce family including, given the dominant legal rules that formally place minor children into parental custody after divorce, questions of access to children. Moreover, the securing of rights was perceived to involve fundamental issues of identity itself: protecting an identity as a masculine self required preserving, or seeking to preserve, sundry spheres of dominance and control (Arendell forthcoming). Men satisfied with their divorce and postdivorce experiences spoke of rights in ways analogous to those men who were intensely dissatisfied with nearly every aspect of their divorce experiences. The principal exceptions were . . . those fathers characterized as androgynous.

The securing of one's rights was a formidable and continuing enterprise, sometimes involving the legal system and typically involving struggles with the estranged spouse: the actions of former wives were often characterized in various ways as "intended to deny me my rights." The primary use of rights entailed a highly personalized contest or competition, involving a polarized outcome of either winning or losing. That rights were to be secured in relationship to another (even if oppositional) demystified the assertions and implications that what was at issue were matters of abstract principles of justice. For example, in discussing his postdivorce experience, a father who had opposed the divorce and subsequently withdrawn from the lives of his five children, recalled,

> When the judge gave my house to my wife [who was living with and caring for the children], he took away my right to manhood. He took it away and left me with nothing. He stripped me clean.

Losing the family home represented for this father the various losses brought about by the divorce; the family home is to many men "a cultural symbol of power and success" (Riessman 1990, 152). Various other fathers explicitly observed that the legal system "had emasculated" them by revoking their rights to control their earnings and ordering them to pay child support. Some men talked of having to fight for their "rights to their parenthood."

The rhetoric of rights fostered an objectifying of relationships and an emphasis on fathering as an achieved status rather than a particular complex of interactional processes and dynamics. Children were often discussed specifically in the context and terms of rights, especially *fathers' rights;* they were defined as a kind of property over which control was to be contested and fought. Contemporary legal dictates and procedures reinforce the stance that children are objects to be fought over by divorcing parents (Emery 1988; Weitzman 1985) although "father right," as traditionally institutionalized in the law, has undergone considerable challenge and change (Polikoff 1983).

Among the fathers using the language of rights to discuss actions taken with regard to their children was this noncustodial father who had his child "snatched from his mother" (his words) by having him taken from a school playground in another state and put on an airplane to New York. He described his efforts and the judicial decision made after 18 months of court hearings and delays:

> My "child snatching" was a huge success. It was a big score. But the judge upheld my ex-wife's custody by ruling that my son was "too young" to be with me. Besides, I didn't really want custody. My son was a kid and needed his mother, I knew that. And I wanted my freedom. But it was all-out divorce war. It was just like an A-bomb: I wanted my son for awhile and got back at her, all in one fell swoop. I legally stole him, just simply resorted to *child snatching.* I'd never really wanted custody. I just needed to flex my *rights.* After all, I still was his father; I wanted to have some say in his life. I needed for her to know that I still had some control around here.

He continued, reflecting on his subsequent withdrawal from his child's life:

> I finally decided that I was putting too much energy into this divorce-war with my ex-wife. We'd played this game for over four years. So I pulled out; she didn't even known when to say "uncle." Someday, if my son wants to get to know me, he can find me.

The entanglement of questions of access to and authority over his child with antagonisms toward and power struggles with his former wife was common among the participants in the study. So too was this father's use of rights as an explanation for aggressive and confrontational actions. Although his tactics were more extreme than most described, eight other fathers reported having engaged in some form of child abduction activities, giving similar explanations for their efforts.[3]

The condition of being a "visiting father," as the noncustodial parents referred to themselves drawing from both legal and popular terminologies, involved extensive interpersonal conflict and emotional turmoil and served as a basis for postdivorce paternal absence.[4] Visitation was an issue of particular significance to the fathers without sole or co-custody because it is the legally defined and protected means of access to one's children after marital dissolution (Novinson 1983). Despite its institutionalization as a parental arrangement, visitation, both as a concept and practice, was engulfed in ambiguity and was a source of extensive dissatisfaction and, for some fathers, particular outrage. Rejecting the status assigned to them in the postdivorce situation as that of visiting father, several withdrew totally from their children's lives. One man characterized his absence as "a response to the condition of 'forced impotence,' a response to the total denial of my rights." Another person, who became an absent father after unsuccessful attempts to obtain shared custody, said,

> I will not be a visiting uncle. I refuse to let some woman [my former wife], judge, attorney, or social worker reduce me to that status. I'm a parent and parents do not "visit" their children. If I see my child only every other weekend, I become nothing more than a visiting uncle. I am a father in name only at this point. Until and unless I can be a father in every sense, I simply refuse to have any part of this. If someone came and stole my child from my house window, there'd be 500 volunteers combing the area. When a wife steals a child, "well, go about your work and if you're really good you'll get every other weekend." These are the things I face, the philosophies that men face. Our rights as fathers are simply negated, erased.

This father's further comments point also to the understanding expressed in numerous ways by these men that the family is predicated on marriage. In this definition of family, relations with children are entwined with those with the former wife and so, likewise, can be transitory.

> Besides, you know, every time I see those children, I am overwhelmed by memories. They are a living reminder of *my* marriage, *my* wife, and the years of pointless effort. Being a father is all tied up with being a husband. That's ended.

Visitation, or the activities and logistics of being a visiting father, typically entailed power struggles between the former spouses; in turn, strategies and tactics aimed at obtaining control usually involved visitation in some way. Most fathers, including those who were involved with their children at relatively high levels, insisted that their former wives had hindered their access to their children. Some fathers encountered persistent interference: scheduled visits were denied, telephone conversations interrupted or prevented, messages not conveyed, and mail intercepted and not given to the child to whom it was sent. The "cat and mouse game," as several fathers described it—going to pick up their children for a visit and finding no one home—was an experience described by nearly half of the fathers and was a recurring one for some. As this father of three recounted,

> Every time I went to pick them up, they'd be gone. I'd drive all the way [40 miles from his residence] over there, as we'd agreed and after I'd called her the night before to remind her, and they'd all be gone, every one of them. Even the neighbors were in cahoots with her; they'd insist they hadn't seen any of them the whole day.

Such actions led some men to initiate court hearings to protest and prohibit the interference. Others, including several who believed that the utility of court proceedings had been exhausted, began to allow longer periods of time to lapse between contacts and visits. Indeed, over half of the factors fully absent from their children's lives became nonvisiting ones over an extended period of time and after a series of confrontations with the former wife.

Failures both of communication processes and imagined alternative lines of action added to the postdivorce interpersonal quagmire. Two absent fathers, for example, noted that they would like to reestablish contact with their children but did not know how to do so. One asked,

> How do I explain to them my absence from them? How do I explain that even though I gave up, I still am their father? I have a difficult time rationalizing that myself.

If visitation was the issue that former wives could use as a tactic of ultimate provocation in the postdivorce hostilities, as alleged by a majority of the fathers, then initiating a challenge to maternal custody was the comparable strategy available to them. Dissension and discord could be continuously played out by challenging or merely threatening to challenge maternal custody status. (Three fathers had fought to obtain custody *during* the divorcing process and had obtained custody; these men like the other three who became sole custodial parents without a custody fight at the time of divorce, were raising their children as single fathers.) More than three-quarters of the fathers without primary custody from the time of separation indicated that they had threatened their former wife with a custody fight since the divorce agreement had become final; more than a quarter had gone through an attorney to make a formal threat. Predominantly at issue in custody challenges, as with the phenomenon of father absence generally, was the relationship with the former spouse and not with the children, and this relationship was fundamentally about issues of power and control. For example, in discussing the problematic relationship between himself and his former wife, this man stated,

> I always know that whenever things get too out of control I can threaten her with a custody challenge. So far, and I've spent over twenty-five thousand dollars, I've lost. But the point is that she and I both know that I can slap a suit against her anytime I want to. She's on the defensive. Economically it kills her. I can make it up in a few good months but she's on a fixed salary.

Another father, discussing his repeated use of custody challenges as a response to his outrage over visitation interference, accounted for his lack of success in changing the custody situation by arguing that a legal bias that overwhelmingly favors women prevails. The belief that mothers are unjustly privileged in custody matters was prevalent among these men, even those who were satisfied with their own situations.

> After that particularly nasty fight over visitation, I took her to court again. I got the usual crap from the judge: "that it's just awful that you fight this way and that you should be ashamed." Judges always look at me when they say that; their eyes are on me. I think what they mean is that "you are the only one here who has rationality, you should be stopping this. Also, you're male so you're inherently wrong, that's why I'm looking at you." And he refused to change the custody, even though he agreed with me that I was correct and she was wrong; he told her not to do it again. Judges would never let men get by with these violations, but women get by with murder in these cases.

Only three fathers had successfully obtained a formal change in child custody and each had subsequently returned the children to their mothers. Two of the three men had then joined the ranks of absent fathers, claiming that they were weary of fighting with the former wife. The primary objective of their court actions had been to torment their former spouses, not to obtain daily parental responsibility for their children.

That a formal custody challenge was sometimes used to harass a former spouse did not exclude other objectives although it did tend to taint them. Several fathers who had lost their custody challenges, including two who had subsequent cases pending, sought custody changes because they believed their children's safety while in their mothers' care was at risk. Confounding their legal cases, however, was the fact that each of these men had been involved previously in extensive and public interpersonal conflict with the former wife; consequently, court personnel viewed these fathers' actions with suspicion and questioned their motives.

Child support was another especially volatile and meaningful issue, and anger about it was standard. Noncompliance with court orders mandating support was not limited to absent fathers, although they had proportionately much higher rates of nonpayment; only one absent father was contributing to the financial maintenance of his children.[5] Refusing to pay child support was described as a legitimate response to unfair treatment from both the legal system and the former wife and was an action typically defined in terms of violated rights. Paying irregularly or not at all was explained also as a way to punish the former wife for various actions and attitudes. Conflict, noncompliance with child support agreements, and father absence were related: the under- or nonpayment of support was an ongoing subject of disagreement contributing to the move to a state of absence. In turn, absence was used to justify nonpayment. One absent father, who was vehement in his rejection of the role assigned to him as a visiting father, explained his refusal to pay child support:

> Why should I have to pay for children who[m] I do not live with and who[m] I do not have a part in raising? By paying child support, I simply reinforce my ex for having left the marriage and denied me my children. What kind of logic is

that? Why should I have to add to my losses by paying out my hard-earned money? So I take my chances that they'll [the legal authorities] throw me in jail. But I refuse to pay.

Alluding to the ambiguity regarding father–child relationships and parental activities and responsibilities in the postdivorce situation, several absent fathers noted that their children were living with stepfathers who, because they "had the pleasure of living with them," should be responsible for their support. For them, since divorce terminated any direct parental involvement, the obligation to provide economic support was ended, as this father suggested:

Not only would I be paying money to her so that she could then spend it on herself, and not the kids, but she has a husband who can afford to support them all. He lives with them; he can support them. We are no longer a family. Why should I support *that* family?

Nearly two-thirds of the men who paid child support regularly expressed an understanding for fathers who refuse to pay support.[6] That child support payments involved a one-directional transfer to the former wife, who was not required to be accountable for its use, was uniformly vexing and viewed as a violation of their rights to control their earnings and have independent lives. Moreover, being obligated by law to provide funds regularly to the former wife was symbolic, representing a shift in the balance of power between them. Child support was defined as constituting "unearned gifts" or "largess" to their former wives rather than as contributions to the support of children. One father described former wives generally as "leeches seeking to drain men dry." Another claimed that "child support taught children, and particularly girls, a welfare mentality in that it tells them that men will always pay their way." Several men insisted that they simply refused to be "money machines" for either their ex-wife or their children.

These divorced men complained that fathers are recognized within the legal system primarily or only as income providers. As evidence, they pointed to the disregard of expenditures made for items other than child support as well as to the restrictions imposed on their access to their children. Expenses incurred while their children were with them, such as those for food, entertainment, and travel, were not recognized by the courts in their cases. Nor was money spent on gifts or clothing acknowledged even when such items were needed by the children; judges dismissed such spending as "voluntary" and not relevant to the discussion of child support levels or payment compliance.

Another issue that infuriated the men was the link made by former wives between child support and visitation. Former wives tried to justify their interference with visitation by leveling charges of inadequate child support payment, although the two are legally defined as distinctive issues. Claims regarding child support were used to harass them in other ways as well. For example, explaining the experiences leading to his absence, this father stated,

She kept insisting that I was hiding money, that I had lied to the judge by filing a false income statement. She knew I was unemployed and that I was only behind in making support payments because I didn't have a cent. Yet she announced that "if you don't pay, you don't get the kids." The fact that I had lost my job was totally meaningless to her. Actually, she believed I quit that job on purpose so that I wouldn't have to pay support. That's what she kept telling her attorney and my kids.

While they complained about the amount of support ordered, arguing that it would have been even higher had they not actively resisted the former wife's demands and her lawyer's efforts, only three fathers indicated that their own financial status was hampered seriously by the support ordered, and two of these fathers were unemployed. All fathers interviewed, however, described men who were impoverished by child support payments. Widespread among these fathers was the belief that financial hardships by divorced men are common but are glossed over in the media and by researchers while those experienced by women are exaggerated. One man characterized the media coverage of the economic outcomes of divorce as "feminist propaganda intended to discredit men."

Largely absent from these fathers' accounts was the scope of their own or other men's participation in the constructing and maintaining of high levels of postdivorce conflict. Tactics and lines of action, such as insistence on achieving a particular outcome, intimidation, and aggressiveness, and seeking to exert control rather than to locate areas for compromise, were defined as necessary responses to particular conditions rather than interactional constructions to which they were party.

Emotion Management

As a strategy of action, absence was used also as a means of emotion management (Hochschild 1983). That is, absence served as a strategy for handling the various and powerful emotions integral to their postdivorce experiences. Emotion management served "to sustain a certain gendered ego ideal" (Hochschild 1990, 24) or to reinforce self-identity. By managing their feelings and directing them along certain avenues, these men empowered themselves in their assertions that they "remained in control," both of themselves and their situations, despite the unfamiliar, complicated, and usually emotionally stressful postdivorce circumstances. Through absence, fathers could distance themselves from the reminders of earlier experiences and limit involvement in situations likely to elicit particular feelings. The avoidance of their children or former wives, or the regular issuing of threats to become absent, were not the only emotion management strategies taken by these men. Other reported actions included drinking excessively; using drugs, particularly cocaine; self-imposed social isolation or, at the other extreme, excessive levels of social activity including the seeking out of a number of women for casual relationships; overt conflicts with friends and other family members; and greatly increased involvement with work. Engaging in any or many of these actions did not exclude also using or threatening to use absence.

Pivotal in many postdivorce conflicts were intense feelings regarding the former wife; anger toward her was a particularly common emotion. Thus absence was often used or considered as a means to preclude further angry and hostile interactions with the former spouse. Absence was also used to limit physical violence: nearly half of the men in the study reported incidents in which they had resorted to violence or had threatened it with serious intent since separating from their wives. "She drove me to it" was a common refrain among the men describing the use of force. Many of those recounting violent episodes insisted that their former wives had "set them up," usually by violating their rights in some deliberate way, seeking to incite violent behavior to justify limiting their access to their children. But even as the men attributed responsibility for their use of force to the former wife, they also acknowledged its efficacy for them. The threat or use of physical force served as a means to reassert control in the immediate situation, to release pent-up feelings of frustration and anger, and to terminate or avoid the search for other problem-solving strategies. The dynamics of interactional violence tended to overshadow all other issues between the man and woman.

Moreover, hostile and aggressive acts also served as a defensive posture against other, usually less familiar or acceptable, feelings. Anger itself was a form of emotion management, and anger, if not already felt, could be triggered readily by defining issues as violated rights. Other feelings, such as sadness and sorrow, loss and pain, and fear, could sometimes be disregarded, denied, or left unexamined by being defined as facets of anger. Anger could be identified, expressed, and directed in ways that other emotions could not be, especially given most men's limited experience in dealing with other divorce-related feelings, such as distress and sorrow. As Riessman (1990) notes, "Men's various manifestations of distress have one thing in common: the distancing of self from feelings of sadness" (p. 153); men "have trained incapacities in the language of feelings . . . they take their distress into the realm of action" (p. 159).

Although given far less attention by the men in their accounts, other emotions contributed to the phenomenon of postdivorce absence and were intertwined with feelings of anger. Unresolved and intense feelings of loss about the ended marriage and the former partner, for example, played a significant part in several men's moves toward ceasing involvement with their children. One father indicated that it was the emotional turmoil prompted by his former wife's remarriage that finally pulled him totally away from his children: "Her marriage was the final nail in the coffin." Another indicated that he saw his children rarely because he had "only finally begun to heal" from the loss of his wife and the end of his marriage. "Seeing my children simply reopens old wounds. It's better to avoid the reminders of the past."

In contrast, several fathers attributed their limited, declining, or total lack of involvement with their children to a gamut of emotions pertaining directly to them.[7] One father, unusual among the participants in the degree to which he distinguished his feelings of sadness from those of anger, discovered that relating to his son sporadically rather than routinely during daily family life prompted unbearable feelings of sorrow and loss. He described his experience:

> Every time I pulled up to the driveway to let him off, it was like part of me was dying all over again. I could barely keep myself together long enough to give him a hug goodbye; I knew it wasn't good for him to leave seeing me so visibly upset each time. He would open the door, step out of the car, and I would feel as if I would never see him again. He would walk up the sidewalk and a sense of grief would utterly overcome me. It would take me several days to pull myself together enough to even function at work. I'd have to keep his bedroom door closed; I couldn't bear to see his empty room. I had to break it off totally just to survive; the visits themselves were terrible because I had this constant unease, knowing what was coming.

Other fathers also found the contacts with their children to be intensely discomforting, though in varying ways. Numerous fathers found themselves confronted with a kind of on-the-job training for which few guidelines were available. Fathers were neither familiar with negotiating the parent–child relationship independent of the children's mother nor were they prepared for the changes prompted by divorce. The ambiguity surrounding what it meant to be a divorced father was only exacerbated by these fathers' feelings of anxiety and sense of uncertainty, particularly during the crucial early months following the marital separation when children looked to their parents for guidance in the altered situation.[8] One father, alluding to the disruption of his former role as the authority figure in the family, said,

> I found that I really couldn't control my boys except by getting angry. They just argued and fought when they were around. Every visit was incredibly tense; we were like coiled rattlesnakes just waiting to strike. I'd end up losing my temper which just made it worse because they treated me like I had no right to punish them.

Most fathers either did not try to establish a routine of family life or were unable to achieve it when their children were with them. Their time together was defined as a "visit." Fathers sought primarily to entertain their children, continuing former family recreational activities or exploring new ones together. Only two fathers enjoyed this approach, one of whom noted,

> We [my two sons and I] play really well together so it works out great. We have a terrific time together. We're like kids together.

While representative in its reference to the awkwardness of visits, the following excerpt is unique in the extent to which the father explicitly expressed empathy for his six-year-old daughter's experience:

> How many times in one day, after all, can I take my daughter to McDonald's or to the park to swing? I just don't know what to do to entertain her. So we end up renting videos and spending hours just sitting in front of the TV screen. I am restless and bored. She is unhappy and bored although she tries, she really does seem to try. She needs to be out playing with her friends, not stuck here with me.

Having been divorced about 18 months, this father continued to be involved with his child but indicated that his contacts with her had been declining steadily over the past year; at the time of the interview, he had neither seen nor talked to her for nearly four weeks. He avoided telephoning her because he anticipated that she would inquire about his plans for the next visit.

Numerous fathers did continue to endure the strains of at least occasional visits. Others, however, viewed the logistical and emotional tensions of visits to be sufficient or further cause for moving toward postdivorce absence. Several, for whom visits were consistently stressful and tense, had begun to deliberately avoid their offspring, screening telephone calls through answering machines and canceling or neglecting to arrange visits.

Interpersonal and emotional conflicts over children also involved the men's feelings of resentment at their perceived marginalization from family relationships. Given their respective primary family activities during marriage, in which they had been primary providers and their wives, even when employed, the primary parent and family caretaker, their parental relationships with their children differed from those of their former wives.[9] Mothers' relationships with children were relatively independent, whereas fathers' relationships with children involved the mother who typically played a mediating role. Consequently, mother–child relations often were viewed as being more durable and able to withstand the transitions and tensions inherent in the divorcing processes. These perceptions were reinforced by the men's place in a gender-structured society, their general acceptance of conventional gender beliefs and stereotypes, knowledge of the widespread prevalence of maternal custody after divorce, and personal experiences with postdivorce relations. After divorce, noncustodial fathers' parental authority and responsibilities in the routines of everyday life decreased, whereas the custodial mothers' increased. Former wives were deeply resented, especially by noncustodial fathers who perceived them to have gained disproportionate and undue amounts of power in the postdivorce family situation.

Absence was a strategy of emotion management for some fathers in response to feeling unappreciated or outrightly rejected by their children, particularly older children who were more likely to enter into the ongoing parental conflicts. Some children "took sides," seeking to explain or defend their mothers' actions. Others expressed their anger or resentment about the economic conditions facing them in the postdivorce situation, implying that their fathers were responsible, sometimes for capricious or vengeful reasons, for the reduced standard of living experienced after divorce. Most commonly, fathers perceived that children were resentful through unspoken messages conveyed, particularly through actions indicating sullen attitudes. On the other hand, a few children behaved "too well," as if they were "guests" during visits; such behaviors were interpreted by fathers to mean that the parent–child relationship was being denied, even if subtly, by the children. Because these fathers typically were unwilling or uncertain about how to engage their children in verbal self-disclosure about their feelings, the tensions between them and the fathers' feelings of rejection increased over time rather than lessened. Absence became a viable response.

Children were active agents in the postdivorce situation in other ways as well. For example, several men's children under the age of 5 actively resisted leaving their homes or their mothers to go for visits, crying and physically fighting when forcibly taken. Although one of these fathers was engaged in a custody challenge, the other two were considering full withdrawal from their children's lives, believing that even at their young ages these children were deliberately rejecting them as fathers. Four older children, ranging from ages 10 to 18, refused to have any contact with their fathers, although, in each of these cases, their siblings remained involved. Merged with these fathers' feelings of anger about the rejection by one of their children was fear that their other children would also come to refuse further contact, influenced by the other sibling. Yet no concrete actions had been taken by these fathers to alter the situation.

Nearly all fathers who were dissatisfied with their visitation experiences, a large majority of those in the study, targeted the former wife as being responsible for the strained father–child relationships, believing that the stress and awkwardness of the visits could be eased by her intervention with the children.[10] The expectation that she facilitate the postdivorce relationship was an extension of the marital division of labor in which women are the emotional caretakers or workers (Hochschild 1983; Riessman 1990). That the former wife refused to continue serving as family mediator as she had during marriage was interpreted by these fathers as evidence of her misuse of power, intended deliberately to undermine or push him outside the bounds of postdivorce family relationships. For fathers who desired a meaningful relationship with their children in the postdivorce situation, the consequences of not having been a primary caretaker or full coparent during the marriage and being relatively inexperienced in handling the emotional dimensions of family life were pronouncedly adverse.

Prior family relationships, however powerful they were in shaping most postdivorce ones, did not inevitably dictate them. While gender is relevant in men's parenting activities in that it places them into one of two possible structural locations, a change in status and thus change in experiences and expectations can lead to a change in parenting behaviors (Cohen 1989; Risman 1989). Ten of the men participating in the study, all but two of whom can be characterized as "androgynous" postdivorce fathers, marveled at how they had "learned to become a father only after divorcing." Each of the ten men had experienced emotional turmoil and confusion in the early days of being a divorced father, like the others. But they, as one noncustodial father of two children described the process, "had refused to drown and instead learned to swim, and even swim well." Meeting the challenges of being a divorced father was rewarding and satisfying; absence was personally inconceivable. Six of these ten fathers noted that the divorce experience had necessitated significant changes in their own behaviors without which they would not have become involved fathers. One said,

> If I hadn't divorced, I probably would have gone on in the same old way, relating to my kids as my father related to me and my sisters. He was just sort of there, in our lives but not relating to us in any important ways. I would probably never have learned how to be a father, how to be a parent.

Most fathers, however, were emotionally disconcerted about the quality of their relations with, or absence from, their children. They felt emotionally isolated, unable or unwilling to express their feelings about their children with others. Men who had remarried felt less emotionally isolated overall but half of those eighteen felt constrained in revealing the scope of their feelings about their children from the prior marriage to their present wives. The men's sense of isolation, together with the gamut of emotions experienced, led many to believe that men are the unrecognized emotional victims of divorce. One father, who had likened the success of his child snatching to the dropping of an A-bomb, bluntly summed up his observation that men are victimized by divorce:

> I learned we men have to be willing to eat the hurt. I'm very strong. But they beat us down so far. And they just keep beating. So the average guy just severs ties and leaves. He stops payments too. Even the payments are too painful a connection. It's just easier to eat the hurt and walk away. These fellahs go, just to survive. That's what I did. I had to survive. I left.

While on the one hand expressing pride in his aggressiveness and interactional tactics and on the other claiming to have been victimized, this father, like most, did not make any significant link between his actions and the persistently high levels of conflict and tension in the postdivorce situation. Nor did he see a relationship between his behaviors and his former wife's enormous distrust of his motives. Absent in a large majority of men's accounts was recognition that the character of the postdivorce situation, which extracted such a high emotional toll, was the outcome of collective activity in which they were primary participants.

Absence as a Standard of Comparison

The second facet of the phenomenon of postdivorce fatherhood was absence as a standard of comparison: absent fathers were the primary comparison group of divorced men for 80 percent of the fathers in this study. That so many fathers perceived absence to be a viable option for themselves partially explains the use of absent fathers as a reference. Also, most knew few if any divorced fathers who were extensively involved with their children; even custodial fathers knew almost no other men whose situations were similar to their own. All the interviewees, however, personally knew men who were totally absent from their children's lives, and their actions and situations were used for comparisons. Any involvement with their own children was evidence of good effort compared to absent fathers as was suggested in the descriptive terms frequently used to refer to absent fathers: *derelict dads, deadbeat dad,* and *bums.* Involvement of any kind rather than absence prompted a stance of self-congratulation among some fathers. For instance,

> I occasionally have to acknowledge that maybe I could have handled most of this better—better for my children—by not insisting on having the last word always with my ex. But at least I haven't checked out like lots of guys in my situation here. I've hung in there and struggled. But most guys just leave, hang it up and leave.

Several men who both used father absence as a reference and expressed an understanding of it, nonetheless condemned the practice. Those who expressed the harshest condemnation of absent fathers, however, were the men who had most successfully dictated the terms of their divorce settlement and child custody arrangements. Able to assert themselves because of their particular access to power and resources, they were critical of other divorced fathers who, they argued, needed to exert more effort and assertiveness to shape the postdivorce situation: they needed simply to be more "manly."

One father whose remarks about absent fathers were consistently derogatory was particularly satisfied with his postdivorce situation; he credited the outcome of his divorce to both his personal competence and assertiveness in dealing with his attorneys and to the coercive and intimidating approach he had used in interacting with his former wife. By threatening her with a legal battle for sole custody and drawing on his extensive business-related legal network, he had obtained shared physical as well as shared legal custody of their four-year-old daughter, an atypical custody arrangement in New York State, especially given the child's age. He had also successfully arranged the property settlement so as "to avoid ever having to pay child support," even though his former wife's income and earning potential were far less than his.

> I refused to let anyone but me set the terms of this divorce. I refused to be a part-time father. And I swore that I absolutely never, never would pay a cent in child support. So I found the hottest lawyer in the state and that's what we got.

The majority of fathers in the study, however, did not have the means to impose their desired divorce outcomes unilaterally and, for the most part, they sympathized, and often identified, with the experiences and choices of fathers who become absent. That absent fathers constituted a primary comparison group for these divorced men suggests basic questions about the meaning of fathering to them in general. Fatherhood as a status within an intact marriage and family had a relatively clear definition, but fathering as an array of activities, interactional processes, and particular kinds of social relations between father and child was ambiguous, at best, to many.

Conclusion

Family transitions prompted gender role and identity transitions in these men's lives. Indeed, much of the apparently intransigent hold of the masculinist discourse of divorce on the majority of men interviewed (many of whom seemed to respond to the stresses and turmoil of divorce by resorting to more traditional or conventional masculine behaviors and perspectives than they had while married) was aimed at reestablishing or buttressing the masculine self as a dominant self amidst the array of changes prompted by divorce. A primary theme of the discourse was a rhetoric of rights through which relationships, actions, and emotions were framed and defined. Absence was understood as a viable response to certain interactional and emotional situations. The possibility that father absence was dysfunctional in that it foreclosed opportunities for negotiating more meaningful relations with children was largely unacknowledged by these fathers, especially since involvement took a variety of forms, many of which were not fully distinct from absence. Rather than being "locked out" of postdivorce relationships with their children by others, as has been suggested by various speculative explanations for father absence (Bauer and Bauer 1985; Doyle 1985), these fathers were more typically "locked in" to particular relational configurations and systems of meanings held by them and shaped by gendered ideology, practices, and social arrangements. These perspectives often countered the development of continuing and reciprocally nurturant postdivorce relationships with their children and contributed, instead, to the acceptance of father absence as an understandable, if sometimes regrettable, practice.

NOTES

1. Other but less common interpersonal conflicts to be avoided by father absence were ones involving a new spouse or companion, and, sometimes, additional children, hers and/or theirs. Logistical issues involving time, money, and geographical distance were also issues contributing to conflict and were to be avoided. Since these issues were typically subsumed under others, especially emotional or interpersonal conflicts with the former wife, they are not more fully developed in this article.
2. According to the men in this study, numerous allegations had been made against them that behaviors toward a child during visits had constituted or approximated sexual molestation. Following formal investigations and hearings, one of these fathers was denied access to his children until they reached age 18 and could choose for themselves whether they wanted contact with their father. Another father gave as his explanation for his absence his fear that if he were again investigated he would be wrongfully convicted of child sexual molestation. Also, two fathers claimed that their children had been sexually molested by friends of the former wife and another by the maternal grandfather. One mother agreed voluntarily to prevent contact between the child and the suspect, and the other mother had denied and been cleared of all charges by an investigation through the New York Family Court. Nearly a fifth of all the men interviewed indicated they suspected that boyfriends of their former wife or her new husband engaged in activities of a sexual character with their children. These findings suggest, at the least, that the exchange of such allegations between divorced parents is widespread and that systematic research exploring this issue is needed.
3. According to a study done for the U.S. Department of Justice (Finkelhor, Hotaling, and Sedlak 1990), "There were an estimated 354,100 *Broad Scope Family Abductions* in 1988," defined as "situations where a family member (1) took a child in violation of a custody agreement or decree, (2) in violation of a custody agreement or decree failed to return a child at the end of a legal or agreed-upon visit, with the child being away at least overnight" (p. ix). "Most of the . . . abductions were perpetrated by men, noncustodial fathers and father figures" (p. xi). In contrast, there were an estimated 3,200 to 4,000 *Legal Definition Non-Family Abductions* known to law enforcement in 1988 (p. xii).
4. Visitation is recognized within the law as the means through which a divorced noncustodial parent retains access to minor children and visitation rights are closely protected by the judicial system in each state (Novinson 1983). A schedule of visitation is typically stipulated in the divorce settlement approved or ordered by the court. Denial or interference with a noncustodial parent's access to minor children by the custodial parent is a violation of law: a noncustodial parent has specified and protected rights to his or her children. Judicial procedures for seeking enforcement of visitation rights and redress for violations are available in all states, and one possible sanction for interfering with visitation is the loss of custody. In contrast to these specified parental rights, there are no legal mandates requiring a parent to become or remain involved with a child (Bruch 1983; Polikoff 1983).
5. A direct relationship exists between the nonpayment of child support and paternal absence: fathers paying no support have particularly low levels of contact with their children. It is the fact of support, not the amount, which is related to maintaining ties with children (Furstenberg et al. 1983, 663; see also Hoffman and Duncan 1988; Weitzman 1985). Payment of child support is not related to socioeconomic or income status (Chambers 1979; Weitzman 1985). Moreover, men's economic situations are not altered markedly by the payment of child support. In sharp contrast to women's—and therefore most children's—postdivorce circumstances, men's standard of living actu-

ally improves after divorce (Corcoran, Duncan, and Hill 1984; Hoffman and Duncan 1988; Weitzman 1985).

6. One aspect of both the widespread noncompliance with child support orders and the sympathy expressed by the men in this study with those who refuse to pay child support has to do with the extensive variability in amounts ordered. As argued by Weitzman (1985) and Glendon (1987), the variability has two sources: the extensive degree of judicial discretion legislatively allowed in the establishing of support orders and the wide latitude that divorcing spouses and their attorneys have in negotiating settlements outside of the judicial system. Fathers are well aware of the differences in child support mandates. As noted by Glendon (1987), "Even though it is not uncommon for a noncustodial parent's child support payments to be lower than his car payments, every support debtor knows someone whose economic circumstances are similar to his, but whose child support payments are lower. Discontent and a sense of the essential unfairness of the system are widespread" (p. 92). Recent federal legislation mandated that all states establish child support guidelines to reduce the variability in awards and to ensure that awards meet certain levels. The results of these efforts are not yet clear, but anger about the recently implemented New York State Child Support Guidelines Act was extensive among the men interviewed; the assertion frequently was made that more fathers will now fight for custody since, by having sole custody, they will retain control of their incomes.
7. In contrast to these particular men's remarks about the emotional effects of absence from their children, Spanier and Thompson (1984) concluded from their research that, "among men, *none* of the child-related variables was a factor in well-being after separation. . . . Even longing for the children—a common form of loneliness among fathers—was unrelated to well-being" (p. 222). Mothers' well-being, in contrast, was found to be dependent on their relationships with their children; moreover, noncustodial mothers maintained significantly more sustained and active relations with their children over time after divorce than did noncustodial fathers (Furstenberg and Nord 1985; Spanier and Thompson 1984).
8. The months immediately following the marital separation are crucial for the negotiation of the post-divorce father–child relationship: studies indicate that father involvement decreases over time for most noncustodial, divorced fathers (Furstenberg and Nord 1985; Furstenberg et al. 1983; Hetherington, Cox, and Cox 1976).
9. Despite some changes in the "good provider role" (Bernard 1981), the increased family involvement of some fathers (Pleck 1985, 1989), and much rhetoric about the "new father," the vast majority of men (in intact families) retain primary responsibility for economic support and have relatively limited involvement with their children (Lamb 1986; Lamb, Pleck, and Levine 1986).
10. Only a small proportion of divorced spouses develop a cooperative, even friendly, relationship with each other, indicating that the "pattern of cooperative parenting, so widely portrayed in the popular media, is, in fact, rather rare" (Furstenberg and Nord 1985, 899). Indeed, former spouses appear to be "reluctant partners in the childrearing process," and children often provide the link between the parents in terms of communication (Furstenberg and Nord 1985, 900).

REFERENCES

Arendell, Terry. Forthcoming. *Fathers and divorce* (tentative title). Berkeley: University of California Press.

Bauer, Bob, and Daphne Bauer. 1985. Visitation lawsuit. In *Men freeing men,* edited by Francis Baumli. Jersey City, NJ: New Atlantis.

Bernard, Jessie. 1981. The good-provider role: Its rise and fall. *American Psychologist* 36: 1–12.

Bruch, Carol. 1983. Developing normative standards for child-support payments: A critique of current practice. In *The parental child-support obligation,* edited by J. Cassetty. Lexington, MA: Lexington Books.

Chambers, David. 1979. *Making fathers pay: The enforcement of child support.* Chicago: University of Chicago Press.

Cohen, Theodore. 1989. Becoming and being husbands and fathers: Work and family conflict for men. In *Gender in intimate relationships,* edited by Barbara Risman and Pepper Schwartz, Belmont, CA: Wadsworth.

Corcoran, Mary, Greg Duncan, and Martha Hill. 1984. The economic fortunes of women and children: Lessons from the panel study of income dynamics. *Signs: Journal of Women in Culture and Society* 10(2): 232–48.

Doyle, Richard. 1985. Divorce. In *Men freeing men,* edited by Francis Baumli. Jersey City, NJ: New Atlantis.

Emery, Robert. 1988. *Marriage, divorce, and children's adjustment.* Newbury Park, CA: Sage.

Finkelhor, David, Gerald Hotaling, and Andrea Sedlak. 1990. *Missing, abducted, runaway, and thrownaway children in America.* Washington, DC: U.S. Department of Justice.

Glendon, Mary Ann. 1987. *Abortion and divorce in Western law: American failures, European challenges.* Cambridge, MA: Harvard University Press.

Hochschild, Arlie. 1983. *The managed heart: Commercialization of human feeling.* Berkeley: University of California Press.

———. 1990. Ideology and emotion management: A perspective and path for future research. Department of Sociology, University of California, Berkeley.

Hoffman, Samuel, and Greg Duncan. 1988. What are the economic consequences of divorce? *Demography* 25: 415–27.

Jaggar, Allison. 1983. *Feminist politics and human nature.* Totowa, NJ: Rowman & Allenheld.

Lamb, Michael. 1986. *The father's role: Applied perspectives.* New York: Wiley.

Lamb, Michael, Joseph Pleck, and James Levine. 1986. Effects of increased paternal involvement on children in two-parent families. In *Men in families,* edited by Robert Lewis and Robert Salt. Beverly Hills, CA: Sage.

Novinson, Steven. 1983. Post-divorce visitation: Untying the triangular knot. *University of Illinois Law Review* 1: 121–200.

Pateman, Carol. 1990. *The disorder of women: Democracy, feminism and political theory.* Stanford, CA: Stanford University Press.

Pleck, Joseph. 1985. *Working wives, working husbands,* Beverly Hills, CA: Sage.

———. 1989. Men's power with women, other men, and society: A men's movement analysis. In *Men's lives,* edited by Michael Kimmel and Michael Messner. New York: Macmillan.

Polikoff, Nancy. 1983. Gender and child-custody determinations: Exploding the myths. In *Families, politics, and public policy,* edited by Irene Diamond. New York: Longman.

Riessman, Catherine Kohler. 1990. *Divorce talk: Women and men make sense of personal relationships.* New Brunswick, NJ: Rutgers University Press.

Risman, Barbara. 1989. Can men "mother"? Life as a single father. In *Gender in intimate relationships,* edited by Barbara Risman and Pepper Schwartz. Belmont, CA: Wadsworth.

Spanier, Graham, and Linda Thompson. 1984. *Parting: The aftermath of separation and divorce.* Beverly Hills, CA: Sage.

Swidler, Ann. 1986. Culture in action—symbols and strategies. *American Sociological Review* 51: 273–86.

Wallerstein, Judith, and Sandra Blakeskee. 1989. *Second chances: Men, women, and children a decade after divorce.* New York: Ticknor & Fields.

Weitzman, Lenore. 1985. *The divorce revolution: The unexpected social and economic consequences for women and children in America.* New York: Free Press.

Divorcing Reality: New State Laws That Slow Down Divorce

Stephanie Coontz

Every time it seems America may finally be coming to terms with how much and how irreversibly our families have changed, a new wave of panic breaks over us. Most recently it's been a rediscovery of the "catastrophe" of divorce. This past summer a new law took effect in Louisiana, giving people the chance to choose a "covenant marriage" in which the state will enforce an agreement not to divorce except for adultery, physical or sexual abuse, alcoholism or a year's abandonment. The sponsor of the bill says he has since received calls from lawmakers all over the country inquiring how to institute similar laws. At least nineteen states already have legislation pending to "slow down" divorce.

Most of the ammunition for this campaign is drawn from Judith Wallerstein's longitudinal study of 131 children whose parents divorced in 1971. In 1989 Wallerstein published a study claiming that almost half had experienced serious long-term psychological problems that interfered with their love and work lives. This summer she released an update based on twenty-six of these young adults, all of whom had been 2–6 years old when their parents separated. They had been extremely vulnerable to drug and alcohol abuse as teens, she reported, and were still plagued in their 20s and 30s by unstable relationships with their fathers, low educational achievement and severe anxieties about commitment.

The media pounced. I found more than 200 newspaper articles and opinion pieces trumpeting the "new" finding that divorce was "worse than we thought," a "catastrophe" for kids. While Wallerstein herself opposes legal restrictions on divorce, she has done little to distance herself from those who cite her work in support of the new crusade. "I've been so misquoted in America," she told *Mother Jones* two years ago. "I cannot worry about it anymore."

But there is good reason to worry about the massive publicity accorded Wallerstein's work. Her estimates of the risks of divorce are more than twice as high as those of any other reputable researcher in the field. Her insistence that the problems she finds were caused by the divorce itself, rather than by pre-existing problems in the marriage, represents an oversimplified notion of cause and effect repudiated by most social scientists and contradicted by her own evidence.

Wallerstein studied sixty Marin County couples, mostly white and affluent, who divorced in 1971. Her sample was drawn from families referred to her clinic because they were already experiencing adjustment problems. Indeed, participants were recruited by the offer of counseling in exchange for commitment to a long-term study. This in itself casts serious doubt on the applicability of Wallerstein's findings. The people most likely to be attracted to an offer of long-term counseling and most likely to stick with it over many years are obviously those most likely to feel they need it. And after twenty-five years in a study about the effects of divorce, the children are unlikely to consider any alternative explanations of the difficulties they have had in their lives.

Wallerstein says she tried to weed out severely disturbed children, yet the appendix to her original study, published in 1980, admits that only one-third of the families she worked with were assessed as having "adequate psychological functioning" prior to the divorce. Half the parents had chronic depression, severe neurotic difficulties or "long-standing problems in controlling their rage or sexual impulses." Nearly a quarter of the couples reported that there had been violence in their marriages. It is thus likely that many of the problems since experienced by their children stemmed from the parents' bad marriages rather than their divorces, and would not have been averted had the couples stayed together. Other researchers studying children who do poorly after divorce have found that behavior problems were often already evident eight to twelve years before the divorce took place, suggesting that both the maladjustment and the divorce were symptoms of more deep-rooted family and parenting issues.

This is not to say that all the problems Wallerstein found can be explained by pre-existing family dynamics. While children in intact families with high levels of conflict usually do worse than children in divorced or never-married families, children's well-being often does deteriorate when a marriage not marked by severe conflict comes to an end. Divorce can trigger new difficulties connected to loss of income, school relocation, constriction of extended family ties or escalation of hostility over issues like custody and finances. (In Wallerstein's sample, many women had not been employed during the marriage; forced entry into the workplace increases the risk of depression and distraction, which can affect the quality of parenting.) Intense conflict after divorce can be even more damaging to children than intense conflict within marriage.

Still, more representative samples of kids from divorced parents yield much lower estimates of risk than Wallerstein's. Paul Amato and Bruce Keith, reviewing nearly every single quantitative study that has been done on divorce, found some clear associations with lower levels of child well-being. But these were, on average, "not large." And the more carefully controlled the studies under review, the smaller were the differences reported.

Interestingly, children whose parents divorced in more recent generations are experiencing less severe problems than those whose parents divorced when laws and social stigmas were stricter. Indeed, a just-published study of 160 Boston-area families conducted by psychologist Abigail Stewart (*Separating Together: How Divorce Transforms Families*) found that while most youngsters had slightly poorer than average mental health a few months after the divorce, their overall mental health had rebounded to average levels after eighteen months.

Wallerstein rejects these studies because they do not take account of what she terms a "sleeper effect," in which problems caused by divorce do not show up until years later. But larger long-term studies do not support this claim, though there may be a sleeper effect for children whose parents continue to battle after the separation. Mavis Hetherington, who has studied more than 1,500 children of divorced parents, reports that the large majority grow up socially and psychologically well-adjusted.

Some past studies have confirmed that children of divorced parents are more likely to get divorced themselves. But another new study shows that even this so-called inheritability of divorce is also on the decline. UCLA researcher Nicholas Wolfinger found that between 1974 and 1993 there was a 50 percent decrease in the tendency for people whose parents had divorced to get divorced themselves.

Family values crusaders often argue explicitly that a little bit of exaggeration, or at least a use of worst-case scenarios, is justified in discussing the effects of divorce because emphasis on children's resilience may lead couples to take divorce too lightly. It is probably true that some people are unwilling to do the hard work of trying to make a relationship succeed, or do not give sufficient thought to the difficulties they or their children may face after divorce. But rising rates of divorce and single parenthood come less from me-first individualism than from long-term historical forces that are not going to be reversed by trying to scare or guilt-trip people into staying married.

If you graph the divorce rate since the 1890s, the current rate is exactly where you'd expect it to be from the trends during the first half of the century. The age of marriage is at an all-time historic high for women; at the other end of the line, a person who reaches age 60 can expect to live, on average, another twenty years. The institution of marriage organizes a smaller portion of people's lives and social roles than ever before.

The economic autonomy of women means that dependence no longer preserves marriages, and the number of people who exist comfortably and happily outside marriage creates an ever-present alternative for people who are unhappy with their mates. No amount of coercion is going to put the toothpaste back in the tube.

In these circumstances, coercion would only make things worse for the very people the antidivorce crusaders say they want to protect. Contrary to conservative rhetoric, women have historically needed the legal protection of divorce more than men have. For centuries, men's greater social and economic power forced many wives to put up with a husband's affairs or his humiliating treatment. Men also had more resources to fight a divorce or penalize a woman for "fault" under older laws. The fact that two-thirds of all divorces today are initiated by women indicates that many women are grateful for the easing of divorce laws.

One group of women has been badly hurt by no-fault divorce in the absence of strong alimony laws: women who played by the old female homemaker rules and whose husbands threw out the rulebook altogether. But making divorce harder and more acrimonious would not protect these women. Would a woman who doesn't want a divorce really be better off if the law says her husband can't divorce her except in case of adultery or violence? What would prevent him from deserting the family, engaging in abuse, provoking her into a compromising situation or even fabricating evidence of her adultery? Better to make sure that strong child-support laws are enforced, and that husbands whose wives sacrificed income and education for the sake of the marriage pay spousal support.

Slowing down divorce is not necessarily in the best interests of children either. If a couple can repair their marriage and develop an effective parental alliance, their kids will certainly benefit. But getting people to "try harder and longer" can make things worse if the marriage does eventually fail. Most studies find that divorces are more damaging for kids when they occur between the ages of 11 and 16 than when they occur between 7 and 11. This doesn't mean parents should rush into divorce, but it does mean that we should beware of frightening or pressuring couples into prolonging a marriage that may well end up being intolerable to one or the other.

We may be able to save more potentially healthy marriages than we currently do, but only by modernizing marriage, not by shoring up a model based on women's self-sacrifice. Modernizing marriage means getting men and women to share child care and housework more equally, helping couples to manage conflict in less destructive ways and building family-friendly workplaces that make it possible to raise children with less stress. (Of course, such measures will also make it easier for divorcing couples, single parents and unmarried partners to raise children.)

It may be true, as conservatives charge, that lessening the stigma and stress attached to single parenting will lead some people to turn to divorce before exploring other options, but it's also true that as divorce has gotten more acceptable it has also gotten less damaging. In 1978, a national sample found that only 50 percent of divorced couples were able to contain or control their anger in a way that allowed them to co-parent effectively. A more recent California study of divorcing couples found that three to four years after separation, only a quarter of divorced parents were engaged in conflict-ridden co-parenting.

Similar progress has occurred in post-divorce parental contact. Surveys at the beginning of the eighties found that more than 50 percent of children living with divorced mothers had not seen their fathers in the preceding year, while only 17 percent reported visiting their fathers weekly. But a 1988 survey found that 25 percent of previously married fathers saw their children at least once a week, and only 18 percent had not visited their children during the past year. As divorce has become more common, more fathers have begun to work out ways of remaining in touch with their children, while more mothers seem willing to encourage such involvement. Researchers can help promote these new trends by explaining what we know both about how to create better marriages and how to parent more effectively after a divorce. . . . In the meantime, parents and the general public should take a hard, critical look at the claims of the antidivorce crusade.

Stepfamilies over the Life Course: Social Support

Lynn White

Most of the research on stepfamilies rightly focuses on the stepfamily with young children. Childrearing and social replacement are clearly the most important functions of families in any society, and knowing how well stepfamilies perform these functions is crucial to understanding how children and society will fare under this increasingly common family form.

Families are much more than childrearing organizations, however. Relationships between parents and children remain important long after co-residence ends and the children are grown. Although the actual quantity of goods and services exchanged between generations may be rather small under routine conditions, family members remain important sources of identity and perceived social support. In the National Survey of Families and Households (NSFH), for example, 80% of adults listed a family member as their first source of assistance in at least one of three emergencies. This perceived supportiveness has been found to be more important than actual exchange for mental health outcomes (House, Umberson, & Landis, 1988).

Thus, one question we need to ask about stepfamilies is whether they will provide these sorts of services to their members. Do people feel confident about relying upon their stepmothers or stepfathers in emergencies? Do stepchildren provide the same kinds of social integration and assistance as biological children?

There has been much speculation on these questions, but few definitive answers. Some scholars have taken a positive stance, arguing that remarriage and stepfamilies enrich the kinship network and provide us with an enlarged pool of potential kin (Furstenberg & Spanier, 1984; Wald, 1981). Others suggest that stepfamily bonds will be weaker than those in biological families, with a resulting shortfall in social support (Bartlema, 1988). In 1979, Troll, Miller, and Atchley (1979) wrote that "we know so little about kinship relations among reconstituted families that we cannot even speculate about them" (p. 69). In the ensuing 15 years, modest steps were taken to fill this gap.

In this [reading], I review theoretical frameworks for stepfamily support, and then examine empirical findings. Because few previous studies have been published, I rely largely on my own NSFH research to answer questions about stepfamily support.

Theoretical Frameworks

I briefly review sociobiological, social psychological, social network, and institutionalization perspectives on stepfamilies as support networks. Each leads to the prediction that stepfamilies will be less supportive than biological families.

Sociobiology: They're Not Carrying My Genes

The sociobiological argument (taken from Daly & Wilson, 1978) suggests that animals are reproductive strategists who maximize the survival of their genes into the next generation, in part by focusing nurturance on their own offspring. Not only is there no genetic predisposition to nurture another's children, such children may be seen as rivals who endanger one's own children's survival. In many species (including some primates), males will kill a mate's young from former unions, increasing the chances that her attention and resources will be devoted to bearing and nurturing his offspring. This argument was noted by some scholars as an explanation for the higher likelihood that stepchildren will be abused or killed (Dobash, Dobash, Wilson, & Daly, 1992; Finkelhor, 1979).

From a sociobiological perspective, not all conflict in stepfamilies will be between stepparent and stepchild. Van den Berghe hypothesized (personal communication, March 5, 1986) that as stepchildren reach puberty, they will become sexual competitors to their parents. In the usual structure of stepfamilies—biological mother/stepfather—his model implies that post-pubertal stepdaughters will be the most troublesome element in stepfamilies and that this family configuration will be most unstable.

Although sociobiology is controversial, I do not introduce it as a straw man. To paraphrase Alice Rossi (1984), failure to recognize the extent to which biology shapes our lives cripples our efforts to understand the role that social structures play. It may be that, given biological predispositions, human stepfamilies are amazingly successful.

Social Psychological Perspectives

Taking a very different approach, scholars who approach the family from a social-psychological orientation argue that what sets parent–child relationships apart from others is the long-term, cumulative nature of the affective bond (Atkinson, 1989). The issue is not whether you like or love your parents/children, but the extent to which your relationship to these persons is a defining characteristic of your identity.

Turner (1970) introduced the concept of *crescive bonds* to describe incrementally built bonds that "link irreplaceable individuals . . . into a continuing relationship" (p. 89). According to Turner, the development of such bonds requires that the relationship: (a) have a long history and an open-ended future orientation, (b) be a label others use to identify you (e.g., "you are John's father"), (c) be a label you use to identify yourself, and (d) provide rewards, particularly enhanced self-esteem. When all criteria are met, the relationship becomes an integral part of identity, and it is maintained even if it ceases to be rewarding.

Empirical evidence about family identities tends to focus on parents rather than children. Research suggests that "parent" is a defining role, a role that sits at or near the top of men's and women's identity salience hierarchies. Parenthood is, after all, an incredibly time-consuming role that, for many years, shapes and constrains lifestyle choices. It would be surprising, indeed, if parenthood was not nearly as important as age and gender in determining our sense of "Who am I?" Stepparenting, however, should be substantially less important to our identities: Stepparenting takes fewer years of our lives, we can walk away, identification is incomplete and tentative ("No I'm not John's father; I'm his stepfather"), and social recognition is weak and inconsistent. As a result, we would anticipate weaker role performance.

In recent research, Thoits (1992) documented the difference in role salience for parenting and stepparenting. Although her results must be tentative because of the small number of stepparent respondents, differences in role salience are more dramatic than those found in behavioral domains. Thoits asked her respondents to rank each of 17 roles (if occupied) on a scale of 0 (respondent held the role but did not list it as one of his or her nine top role identities) to 3 (most important identity). Although parenthood received ratings of 2.7–2.9 (the highest ratings of any role), the stepparent role ranged in salience from only 0.2–0.6. Nearly a dozen roles—including neighbor, churchgoer, in-law, relative, friend—intervened between parenthood and stepparenthood, and in general, being a stepparent was about as important as belonging to voluntary associations or having a hobby.

Thus, social psychological theory and evidence suggest that stepfamily roles are less salient to parents than are biological family roles. We might expect that this salience would be significantly greater when the stepfamily relationship had lasted longer and been formed earlier in the child's life. It is startling in Thoits' data, however, to note that stepparent roles were only marginally more salient for married than for divorced respondents.

The consequences of low role salience remain to be established. It seems likely that social support to stepchildren will be far higher than these salience rankings suggest—at least as long as the marriage creating the stepfamily lasts. Stepparents may aid stepchildren in order to please their spouse, rather than because of their own relationship to the stepchildren; likewise, stepchildren may maintain cordial relationships with the stepparent in order to please their parent. In the face of such very low role salience, the enactment of role obligations must depend on external forces—in this case, the biological parent.

Social Network Perspectives

In the last decade, a body of work has developed around the notion of social network—an interacting group whose exchanges are regulated by the density and duration of social relationships. Although the family is distinct from voluntary groups in many ways, the structural factors that operate to make other groups cohesive also apply to the family. Briefly, social network characteristics may partly explain why the family is more cohesive than other social groups: The family has high levels of density (everyone knows one another) and so has higher social control capacity (Fischer, 1982); long and open-ended interaction, and so greater likelihood of cooperation and diffuse reciprocity (Axelrod, 1984; Carley, 1991; Ekeh, 1974); and strong similarity, which reduces the cost of exchange (Carley 1991; Knoke, 1990). In every case, the stepfamily is structurally weaker than the biological family. The uniformly dense network of the biological family breaks into cliques where stepchildren and stepparent may have nonoverlapping sets of ties (to noncustodial fathers, noncustodial children, former affines), ties are of shorter duration and not so open-ended (stepparents can walk away), and genetic and, perhaps, even social similarity is less.

The structural weakness of the stepfamily implies that stepfamilies will be less able to maintain and enforce obligations to one another. They may be more apt to operate as a voluntary group, characterized by demands for immediate as opposed to diffuse reciprocity, and more dependent on affect. This line of reasoning leads to the direct hypothesis that stepfamilies will be weaker support mechanisms than biological families. It also leads to subsidiary hypotheses that exchange within stepfamilies will be greater when the stepfamily has lasted longer and when there are fewer competing ties (stepparent has no own children, noncustodial parent is dead or absent); it also implies that affect will have a greater effect on exchange within stepfamilies than biological families.

Institutionalization Perspectives

Many family scholars argue that what sets the family apart from other collections of individuals is not genetics, crescive bonds, or density of networks, but institutionalized norms of obligation—norms that hold us responsible for our children or parents even if we don't like them (Finch, 1989).

Norms of family obligation are weakening in contemporary society for blood relatives. Certainly, many fathers have found ways to walk away from their children without being, apparently, overwhelmed with guilt or ostracized by polite society. As Cherlin (1978) noted, these norms are even weaker for stepfamilies, and it is not at all clear how much obligation stepparents and stepchildren owe one another. How much guilt do stepparents/children impose on themselves if they fail to meet a need in the other? How much censure will others impose on the individual? Two kinds of evidence can be brought to bear on these questions: evolving legal obligation among stepfamily members and survey evidence about perceived normative obligation.

Legal Obligations Laws about family obligation reflect, sustain, and create behavior. Thus, laws about stepfamilies are informative about how our society views these obligations. Two issues are of particular relevance: child support and inheritance. Fine and Fine (1992) provided important insight into stepparent/stepchild relationships by reporting on legal rights and responsibilities in these areas.

In 45 states, stepparents have no obligation to support minor stepchildren even while they live with the child's parent; in no state are they required to continue providing support if their marriage to the child's parent ends or they cease to live with the child.

In regard to inheritance, stepparents *may* choose to specify that their stepchildren receive a share of their estate, but in cases where they die without a will, the courts will assume that they intended to omit their stepchildren. If a will directs that the estate is to be divided among "my children," the courts will interpret that phrase as excluding stepchildren. No evidence exists to support or refute the assumption of such an intention on the part of stepparents.

California has recently permitted stepchildren to inherit if "the relationship of parent and child [began] during the person's minority and continued throughout the parties' joint lifetimes [and if] it is established by clear and convincing evidence that the foster parent or stepparent would have adopted the person but for a legal barrier" (Waggoner, Wellman, Alexander, & Fellows, 1991, p. 131). The problem with this doctrine is that very few stepparents formally adopt their stepchildren, even though (in the many cases where the noncustodial parent disappears from the child's life and provides no support) such adoptions could probably be achieved even over the noncustodial parent's objections (Waggoner et al., 1991). Failure to adopt signals to the courts an absence of commitment. Does it carry the same signal to the child or the child's parent? Social science research could make a useful contribution, here.

From their review of statutes and judicial findings, the Fines deduce a judicial assumption that "stepfamily status entails no rights or obligations" (Fine & Fine, 1992, p. 335). If this assumption accurately reflects the feelings of stepfamily members—and little evidence exists one way or the other—then it is an inescapable conclusion that stepfamilies are seriously weak organizations whose roles lack substance.

Normative Obligations to Stepfamily Members

Bartlema (1988) argued that as serial monogamy replaces monogamy, we will find kinship networks filled with "all kinds of affinal kin of varying degree of relationships" (p. 218). In many cases, we may not even be sure that these people are, in fact, relatives or what we should call them (Day, 1988). Such confusion "leaves the individual free to fill in the nature and degree of affinity of such relationships according to his or her own needs and preferences" (Bartlema, 1988, p. 218). The result is less normative pressure and more dependence on individual characteristics, such as affection, affluence, and proximity.

Rossi and Rossi's (1990) study of three-generational family relationships in the Boston area provided unique information about norms of stepfamily obligations. Rossi and Rossi gave each of their respondents a set of 32 vignettes that randomized hundreds of combinations of situation, relationship, and gender. For each vignette, respondents were asked how obligated they would feel to provide the assistance specified to the kin member named. Stepparents and adult stepchildren were among the relationships examined.

The vignettes represented hypothetical situations, and the vast majority of respondents reporting on kinship obligations to adult stepchildren and stepparents had no such relatives in their own families. Evidence that those with stepfamily experience are less likely to carry negative stereotypes about stepchildren/parents than others (Fine, 1986; Fluitt & Paradise, 1991) suggests that these hypothetical reports may underestimate actual feelings of obligation among stepfamily members. On the other hand, Thoits' data showing extraordinarily low salience of stepparent roles suggests that stepparents may distance themselves from their stepchildren even more than the public anticipates. Because generalized norms affect the degree to which stepfamily members will be sanctioned for failing to meet obligations, however, they are important.

Rossi and Rossi (1990) showed that normative obligations to stepchildren fall very materially below those to own children, but they are nevertheless substantial.[1] Obligation was rated on an 11-point scale, where 0 represented *no obligation at all* and 10, *a very*

---

[1]The same general finding is reported in Schwebel, Fine, and Renner's (1991) small study of college students' perceptions of stepparent obligations to school-age children. Averaging across several vignettes in which the child needed some sort of assistance or support, 80% of the students thought stepparents ought to provide assistance, compared to 83% of biological parents—a significant, but quite small difference. These authors note that standard deviations differ more across categories than do the means, suggesting less consensus on norms about stepfamily obligations.

strong obligation. Averaging obligation ratings across the four crisis situations for which the Rossis provide tabular data (p. 173), the average obligation rating was 8.3 for own children and 6.4 for stepchildren. These figures suggest a public perception that stepparents ought to feel a moderate obligation to assist adult stepchildren, and that, although the obligation to assist stepchildren falls materially below the obligation to help own children, obligations to stepchildren are rated higher than those to nieces/nephews, aunts/uncles, cousins, friends, and neighbors. Stepparents fared somewhat less well than stepchildren, and the gap between biological and stepparents was greater than that between biological children and stepchildren. Stepparents rated below friends, but ahead of neighbors, nieces/nephews, aunts/uncles, and cousins. Average obligation to stepparents was 5.8, compared to 8.3 for own parents.

The Rossi and Rossi data suggest that public opinion (presumably reflected in internalized normative pressure) would support giving less to stepparents and stepchildren than to biological parents and children. Public opinion would condemn those who failed to provide comfort or support in times of crisis, or who failed to keep up normal intercourse with stepfamily members.

A question unanswered by the Rossis' hypothetical data is whether obligations to the two types of kin might be more or less determined by circumstances. It seems likely that the normative obligations between stepchildren and stepparent like the very modest legal obligations discussed earlier, depend on an intact marriage between parent and stepparent. Normative obligations probably also depend on such factors as duration of the stepfamily relationships, the age of the child when the stepfamily was formed, length of co-residence with the stepparent, and whether one has relatives in the higher ranked categories (e.g., whether one has own children).

Whether normative and legal obligations extend beyond the marriage between stepparent and parent brings up an even more basic question: Does the relationship itself extend beyond the marriage? Is John still your stepson if you divorce his parent? Day (1988) noted that we have no readily agreed-upon terms to refer to such people, and that the vagueness of the role relationship is likely to be paralleled by the vagueness and ambiguity of the role expectations. In this regard, however, it is interesting that Thoits' research on role identities found that 11% of divorced men and 3% of divorced women (compared to 19% of married women and men) reported being a stepparent—and that the majority of these divorced stepparents claimed the role by saying they thought of themselves as stepparents. These data imply that role relationships, and perhaps, role obligations between stepparents and stepchildren may have a lasting quality not entirely embedded in the marriage between parent and stepparent.

The extent and determinants of normative obligation in stepfamilies are empirical questions worth further study. Both theory and the limited existing evidence, however, suggest that members of stepfamilies will feel less normative obligation (internal and external) to provide social support.

Conclusion

From each of the theoretical perspectives reviewed here, we come to the hypothesis that stepfamilies will not be as effective as biological families in providing social support to members. The extent of supportiveness in stepfamilies is likely to be enhanced by longer duration of childhood co-residence, weakness of competing ties, and especially by the continued marriage of the stepparent to the biological parent. In the absence of this last critical link and the absence of strong institutionalized support, exchange among stepparents and stepchildren is likely to be fragile. . . .

Overall, I think stepfamily members are best viewed as affines, much like parents-in-law or children-in-law. The relationship is created and sustained by a marriage; although affectionate relationships may develop, the primary impetus for exchange with the affine is to benefit or please the biological relative. When the marriage ends, the relationship falters or dies.

The fact that stepfamily ties are not interchangeable with biological ties does not mean that stepfamily ties are useless as sources of support. As Rossi and Rossi's (1990) data on perceived normative obligation to stepparents and stepchildren suggest, people in stepfamilies have weaker, but still important, family ties. Stepfamily members continue to keep track of one another and to exchange at above minimum level. Ties may be weak rather than strong, but a growing literature shows that we should not underestimate "the strength of weak ties" (Granovetter, 1973).

The question remains as to whether or not the proliferation of stepfamilies in contemporary society actually reduces family support. I think the general answer is yes, but that the strength of effect differs substantially by gender. Fathers and stepfathers are probably seriously disadvantaged. Divorced fathers tend to lose contact with their own children, yet fail to form enduring bonds with their stepchildren. In general, stepfathers will remain a part of their stepchildren's lives only as long as they remain married to the children's mother.

Stepmothers and their stepchildren are probably the most serious losers in stepfamily relationships. From both child's and parent's points of view, the family with a stepmother fares worse than that with a stepfather. Several explanations can be offered for this finding. Stepmother families may have a weaker foundation because of the generally poorer relationship between the child and the custodial parent.

A second possibility is that it is harder to fill the role of mother than father. Although one can be a decent father by being a good provider and a nice guy, filling an absent mother's shoes requires establishing gut-level empathy and attachment that may be more difficult to develop or fake. I would also like to raise the possibility that, although sociobiologists have focused on stepfathers because female animals seldom inherit their mates' offspring from previous unions, it is females more than males for whom the distinction between own and step is most critical.

Those least affected by stepfamily proliferation are biological, custodial mothers and their own children. Mothers who have remarried report moderately weaker bonds to their own adult children than mothers in intact marriages (White, 1992; ...), but their relationships are least affected by marital disruption and remarriage....

REFERENCES

Atkinson, M. (1989). Conceptualizations of the parent–child relationship: Solidarity, attachment, crescive bonds, and identity salience. In J. Mancini (Ed.), *Aging parents and adult children* (pp. 81–97). Lexington, MA: Lexington.

Axelrod, R. (1984). *The evolution of cooperation.* New York: Basic Books.

Bartlema, J. (1988). Modeling step-families: Exploratory findings. *European Journal of Population, 4,* 197–221.

Carley, K. (1991). A theory of group stability. *American Sociology Review, 56,* 331–354.

Cherlin, A. (1978). Remarriage as an incomplete institution. *American Journal of Sociology, 84,* 634–649.

Daly, M., & Wilson, M. (1978). *Sex, evolution, and behavior.* North Scituate, MA: Duxbury.

Day, A. T. (1988). Kinship networks and informal support in the later years. In E. Grebenik, C. Hohn, & R. Mackensen (Eds.), *Later phases of the family cycle: Demographic aspects* (pp. 184–207). Oxford, England: Clarendon.

Dobash, R. P., Dobash, R. E., Wilson, M., & Daly, M. (1992). The myth of sexual symmetry in marital violence. *Social Problems, 39,* 71–91.

Ekeh, P. (1974). *Social exchange theory: The two traditions.* Cambridge, MA: Harvard University Press.

Finch, J. (1989). *Family obligations and social change.* Oxford, England: Polity Press.

Fine, M. A. (1986). Perceptions of stepparents: Variation in stereotypes as a function of current family structure. *Journal of Marriage and the Family, 48,* 537–543.

Fine, M. A., & Fine, D. R. (1992). Recent changes in laws affecting stepfamilies: Suggestions for legal reform. *Family Relations, 44,* 334–340.

Finkelhor, D. (1979). *Sexually victimized children.* New York: The Free Press.

Fischer, C. (1982). *To dwell among friends: Personal networks in town and city.* Chicago: University of Chicago Press.

Fluitt, M. S., & Paradise, L. V. (1991). The relationship of current family structure to young adults' perceptions of stepparents. *Journal of Divorce and Remarriage, 15,* 159–173.

Furstenberg, F. F., Jr., & Spanier, G. (1984). *Recycling the family: Remarriage after divorce.* Beverly Hills, CA: Sage.

Granovetter, M. (1973). The strength of weak ties. *American Journal of Sociology, 78,* 1360–1380.

House, J. S., Umberson, D., & Landis, K. R. (1988). Structures and processes of social support. *Annual Review of Sociology, 14,* 293–318.

Knoke, D. (1990). Networks of political action: Toward theory construction. *Social Forces, 68,* 1041–1063.

Rossi, A. S. (1984). Gender and parenthood. *American Sociological Review, 49,* 1–18.

Rossi, A. S., & Rossi, P. H. (1990). *Of human bonding: Parent–child relations across the life course.* Hawthorne, NY: Aldine de Gruyter.

Schwebel, A. I., Fine, M. A., & Renner, M. A. (1991). A study of perceptions of the stepparent role. *Journal of Family Issues, 12,* 43–57.

Thoits, P. A. (1992). Identity structures and psychological well-being: Gender and marital status comparisons. *Social Psychology Quarterly, 55,* 236–256.

Troll, L., Miller, S. J., & Atchley, R. C. (1979). *Families in later life.* Belmont, CA: Wadsworth.

Turner, R. H. (1970). *Family interaction.* New York: Wiley.

Waggoner, L. W., Wellman, R. V., Alexander, G. S., & Fellows, M. L. (1991). *Family property law: Cases and materials on wills, trusts, and future interests.* Westbury, NY: Foundation Press.

Wald, E. (1981). *The remarried family: Challenge and promise.* New York: Family Service Association of America.

White, L. K. (1992). The effect of parental divorce and remarriage on parental support for adult children. *Journal of Family Issues, 13,* 234–250.

White, L. K., & Riedmann, A. C. (1992). When the Brady Bunch grows up: Relations between fullsiblings and stepsiblings in adulthood. *Journal of Marriage and the Family, 54,* 197–208.

The New Stepfamily Requires a New Public Policy

Mary Ann Mason and Jane Mauldon

Stepparents presently exist in a kind of legal and social limbo, rarely recognized as "real" fathers or mothers, yet expected to adopt some elements and obligations of the parental role. While the act of marriage creates legal rights and obligations between stepparents and their spouses, stepparents in most respects are without legal obligations or entitlements with regard to their stepchildren. During marriage, stepparents usually have no obligation to support their stepchildren, yet neither do they enjoy any right of custody or control. If the marriage terminates through divorce or death, stepparents have no rights of custody or visitation, and conversely, no obligation to pay child support however long-standing their relationship with their stepchildren. Stepchildren, in turn, have no right of inheritance in the event of a stepparent's death (Mason & Simon, 1995).

The legal invisibility of stepparental rights and obligations can be traced to the traditional notion of parenthood, which is biologically based and which bars the possibility of one having two mothers or two fathers (Mason, 1994). For that reason also, the legal option of adoption is not available to most stepfamilies since the natural noncustodial parent is usually still living and not willing to relinquish parental rights. Indeed, the contrast between the legal status of stepparents and the presumptive rights and obligations of natural parents is remarkable. Child support obligations, custody rights, and inheritance rights exist between children and their natural parents by virtue of a biological tie alone, regardless of the quality of social or emotional bonds between parent and child, and regardless of whether the parents are married. In recent years federal and state policies have strengthened the rights and obligations of natural parents, particularly in regard to unwed and divorced parents, but have not advanced with regard to stepparents....

The Face of Today's Stepfamilies

In 1990, about 11% of all U.S. children (7.2 million children) were living with a stepparent, while 58% were living with both biological parents, 25% with one biological parent, and 6% with neither parent or were "unknown or unaccounted for" (U.S. Bureau of the Census, 1994). However, many of the children who in 1990 were still living with both biological parents or with an unmarried mother could expect, before they reached adulthood, to acquire a stepparent (usually a stepfather) by virtue of their mothers' remarriages. In all, some 25% of children born in the 1980s are expected to reside for some time with a stepparent (Furstenberg, 1987).

More than one-fifth of all married-couple families with children are stepfamilies (step*families* are more common than are step*children* in part because some stepfamilies also include children who are biological offspring of both parents). Nearly 80% of stepfamilies are stepfather–custodial mother families, while about 13% are stepmother–custodial father, and fewer than 10% are "complex" stepfamilies resulting from the marriage of two custodial parents. While most custodial parents in stepfamilies have been married before, the fraction of stepfamilies formed by the first marriage of a never-married woman is likely to increase quite sharply in the future as more children are born to unmarried women. This trend may also make stepfamilies more common in the future. In 1993, 30 percent of all births were to unmarried women, a fraction that had more than doubled since 1973. . . .

[A detailed portrait of stepfamilies developed from data collected in the 1987–1988 National Survey of Families and Households (NSFH) reveals that] stepfathers' incomes [are] essential in preventing or ending poverty among custodial mothers and their children. Consequently, our discussion will concentrate on stepfather–custodial mother families, since these are the most common form by far. Custodial mothers in stepfamilies had similar incomes to single mothers, earning about $12,000 in 1987 [see Table 1]. If, as seems plausible, their personal incomes were about the same before they married as after, then marriage had increased their household incomes more than threefold, even though the stepfathers themselves contributed about $6000 less to the family economy than did husbands in nuclear families. Stepfathers' incomes were, on average, more than twice as great as their wives', and this accounts for nearly three-fourths of the family's income.

In contrast to custodial stepparents, absent biological parents only rarely provide much financial or other help to their children. Some do not because they are dead or cannot be found: about 26% of custodial remarried mothers and 28% of single mothers report that their child's father is deceased or of unknown whereabouts.

Yet even in the three-quarters of families where the noncustodial parent's whereabouts are known, contact between children and their natural father is generally very limited (see Table 2). Only about one-third of all custodial mothers (whether single or remarried) receive child support or alimony from former spouses, and the amounts involved are small compared to the cost of raising children. Remarried women with support awards had received on average $1780 in the preceding year, while single mothers received $1383. Clearly, former spouses should not be relied on to lift custodial mothers and their children out of poverty.

Personal contact between absent parents and children is only slightly more common than is payment of child support. Half of all children living apart from their fathers see them less often than once a year (although about half of this number have fathers who are dead or of unknown whereabouts). About 28% of absent fathers see their children at least once a month, and roughly 15% see their children once a week or more. Contact with children is slightly more frequent for children in single-mother families than for children of remarried mothers. For example, 25% of children in stepfamilies see their natural fathers at least monthly, while 33% of children of single mothers see their fathers that often.

These data on payments and visits fall into four broad patterns of contact between stepchildren and their absent natural fathers. Roughly one quarter of all stepchildren have no contact at all with their fathers and receive no child support; one quarter see their fathers only once a year or less often and receive no child support; one quarter have intermittent contact or receive some child support; and one quarter may or may not receive child support but have fairly regular contact, seeing their fathers once a month or more. Using these data as guides to the quality and intensity of the father–child relationship, it appears that relatively few stepchildren are close to their fathers or have enough contact with them to permit their fathers to play a prominent role in the children's upbringing. In contrast, at least half of natural fathers play an active role in their children's lives, to some degree.

Some noncustodial fathers have remarried and have stepchildren themselves, and these relationships too are evident in the NSFH data. As the lower panel of Table 2 illustrates, nearly one-quarter (23%) of stepfathers have minor children from former relationships living elsewhere. One-sixth (16%) report paying child support for those children. This overlap between the categories of stepparent and noncustodial parent has implications for policies governing child support, and we return to this matter, below.

While it is plausible to assume that stepfathers' substantial contributions to family income improve their stepchildren's material well-being by helping to cover basic living costs, the impact of the step relationship on children's emotional and intellectual

TABLE 1 *Income and Education Levels of Parents in Original Two-Parent, Stepparent, and Single-Parent Families*

	Types of Families[a]			Types of Stepfamilies		
	Original Two-Parent Family	Single Mother (Not Cohabiting)	Stepfamilies (All Types)	Stepfather–Custodial Mother	Stepmother–Custodial Father	Both Parents Are Step and Custodial
Family type as a percent of all families with children (1990)	60%	17%	16%	13%	2%	1%
Father's education (years)	13.5	—	12.9	12.9	13.2	13.0
Mother's education (years)	13.1	12.1	12.8	12.8	12.9	12.9
Father's income ($1,000s)	33.3	—	29.4	27.0	43.6	29.2
Mother's income ($1,000s)	9.5	11.8	12.3	12.0	13.8	12.3
Couple's Income[b] ($1,000s)	44.6	11.8	43.1	40.3	59.4	43.8
Number of families in sample	3021	1615	678	539	92	47

Notes. Income and education data are from the 1987–1988 National Survey of Families and Households (weighted estimates), while the percentages of types of households across types are also based on 1990 Current Population Survey data reported in the 1993 *Statistical Abstract of the United States.*

[a] The family types described here account for 93% of all families with children. In addition to these, 2% of families are headed by a single male, and 5% of families are cohabiting families (almost all of them headed by a single biological parent and her or his partner). Such cohabiting unmarried families are considered stepfamilies by some, single-parent families by others. We exclude them from these analyses.

[b] The couples' incomes exceed the sum of husbands' and wives' incomes because couples have sources of income that are joint and not counted in the individual incomes. In addition, couple income is only computed where complete income data is reported for both spouses; those couples tend to have slightly higher individual incomes than average.

TABLE 2 *Contact Between Absent Parents and Children: Child Support and Visits*

	Types of Families		
	Original Two-Parent Family	**Single Mother (Not Cohabiting)**	**Stepfather–Custodial Mother**
Children's contact with noncustodial fathers			
% of custodial mothers who got any child support in the preceding year	—	36%	36%
Average child support paid in the preceding year to mothers with awards	—	$1,383	$1,780
% in which absent father[a] sees child yearly or more often	—	50%	45%
% in which absent father sees child monthly or more often	—	33%	25%
% in which absent father pays any support or sees child more than yearly	—	53%	47%
Parent's contact with nonresident children			
% of fathers with minor children elsewhere ("nonresident children")	5%	—	23%
% of fathers paying support for nonresident children	3%	—	16%

[a] "Absent fathers" may be dead or of unknown whereabouts.

development is less certain. Stepfathers vary in how enthusiastically and effectively they parent their stepchildren, and stepchildren also vary in how willingly they permit a parental relationship to develop. Indeed, many stepfather–stepchild relationships are not emotionally close. Hetherington and Clingempeel found that the most common parenting style found among stepfathers was disengaged, characterized by low levels of communication and warmth, and a lack of control, discipline, and monitoring of the stepchild's behavior and activities (Hetherington & Clingempeel, 1992; Thomson, McLanahan, & Curtin, 1992). The fragility of the stepfamily relationship is evident in the high divorce rate as well. About one-quarter of all remarrying women separate from their new spouses within five years of the second marriage, and the figure is higher for women with children from prior relationships (Bumpass & Sweet, 1987). Overall, we estimate conservatively that between 20 and 30% of stepchildren will see their custodial parent and stepparent divorce, before they themselves turn 18. Given this relatively high rate of divorce, public policy needs to shape stepchild–stepparent relations at the time of and following divorce in ways that protect children from falling into poverty.

The detached parenting style that characterizes many stepparents, and their above-average risk of divorce, may result in part from the absence of a clearly defined social or familial role for stepparents. Cherlin (1978) proposed that, whereas strong norms exist governing relations between immediate family members, the norms governing relations between stepkin are weak or absent. As we shall see, our review of the relevant state and federal laws supports this contention; with few exceptions, stepparents have none of the legal rights or responsibilities of parents vis-à-vis their stepchildren. . . .

Discussion and Conclusion

Federal and state policy do not reflect the modern reality of the large numbers of stepfamilies in which children are actually dependent upon their stepparent for support. Our analysis has shown that families with a residential stepfather have a much higher family income than do mother-headed single families; indeed, their household incomes look much like nuclear families. However, research also demonstrates that stepfamilies are emotionally fragile, and are at least as likely to terminate in divorce as are nuclear families.

The failure of state and federal policy to address the financial and emotional interdependencies of step relationships can be detrimental to children in at least two ways. First, children dependent upon a residential stepparent may not receive adequate support or benefits from that parent during the marriage, and they may not be protected economically in the event of divorce or parental death. Second, residential stepparents are not given the legal status and authority that may be necessary to effectively parent their stepchildren. Moreover, the lack of legal recognition may influence social expectations. Stepparents may not have a clear understanding of their role because policies and laws are ambiguous or nonexistent.

We propose a new conceptualization of stepparents that will clarify existing policy and strengthen the legal position of those stepparents who are, in many ways, acting as parents. This conceptualization requires dividing stepparents into two subclasses: those who are "de facto" parents and those who are not. De facto parents would be defined as "those stepparents legally married to a natural parent who primarily reside with their stepchildren, or who provide at least 50% of the stepchild's financial support."

For the purposes of federal and state policy, under our scheme, a de facto parent would be treated virtually the same as a natural parent during the marriage; the same rights, obligations and presumptions would attach vis-à-vis their stepchildren. These rights and duties could continue in some form, perhaps based on the length of the marriage, following the natural parent's death or divorce from the stepparent, or the death of the stepparent. Stepparents who do not meet the de facto parent requirements would, in all important respects, disappear from policy.

Creating a de facto parent category for stepparents would not invalidate the existing rights and obligations of custodial and noncustodial natural parents. Rather, our proposal would empower a stepparent as an additional parent. If the stepfamily marriage terminated through death or divorce while any stepchildren were still minors, the stepparent would have standing to seek custody or visitation. The stepparent could also be obligated for child support, prorated according to the length of the marriage. If conflicts regarding children arose between any of the natural or stepparents, they could be resolved on their merits by mediation or in court based on clear guidelines (which might include the duration of the stepparents' de facto parenthood as a criterion) just as current conflicts between natural parents are resolved.

A clear definition of stepparents as de facto parents would eliminate the inconsistencies regarding stepparents that plague current federal policies and would clarify the role of the residential stepparent. For the duration of the marriage a stepchild would be treated as a natural child for purposes of support and the receipt of federal benefits. This treatment would persist in the event of the death of the stepparent. When a stepparent dies the child should receive all the survivor and death benefits that would accrue to a natural child (Mason & Simon, 1995).

In the case of divorce, the issue is more complicated. We do not argue that stepchildren and natural children should have identical coverage for federal benefits following divorce, but neither do we believe it is good policy to cut off summarily children who have been dependent, sometimes for many years, on the de facto parent. A better policy is to extend federal benefits for a period following divorce, possibly based on a formula that matches the number of years of dependency (Mason & Simon, 1995). For instance, if the stepparent resided with the stepchild for four years, the child would be covered by Social Security survivor benefits and other federal benefits for a period of four years following the divorce. This solution would serve child welfare by at least providing a transitional cushion. It would also be relatively easy to administer. In the case of the death of the natural parent benefits similarly could be extended, or continued indefinitely as long as the child remains in the sole custody of the stepparent.

Federal policy, however, is limited in promoting increased rights and responsibilities for de facto parents. Critical matters of everyday custody and control and rights to custody or visitation following a divorce from or death of the custodial parent are determined by the laws of the individual states, not the federal government. So too are the crucial issues of child support following divorce and inheritance rights.

Since state law governs these and most areas of family law, a complete legal role for the stepparent as a de facto parent can only be established by encouraging reform in this confused arena. The first step in promoting this new policy is to insist all states pass stepparent general support obligation laws that would require stepparents acting as de facto parents to support their stepchildren as they do their natural children. Federal policy already assumes this support in figuring eligibility in many programs, but it has not insisted that states change their laws. This goal could be accomplished by making stepparent general support obligation laws a prerequisite for receiving federal AFDC grants. Precedent for this strategy has been set by the Family Support Acts of 1988, which mandated that states set up strict child support enforcement laws for divorced parents and unwed fathers at AFDC levels in order to secure AFDC funding (100 P.L. 485; 102 Stat. 2343, 1988). (See also Mason & Simon, 1995.)

Requiring stepparent support, however, raises a central issue of fairness. If the stepparent is indeed supporting the child, there is a question about the support obligations of the noncustodial parent. Traditionally, most states have not recognized the stepparent contribution as an offset to child support (Ramsey & Masson, 1985; Levy, 1966). While this policy promotes administrative efficiency, it may not be fair to the noncustodial parent. An important advance in recognizing the existence of multiple parents in the nonlinear family is to recognize multiple support obligations. The few states that require stepparent obligation have given limited attention to apportionment of child support obligations, offering no clear guidelines. We propose that state statutory requirements for stepparent obligation as de facto parents also include clear guidelines for apportioning child support between the noncustodial natural parent and the stepparent.

Furthermore, if stepparents are required to accept parental support obligations, equal protection, and fairness concerns dictate that they must also be given parental rights. Currently state laws recognizes only natural or adoptive parents; a stepparent has no legal authority over a stepchild. In some cases, such as when the parents have shared legal custody, the law would be recognizing the parental rights of three parents, rather than two. This sounds unusual, yet it is an accurate reflection of how many families now raise their children. Most often, however, it would be the custodial parent and his or her spouse, the de facto parent, who would have authority to make decisions for the children in their home.

The next logical, albeit more radical step, would be for states to impose child support obligations on stepparents following divorce from the custodial parent. This approach would conform with our policy proposal extending federal benefits following divorce. If federal policy took the lead in extending benefits to children following divorce where the parent qualified as a de facto parent, states might be encouraged to follow a similar scheme regarding child support. Only a few courts have ruled in favor of support payments following divorce, and each of these cases has been decided on individual basis (Clevenger v. Clevenger, 1961). Only one state, Missouri, statutorily continues stepparent support obligations following divorce (Vernon's Annotated Missouri Statutes 453.400, 1994).

It would be in the fiscal best interests of both state and federal governments to encourage an extended period of support following divorce. It is likely that a significant number of children now turn to AFDC when their stepfamilies dissolve through divorce and they lose access to their stepparent's income. A formula similar to that suggested for federal benefits could be appropriate. A stepparent who qualified as a de facto parent for at least one year would contribute child support for the number of years of dependency until the child reached majority. If the natural noncustodial parent were still paying support payments, the amount could be apportioned. While it may be said that this policy would discourage marriage, it could also be said to discourage divorce. Stepparents might consider working harder at maintaining a marriage if divorce had some real costs.

If policy mandates child support following divorce, the divorced or widowed stepparent must be able to pursue visitation or custody rights. The relationship between stepparent and stepchild should not necessarily terminate with the marriage relationship. Likewise, the death of the stepparent should not end the relationship. State inheritance laws are frequently archaic and notoriously difficult to modify. Yet a consistent reformation of the stepparent–stepchild relationship must include strengthening the rights of dependent stepchildren to inherit when there is no will. While Social Security and other federal survivor benefits are based on the premise that a child relies on the support of the residential stepparent and will suffer the same hardship as natural children if the stepparent dies, state inheritance laws decree that only biology, not dependency, counts. The existing model is inadequate; state laws should assume that de facto parents would wish to have all their dependents receive a share of their estates if they died without a will. The same assumption should prevail for compensation claims following an accidental death. A dependent stepchild, just as a natural child, should have the right to sue for loss of support.

Critics of this scheme may argue that adoption, not the creation of the legal status of de facto parent, is the appropriate vehicle for granting a stepparent full parental rights and responsibilities (Hollinger, 1988). If, as we have described, nearly two-thirds of stepchildren are not being supported by their noncustodial parents, policy initiatives could be directed to terminating the nonpaying parents' rights and promoting stepparent adoption. Adoption is not possible, however, unless the parental rights of the absent natural parent have been terminated—a difficult procedure against a reluctant parent. Normally, the rights of a parent who maintains contact with their child cannot be terminated even if that parent is not contributing child support. And when parental rights are terminated, visitation rights are terminated as well in most states. It is by no means clear that it is in the best interests of children to terminate contact with a natural parent, even if the parent is not meeting his or her obligation to support (Bartlett, 1984).

Stepfamilies compose a large and growing sector of American families that is largely ignored by public policy. In this article we have demonstrated that stepfamily relationships are important in lifting single-parent families out of poverty. When single or divorced mothers marry, the household income increases by more than threefold, rising to roughly the same level as nuclear families. Our data show that in at least half these stepfamilies, the noncustodial parent neither pays child support nor has contact with the children. A substantial portion of these families experience divorce, placing the stepchildren at risk of falling back into poverty. It makes good public policy sense both to strengthen these stepfamily relationships and to cushion the transition for stepchildren should the relationship end.

REFERENCES

Bartlett, K. (1984). Re-thinking parenthood as an exclusive status: The need for alternatives when the premise of the nuclear family has failed. *Virginia Law Review, 70,* 879–903.

Bumpass, L., & Sweet J. (1987). *American families and households.* New York: Russell Sage Foundation.

Cherlin, A. (1978). Remarriage as an incomplete institution. *American Journal of Sociology, 84,* 634–649.

Clevenger v. Clevenger (1961). 189 Cal. App. 2d, 677.

Code of Federal Regulations (C.F.R.) (1994).

Furstenberg, F. (1987). The new extended family: The experience of parents and children after remarriage. In K. Pasley & M. Ihinger-Tallman (Eds.), *Remarriage and stepparenting: Current research and theory* (pp. 185–196). New York: Guilford Press.

Hawaii Revised Statutes Annotated (1994).

Hetherington, E. M., & Clingempeel, W. G. (1992). Coping with marital transitions: A family systems perspective. *Monographs of the Society for Research in Child Development, 57*(2–3), Serial No. 227.

Hollinger, J. (Ed. in Chief). (1988). *Adoption law and practice.* New York: M. Bender.

Levy, R. J. (1966). Family law and welfare policies: The case for "dual systems." *California Law Review, 54,* 748–798.

Mahoney, R. (1987). Stepfamilies in the federal law. *University of Pittsburgh Law Review, 40,* 480–495.

Mason, M. A. (1994). *From father's property to children's rights: A history of child custody in the United States.* New York: Columbia University Press.

Mason, M. A., & Simon, D. (1995). The ambiguous stepparent: Federal legislation in search of a model. *Family Law Quarterly, 3,* 445–483.

Ramsey, S., & Masson, J. (1985). Stepparent support of stepchildren: A comparative analysis of policies and problems in the American and British experience. *Syracuse Law Review 36*, 649–666.

Thomson, E., McLanahan, S. S., & Curtin, R. B. (1992). Family structure, gender, and parental socialization. *Journal of Marriage and the Family, 54*, 368–378.

Trudell v. Leatherby. (1931). 300 P. (Supreme Court of California) pp. 7–15.

U.S. Bureau of the Census. (1994). *1993 Statistical Abstract of the United States*. Washington, DC: U.S. Environment Printing Office.

United States Code (1994).

Vernon's Annotated Missouri Statutes (1994).

CHAPTER 7 Questions for Writing, Reflection, and Debate

READING 35

Uncoupling: The Breakdown of the Cover-Up • *Diane Vaughan*

1. In the initial stages of dissolving a relationship, how do initiators of a breakup weigh the costs and benefits of their actions?
2. How do initiators of a breakup experience the dissolution of a relationship differently from their partners? Who has the most power? Why?
3. Why might an initiator rely on indirect methods of initiating a breakup?

READING 36

The Economic Consequences of Divorce • *Frank F. Furstenberg, Jr., and Andrew J. Cherlin*

4. How does gender affect the economic impact of divorce? In what ways do institutions such as the courts and the labor market exacerbate these results? Why is child support often not obtained or enforced, or set at a low amount? What motivates men not to pay court-ordered child support?
5. Why is losing the family home so devastating?

READING 37

After Divorce: Investigations into Father Absence • *Terry Arendell*

6. How do current trends in custody and child support encourage men to distance themselves from their children?
7. In what ways do fathers use absence as a means of exerting power and managing emotion?
8. What are the consequences of viewing fatherhood as a status within an intact marriage rather than as a relationship between a man and his children?

READING 38

Divorcing Reality: New State Laws That Slow Down Divorce • *Stephanie Coontz*

9. What might be the positive and negative effects of instituting laws that make divorce more difficult to obtain?
10. Some people advocate making it harder to get married rather than making it harder to get a divorce. How could this be done? How might it affect family life?

READING 39

Stepfamilies over the Life Course: Social Support • *Lynn White*

11. What are the broad social implications of the fact that social support in stepfamilies is not as strong as it is in biological families?
12. In a society where divorce and remarriage are becoming the norm, how does the impact of this loss of social support mesh with the withdrawal of financial support?

READING 40

The New Stepfamily Requires a New Public Policy • *Mary Ann Mason and Jane Mauldon*

13. Can a society legislate to create stronger social bonds? Would legal recognition influence social expectations for the stepparent role? What would the impact be on children, parents, and grandparents of creating a de facto, third, or additional parent role?

8

Family Transitions

How will family life change as the U.S. population ages?

My mother is 81 years old. Her mother was widowed in her forties, left with five children to raise. She took in laundry to keep food on the table and pay the rent, but she wasn't the only income earner in the family. The older children took whatever jobs they could get, and their pay contributed to the family's income. That pattern established a lifetime of mutual support. At 31 years old, my mother left her mother's house to get married. Grandma still had adult children living with her, and their wages helped pay the rent and put food on the table. When the last child left home to join the Army, part of his pay joined contributions from his siblings to support Grandma.

My mother's story is not an uncommon one for families of that time. However, for those of you in your twenties, who couldn't or can't wait to be independent from your parents, this story has an old-fashioned ring. As our country has become more affluent and poverty among the elderly has decreased, both parents and their children have developed a preference for an independent, although still interdependent, lifestyle. Like my mother, I also married when I was 31. But I had long before left my parents' home to live and work in faraway lands. My wages were spent on traveling, a new car, and electronic entertainment systems, not to support my parents. My story is not an uncommon one for families of my time. But the times are changing again, and some people predict that these changes will call for new interdependencies between the generations. Some people think the current economic situation is returning us to a lifestyle where adult children live with their parents, pooling their meager earnings to survive. Whether the parents or the children will make the bigger contribution to the pooled resources is a matter of debate.

As my family's story suggests, there is little doubt that the relationships between the generations have changed and will continue to change. The economic climate and cultural values are one aspect of this change, but at the core is one inescapable fact: The dramatic increases in life expectancy in the recent past have forced us to adapt how we interact with our parents and grandparents because they are around longer than in the past. In going through some old family albums with my mother when I was about 8 years old, I remember her pointing excitedly to a photograph of three generations of some branch of the family. At the time, only my maternal grandmother was alive. Both grandfathers died before I was born, and my paternal grandmother died before I started school. My two younger brothers never had a "three generations" picture, and I have very few. But today my students talk of four-generation families as the norm, and five generations are not that rare today. The longer people live, the more time we have to figure out how to work together as a family, and the more generations there are to consider.

When I raise these issues in my Sociology of the Family class, students seem to dwell on the negative aspects of an aging population. I do not discount the importance of considering how we will care for aging family members afflicted by illness and poverty. However, I also believe there are positive aspects for family life of the increasing life span. Maybe my feelings are influenced by the fact that I never knew my grandfathers and have only a dim recollection of the grandmother who died when I was 5 years old. As I selected these readings, I became determined that we should consider both the good and the bad that are likely to accompany the aging of America. In addition, the sociologist in me believes that whether we experience these changes as costs or benefits to our families will be influenced by our cultural values and the support we get from social institutions.

I open these readings with Harriet Presser's "Some Economic Complexities of Child Care Provided by Grandmothers." Child care is an interesting topic through which to begin thinking about interactions between the generations, because it involves at least three generations and is an area commonly mentioned by my students when I ask them to consider how grandparents can contribute to the lives of their children and grandchildren. Politicians today talk about the role of elder family members in providing child care. Listening to the picture they paint, you would think that the majority of American families have a cadre of older members, with time on their hands, who live right around the corner from their children, and adore filling their lonely days with activities with their grandchildren. However, as this piece shows, the realities of this kind of exchange between generations are much more economically, socially, and emotionally complex.

The complexity of meshing two busy work lives in order for grandparents to care for grandchildren gives us insights into the busy, active lives most grandparents have today, and into one type of economic partnership developed between generations. A less common but equally interesting partnership is found on family farms and businesses. In "Legacy, Aging, and Succession in Farm Families," Norah Keating offers great insight into the family dynamics of choosing a successor for a family business, and the process involved in transferring the farm. My first teaching position was in the Midwest, and I now work in western New York. In both locations I have learned the unique problems presented farm families in negotiating the transfer of assets that have great emotional, historic, and economic implications. This article captures many of the issues my students have conveyed over the years. Of interest to this chapter are the changes in this transition caused by the older generation living longer and in an active, healthy state. As Keating suggests, this process of succession may need adjusting to accommodate this demographic change.

Keating's article ends on a note of the struggle within families to allocate resources for the care of the oldest generation when they can no longer contribute to the productivity of the family business. Mal Schechter's piece, "Social Security and Medicare Policy: A Personal, Intergenerational Story" provides insight into the degree to which some families depend on social welfare programs when parents reach that stage. As you read this very personal account, ask yourself how the lives of the two generations in these families would have been different if government programs were more generous or less generous for the oldest generation.

One of the main arguments for increasing any benefits to the elderly is the demographic projection of the changes in population. We read and hear daily about the growing numbers of elderly and the shrinking numbers of people earning the paychecks that government must tap to provide funds for assistance programs. How real is this problem? Is it a national or an international phenomenon? Kevin Kinsella addresses these questions in "Aging and the Family: Present and Future

Demographic Issues" and provides many other interesting demographic trends worth considering to understand the global changes affecting the multiple generations in families. As you read Kinsella's work, ask yourself how you would use some of these trends to develop social policies to ease the effect of these population transitions on families.

In "Programming for Family Care of the Elderly Dependents: Mandates, Incentives, and Service Rationing," Amanda Barusch reviews several options in policy. As the title suggests, Barusch surveys three policy options implemented in various countries around the world. She does not extensively evaluate the effectiveness of these options, but does a good job of challenging us to think about how effective they might be in our own country.

These readings provide some very personal accounts of the struggles families go through in negotiating roles as the oldest generation moves from a productive contribution to the family to a more dependent role. They also suggest ways in which government policies may facilitate or complicate those transitions. Some of you may be facing these issues in your families now. Others of you may have an oldest generation that has not reached this transition stage yet. Regardless of your personal situation, the demographics imply that your generation will be making the policies that affect families facing these decisions in our country in the near future. I hope the material presented here informs your decisions.

Some Economic Complexities of Child Care Provided by Grandmothers

Harriet B. Presser

The dramatic increase in the employment of mothers with young children in recent decades is well documented. In 1970, 30.3% of mothers with children under age 6 were in the labor force; by 1987, the percentage had risen to 56.7 (Hayghe, 1986, Table 3; U.S. Department of Labor, 1987b, Table 3). The rapid increase in employment is evident for mothers of children of all preschool ages but is greatest among those with infants and toddlers. By 1987, 51.9% of mothers with children one year of age or younger were in the labor force, and 58.5% of those with children who were age 2 (U.S. Department of Labor, 1987b, Table 1). Increasingly, women are entering or returning to work very soon after the birth of their children. This phenomenon is generating substantial increases both in the absolute numbers of preschool-age children who require alternative child care and in the proportion of the preschool-age child care population composed of infants and toddlers. Who are the people caring for these increasingly younger preschoolers when mothers are employed?

Most of them are relatives, the primary relative being the grandmother. The younger the child, the more likely are grandmother care arrangements (U.S. Bureau of the Census, 1983, Table 2).[1] However, the participation of grandmothers in child care seems to be on the decline at a time when infants and toddlers in need of care are increasing in number. Trend data specifically on grandmother (or grandparent) care are not available, but more generally we know that care by relatives (excluding fathers) has been decreasing since 1958 for both children under age 3 and for 3-to-4-year-olds (U.S. Bureau of the Census, 1982, Table A-3; U.S. Bureau of the Census, 1983, Table 2; 1958 is the earliest year for such a trend analysis). The implied decline in grandmother care may be anticipated by the rising employment rates of older as well as younger women (U.S. Department of Labor, 1987a, Table A), which suggest that grandmothers as well as mothers are increasingly likely to be employed at the same time.[2] Just as the rise in real wages or "opportunity costs" for mothers has increased their labor force participation (Blau and Ferber, 1986), so too, grandmothers may be less willing (or financially able) to forgo improved job opportunities in the wage economy by providing child care for their grandchildren. But are grandmothers who provide such care in fact forgoing such opportunities?

This need not be the case; that is, if grandmothers are geographically available and willing to provide child care for their grandchildren, they may be able to do so and be employed as well. Indeed, the growing demand for the care of very young children when mothers are employed may encourage some adaptations that would enable mothers and grandmothers both to work and to share child care: (*a*) the grandmother and mother may work different hours (e.g., one during the day and the other during the evening); and/or (*b*) the mother may pay the grandmother for child care to compensate the grandmother, at least partly, for not seeking other employment during the hours that child care is needed.

Previous research has shown that among married dual-earner couples, a substantial minority of husbands and wives work very different hours from each other; when this occurs, almost all spouses jointly participate in child care (Presser, 1988; Presser and Cain, 1983). Grandmothers, too, may be employed and provide child care by working different hours from employed mothers; this arrangement may be especially likely when mothers are not married and thus do not have husbands available for sharing child care.

To explore this possibility, ideally we would want to examine the employment status and work schedules of all grandmothers who reside near or with their grandchildren, and to assess the likelihood of grandmothers' participation in child care according to the employment characteristics of both grandmothers and mothers.[3] Unfortunately, such data on all geographically accessible grandmothers are not available. But, as we shall see, we have a unique opportunity to assess the employment characteristics of grandmothers *who provide child care* when mothers are employed. The neglect of this issue in child care research reflects an implicit assumption that grandmothers who provide such care are not otherwise employed. Moreover, the fact that some grandmothers are paid for child care has received only minimal attention. It is the purpose of this study to examine both issues in some detail and to demonstrate that caregiving by grandmothers is more complex than is generally acknowledged.

Sample Description

The data source for this study is the Youth Cohort of the National Longitudinal Survey of Labor Market Experience (NLSY), a national probability sample of both males and females who were 14 to 21 years old as of 1 January 1979. The fieldwork was conducted by the National Opinion Research Center for the Center for Human Resource Research at Ohio State University. Blacks, Hispanics, and economically disadvantaged whites (in 1979) were oversampled; a weighting procedure adjusts for this (as well as for different response rates and attrition). A sample of persons serving in the Armed Forces was also included. In all, 12,686 persons were interviewed in 1979, with annual reinterviews thereafter. The field completion rate for the first interview was 85%. The attrition rate between 1979 and 1984 was only 5% for the entire cohort, with little variation for subgroups. (For further sample details, see Center for Human Resource Research, 1986.)

The present analysis focuses on the 1984 wave of the NLSY, when all respondents were age 19 to 26 as of 1 January 1984. The sample is further restricted to employed mothers (married and unmarried) with children under 5 years of age. There are 796 eligible cases with data on the employment and child care variables under study. (The actual number of cases is shown in the tables, but the analyses are based on weighted values of these respondents, divided by the average sample weight.) The mean age of these employed mothers is 23.7; 64.6% have only one child; and 43.0% have a child who is less than 2 years old.

The 1984 wave of the NLSY includes data on the work schedules of employed respondents (and their spouses if married) and the type of child care arrangements they made for the youngest child while employed. This is the only wave of the NLSY with data on the employment status of grandmothers who provide child care and their work schedules if employed. Moreover, no other national survey contains such data. Thus, the present study yields the only estimates of the extent to which American grandmothers who provide child care to their own grandchildren are otherwise employed and how their hours of employment relate to their hours of child care and the work hours of employed mothers. The 1984 wave also includes data on whether (but not how much) payment is made for child care and whether the type of payment is cash and/or non-cash.

The Prevalence and Extent of Grandmother Care

About one-half (52.9%) of employed mothers in the sample rely upon relatives (including fathers) as the principal providers of care for their preschool-age children. As may be seen in Table 1, grandmother care is the most common type of care by a relative (23.9% of all care); care by fathers is the next most common (15.5%), with the remainder being by other relatives.[4]

TABLE 1 *Percentage Distribution of Principal Child Care Providers When Mothers Are Employed, for Youngest Child Under 5 Years of Age, by Marital Status: U.S. Sample of Mothers Aged 19 to 26 in 1984*

		Marital Status	
Principal Child Care Provider	**Total[a]**	**Married, Husband Present**	**Other[b]**
Mother	7.0	8.5	3.6
Father	15.5	20.3	4.8
Grandmother	23.9	21.8	28.7
Other relative	13.4	11.2	18.4
Nonrelative	40.1	38.2	44.5
Total %	100.0	100.0	100.0
$(n)^c$	(796)	(501)	(295)

Note: p (chi-square) < .001.
[a] Percentages may not total 100.0 because of rounding.
[b] Includes never-married, separated, divorced, and widowed.
[c] *n*'s are unweighted; percentages are based on weighted *n*'s.

Two-fifths (40.1%) of the mothers rely upon nonrelatives for principal care, and an additional 7.0% care for the child themselves while employed (at home or elsewhere).

The distribution of child care by provider differs significantly by marital status. The majority of both those married (53.3%) and those not married (51.9%) rely on relatives; however, as might be expected, there are differences by type of relative (Table 1). Father care is rarely reported for unmarried mothers, but this is compensated for by higher rates of care by grandmothers and other relatives.

One-fourth of the mothers (23.7%) report two or more types of child care arrangements for their youngest child when they are employed (22.2%, two types; 1.5%, three types). There is little difference in the number of arrangements by who is the principal provider: grandmother, other relative, or nonrelative. (When there are two or more types of arrangements, principal care is defined as the one involving the longest number of hours.) Whereas 23.9% of employed mothers rely on grandmothers as principal caregivers, 7.1% rely on them as secondary caregivers. Moreover, there are some cases in which grandmothers are both the principal and secondary caregivers (1.0% of the sample); presumably, they are different grandmothers. Thus, grandmothers are the principal or secondary caregivers for 30.0% of the sample.[5] (Detailed questions on the caregiver were not asked for third type of care.)

Not only are grandmothers a major source of caregiving for this sample of young employed mothers, but the number of hours they provide care is substantial: the mean number of hours per week is 27.1. Whether the grandmother is the principal or secondary caregiver is clearly relevant: 31.1 versus 11.4 hours, respectively ($p < .001$). Also relevant is the number of hours the mother is employed (an indicator of the extent of need for child care): 31.6 hours when mothers are employed full-time (35 hours or more) and 19.3 hours when they are employed part-time ($p < .001$). When mothers are employed full-time and grandmothers are the principal and sole caregivers (38.5% of the grandmother caregivers), the mean number of hours of child care they provide is 38.9.

Employment Status and Hours of Caregiving and Employment

For the remainder of the article, we focus specifically on the grandmothers who provide child care and consider our central interest: the economic complexities of such care.[6] This will provide some insight as to the extent to which grandmothers as well as mothers are juggling work and family roles to provide for the care of children.

We begin with the grandmother's employment status. The survey asked whether the grandmother who provided child care (principal or secondary) worked in the last 4 weeks. Over one-third (35.2%) were reported as working during this period: 31.4% of grandmothers who are the principal caregivers and 50.2% of grandmothers who are the secondary caregivers ($p < .05$).[7]

The intent of the question was to identify work other than providing child care for the youngest child, although this was not explicitly stated in the question and requires cross-checking for validation. The grandmother, however, would be classified appropriately as working if she were a child care provider for other children as well as her grandchild. The occupations of employed grandmothers were not asked, but we can compare the hours that grandmothers usually worked in the last 4 weeks with the hours they usually provided child care during this same period. To the extent that there is some or complete overlap in hours, grandmothers may be child care providers or have other occupations that permit the presence of children on the job—or child care hours may have been added inappropriately as hours of (other) employment.

Only 4 of the 78 employed grandmothers are reported to have worked hours that match identically with their child care hours (considering both beginning and ending time, and allowing for 1/2 hour variability in either). This suggests that very few grandmothers are inappropriately classified as working when the work is solely the care of their grandchild. The grandmother's work hours are completely different from their child care hours for about half the grandmothers—no overlap at all—and the others show work hours that exceed child care hours and/or child care hours that exceed work hours.

The non-overlapping hours in child care and grandmother's employment are clearly related to the non-overlapping hours in grandmother's and mother's work schedules. For 36.6% of all employed caregiving grandmothers, there is no overlap whatsoever between their work hours and those of employed mothers, and for an additional 22.0%, there are less than 3 hours of overlap. The difference by mother's marital status is significant: 50.5% of employed unmarried mothers have no overlap in work hours with employed grandmothers who provide child care, as compared to 27.2% of married mothers ($p < .05$).

Although grandmothers and mothers could both be working different daytime hours, especially if both are employed part-time, an alternative pattern is of one working days and the other non-days. Who is more likely to be the one working evenings, nights, or on a rotating shift: the grandmother or the mother? We would expect it to be the mother, since a previous study has shown that the prevalence of non-day employment for women (and men) decreases with age (Presser, 1987). These data support this expectation: among employed mothers who rely on grandmothers for child care, 26.0% begin work between 4 p.m. and 4 a.m. or are rotators, as compared to 10.7% of employed caregiving grandmothers. We do not know the marital status of the grandmothers, but the mothers are more likely to work non-days if they are unmarried (32.7%) than if married (22.5%)—although the difference is not significant ($p = .12$).[8]

We see, then, that grandmothers who are employed and also provide child care often work different schedules than the mothers, facilitating the sharing of child care—a pattern previously documented for dual-earner spouses. This arrangement may be especially likely among unmarried mothers because they do not have the option of negotiating this type of arrangement with husbands. This does not mean, however, that all unmarried—or married—women who rely on grandmother care choose their work schedules for this reason; for some, particularly those with minimal skills, the more relevant factor may be that their daytime job opportunities are very limited, even though they may have the side benefit of the grandmother's participation in child care. This situation may apply to some married mothers as well. Unfortunately these suppositions cannot be assessed empirically, since the survey did not ask the reasons mothers or grandmothers were working particular hours.

Payment for Grandmother Care

Grandmother care of children may have its benefits, but it also may have costs—social, psychological, and economic. As Carol Stack (1975) has noted for black women, child care by kin may be a means of obligating the child's mother for future needs. Economically, there may be immediate payment as well. The particular aspects of immediate payment we are able to consider in this study are payment in cash and/or service to grandmothers.[9]

We find that close to one-third (31.1%) of all grandmothers who provide child care receive cash payments, and an additional 11.7% receive noncash payments, such as meals, transportation, or exchanging other services.[10] (Some who receive cash payments also receive these other services; the amount of the cash payment was not asked, nor the specific type of other services provided.)

What determines whether the grandmother is paid either with cash or services for child care? This question can be addressed from at least three different perspectives: the grandmother's need for (additional) income, the relative burden of the caregiving job, and the mother's ability to pay.[11]

With regard to the grandmother's need for (additional) income, we have no direct measures in this data set. Her employment status may be an indirect indicator, but whether it should be positively or negatively associated with payment for child care is not obvious. Income from employment might reduce the need for additional income—and thus payment—for child care relative to nonemployed grandmothers, or employment itself might reflect minimal economic resources (perhaps an unemployed husband or none, about which data are not available), and thus a greater need for—and receipt of—child care income among employed compared to nonemployed grandmothers.

As for the relative burden of the caregiving job, the number of hours the grandmother provides care may be relevant, as well as which hours they are provided (mother employed during daytime versus evening, night, or rotating shift). We would expect that the fewer the child care hours, the less likely grandmothers would be to receive payment. We would also expect grandmothers who provide care during the evening or night (when the young child is more apt to be asleep) to be less likely to receive payment than those providing daytime care. (The expectation when hours vary because of a rotating shift is unclear, but this category is small and is grouped with evening and night shifts.)

The age of the youngest child and the number of children in the household also may be regarded as indicators of the burden of care. Generally, the younger the child, the more intensive the care (which is why child care centers typically have higher staff/child ratios for infants than for older children). Thus we would expect that the younger the child, the more likely the grandmother is to receive payment for care. We would also expect that the more children under age 5 in the household, the more likely it is that the grandmother will receive payment, since it is highly likely she would be caring for more than one child. The present study provides information on the care of the youngest child only, but a 1977 national study revealed that when there were two children under age 5 in the household—and an employed mother—the same type(s) of care were provided to both children in the large majority of cases (Presser, 1982).

As for the mother's ability to pay (although she need not always be the one who pays), one factor is the number of child care providers involved. We have such data for the youngest child only. (Two-thirds of the employed mothers who report grandmother care have only one child.) We would expect that those who use another provider in addition to the grandmother are less likely to pay the grandmother for child care than those who rely solely on the grandmother.

Another important aspect of ability to pay is the current hourly wage of the mother. We would expect that the higher the wage, the more likely it is that a grandmother would be paid for child care. Since the income of a husband might also be relevant, the mother's marital status is another consideration. We expect that payment for grandmother care is more likely for the children of married than of unmarried employed women.

There is also the expectation of differences in the prevalence of payment for grandmother care by race and ethnicity, reflecting cultural differences in the role of grandmother. For example, Cherlin and Furstenberg (1986) found that black grandparents were much more likely than white grandparents to take on a "parentlike" role with their grandchildren.[12] The assumption of this role may minimize the expectation of payment on the part of both mothers and grandmothers. Whether this is relevant for Hispanics, we do not know. However, it may be that the lower opportunity costs of Hispanic grandmothers in providing child care (given their low employment rates in comparison to blacks and whites; U.S. Department of Labor, 1988) may also minimize the expectation of payment for such care.

Table 2 shows the results of a multiple classification analysis (MCA) of payment for grandmother care (dependent variable), specified in two ways: (*a*) cash

TABLE 2 *Multiple Classification Analysis of Percentage of Grandmothers Providing Child Care Who Receive Payment for Care, by Selected Independent Variables: U.S. Sample of Mothers Aged 19 to 26 in 1984 with Child Less Than Age 5 (n = 228)*

		Cash Payment ($\bar{x}$ = 32.6)				Cash and/or Noncash Payment ($\bar{x}$ = 44.3)			
		Unadjusted		Adjusted[a]		Unadjusted		Adjusted[a]	
Independent Variables	n^b	%	Eta	%	Beta	%	Eta	%	Beta
Grandmother employed									
No	(153)	40.9		36.7		53.5		50.4	
Yes	(75)	16.9		25.0		26.9		32.8	
			.24		.12*		.25		.17**
Number of hours of grandmother care									
< 20	(53)	13.1		15.5		28.2		32.9	
20–39	(77)	26.2		28.5		39.2		40.4	
40 +	(98)	49.5		46.2		58.1		54.3	
			.32		.26***		.25		.18*
Mother's work shift									
Day	(166)	37.4		34.6		49.9		48.1	
Nonday	(62)	19.6		27.1		29.0		33.7	
			.17		.07		.19		.13*
Number of children in household									
1	(147)	28.2		29.3		39.7		40.5	
2 +	(81)	41.0		38.9		53.0		51.5	
			.13		.10		.13		.11
Age of youngest child									
< 1	(49)	23.0		20.8		35.6		34.5	
1	(57)	42.4		43.1		49.1		51.2	
2	(59)	37.1		36.2		47.8		45.6	
3 +	(63)	25.5		28.0		43.2		44.5	
			.17		.18*		.10		.12
Number of types of child care									
1	(151)	41.1		37.1		49.5		45.4	
2 +	(77)	18.8		25.3		35.8		42.5	
			.23		.12*		.13		.03
Mother's hourly wage									
< $4.00	(79)	31.4		27.2		42.8		38.6	
$4.00–5.49	(68)	38.8		44.3		48.2		52.7	
$5.50 +	(81)	28.1		27.2		42.1		42.2	
			.10		.17**		.05		.12
Mother's marital status									
Married, spouse present	(128)	33.5		32.0		43.3		40.7	
Other	(100)	30.8		33.8		46.2		51.1	
			.03		.02		.03		.10
Mother's race/ethnicity									
Hispanic	(55)	41.0		39.4		49.3		46.0	
Black	(70)	32.5		26.4		47.2		39.6	
Other	(103)	31.2		33.2		42.7		45.3	
			.07		.08		.05		.05
Multiple *R* squared		.212				.147			
Multiple *R*		.461				.384			

[a]Adjusted for all other independent variables.
[b]*n*'s are unweighted; percentages based on weighted *n*'s.
*F ratio significant at .10.
**F ratio significant at .05.
***F ratio significant at .01.

payment only and (*b*) cash and noncash payment combined. The independent variables for this analysis are those discussed above. MCA is a multiple regression technique that estimates the mean value of the dependent variable for each category of the independent variable. The "unadjusted" columns show the simple two-way relationships between each of the independent variables and the dependent variable. The "adjusted" columns show the two-way relationships net of the influence of the other independent variables in the table.

Looking first at cash payment for grandmother care as the dependent variable, the unadjusted column shows that grandmothers are more likely to receive cash payment for child care when (*a*) they are not employed; (*b*) they provide care for a relatively long number of hours per week; (*c*) they provide care in the daytime rather than evening, night, or at varying times; (*d*) there are two or more children under age 5 in the household; (*e*) the youngest child is age 1 or 2 rather than 0 or 3; (*f*) the youngest child is cared for only by the grandmother; (*g*) the mother's hourly wage is between $4.00 and $5.49; (*h*) the mother is married; and (*i*) the mother is Hispanic.

The adjusted column reveals that the only statistically significant relationships ($p < .05$), when the other independent variables are controlled, are between the number of hours the grandmother provides child care and cash payment (positive association) and between the mother's hourly wage and cash payment (curvilinear association). The positive relationship for number of hours was predicted, but the curvilinear association for mother's hourly wage was not. It may be that when mothers make a very low income (less than $4.00 an hour), they cannot afford to pay anyone for child care since their net gain from employment would be close to nil; when they have a relatively high income ($5.50 + an hour), they may be more likely than those receiving $4.00–5.49 an hour to have grandmothers who are not in need of income, thereby reducing the likelihood of their being paid for child care.

Near statistical significance ($p < .10$) is evident for the relationship between cash payment and grandmother's employment status (negative association), age of youngest child (curvilinear association), and number of types of care for the youngest child (negative association). Although these findings are only suggestive, it seems that grandmother's employment may reduce her economic need for income from child care. It is more difficult to interpret the curvilinear association between age of youngest child and cash payment; it may be that many of these grandmothers view infant care as a temporary phenomenon until the daughter finds an alternative arrangement, and therefore they are less inclined than grandmothers caring for older children to want cash payment. Excluding the category of children less than one year old, there is the expected negative relationship between age of child and payment for grandmother care. The negative association between number of types of child care and cash payment is in the predicted direction. These relationships are only near statistical significance, despite substantial percentage differences between categories, because of the small subsample size. Future studies in this area should consider the possible relevance of these variables.

The other independent variables in this multivariate analysis seem less important; it is particularly interesting that marital status shows very little association with cash payment for grandmother care. All of the independent variables together explain 21.2% of the total variation in cash payment.

The last two columns in Table 2 consider payment in cash and/or noncash forms as the dependent variable. The unadjusted column shows two-way relationships generally similar to those for cash payment only. The adjusted column, however, shows only one variable that is statistically significant ($p < .05$): grandmother's employment status. Those who are employed are less likely to receive payment. The greater importance of this variable in this analysis indicates that, when the other independent variables are controlled, grandmother's employment discriminates more whether noncash payments are received than whether cash payments are received (employment decreasing the likelihood of payment in both cases). The number of hours of grandmother care and the mother's work schedule both show relationships with payment in the predicted direction that are near significance. The total variation explained in payment to grandmothers (cash and noncash combined) by all the independent variables is 14.7%, less than the 21.2% when cash payment only is the dependent variable.

Summary and Implications

We have seen that over one-half of young mothers with young children rely upon relatives for child care when they are employed, and the primary relative is the grandmother. Moreover, the number of hours per week provided by grandmothers is substantial. The reliance on relatives generally, and on grandmothers in particular, is implicitly regarded as a superior alternative to care by nonrelatives in much of the literature—particularly for infants and toddlers (although father care is more often considered "ideal" by mothers than is care by any other relative for preschool-age children of all ages when mothers are employed, but second to formal care for 4-year-olds; Mason and Kuhlthau, 1988). The underlying assumption is that a relative is more emotionally committed to the child and will provide more loving care than a nonrelative. This may be true in most cases—although never empirically tested—but it is only one factor of many that should be considered when assessing care provided by a relative.

Our findings suggest that there may be a more complicated negotiation of work and family roles between grandmothers who provide child care and employed mothers who rely on such care than is generally acknowledged in the literature. About one-third of grandmothers who care for children are otherwise employed, although some of this employment may consist of caregiving for other children. Among employed grandmothers who provide child care, over one-third work completely different hours than the child's mother and more than one-fifth have only a few overlapping work hours. The comparison of the child care hours of employed grandmothers with their work hours suggests that there is considerable juggling of time demands on the part of grandmothers—not just mothers—to enable their participation in child care.

Although the participation of grandmothers in child care may have many benefits, it is not without its costs. We were only able to investigate immediate economic costs, specifically whether there was cash payment (not the amount) and/or noncash payment. We found that close to one-third of the grandmothers receive cash payment, sometimes along with payment of other kinds. An additional one-tenth of grandmothers in our survey receive only noncash payment in the form of meals, transportation, and/or the exchange of other services. A multivariate analysis revealed two significant relationships: cash payment for grandmother care is more likely the longer the number of hours in which grandmothers provide child care, and it is more likely when mothers' hourly wages are in the middle range of the distribution for this sample (between $4.00 and $5.49 an hour). Some dimensions of the relative burden to the grandmother and the ability of the mother to pay thus seem to be relevant determinants of cash payment (although many predicted relationships did not obtain). Moreover, the curvilinear association of mother's current hourly wage and cash payment suggests that the grandmother's need for income may be salient. This interpretation is further supported by the finding that when both cash and noncash payment for child care are combined as the dependent variable, the grandmother's employment status is the significant source of variation; when she is employed (and thus has other income), she is less likely to receive payment.

As we noted earlier, the demand for infant and toddler care is increasing as the availability of grandmothers to provide such care seems to be on the decline. This research suggests that families are responding in complex ways to facilitate the employment of mothers. Many mothers and grandmothers, as well as husbands and wives, are working different hours and sharing child care. How satisfactory these adaptations are, however, is open to question. An earlier analysis of this data source (based on dual-earner married couples; Presser, 1988) has shown that child care by relatives (including grandmothers and fathers) is especially constraining on the number of hours mothers work; part-time employed mothers who rely on relatives for child care are more likely to report that they would work more hours if satisfactory and affordable child care were available, compared to part-time employed mothers who rely on nonrelatives.

To conclude, child care researchers, who tend to focus on the problems of family day care providers and child care centers, need to give more attention to care by relatives in general, and by grandmothers in particular. The implicit assumption that care by relatives is less problematical than care by nonrelatives needs to be investigated empirically. To do so requires that we further explore a neglected dimension of such care—namely, the employment and work schedules of caregiving relatives—and include detailed information in our studies on the nature and amount of payment for such care, as well as other forms of indebtedness that may be involved (social and psychological as well as economic). A longitudinal perspective on these issues is needed. The extent to which child care by relatives is stable (or unstable) over time, as compared with care by nonrelatives, needs to be investigated, with special attention given to the reasons for discontinuity, including alternative job opportunities for the caregivers. Stability of care is an important component of quality of care (Howes, in press; Phillips, 1988) and thus has implications for child development as well as for the employment of mothers and fathers. Many aspects of this important social issue merit further study.

NOTES

This research was funded by a grant from the National Institute of Child Health and Human Development (R01 HD-20187). The paper was presented at the annual meeting of the Southern Sociological Society in Norfolk, Virginia, 14–16 April 1989. The author gratefully acknowledges the helpful comments of Andrew Cherlin, Sandra Hofferth, and Joan Kahn on an earlier draft.

1. These data are for grandparents, not specifically grandmothers, but as may be seen in Table 1, most grandparent care is by grandmothers.
2. Another indicator of the declining availability of grandmothers for child care is the modest decrease in three-generation households. Between 1940 and 1980, the percentage of white children who lived with their grandparents declined from 10% to 5%; for black children it declined from 17% to 12% (Hernandez and Myers, 1987). It could also be that grandmothers are less likely to be residing near their grandchildren than in the past, but unfortunately there are no national data on proximity of grandmothers to assess this. For the U.S. population as a whole, geographical mobility between counties (including between states) has been fairly stable over the period from 1960–61 to 1984–85 (an average annual rate ranging from 6.0% to 6.5%; U.S. Bureau of the Census, 1985).
3. Such an analysis would relate closely to the existing literature on the relevance of kin availability for participation in kin networks (see, for example, Gibson, 1972; Lee, 1980).
4. Grandfather care is included in care by other relatives; such care represented only 2.0% of all principal care.
5. These estimates are higher than for U.S. employed mothers aged 18 and over with children less than 5 years of age. In 1982, 17.2% of all principal child care was provided by grandparents, presumably mostly grandmothers (U.S. Bureau of the Census, 1983, Table 2). We cannot determine in either data source whether the grandmother providing care is the maternal or paternal grandmother. We would expect the maternal grandmother to participate in such care more than the paternal grandmother, particularly if the mother is not currently married; the assumption is that the bonds are stronger between mother and daughter than between (ex)mother-in-law and (ex)daughter-in-law, and that grandmothers perceive providing child care as helping the child's mother more than the child's father.
6. The determinants of grandmother care cannot be assessed adequately with this data set, given the absence of data on the social and economic characteristics of all grandmothers (not only those who provide child care) as well as their physical proximity to the mother. Whether grandmothers reside in the household is known, but this is too restricted a measure of physical proximity. Dennis Hogan and his colleagues at Pennsylvania State University are preparing a public use tape for the 1983, 1984, and 1985 NLSY with derived data on kin access. These data are based on a series of responses on the survey locator sheets to questions on the respondent's mother and father and two other relatives with whom they are frequently in touch (questions on mothers-in-law were not directly asked but could be included in the latter). Hogan, Hao, and Parish (1988) report that 10.5% of white mothers and 19.1% of black mothers did not provide information on their mother (some of these grandmothers were deceased); an additional 16.9% of black mothers and 19.4% of white mothers reported that their mothers lived more than 50 miles from them.

7. If two grandmothers participated in child care, employment questions were asked of the grandmother who provided principal care.
8. Among those employed full-time, unmarried mothers with preschool-age children are in general—without consideration of type of child care—much more likely to work non-days than their married counterparts: 20.6% versus 12.3%, respectively; among those employed part-time, there is little difference by marital status (Presser, 1986).
9. Parents may use cash payments to grandmothers (and other relatives) for the child care credit allowed by the Internal Revenue Service when filing the federal tax return (effective 1976). The only limitation is that the person providing the care is not also a dependent for whom a personal deduction is taken. State welfare programs differ in their regulations concerning payment for care of children by relatives, but this consideration would be relevant for very few women in our sample.
10. Cash payment for grandmother care in this young sample approximates that for U.S. employed mothers aged 18 and over in 1982 with children under age 5, although it is lower on noncash payment. For the 1982 sample, 28.1% of grandmothers who provided either principal or secondary child care received cash payment (some noncash payment as well); an additional 25.4% received only noncash payments, 45.2% received neither, and 1.4% gave no answer. Noncash payment was defined in this study to include "lunches provided for sitters, and exchange of child care services, or other similar inkind arrangements" (U.S. Bureau of the Census, 1983: 3; data from Table 2).
11. The question asks the mother: "Did you or someone else in the family usually pay for this [principal/secondary] care either in cash or in a noncash arrangement such as providing meals, transportation, or exchanging other services?" Thus, it may not always be the mother who pays for grandmother care; for example, it may be the father. Typically, however, the employed mother regards her salary as the source of payment for child care, even though such care may facilitate the dual employment of spouses.
12. They state that the assumption of this instrumental role "is not a style of grandparenting freely chosen; rather it is a style adopted under duress. It is rooted in past and present experiences with hard times. It reflects the continuing instabilities that separation, divorce, unemployment, illness, and death can bring to black family life" (Cherlin and Furstenberg, 1986: 130).

REFERENCES

Blau, Francine D., and Marianne Ferber, 1986. *The Economics of Women, Men, and Work.* Englewood Cliffs, NJ: Prentice-Hall.

Center for Human Resource Research. 1986. *The National Longitudinal Surveys Handbook.* Columbus: Ohio State University.

Cherlin, Andrew J., and Frank F. Furstenberg, Jr. 1986. *The New American Grandparent.* New York: Basic Books.

Gibson, Geoffrey. 1972. "Kin family network: Overheralded structure in past conceptualizations of family functioning." *Journal of Marriage and the Family 34:* 13–23.

Hayghe, Howard. 1986. "Rise in mothers' labor force activity includes infants." *Monthly Labor Review 109*(2): 43–45.

Hernandez, Donald J., and David E. Myers. 1987. "Children and their extended families since the Great Depression." Paper presented at the annual meeting of the Population Association of America, Chicago.

Hogan, Dennis P., Ling-zin Hao, and William L. Parish. 1988. "Race, kin networks, and assistance to mother-headed families." Unpublished manuscript.

Howes, Carollee. In press. "Current research on early day care: A review." In S. Chehrazi (ed.), *Balancing Working and Parenting: Psychological and Developmental Implications of Child Care.* New York: American Psychiatric Press.

Lee, Gary B. 1980. "Kinship in the seventies: A decade review of research and theory." *Journal of Marriage and the Family 42*: 923–934.

Mason, Karen Oppenheim, and Karen Kuhlthau. 1988. "Determinants of child care ideals among mothers of preschool-aged children." Paper presented at the annual meeting of the American Sociological Association, Atlanta.

Phillips, Deborah A. 1988. "Quality in child care: Definitions and dilemmas." Paper presented at the A. L. Mailman Family Foundation Symposium on Dimensions of Quality in Programs for Children, White Plains, NY.

Presser, Harriet B. 1982. "Working women and child care." Pp. 237–249 in P. W. Berman and E. R. Ramey (eds.), *Women: A Development Perspective.* NIH Publication No. 82-2298, U.S. Department of Health and Human Services.

Presser, Harriet B. 1986. "Shift work among American women and child care." *Journal of Marriage and the Family 48*: 551–564.

Presser, Harriet B. 1987. "Work shifts of full-time dual-earner couples: Patterns and contrasts by sex of spouse." *Demography 24*: 99–112.

Presser, Harriet B. 1988. "Shift work and child care among young dual-earner American parents." *Journal of Marriage and the Family 50*: 133–148.

Presser, Harriet B., and Virginia S. Cain. 1983. "Shift work among dual-earner couples with children." *Science 219*: 876–879.

Stack, Carol B. 1975. *All Our Kin.* New York: Harper & Row.

U.S. Bureau of the Census. 1982. *Trends in Child Care Arrangements of Working Mothers.* Current Population Reports, Series P-23, No. 117. Washington, DC: Government Printing Office.

U.S. Bureau of the Census. 1983. *Child Care Arrangements of Working Mothers: June 1982.* Current Population Reports, Series P-23, No. 129. Washington, DC: Government Printing Office.

U.S. Bureau of the Census. 1985. *Geographical Mobility: 1985.* Current Population Reports, Population Characteristics. Washington, DC: Government Printing Office.

U.S. Department of Labor. 1987a. *Employment in Perspective: Women in the Labor Force.* Report 749.

U.S. Department of Labor. 1987b. *News.* USDL 87-343, Bureau of Labor Statistics (August 12).

U.S. Department of Labor. 1988. *Employment and Earnings, April.* Washington, DC: Government Printing Office.

Legacy, Aging, and Succession in Farm Families

Norah C. Keating

Farm families provide an interesting forum for the study of legacies. They are bound together through kinship, through business relationships, and through connections to the land that may span several generations. One important farming legacy is the capital asset of farm land, buildings, and equipment. Yet farming legacies include more than property. They include succession to an occupation, to a business, and to a way of life—all effected through generational connections. For many farm families, the choice and placement of a family successor and the maintenance of family ownership of farm property are the means of transferring these legacies.

Generational transfer in farming is a living legacy. For the most part, farming legacies are handed down through an active process of negotiation between generations that may begin when children are young and continue well into parents' old age. Younger and older generations in farming families must create the process of passing on the legacies and live with the outcome. The success of this endeavor affects the quality of retirement of the older generation, the nature of generational relationships in the family, and the viability of the farm into the next generation.

The Generations Project

The Generations project comprises several studies on generational relationships in farm families, conducted in Canada, New Zealand, and the United States from the mid 1980s to the mid 1990s. The project was designed to increase understanding of generational relationships in two-generation families who were at a life stage in which transfer of the business was an issue. The project was undertaken during a period of rapid change in rural communities. The United States was in the midst of a farming crisis in which land prices had escalated, commodity prices had dropped, and many farmers were being forced out of business. In New Zealand, agriculture was deregulated with the withdrawal of subsidies and incorporation of free trade of agricultural commodities. Farm incomes declined, while interest rates rose. In Canada, numbers of farms continued to decline, and average farm size on the Canadian prairies rose as smaller operators left farming.

During the restructuring of farming in each of the three countries, concern about the continuation of generational succession in farming was acute. During this period, farms were becoming even more family oriented. Because of difficult economic times, they relied more on family labor than hired labor, and some provided an employment refuge for younger family members who could not find off-farm jobs. At the same time, farms carrying higher debt loads were becoming less able to support more than one successor. What's more, new attitudes toward gender equity resulted in a larger pool of eligible successors that now included daughters as well as sons. High business demands and new family issues meant that farming legacies might be played out in new ways.

Transferring the Legacies

Legacies in these farming families were transferred through both succession (transfer of management of the business) and inheritance (transfer of ownership of assets) (Fennell, 1981; Rogers and Salamon, 1983; Gasson and Errington, 1992). In each of the Generations countries, succession occurred before inheritance. Succession involved gradual transfer of management of key business decisions. It began with successor children assuming production decisions about such matters as the timing of planting and harvesting. The child who made good production decisions was next allowed to make marketing decisions—how to market farm commodities, for example. Financial management was transferred last. It included developing business plans, deciding on loans and mortgages, and buying or leasing land. Financial decisions, which have more long-term impact on the business, symbolized the most important element of management control. Fathers in their late fifties generally shared production and marketing decisions with successor children and relinquished decision-making in their sixties. In contrast, many fathers were still actively involved in financial management into their seventies. The process of transfer of management often spanned several decades.

Transfer of ownership of farm property and other assets began in some families at about the same time as succession. However, many farmers in their seventies still owned the majority of the farm, which would not be wholly transferred until the death of the parents. Inheritance in these families is similar to inheritance in nonbusiness families in which assets of the previous generation are passed along at their death. Yet inheritance of farm assets differs in important ways. Farm ownership is of practical and symbolic importance to older parents. It gives them control over a major source of their retirement income as well as a reason to continue to have an active interest in the farm and a status in the community.

Parents did not see posthumous transfer of land as an impediment to succession. Older farmers believed that transfer of management allowed successor children to be farmers while relieving the retiring generation of ongoing responsibility for the farm operation. Surprisingly, successors concurred that they did not need to own all of the farm assets in order to farm. They felt that a combination of management control and control over sufficient assets to operate the business was sufficient.

Generations in these families were highly interdependent because of these ongoing ties to business, family, and ownership. Thus it is not surprising that they believed that a successful farm succession had two components: amicable family relationships and a viable farm operation (Stalker, McGregor, and Rock, 1996). The first indicator of successful succession was that families came through the process with open communication among family members. They believed that the transfer had been successful if all family members including nonfarming children had been included in transition decision-making and had been fully informed of decisions. Yet, if the succession process was to proceed smoothly, there also had to be family consensus on the appropriateness of the choice. In general, members of both generations accepted the principle that it was impossible to treat children equally and still retain the farm in the family. In most families, ineligible children moved into other spheres and allowed the farming child to begin. However, acknowledging the successor and relinquishing claims to the farm did not always occur. One man felt that his sisters "had been robbed" in order for him to become the farmer. Another spent years of negotiating with his siblings, all of whom saw themselves as successors.

The second criterion for a viable farm operation was that the farm maintain its value and profitability after the transfer. In order to ease the entry process for family successors, many farming parents transferred the farm at below market value. However, volatile real estate markets meant that in some cases, land values were soon substantially lower than the transfer price. Thus, successor children were locked into financial commitments that sometimes exceeded the value of the asset. Encumbrances on the farm affected the burden carried by the farming child and the perceived equity or inequity of the division of family assets among the siblings. To be viable, most farms needed to support the retiring and the succeeding generations. Parents who had assets outside the farm were in the best position to allow successor children to manage without placing additional demands on farm assets.

Gender Issues in Transfer and Succession

Women are often seen as invisible farmers who labor in their farms but receive little public recognition for their work (Keating and Munro, 1988). Certainly the language of farming is "gendered." We have had ongoing conversations with our British colleagues (Gasson and Errington, 1993) about their use of the terms farmer and farm wife rather than more neutral terms of farm women and farm men. However, given the place of women in farm succession and inheritance, it may be necessary to concede that their perspective is accurate.

Many researchers believe that girls are not viewed as eligible successors, especially in families in which there are both daughters and sons (Moss and Abramowitz, 1987; Fink, 1986). Others believe that the rule that girls are ineligible may be breaking down. Gasson et al. (1988) suggest that increasing[ly favorable] attitudes toward equality of the sexes make girls eligible as farm successors. This new attitude toward gender equity was evident among middle-aged women and men in the Generations project. Many of these parents viewed their daughters as possible successors. Yet, beliefs in gender equity did not translate into action. Parents did not actively help daughters become family successors. Daughters who wanted to stay in farming were believed to have the option to do so by marrying a "farmer." Parents attempted to be equitable by providing ineligible girls with resources such as career training or a down payment on a house in lieu of giving them equal shares of the farm. Nonetheless, the choice of sons as next-generation successors suggests that there is a gap between attitudes toward equity and actual behavior, in which chosen successors are male. Gender-related traditions in farm succession have not disappeared. Overall, New Zealand women were less involved in management than their North American counterparts.

Generational Issues in Succession (the Prince Charles Syndrome)

A potential family difficulty in passing on the legacy is that when children are ready to begin their careers, parents are not always ready to pass along the farm. In Canada, New Zealand, and the United States, adult children expected to be fully into their farming careers by their mid thirties. Yet while some parents planned to retire in their late sixties, many did not expect to retire at all. Farming parents did not view their continued involvement in the business as an impediment to their farming children's careers. "From the view of the parent, retirement from management responsibility and lightening one's physical work load affords adult children the opportunity to be more autonomous while under the guidance of parents, who still own at least part of the enterprise" (Marotz-Baden, Keating, and Munro, 1995, p. 41). Continued parental guidance is not exactly what farming children had in mind, and many said that it took twenty years for them to assume management control.

This long period of family apprenticeship is analogous to that of Prince Charles, heir to the British throne. The prince is working in the family business but has not taken it over, even though he is approaching so. His mother is still in control, and even his grandmother is still involved. Given Charles's apparently poor decision-making ability and divorce, it may be that his parents do not believe that the chosen successor has the skills necessary to succeed. It seems likely that by the time the prince becomes king, he may be at an age when others would already be considering retirement. By the time Charles succeeds to the throne, his own son will be ready to assume adult roles. Yet Prince William may himself spend most of his adult life as an apprentice in the family business.

Family succession may have made more generational sense when life expectancies were shorter and farming required a young, physically strong person. Today, people can farm well into their seventies, while adult children are still expected to assume occupations in their twenties. It might make more sense for alternate generations to be in charge of the farm (or the country). At least then, alternate generations would have a reasonably long career in their chosen profession. And the middle generations would know that they were just keepers of the land (or the kingdom) until the succeeding generation was ready to take over.

Gender and the Next Generation

The model of the male successor who operates the farm with his wife as helpmate may be making way for a new generation in which married couples are partners in farming. In the Generations project, younger couples were asked what level of mid-life involvement each expected to have in their farm business. In almost all areas of work, management, and ownership, there were two consistent findings. The first is that younger women expect to be more involved in the business than their mothers-in-law have been. The second is that younger men expected to be less involved than their fathers have been.

It may be that younger couples are expecting to move toward more equity in running the business. Younger women want to have more control over the source of their livelihood (Keating and Little, 1994), while younger men have been influenced by notions of gender equity and look forward to having their wives as partners. Many younger men also realize that they may need assistance in running the farm, since they may have to take an off-farm job to increase farm income. Another possible interpretation of younger men's expectation of lower involvement is that they think that their fathers will not retire. They may see shared responsibility with their fathers, not with their wives. An anecdote from Generations Alberta illustrates this point. A farm man in his [fifties] recounted a story of his father who had "retired" just before his 80th birthday. The next summer, the father purchased a new combine, so that he would be available to help during busy times. If family farms continue to be run by generations of farm men, younger women's fantasies of higher levels of farm involvement may not be fulfilled until very late in their lives.

Living with the Legacy

The Generations project was focused on families in the midst of transfer of their farm. They were just embarking on a phase in which parents in later life and their adult children would live with the succession legacy. The movement of parents out of the business and a successor child into the business will change the nature of family relationships in ways that have yet to be explored. Important issues in these later-life families include the nature of relationships with successor and nonsuccessor children and the way in which older parents deal with exit from a way of life.

In the Generations project, the voices of nonsuccessor children were heard only through their successor siblings and their parents. Most retiring-generation parents felt that their farm business could only support one successor, and they attempted to compensate nonsuccessor children through the provision of career training or nonfarm assets such as houses in town. Families who were seen as models of successful succession asserted that their nonsuccessor children had been involved in farming decisions and were supportive of the choice of the successor. Yet little is known about their level of support for their parents' choice of the successor. Some nonsuccessor children may mourn loss of a way of life. Others may resent a sibling being the chosen successor. Some may feel free from the burden of living up to a generational legacy. Each of these possible scenarios has implications for family interactions once the family legacy has been passed to one child.

Overseeing the placement of a family successor is part of the living legacy in farming families. However, for some older farmers, exit from farming marks loss of a way of life. Farmers have different orientations to the centrality of farming in their lives. For some, life is very focused on the farm. These dedicated producers are highly committed to farming as an occupation, to farm work, and to the farming life. For others, farming is a means to an end—a desirable lifestyle (Fairweather and Keating, 1994). It is the latter group that seems most likely to be able to retire, to achieve some emotional distance from the farm, and to maintain some intimacy at a distance from their farming children.

Successor children and their aging parents will continue to be interdependent. Most farm couples need income from the farm to support their retirement. Assets may be taken from the business when the parents retire, although ongoing financial support to parents may be part of the contract with the successor child. This part of the legacy puts pressure on the business to support two families, on generational relationships because parents have a continuing stake in the success of the business, and on successor children who may reach middle age feeling as if they do not have control.

Successor children are tied to their aging parents through commitments to the farm, geographic proximity, and economic interdependence. These ties may also make successor children the most likely candidates for family caregivers as parents grow older. Little is known about how much the economic ties between these children and their parents bring with them implied caregiving responsibilities as parents grow older. Are successor children the ones who are the first line of defense when parents need care? Do continuing business ties create a level of interdependence that fosters or impedes caring relationships? Do nonsuccessor children believe that successor children should be responsible for caring for parents, since successor children were most advantaged by the parents' legacy?

Farming legacies do not just fall into place. Succession is a long process of choosing a family successor, attempting to be equitable in dealing with other children, and moving away from control of the business. The farming successor usually becomes the owner of the farm. However, complex arrangements with siblings and with parents may place this legacy at risk. Finally, losing a way of life may be especially important for this cohort of older farmers. We do not yet know how the legacies in these families will influence their ongoing relationships into the parents' later years and the entry period of the next generation.

REFERENCES

Bennett, J., and Kohl, S. 1982. "The Agrifamily System." In J. W. Bennett, ed., *Of Time and the Enterprise.* Minneapolis: University of Minnesota Press.

Fairweather, J., and Keating, N. 1994. "Goals and Management Styles of New Zealand Farmers." *Agricultural Systems 44*: 181–200.

Fennell, R. 1981. "Farm Succession in the European Community." *Sociologia Ruralis 21*: 19–41.

Fink, D. 1986. "Constructing Rural Culture: Family and Land in Iowa." *Agriculture and Human Values* (4): 43–53.

Gasson, R., and Errington, A. 1992. *The Farm Family Business.* Wallingford, U.K.: CAB International.

Gasson, R., et al. 1988. "The Farm as a Family Business: A Review." *Journal of Agricultural Economics 39*(1): 1–41.

Keating, N., and Little, H. 1994. "Getting into It: Farm Roles and Careers of New Zealand Women." *Rural Sociology 59*: 720–36.

Keating, N., and Munro, B. 1988. "Farm Women/Farm Work." *Sex Roles 19*: 155–68.

Marotz-Baden, R., Keating, N., and Munro, B. 1995. "Generational Differences in the Meaning of Retirement from Farming." *Family and Consumer Science Research Journal 24*: 29–46.

Moss, V. E., and Abramowitz, S. I. 1982. "Beyond Deficit-Filling and Developmental Stakes: Cross-Disciplinary Perspectives on Parental Heritage." *Journal of Marriage and the Family 44*: 357–66.

Rogers, S., and Salamon, S. 1983. "Inheritance and Social Organization among Family Farmers." *American Ethnologist 10*: 529–50.

Stalker, N., McGregor, J., and Rock, G. 1996. *From One Generation to the Next: Successful Farm Transfer from the Perspectives of Retiring and Successor Farm Family Members.* Report to the Rural Education Development Association. Edmonton, Alberta, Canada.

Social Security and Medicare Policy: A Personal, Intergenerational Story

Mal Schechter

My father, an expatriate from Rumanian poverty, supported Social Security legislation in 1935. I don't know what my mother's opinion was. She came to America from Poland with her family. They met on New York's Lower East Side; he was looking for a room, and the person he asked was my maternal grandmother. He got a room in a building with my aunt and two uncles and my grandfather.

In 1935, when Social Security became law, Pop was 35, Mom 29, and I, a born American, age 4. By the time the United States entered World War 11, our family included my brother and sister. The idea that I might reach 65 one day was as inconceivable to me as was the idea that my parents were mortal. But they have died, and I am now "elderly."

Medicare was enacted in 1965, when Pop was 65 and Mom 59. My wife and I, 32 and 34, respectively, had presented them with two grandchildren, ages 7 and 1. A third arrived in 1967, a Generation X-er among two baby boomers. Medicare proved to be important to us all. By protecting our parents against major (though not all) costs of sickness, Medicare helped the rest of the family. We could venture more spending for ourselves and the children without worrying as much about sickness expenses for Mom and Pop. In 1967, my wife and I bought a $34,000 house, applying my veterans' benefits. Our savings were minimal. From time to time, both sets of grandparents gave gifts, including money to help us along. At the time, I had job-based health insurance but no job-based pension plan. I had some commercial life insurance in addition to Social Security life insurance (survivor's benefits). Retirement, a distant prospect, would be financed later on, we hoped.

In 1969, when I was 38, my father died of lung cancer. A spot was discovered on an x-ray film taken routinely by the Baltimore Longitudinal Study of Aging, which the two of us had joined. The cause of the lung cancer was surely heavy cigarette smoking. I blanch today when thinking of the times I, not even in my teens, returned from a drugstore with packs of Lucky Strikes. In 1964, as a Washington reporter, I covered the press conference given by Surgeon General Luther Terry proclaiming the hazards of smoking, especially lung cancer. Pop cut down, but he didn't quit until the diagnosis was made in 1967. An operation followed for removal of a third of one lung. The lung specialist told us he might have escaped metastasis. That hope expired in a year. The rest of Pop's life was awful for Mom and the rest of the family.

Mom couldn't have covered the bills, exceeding $50,000 from a hospital and his doctors. Medicare paid almost all of them. It was a relief to her and to my brother, sister, and me—all on financial tightropes with growing families—that Mom's finances were protected. As Medicare protected her, it also protected the education, housing, and career aspirations of her children and grandchildren.

Mom survived on her Social Security, investments, and cash savings until 1991. At age 85, she had a stroke. Medicare paid her bills for doctors, the hospital, and the nursing home (briefly, for convalescence and rehabilitation). But when her acute care and rehabilitation needs were deemed to have been met, Medicare stopped paying. With severe memory loss and other limitations, Mom never went home. She became a long-term-care patient, receiving "custodial care." Medicare by statute did not, and does not, pay for such care, as essential as it was to her survival. In three months, her assets were wiped out. Her $600-a-month check from Social Security went almost entirely to the nursing home. Because she qualified as poor, Medicaid paid the rest of the bill, about $3,000 a month.

Last summer—five years later—she died, thirty days short of her 90th birthday, in the financial "arms" of Medicaid and the Medicare hospice program for terminally ill persons. A month before she died, I showed her my Medicare card. I had recently passed my 65th birthday. I thought she would want to see the card. Had she been in her old mood, I am sure she would have accused me of a joke, as her way of denying her own age. But Mom was uncommunicative. She no longer spoke or walked. She seldom ate.

I like to think she would have been glad to see that I had become a Medicare beneficiary. I had paid my way. My parents never wanted handouts. They believed as did the philosopher Maimonides that the best form of charity was to help someone obtain the tools to be self-reliant. Uncle Sam had provided a way for them to pay their way for protections against poverty and medical poverty in old age. Dignity was important. Medicare and Social Security allowed that for Mom, up to a point.

My maternal uncle died at age 86 in a hospital a few months after Mom. He, too, was a Social Security and Medicare beneficiary. His last years were lived on a stringent budget. Rent and the monthly cost of multiple prescribed drugs taken at home, which Medicare does not cover, quickly exhausted his Social Security check and a small pension. His wife, as frail as he, managed his complex drug regimen. Neither he nor she would agree to apply for Medicaid, even though those resources seemed to me to be essential.

Mom's dementia protected her from understanding that she was, officially, a pauper. That would have pained her as much as any appearance of being a burden to her children. The same with Pop. They had saved for "rainy days," but they also knew this was not enough. Who can predict cancer, stroke, and the costs of these conditions? Who can predict economic depression and inflation? My parents appreciated having a low-overhead, publicly accountable system made affordable by people joining together for mutual support.

But something has gone away. The commitment to social insurance as the preferred guarantee of dignity in old age is weakening, and despite the country's distaste for "welfare," public assistance—Medicaid for long-term care—is emphasized in old age, especially for women, who generally out-live their health, wealth, and menfolk. In addition, the concept of security as the product of a three-legged stool—Social Security, employment benefits, and savings—seems to be fading.

My analysis is that while the Social Security system is incomplete, it is fixable. However, the will to shore it up seems relatively feeble. It is obvious that pension and health insurance as employment benefits are eroding, exposing breadwinners to potential disasters now and in the future; and that the average individual savings rate appears anemic, partly because many Americans have had stagnant or only slowly growing earnings over the past 20 years (Mishel, Bernstein, and Schmitt, 1996).(1) Corporate "downsizing" of workforces and benefits, the "globalization" of national economies, and shifts in structure and distribution of job opportunities in our economy, are not necessarily ominous. But what makes these factors truly ominous to me is the loss of a collective-security vision and ethic among our political and corporate leaders.

One has only to imagine the executive, technician, or laborer in a downsizing corporation. Income is insufficient for current household needs, including college education of children. Social Security and private pension credits are not accumulating during unemployment between jobs or when the last good job ends a few years before one is qualified for Social Security. Company-paid life insurance, health, and retirement benefits (all of which are government-subsidized) are not there for a workforce newly emphasizing "outsourced" and part-time workers. These staples of the private leg of the three-legged stool become more "iffy" in a restless, technology-driven economy. The three-legged stool is no fit place for an American, young or old, to sit. Being an "American" gets no points from the plutocrats and macroeconomists.

As a study by Chen and Leavitt (1997) indicates, for most of the period since Medicare was enacted, Americans' sense of income security weakened, both in the present and in their expectations for their old age. Some 41 million Americans today lack health insurance. Wages have stagnated for many workers over the past 20 years, and the percentage of workers covered by private pension plans and health insurance has peaked (at about 50 percent). The percentage with neither grew to 25 percent from 22 in the period from 1988 to 1993, and those with either health or pension coverage available diminished to 24 percent from 27 percent. Chen and Leavitt note that "the percentage of insecure workers increased in every worker subgroup, but the magnitude of the increase varied widely among these subgroups."

Assuming that the trends noted above strengthen, I would expect an enlarged and older population on public assistance or doing without needed healthcare because of unaffordable expenses, including those expenses covered by a shrinking Medicare and those excluded (for example, prescribed drugs and long-term care). Family demoralization can be expected to flow from old to young and back again. The politics of savaging the elderly, portraying them as siphoning away healthcare and education resources for children and as destabilizers of our economy, (2) will be offered in place of the politics of truly shared responsibility according to ability to pay. Some, of course, are fighting this trend. (See, for example, *The Economist* [1997], which recently simplified a core issue in population aging—not as old against young—but rich against poor.) Arthur Flemming, in a rousing extemporaneous speech to the 1995 White House Conference on Aging, rose from a wheelchair to talk about America, the community. "America, Inc." may have its place, but America the community had a greater place, he seemed to suggest. After his talk, President Clinton entered the hall and spoke about protecting and cutting Medicare to balance the federal budget. I couldn't applaud. We were back at America, Inc.

Meanwhile, the top 10 percent of households control two-thirds of America's wealth, and the polarization of income and wealth accelerates to proportions not seen since just before the Great Depression. The top 1 percent of households control 26 percent of wealth, we cut progressive income taxes, we raise payroll taxes on workers, we borrow billions from an overflowing Social Security trust fund, and, while the Dow crests at 7,000 in the New York Stock Exchange, we worry about how to support the oncoming wave of baby boomer retirements, starting in 2011. That's when I'll be 80, possibly if not probably a Greedy Geezer falling through the "safety net."

My fate depends on how well our leaders, the media, and the public explore the causes of social insecurity in the interests of a humane and just society. Population aging is a challenge to all our institutions, public and private. Part of the remedy may be more democracy and less plutocracy, less insurance and more public health approaches, less profit in healthcare and more professionalism—but not less care for the elderly to finance more care for children. If poor children are to share in the longevity potentials of our society, then they cannot be isolated from healthcare because their families lack money. Long-term care is not insurable, but it can be prepaid, just as hospital coverage under Medicare is prepaid over the "work span." The affordability of enlarging Medicare to do this—that is, the ability of breadwinners to cover necessary contributions from their pay—depends on how much they get paid. Just as the markets of America depend on consumer income, so do the Social Security and Medicare systems.

If our political and economic leaders don't realize that their labor policies leave many American families short of the money for a decent standard of living and security over the longer average lifetime, these leaders will be courting social as well as marketplace disasters.

My parents should have been rabidly contemptuous of government. Governments rarely had a good name among Eastern European expatriates like my parents. But they recognized Social Security as "good," necessarily compulsory, and something they paid the government to do, specifically. President Roosevelt linked Social Security to earmarked payroll taxes, making benefits an earned right that "no damn politician" would take away, as could easily be the case with a handout. FDR surely would have said the same thing about Medicare, an earned right.

A child born today can expect to live almost to age 76, about twenty-eight years longer than a child born in the year of my father's birth, 1900. The population over age 6, will double to 70 million by 2030. The future elderly will constitute 20 percent of the U.S. population, up from about 13 percent today. A greater proportion will be over age 80 than is the case among today's elders. For the longer lifespan, including a longer work span, America needs a philosophy and a program containing assurances for enough protected income to support a decent old age. We should make good on the promise of enabling all Americans to capitalize on the vaunted opportunities of our economy.

We will have to spend more money on health, education, skill development, and public and private insurance and pension programs. If we are concerned about the next generations, let's stop kidding ourselves: A long-lived society can't reexamine its social insurance programs while keeping off the table discussion of employment policy, income and other taxes, child health coverage, and long-term care. The private sector, uneasy with its load of job-based benefits, might find relief and freedom in passing the burdens to an enhanced Social Security system. The enhancements could include provisions for an individual's limited borrowing against retirement benefits in order to finance further education and career training to keep up with technological change and maturing personal objectives. With the resources to improve primary and secondary schooling, such a change may improve national productivity, earnings, and savings.

Rather than castigating social insurance, our society should look to it as a tool of social adaptation to the longevity revolution. In this way, we would continue the work of our parents and grandparents. We might even feel good about it.

NOTES

1. Mishel, Bernstein, and Schmitt (1996) summarize their work as follows: "As this book goes to press, the economy is in an expansion, but many of the economic problems first evident in the 1980s continue to be felt. For example, despite growth in both gross domestic product and employment between 1989 and 1994, median family income in 1994 was still $2,168 lower than it was in 1989, suggesting that overall growth does not, under current economic circumstances, lead to improved economic well-being for typical families. The 1980's trends toward "greater income inequality and a tighter squeeze on the middle class show clear signs of continuing in the 1990's."
2. See, for example, Lester Thurow (1996), "The Birth of a Revolutionary Class." Thurow warns of impending generational conflict due to the growth of government benefits for the elderly. He blames the elderly for "bringing down the social welfare state, destroying government finances . . . and threatening the investments that all societies need to make to have a successful future." A sophisticated economist rests his arguments on a surprisingly narrow interpretation of government roles. For example, Thurow sees the federal budget unbalanced in favor of older people, but ignores the role of states in providing benefits to children, part of a federal division of labor easily ignored in the quest for a balanced U.S. budget. That only 6 percent of America's elderly receive $50,000 or more in income annually is unnoted; the elderly population mirrors the income polarization seen in the general population. Finally, Thurow claims that the government can use money collected through designated payroll taxes for any purpose.

REFERENCES

Chen, Y-P., and Leavitt, T. D. 1997. "Economic Security of American Workers." *Contingencies* (Jan.–Feb.): 27–9.

Economist. 1997. "Not Young Against Old, But Rich Against Poor" (Jan. 4): p. 27.

Mishel, L., Bernstein, J., and Schmitt, J. 1996. *The State of Working America: 1967–1997.* Washington, D.C.: Economic Policy Institute.

Thurow, L. 1996. "The Birth of a Revolutionary Class." *New York Times* Magazine (May 19), p. 46.

Aging and the Family: Present and Future Demographic Issues

Kevin Kinsella

In the early 1980s, Myers and Nathanson (1982) framed three paramount issues regarding population aging and the family. These concerns involve (1) the extent to which changing concepts of social duties and responsibilities alter traditional ways of providing care for aged people within the family context; (2) the potential social support burden arising from reduced economic self-sufficiency of aged people, longer life expectancy that might involve prolonged episodes of chronic disease morbidity and functional impairment, and increased complexity of social life resulting from urbanization and modernization; and (3) the processes by which countries determine funding priorities for national care systems given competing demands for scarce resources.

The salience of these issues intensified during the 1980s and early 1990s. Governments are concerned not only about the need to provide care and services to growing numbers of aged individuals but about the mix of resources among generations. Short of a revolution in cultural norms, the broad growth of elderly populations ensures that an increased proportion of Social Security resources and expenditures will be allocated to maintaining an adequate income for the elderly. Such a fiscal increase might compete with finances available for traditional family support. A central political tension in developed countries arises from the tendency to withdraw resources from the direct support of families, while promoting an implicit expectation for families to increase support for an elderly generation (ILO, 1989; . . .)

The purpose of this [reading] is to outline several interrelated demographic trends that affect both the structure and functioning of the family institution worldwide. These trends . . . will significantly shape the decisions that families and governments make with regard to care of, and services for, older constituents. The plethora of family types among and within societies precludes any detailed analysis of global familial evolution.[1] Rather, I seek to highlight distinctions between developed and developing regions of the world. Although these overarching categories[2] mask many differences among nations, they do serve as useful foci for identifying general phenomena that affect specific nations to a greater or lesser degree.

Demographic Factors Affecting Family Structure

Evolving Population Age Structure

Underlying many of the changes in family relationships are shifts in population age structure. Population aging refers most simply to an increasing proportion of elderly[3] persons within an overall population. In most countries today, the aging process is determined primarily by fertility rates and secondarily by mortality rates, so that populations with high fertility tend to have low proportions of older persons and vice versa. Demographers use the term *demographic transition* to refer to a gradual process wherein a society moves from a situation of high rates of fertility and mortality to one of low rates of fertility and mortality. The initial stage of this transition is characterized by declines in infant and childhood mortality as infectious and parasitic diseases are eradicated. The resulting improvement in life expectancy at birth occurs while fertility tends to remain high, thereby producing large birth cohorts and an expanding proportion of children relative to adults.

Whole populations begin to age when fertility rates decline and mortality rates at all ages improve. Successive birth cohorts might eventually become smaller and smaller, although many countries experience a "baby boom echo" as women from prior large birth cohorts reach childbearing age. International migration usually does not play a major role in the population aging process but can be important in small nations. Certain island nations, for example, have experienced a combination of emigration of working-age adults, immigration of elderly retirees from other countries, and return migration of former emigrants who are above the average population age; all three factors contribute to population aging.

Figure 1 illustrates the historical and projected aggregate population age-structure transition in developed versus developing countries. At one time, most countries had a youthful age structure similar to that of developing countries as a whole in 1950, with a large percentage of the entire population under the age of 15. Given the relatively high rates of fertility that prevailed in most developing countries from 1950 through the early 1970s, the overall pyramid shape had changed very little by 1990. The effects of fertility decline can be seen in the projected[4] pyramid for 2025, however, in which the strictly triangular shape changes as the elderly portion of the total population increases.

The picture in developed countries has been, and will be, quite different. In 1950, relatively little variation existed in the size of 5-year groups between the ages of 5 and 24. The beginnings of the post–World War II high-fertility baby boom can be seen in the 0-to-4-year age group. By 1990, the baby boom cohorts were 25 to 44 years old, and the cohorts under age 25 were becoming successively smaller. Several developed countries (e.g., Italy, Germany, Hungary, Sweden) have had total fertility rates below the natural replacement level of 2.1 children per woman for some time. Successive small birth cohorts have contributed to the large proportions of elderly people in these societies. If fertility rates remain relatively low through 2025, the aggregate developed-country pyramid will start to invert, with more weight on the top than on the bottom.

Within aging populations, older age groups tend to grow at different rates. The average age of a nation's elderly population often increases because the oldest-old (persons aged 80 years and over) are the fastest growing portion of many elderly populations worldwide. This oldest group constituted 16% of the world's elderly in 1992: 22% in developed countries and 12% in developing countries. As seen in Figure 1, the prominence of the oldest-old—especially women—will increase over time.

Lengthening Life Expectancy

Developed countries have made enormous strides in extending life expectancy since the beginning of this century. From 1900 to 1950, many Western nations were able to add 20 years or more to their average life expectancy at birth. In some countries, life expectancy more than doubled from 1900 to 1990 (Table 1). In the eighteenth and nineteenth centuries, low life expectancy meant that persons spent a relatively short amount of time in a multigenerational family (UNDIESA, 1990). Although most persons lived their older years with family members, time spent in extended families was limited because the average individual died shortly after becoming a grandparent (. . . Ruggles, 1987).

Declining fertility and increased longevity have enhanced the joint survival of different generations. In developed countries, this has led to the emergence of the *beanpole family* (Bengtson, Rosenthal, & Burton, 1990). This vertical extension of family structure is characterized by an increase in the number of living generations within a lineage and a decrease in the number of members within each generation. If fertility rates continue to decline or remain at low levels, four-generation families may soon become the norm in developed countries. The post-World War II baby boom generation could become the great-grandparent boom of the 2020s (Taeuber, 1992).

FIGURE 1 *Population, by Age and Sex: 1950, 1990, and 2025 (in millions)*

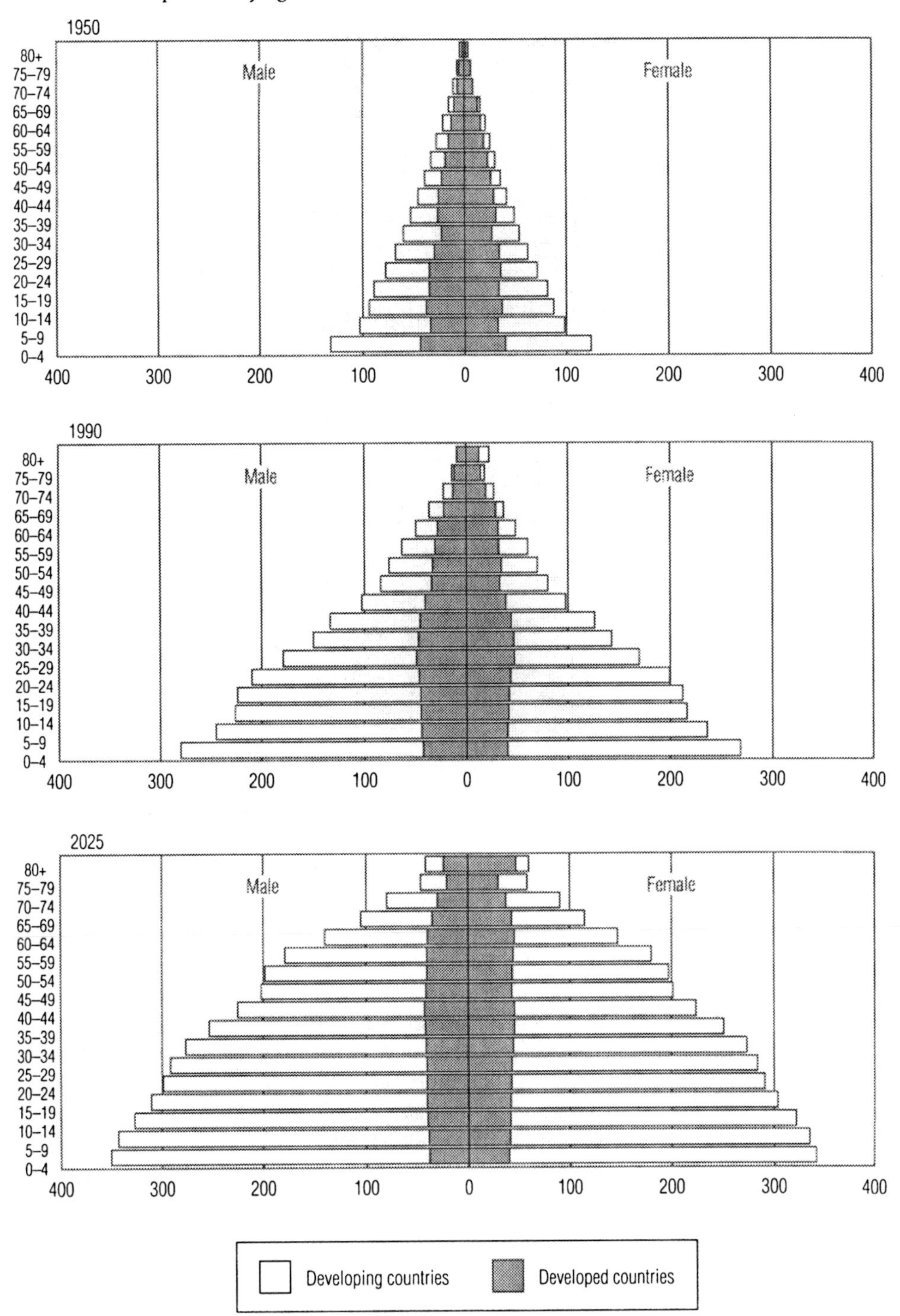

TABLE 1 *Life Expectancy at Birth for Selected Developed Countries: 1900–1990*

	Circa 1900		Circa 1950		1990	
Region/Country	**Male**	**Female**	**Male**	**Female**	**Male**	**Female**
Western Europe						
Austria	37.8	39.9	62.0	67.0	73.5	80.4
Belgium	45.4	48.9	62.1	67.4	73.4	80.4
Denmark	51.6	54.8	68.9	71.5	72.6	78.8
England/Wales	46.4	50.1	66.2	71.1	73.3	79.2
France	45.3	48.7	63.7	69.4	73.4	81.9
Germany	43.8	46.6	64.6	68.5	73.4	80.6
Italy	42.9	43.2	63.7	67.2	74.5	81.4
Norway	52.3	55.8	70.3	73.8	73.3	80.8
Sweden	52.8	55.3	69.9	72.6	74.7	80.7
Southern and Eastern Europe						
Czechoslovakia	38.9	41.7	60.9	65.5	68.7	76.5
Greece	38.1	39.7	63.4	66.7	75.0	80.2
Hungary	36.6	38.2	59.3	63.4	67.2	75.4
Poland	n/a	n/a	57.2	62.8	68.2	76.7
Spain	33.9	35.7	59.8	64.3	74.8	81.6
Other						
Australia	53.2	56.8	66.7	71.8	73.5	79.8
Canada	n/a	n/a	66.4	70.9	74.0	80.7
Japan	42.8	44.3	59.6	63.1	76.4	82.1
New Zealand	n/a	n/a	67.2	71.3	72.2	78.4
United States	48.3	51.1	66.0	71.7	72.1	79.0

Sources: UNDIESA, 1988; Siampos, 1990; U.S. Bureau of the Census, unpublished estimates/projections; and various national sources.
Note: Figures for Germany refer to what was West Germany.

The widening gender differential in life expectancy has been a central feature of mortality trends in the twentieth century. In 1900, the gender gap in life expectancy was typically 2 to 3 years in Europe and North America. Today, women in most developed countries outlive men by 5 to 9 years, reflecting the fact that females have lower mortality than males in every age group and for most causes of death. Average female life expectancy now exceeds 80 years in at least 15 countries and is approaching this threshold in many other nations. The gender differential is usually smaller in developing countries and is even reversed in some South Asian and Middle East societies where cultural factors (low female social status, preference for male rather than female offspring) are thought to contribute to higher male than female life expectancy at birth.

Although the effect of fertility decline is usually the driving force behind changing population age structure, current and future changes in mortality (mostly at older ages) assume greater weight in countries that already have high proportions of elderly citizens. Caselli, Vallin, Vaupel, and Yashin (1987) demonstrated the growing impact of mortality change in population projections for France and Italy. Even if Italian fertility is held to a very low level of 1.4 children per woman through the year 2040, more than half the increase in the proportion of population aged 60 and over will be due to mortality change and less than half to fertility change.

The Epidemiologic Transition

The term *epidemiologic transition,* coined by Omran (1971) and now in its third decade of use, refers to a long-term change in leading causes of death from infectious and acute to chronic and degenerative diseases. In general, the epidemiologic transition is related to, but lags behind, the demographic transition. The initial mortality declines that characterize the demographic transition result largely from reductions in infectious diseases at young ages. As children survive and age, they are increasingly exposed to risk factors linked to chronic disease and accidents. As fertility declines begin to induce population aging, growing numbers of older persons shift national morbidity profiles toward a greater incidence of continuous and degenerative ailments (Frenk, Frejka, Bobadilla, Stern, Sepulveda, & Jose, 1989). Such conditions exact a toll not only on those who suffer from them but on family members who must respond to long-term care needs.

Although comparable cross-national morbidity data are scarce, the implications of epidemiologic change for individual and family care can be ascertained from mortality statistics. Obstructive heart disease has been, and remains, the leading cause of adult mortality in developed (and some developing) countries, though cancers now rank a very close second. In North America, death rates from heart disease peaked in the 1960s and have fallen by almost 50% since that time. Many Western European nations have recorded more modest declines of 10% to 20%, whereas mortality rates from heart disease have increased in several Eastern European nations. Overall age-standardized death rates for malignant neoplasms (cancer) in developed countries have risen 30% to 50% among men since 1950 and fallen by about 10% among women. Such broad trends, however, often are the net result of quite different changes in mortality for the leading sites of disease. In the United States and Western Europe, stomach cancer has been declining steadily since the 1930s, a decline clearly attributed to nutritional change (i.e., a reduction in the salt content of food, especially in preserved food) (Lopez, 1990). On the other hand, a dramatic rise in lung cancer has occurred since World War II, initially among men but now increasingly among women because of increased tobacco use.

The pace of epidemiologic transition varies throughout the developing world. The situation in much of Latin America and the Caribbean is similar to that of developed countries. Data from the Pan American Health Organization (1990) indicate that cardiovascular diseases are the principal cause of death in the populations of 27 of the 37 countries of the Americas for which recent mortality data are available. In 6 of the remaining 10 countries, either cancer or cerebrovascular disease (stroke) is the leading killer. The proportion of deaths attributable to cardiovascular diseases increased from 27% in 1975 to 33% in 1985; if North American data were excluded, the increase would be larger still. The proportion of mortality due to cancer also increased, although the percentage of deaths from strokes declined during the decade.

In several East and Southeast Asian nations the pace of epidemiologic change appears quite rapid. In Singapore, for example, life expectancy at birth rose 30 years in barely one generation, from 40 years in 1948 to 70 years in 1979. During the same period, deaths due to infectious diseases declined from 40% to 12% of all deaths, whereas the share of cardiovascular deaths rose from 5% to 32%. Recent data from China indicate that cardiovascular diseases are often the primary killers in both urban and rural locales; heart disease and cancer together account for 59% of reported deaths in cities and 46% in rural areas. Similar patterns have been reported for Turkey and Sri Lanka, but to date, comparable indicators for the majority of Southern and Western Asia and for Africa are not available.

Overall gains in life expectancy imply, other factors being equal, a greater potential for coresident multigenerational families and an enhanced opportunity to provide care for older family members....

The experience of many developed countries, however, has not substantiated the implication. Numerous researchers (e.g., Rice, 1984; Liu & Manton, 1985; Doty, 1988) documented the direct relationship between population age-sex structure, age-sex-specific rates of chronic disease and disability, and the need for long-term care. But the confluence of several macrotrends in developed countries—older population age structures, higher incidence of noncommunicable diseases, lowered fertility, increased geographical mobility, and rapid advances in medical technology—encouraged the formation of an institutionalized response to population aging. As families no longer could, nor desired, to provide direct care for needy elderly members, medical and nonmedical facilities adopted this role.

The highest rates of institutional use are found in many of the world's oldest countries (as measured by the percentage of the population aged 65 and over), and absolute numbers of users have tended to expand in spite of efforts to enhance community-based services and to avoid or greatly reduce levels of institutionalization. By the early 1990s, however, signs of change appeared. Rates of institutional use were declining in the United States and several other developed countries.... At the same time, age-specific disability rates seem to be decreasing, at least in the United States (Corder, 1992), such that some elderly populations might be spending a greater proportion of their remaining years in a healthy, rather than in a disabled, state (Robine, 1991). If real and sustained, this trend will have enormous import for the well-being of elderly families and younger relatives and for social costs associated with treatment and long-term care. It remains the task of researchers to determine the prevalence of such change and the extent to which it is due to cohort effects.

Effects of Demographic Change on Families and Households

Changing Marital Status

The marital status of older men is very different from that of older women throughout the world. Although widowhood rates rise with age for both sexes, a large majority of men aged 65 and over are married. Even at ages 75 and over, married men usually outnumber widowers. Quite the opposite is true for women; in many countries, the percentage of widows among elderly women exceeds 50%. Several reasons for the gender disparity in widowhood emerge in both developing and developed nations. The most obvious factor is simply that women live longer on average than do men. Also, the nearly universal tendency for women to marry men older than themselves compounds the likelihood of their outliving their spouses. Finally, widowers are much more likely than widows to remarry; in the United States, for example, elderly widowed men have remarriage rates over eight times higher than those of elderly women.

Recent trends for developed countries indicate that growing proportions of older populations are married, declining proportions are widowed, and the percentage of older people who are divorced or separated is small but steadily rising. One might expect that the wide difference in life expectancy that favors women would lead to the increased probability of widowhood. Myers (1990) suggested, however, that the increase in joint survival has meant that higher proportions of husband–wife families reach age 65 and continue intact for some time. Thus, in the United States, actuarial data show that a couple could anticipate 47 years of married life in the mid-1980s, compared with 44.5 years around 1970 and fewer than 35 years in the early 1900s.

In one sense, then, concerns about growing proportions of elderly widows might be less onerous than previously thought. It is crucial, however, to consider absolute numbers as well as proportions. Numbers of older widows, especially in large developing countries, often are increasing rapidly (Figure 2). In Indonesia, for example, the number of elderly widows rose from 1.4 million in 1976 to 2.2 million in 1985. If the gender trend in life expectancy in developing countries follows the historical pattern in developed countries, an increasing share of their older populations will be women. The net result might be an overall improvement in status for women under the age of 65 or 70 due to spousal survival but a worsening of the situation for older women, who are at greater risk in terms of loss of spouse, lack of economic resources, and frail health.

FIGURE 2 *Percent Increase in Number of Widows Age 65 and Older in Selected Countries*

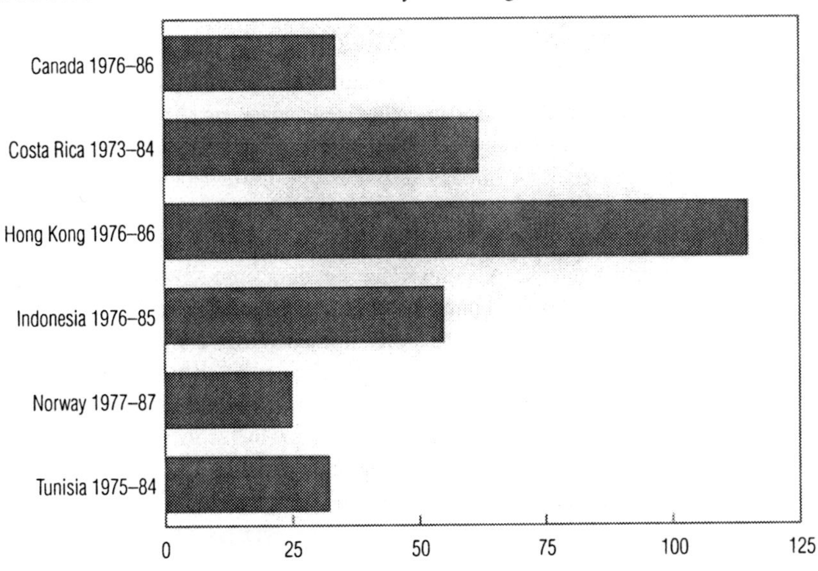

In most countries, more women than men are divorced or separated, probably reflecting gender differences in remarriage rates. Increases in numbers of divorced and separated persons have been observed among the older populations of virtually all countries (Kinsella & Taeuber, 1993). About 5% of both elderly men and elderly women in the United States were divorced (and had not remarried) in 1990, compared with fewer than 2% in 1960. Projections of the U.S. Social Security Administration indicate a continuation of this trend, implying that 8% of elderly men and 14% of elderly women will be divorced by 2020 (Taeuber, 1992).

Proportions of persons never married are likely to become much more significant to the analysis of family structure. A study of 14 European countries (Gonnot & Vukovich, 1989) showed that proportions of never-married men and women increased during the 1980s in all five-year age groups between ages 20 and 40 (with two exceptions) in all countries. The highest such proportions ever recorded in Europe are in Sweden: as of 1985, 39% of all men and 30% of all women had never married.[5] Percentages of never-married Swedish elderly are much lower—roughly 12% for men and women—but still high by international standards. Double-digit percentages of never-married people also obtain among the elderly of both sexes in Norway and among elderly women in several other developed countries.

Rising divorce and nonmarriage rates raise important questions about the likelihood of changing family networks and future social support. Part of the decline in European nuptiality is related to the increasing prevalence of consensual unions, particularly among young adults. The duration of such unions and their implications for later life are as yet unclear. Also unclear is the extent to which sibling interaction and support are substitutes for spouse/child support among both never-married and divorced persons....

Declining Household Size and Altered Living Arrangements

The intersection of the demographic transition and other social forces—urbanization, economic growth, increased female education and labor force participation—produces changing household structures. The enormous worldwide variability in family and household structure makes it difficult to generalize about patterns of change. Nevertheless, it is fair to say that the latter half of the twentieth century has been characterized by declining household size and a trend toward the nuclear family.

Lowered fertility and increased migration have reduced average household size in both developed and developing countries and led to more dispersed family networks. At the same time, increasing proportions of persons are living alone in single-person households. This is partially due to normative changes—delayed marriage and changing gender roles—but is also related to higher rates of marital dissolution and growing numbers of elderly persons whose spouses have died.... The end result is that the number of households in many countries is growing faster than the total population.

The most prominent change in household profiles in developed countries has been the trend toward single-person households. In Sweden, more than a third of all households consist of a single individual. In rapidly aging Japan, the percentage of single-person households rose from 5% in 1960 to 21% in 1985 (UNDIESA, 1990). In addition to having an effect on aggregate saving and spending patterns in a given society, the rise in single-person households shifts a society's caregiving equation as fewer and geographically dispersed children are available for traditionally family support functions.

Table 2 presents proportions of persons aged 65 years and over (unless otherwise noted) who live alone in 57 countries. These figures refer to persons living in what typically are private, noninstitutional households. In developed countries, the proportions range from a low of 10% in Japan to more than 38% in Denmark, Sweden, and what was West Germany. Considerable variation exists within Europe, with the younger countries (in terms of percentage of total population aged 65 years and over)—Spain; Portugal; Ireland; and, to a lesser extent, Greece—exhibiting the lowest rates of single habitation.

Among older individuals, living alone is most often the result of having outlived a spouse and even children or siblings (Kasper, 1988). Consequently, the likelihood of living alone increases with age, although a decline might occur at the oldest ages, especially among women. The decline might be explained by the fact that some of the oldest-old obtain care within institutions, whereas others might seek additional income and/or assistance in maintaining a home by taking in a companion or boarder. Because women on average outlive men and tend to be younger than their spouses, it is not surprising to find that in all older age groups the percentage of women living alone is usually much higher than that of men. Cross-tabulations of marital status and living arrangements are not commonplace, but data from New Zealand and the United States reinforce the intuitive impression that most elderly women living alone are widows—roughly 80% in both countries. Hence, it has become a truism that in most developed countries, women should anticipate a period of living alone at some point during their older years.

Both numbers and proportions of elderly people living alone have risen sharply during the past three decades. Data from Canada illustrate the trend common to most developed countries that the increase in absolute numbers largely consists of women. The 1986 Canadian census recorded 526,000 elderly women living alone, or 200,000 more than a decade earlier. Put in different terms, the number of women living alone grew at an average annual rate of 6.3% from 1961 to 1986. The rate of growth for the entire Canadian population during this period was 1.3%.

In spite of the frequently high proportions of elderly who live alone in developed countries, the fact remains that a majority of those aged 65 years and over live with other persons. Comparable data for 12 Western and Southern European nations from the early 1980s showed that the proportion of elderly that lived with one other elderly person only (in most cases, a spouse) was higher than the proportion that lived singly. The share of elderly population residing with only one other elderly person varied widely, from just under 22% in Ireland to 41% in the United Kingdom. Between 10% and 15% of elderly adults in the 12 European nations lived with one other person younger than 65 years of age; many of these elders were likely to be either men living with a younger spouse or widowed and divorced persons living with a child. The next most common household arrangement was a single elderly person residing with two or three other

TABLE 2 *Percent of Household Population Age 65 Years and Older (Unless Noted) Living Alone: Latest Available Data, 1980–1990*

Europe		*Other Developed*	
Austria, 1980 (60+)	30.9	Australia, 1981	26.2
Belgium, 1981	31.9	Canada, 1986	27.7
Czechoslovakia, 1983 (60+)	32.4	Japan, 1985	9.7
Denmark, 1981	38.3	New Zealand, 1981	26.4
France, 1982	32.6	United States, 1990	31.0
Germany (West), 1982	38.9		
Greece, 1981	14.7	*Asia*	
Hungary, 1984	24.8	China (PRC), 1987 (60+)	3.4
Ireland, 1981	20.1	Indonesia, 1986 (60+)	8.0
Italy, 1981	25.0	Israel, 1985	26.1
Luxembourg, 1981	22.6	Korea, Rep. 1984 (60+)	2.2
Malta, 1980 (60+)	10.5	Malaysia, 1986 (60+)	6.4
Netherlands, 1982	31.3	Philippines, 1984 (60+)	3.0
Poland, 1988 (60+)	36.2	Singapore, 1986 (60+)	2.3
Portugal, 1981	17.7	Sri Lanka, 1987 (60+)	7.6
Spain, 1981	14.1	Taiwan, 1989	8.9
Sweden, 1982	40.0	Thailand, 1986	6.4
United Kingdom, 1989	35.8		
		Central/South America	
Caribbean		Argentina, 1980	12.0
Barbados, 1982	27.1	Brazil, 1980	9.8
British Virgin Is., 1980	20.4	Chile, 1984–85 (60+)	7.0
Cuba, 1981	10.0	Costa Rica, 1985–86	6.9
Dominica, 1980	18.6	French Guiana, 1982	40.0
Grenada, 1981	21.0	Mexico, 1981 (60+)	6.4
Guadeloupe, 1982	32.4	Uruguay, 1985	16.2
Jamaica, 1984	23.0		
Martinique, 1982	30.6	*Other*	
Montserrat, 1980	25.2	Cote d'Ivoire, 1986	2.8
St. Lucia, 1980	19.7	Fiji, 1984 (60+)	2.0
St. Vincent, 1980	16.5	Kenya, 1983 (50+)	16.1
Trinidad/Tobago, 1985 (60+)	13.6	Reunion, 1982	23.3
Turks and Caicos, 1980	17.9		

Source: Compiled at the Center for International Research, U.S. Bureau of the Census from primary census and survey volumes, international compendia, and published research.

Notes: Chile and Czechoslovakia—refer to urban areas; rural Czech percentage is 24.5. *Hungary* refers to pensioners and persons of retirement age. *Sweden* refers to pensioners, with usual pension age being 65 years. *United Kingdom* refers to men 65 years and over, women 60 years and over. *Costa Rica* refers to two cantons only. *Mexico* refers to urban and suburban elderly in four states. *Jamaica* refers to a single urban community of Kingston. *Indonesia* refers to the island of Java. *Malaysia* refers to three Peninsular states. *Kenya* refers to three districts only (Nairobi, Kakamega, Machakos).

persons under age 65 years; this group ranged from only 3% or 4% in Denmark and the Netherlands to 14% in Greece. Small proportions of elderly people (under 3%) lived in three-person households where all members were aged 65 years or over.

Although many factors are related to the changing likelihood of elderly persons' living alone or with their spouse only, rising income appears to be the primary vehicle that affords older individuals the opportunity to maintain households apart from younger family members. The desire for intimacy at a distance is frequently cited in Western gerontological literature. One cross-national study of the propensity of older women to live alone confirms the importance of income and posits that economic factors (income per capita and housing stock per capita) are more pivotal than demographic forces in producing change in residence patterns over time (Wolf, 1990).

Japan is unique among developed countries insofar as a high proportion of older adults—65% in 1985—reside with one or more of their offspring. Three national sample surveys taken during the 1980s showed that roughly one-half of persons aged 60 and over lived with married children, another 20% lived with unmarried children, 22% resided with their spouse only, and the remaining 10% lived alone. The high prevalence of cohabitation with married children is an indication of Japan's traditional *stem family*, which in classic form consists of a married couple living with their unmarried children, the eldest son, his wife, and their grandchildren. In rural areas, such an arrangement is still typical (Kamo, 1988).

Most nonstatistical writing on living arrangements of the elderly in developing countries asserts that relatively few elderly individuals live alone. Although this assertion appears to be true on balance, regional differences are apparent from Table 2. In the Caribbean, which in toto is the oldest of the world's developing regions, between one-fifth and one-third of many elderly populations live alone. Asian countries have higher proportions of older women than men living alone, but in the Caribbean, older men are more likely than older women to live singly. This difference might be related to patterns of migration and marital status unique to parts of the Caribbean region.

In all other developing regions of the world, the paramount living arrangement for elderly persons is with children, with or without grandchildren. Available Latin American data from the 1970s and 1980s indicate that a majority of persons aged 60 years and over lived in *complex family* households consisting of members who belong to more than one conjugal unit (e.g., an older couple, a married child, and grandchildren). Roughly one-fourth to one-fifth of older persons lived in *simple family* households (a married couple or unmarried individual living with unmarried child(ren)), although in Uruguay and urban Chile this proportion was as high as for complex households. In several Latin American nations, time series data reveal declining proportions of elderly in complex (extended) households and greater shares of persons living alone or in nuclear families, reminiscent of the historic trend observed in developed countries.

Surveys in Asia and the Pacific show aggregate residence patterns similar to those observed in Latin America. According to findings from 1984 World Health Organization surveys, in Malaysia, the Philippines, Fiji, and the Republic of Korea, between 72% and 79% of older respondents (aged 60 and over) lived with children (Andrews, Esterman, Braunack-Mayer, & Rungie, 1986). The figures for Malaysia and the Philippines were reconfirmed in later surveys, and similar results have been observed in Indonesia and Singapore. The percentage living with spouse ranged from only 6% in the Philippines to 15% in Indonesia (ASEAN, 1988).

With regard to determinants of living arrangements, data from both Latin America and Asia suggest that marital status is the strongest demographic determinant of whether an older person lives in a complex family household. As might be expected, spouse survival reduces the likelihood of living with children. At the same time, availability of children reduces the likelihood of spouse-only arrangements. Of more interest are associations involving age, sex, education, and urban versus rural residence. The effects of these variables on living arrangements in Latin America appear ambiguous, with no clear trends evident among countries (De Vos, 1990; Christenson & Hermalin, 1989). In Asian nations, males and young-old adults were generally more likely to live with their children than were females or old-old persons (Martin, 1988). This is surprising in view of the commonly held notion that the latter two categories of the elderly population tend to be the most vulnerable in terms of spousal availability and economic resources and, therefore, most in need of support from offspring (unless other relatives provide support). Where urban residence influenced living arrangements, it did so by increasing coresidence with children. This has been observed in other studies (UNDIESA, 1985) and might be related to shortages and high costs of housing in urban areas of developing countries.

Such information hints at the diversity of living arrangements found throughout the developing world. A more detailed portrayal of differences emerges from a comparison of communities in seven countries, part of a United Nations University project (Social Support Systems in Transition) that assembles comparative data on living arrangements in disparate contexts. Although the data refer only to single communities in each nation, the sites are at different stages of urbanization and might reveal differences in patterns of adaptation to the urbanization process as a result of differing social and historical conditions (Hashimoto, 1991).

Several observations arise from a comparative view of the seven communities. The importance of coresidence with family members in both urban and rural settings is reaffirmed, but the proportion of elderly persons (aged 60 years and over) who live with family members of direct descent is highest in the four Asian countries (India, the Republic of Korea, Singapore, and Thailand), more so than in Egypt, Zimbabwe, and Brazil. The stem family predominates in the Asian communities, whereas in Zimbabwe the *skip-generation* arrangement (elderly without children but with grandchildren) is most common. Nuclear households predominate in urban Egyptian and Brazilian communities, and the relatively high share of single and conjugal (spouse-only) households in the Brazilian case indicates a pattern very different from that of the other samples. As might be expected, the likelihood of residing with a married child increases with age in all seven communities. Not as expectedly, the percentage of elderly living alone also increases with age.

The foregoing summary reflects the importance of cultural norms that define family types and the resulting effects on living arrangements. In Cote d'Ivoire, very few old men or women live alone, but for very different reasons. For men, it is difficult to become a widower because polygamy is increasingly common across age groups until late in life; nearly a third of men aged 65 to 74 years have more than one wife, and 13% of men aged 70 to 74 years have three or more wives (Deaton & Paxson, 1991). Hence, more than 80% of elderly men live in households with at least one spouse. In stark contrast, 71% of elderly women are widowed and live either with their children, with a brother's family, or with more distantly related relatives.

Data for the developing world generally are insufficient for documenting changes over time in living arrangements of the elderly. Existing information tends to support the assumption that the family (in its various cultural forms) provides direct support for the vast majority of older persons. A commonly voiced concern in developing nations, however, is that the twin processes of modernization and industrialization are shaking traditional family structures and threatening to create, as a by-product, a marginalized class of older citizens. We know that in many countries, rural areas have become disproportionately older as young adults migrate to urban centers in search of employment (Kinsella, 1988). Beyond anecdotal information, however, it is not yet clear what impact this migration of the young has on older rural residents.

Although the case of Japan does not appear immediately relevant to the situation of developing countries, the extended family structure common to developing countries historically has been a feature of Japanese society. Even after the rapid post–World War II period of economic development and subsequent fertility decline, the large numbers of elderly Japanese persons living with their married children challenge the contention that the nuclear family might be an inevitable product of industrialization. Nevertheless, time series data clearly show that the number and proportion of extended-family households are decreasing (Way, 1984; Wada, 1988), whereas the proportion of childless elderly couples is rising—from 7% in 1960 to 18% in 1985 to a projected level of 30% in the early twenty-first century (Sodei, 1991). These trends have led one author to suggest that the effects of industrialization are so strong that the indigenous culture of Japan vis-à-vis the status of elderly citizens is steadily being undermined in favor of nuclear families (Kamo, 1988). One result of this nuclearization process was identified by Burgess (1986), who reported that significant numbers of elderly people lead "destitute, solitary lives. More than 900,000 women aged 65 years and over live alone in Japan, many of them scraping by on meager pensions, doing menial work, dreaming of getting into shabby government nursing homes" (p. A21).

Changing Social Support Ratios

The notion of social support is at the heart of policy planning for aging populations....

Demographic assessments of intergenerational support often have focused on social support ratios, also known as dependency ratios. Such ratios are seen as indicative of economic dependency within a society and of potential problems concerning provision of health and social services, pension benefits, adequate housing, and (indirectly) family relationships. One such measure, the *elderly support ratio,* relates in a crude fashion the number of persons aged 65 and over to the size of the working-age population aged 20 to 64 years. The combination of declining fertility and increasing longevity has produced rising elderly support ratios in most developed countries. Because the large baby boom cohorts are still of working age, however, the rise has been, and will continue to be, modest in most countries until after the turn of the century. In France, the United Kingdom, Australia, the United States, and several other developed nations, elderly support ratios of the 1990s will increase 10% to 15% by the year 2010. However, between 2010 and 2025 the rise will be 35% to 50% as the large working-age cohorts begin to retire (Figure 3). By 2025, Japan's elderly support ratio will be nearing 50, and many European countries are projected to be in the 40-to-46 range.

In contrast, the majority of developing countries will experience little, if any, change in their elderly support ratios during the next 20 years. This is because the high-fertility cohorts of the 1960s and 1970s will still be in the denominator of the ratio (i.e., under the age of 65). In countries as varied as Zimbabwe, Jamaica, Israel, and Pakistan, the elderly support ratio is expected to decline by 2010, even though the absolute numbers of elderly population are increasing. Eastern and Southeastern Asia and parts of Latin America, on the other hand, will see real change over the next three to four decades. The elderly support ratio is projected to double between now and 2025 in China, Indonesia, Thailand, Brazil, Colombia, and Costa Rica and to more than triple in South Korea and Singapore.

The major flaw in standard dependency ratios is the implicit assumption that age itself indicates need for support. Although this assumption essentially might be true at younger ages, concepts of the traditional needs of elderly adults are being challenged by prolonged health, enhanced financial security, and new forms of social organization. In countries with well-established pension and Social Security programs, many elderly adults are now providing increasing support—instrumental and emotional—to adult children who have run afoul of economic recession, broken marriages, and lack of affordable housing. In developing countries where elderly citizens' economic assets are minimal, aged adults contribute to family maintenance in ways ranging from socialization, housekeeping, and child care to sometimes-exclusive responsibility for child rearing. Such activities free younger adult women for both modern-sector and agricultural production (Hashimoto, Kendig, & Coppard, 1992). Thus it is important to realize that many elderly people are not dependent on younger ones, and when dependency does occur, it may be episodic rather than permanent.

Caretaker Ratios

Given the affective nature of family relations and embedded conceptions of member support, demographic analyses have moved beyond the crude category of elderly support ratios. An alternative

FIGURE 3 *Elderly Support Ratios*

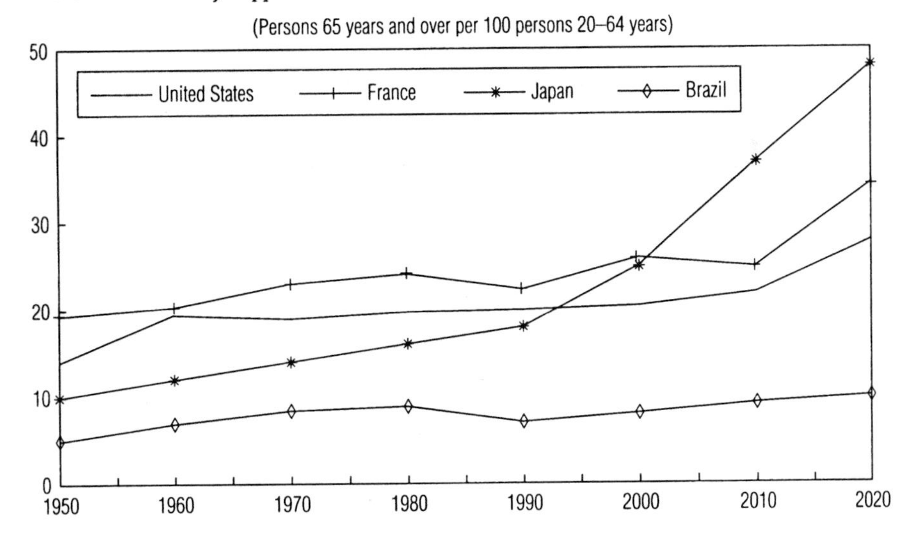

approach examines *caretaker* or *familial support ratios,* measures that relate one generation to another. Several such ratios can be constructed across an age spectrum, depending on the generations under study. For purposes of this [reading], the caretaker ratio is the ratio of the population aged 80 years and over—those most likely to be in need of long-term care and economic support—to the population aged 50 to 64 years, which, in a general sense, relates the oldest-old to their offspring who were born when most of the oldest-old were 20 to 35 years old.

Potential caretaker ratios evolve very differently both within and among world regions. In most developing countries, the ratio of persons aged 80 years and over to those aged 50 to 64 years was between 3% and 6% in 1950. In East and Southeast Asia, this ratio has risen slowly during the past 40 years but will increase significantly in the next 35 years; by 2025, Japan's ratio will be nearing 50. In South Asia, however, the rise has been, and will continue to be, slight. In Latin America, the historical and projected rise appears more constant across time, whereas in Africa, most countries will see little future change in the ratio, even though oldest-old populations are growing rapidly. The most dramatic changes have occurred in the industrialized world; in the Scandinavian and several other countries (including the United States), the ratio is now greater than 20. In some nations, especially in Eastern Europe, the ratio will continue to rise sharply through 2025, whereas other developed countries will see little change over the next three decades.

Women traditionally have been the primary care providers for elderly parents, and this remains true (UNDIESA, 1990; Sundstrom, 1986). Examining the ratio of persons aged 80 years and over to *women* aged 50 to 64 years reveals patterns similar to those just mentioned, but the ratios are, of course, much higher. Projections to the year 2025 suggest that Japan will have 91 persons aged 80 years and over per 100 women aged 50 to 64 years, with most other developed-country ratios in the 60-to-76 range.

Reduced Kin Availability

Living with other people reduces the likelihood of using formal medical care and increases the use of informal care, at least in the U.S. context (Cafferata, 1988). Because most physical, emotional, and economic care to older individuals is provided by family members, the demography of population aging is increasingly concerned with understanding and modeling *kin availability* (the number of family members that will potentially be available to elderly individuals if and when various forms of care are needed).[6] Although smaller families obviously imply fewer potential caregivers, this is offset to some extent by increased population longevity and caregiver survival. Modeling is further complicated by the fact that whereas demographic forces impose constraints on family, household, and kin structures, these structures also are determined by social and cultural factors (Myers, 1992).

The consensus to date foresees a declining biological kinship support network for elderly persons in developed and some developing countries. In addition, future older cohorts are likely to live at greater distances from their children, perhaps resulting in decreased face-to-face interaction (Crimmins & Ingegneri, 1990). We know, however, that decreased coresidence has not necessarily led to less familial interaction; patterns of reciprocal support continue to characterize older people and their children (Mancini & Blieszner, 1989).

Research on prospective kin availability is now expanding in the context of developing countries. This is particularly true in East and Southeast Asian countries, driven in large part by the rapid declines in fertility that have greatly reduced the average family size of young-adult cohorts. The complex interplay of demographic and cultural factors is illustrated by the case of the Republic of (South) Korea, where two-thirds of the elderly are economically dependent on their adult children (Korea Institute for Health and Social Affairs, 1991) and where cultural norms dictate that sons provide economic support for elderly women who have lost their husbands.

Lee and Palloni (1992) showed that although declining fertility results in an increase in the proportion of South Korean women with no surviving son, increased male longevity means that the proportion of elderly widows also will decline (their husbands will live longer). Thus from the elderly woman's point of view, family status will not deteriorate significantly in the coming years. From society's perspective, however, the demand for support of elderly women will increase, because the momentum of rapid population aging means that the fraction of the overall population that is elderly women (especially sonless and childless widows) will increase among successive cohorts. Given the strong trend toward nuclearization of family structure in the Republic of Korea and the traditional absence of state involvement in socioeconomic support, the future standard of living for a growing number of elderly widows is tenuous. A similar prospect looms in Taiwan and Japan (Tu & Liang, 1988; Hermalin, Ofstedal, & Chang, 1992; Burgess, 1986) and, to a lesser extent, in numerous developing nations worldwide.

In spite of gloomy predictions regarding family support in developing countries, most nations will have favorable demographic support ratios for the next several decades. High-fertility population cohorts of the recent past will swell the ranks of working-age populations in many countries at, or shortly after, the turn of the century, thereby providing a large potential tax base for social support at both ends of the age spectrum. In theory, rising family incomes will allow adult children to fulfill obligations to aged parents even if extended family structures become less prevalent and economies modernize.

Directions for Future Research

As populations age, the economic dimensions of social support take on added importance. At the heart of many current debates is the issue of the proper mix of public versus private economic support for older citizens. Researchers of aging and the family should seek to inform the policy debate via an integration of demographic and economic perspectives. For example, studies show that patterns of reciprocity and support exist among family members in all societies. More information is needed, however, about the economic sufficiency of such patterns (for all parties involved), as well as their variability within a given society. One overarching research goal should be the identification of patterns of economic vulnerability in old age and the social aspects (e.g., class, gender, economic status, parental status) that correlate with individual and familial vulnerability.

In addition to identifying vulnerable family types, research needs to anticipate the changing profile of future families. Wider application of family-status life tables (e.g., Yi, 1988) and microsimulation methodologies (e.g., Yang, 1992; UNDIESA, 1990) could give policy planners concrete scenarios of expected kinship structures and living arrangements. Another, even more promising methodology is the cohort approach to population-based forecasts (e.g., Stone & Ng, 1992), wherein inter- and intracohort heterogeneity can be measured, analyzed, and accounted for when projecting variables such as joint kin survival and health service use requirements.

In addition, it is important to link demographic trends in population aging with an analysis of housing options, housing availability, and the housing needs and desires of end users. Housing is an important variable in the income support and care of older persons in the community, yet housing outcomes in old age are largely dictated by economic and planning decisions not directed toward older people (Kendig, 1990).

Finally, policy options and outcomes that already exist must be used in the preparation for demographic change. Although new programs will surely evolve, they should emerge from assessments and evaluations of extant efforts (e.g., tax incentives, home assistance schemes, self-help programs) to assist families vis-à-vis support and care of elderly members. As has been noted in a variety of cultural contexts, current evaluations of outcomes are based more on expectations and hopes of policymakers than on objective, scientific evidence (Hashimoto et al., 1992). This is especially true in many developing countries where official pronouncements routinely refer to the need to strengthen traditional family structures but ignore the indisputable social forces that have altered such structures throughout the world.

NOTES

The views expressed in this [reading] are those of the author and do not necessarily represent the views of the U.S. Census Bureau.

1. For a discussion of difficulties involved in cross-national comparisons of the family, see Myers (1992). For an excellent summary of diversity in family structure within a single nation—the United States—see Bengtson, Rosenthal, and Burton (1990).
2. The "developed" and "developing" country categories used in this chapter correspond directly to the "more developed" and "less developed" classification employed by the United Nations. Developed countries comprise all nations in Europe (including the former Soviet Union) and North America, plus Australia, Japan, and New Zealand. The remaining nations of the world are considered to be developing countries.
3. To facilitate comparisons among countries, the admittedly heterogeneous term *elderly* refers to all persons aged 65 years or over.
4. Unless otherwise noted, projected population figures in this [reading] are those of the Center for International Research, U.S. Bureau of the Census, as of mid-year 1992.
5. Overall figures in the United States are much lower, although among black women aged 35 to 39, 25% had never married as of 1990.
6. Numerous analytic and simulation models of kinship structure have been generated since the mid-1970s. See Wolf (1988) for an overview of such models and a microsimulation methodology applied to developed countries and Yang (1992) for a developing country example (rural China).

REFERENCES

In addition to the following sources, primary census and survey publications of individual nations were used to calculate many of the statistics in this [reading]. The majority of these data are contained in an International Data Base on Aging; see Kinsella & Taeuber (1993) for further information.

Andrews, G., Esterman, A. J., Braunack-Mayer, A. J., & Rungie, C. M. (1986). *Aging in the Western Pacific; A four country study* (Western Pacific Reports and Studies No. 1). Manila: World Health Organization Regional Office for the Western Pacific.

Antonucci, T. C. (1990). Social supports and social relationships. In R. H. Binstock & L. K. George (Eds.), *Handbook of aging and the social sciences* (3rd ed., pp. 205–226). San Diego: Academic Press.

Association of South East Asian Nations (ASEAN). (1988). *Socioeconomic consequences of the ageing of the population* (Phase III Population Project, Inter-Country Final Report.) Jakarta: ASEAN.

Bengtson, V., Rosenthal, C., & Burton, L. (1990). Families and aging: diversity and heterogeneity. In R. H.

Binstock & L. K. George (Eds.), *Handbook of aging and the social sciences* (3rd ed., pp. 263–287). San Diego: Academic Press.

Burgess, J. (1986, October 12). Japan faces "graying" of population. *Washington Post,* A21.

Cafferata, G. (1988). Marital status, living arrangements, and the use of health services by elderly persons. *Journal of Gerontology: Social Sciences, 42,* S613–618.

Caselli, G., Vallin, J., Vaupel, J. W., & Yashin, A. (1987). Age-specific mortality trends in France and Italy since 1900: Period and cohort effects. *European Journal of Population, 3,* 33–60.

Christenson, B., & Hermalin, A. (1989, November). *A demographic decomposition of elderly living arrangements: A Mexican example.* Paper presented at the 42nd Annual Scientific Meeting of the Gerontological Society of America, Minneapolis.

Corder, L. (1992, June). *National long term care survey.* Paper presented to the Inter-Agency Forum on Aging Related Statistics, U.S. National Institute on Aging, Bethesda, MD.

Crimmins, E. M., & Ingegneri, D. G. (1990). Interaction and living arrangements of older parents and their children. *Research on Aging, 12,* 3–35.

Deaton, A., & Paxson, C. (1991). *Patterns of aging in Thailand and Cote d'Ivoire* (Living Standards Measurement Study Working Paper No. 81). Washington, DC: World Bank.

De Vos, S. (1990). Extended family living among older people in six Latin American countries. *Journal of Gerontology: Social Sciences, 45,* S87–94.

Doty, P. (1988). Long-term care in international perspective. *Health Care Financing Review* (Annual Supplement), 145–155.

Frenk, J., Frejka, T., Bobadilla, J. L, Stern, C., Sepulveda, J., & Jose, M. (1989). The epidemiologic transition in Latin America. In *International Population Conference, New Delhi* (Vol. 1, pp. 419–431). Liege, Belgium: International Union for the Scientific Study of Population.

Gonnot, J.-P., & Vukovich, G. (1989). *Recent trends in living arrangements in fourteen industrialized countries* (Working paper 89–34). Laxenburg, Austria: International Institute for Applied Systems Analysis.

Hashimoto, A. (1991). Urbanization and changes in living arrangements of the elderly. In *Ageing and urbanization* (pp. 307–328). New York: United Nations. (ST/ESA/SER.R/109)

Hashimoto, A., Kendig, H. L., & Coppard, L. (1992). Family support to the elderly in international perspective. In H. L. Kendig, A. Hashimoto, & L. Coppard (Eds.), *Family support for the elderly: The international experience* (pp. 293–308). Oxford: Oxford University Press.

Hermalin, A., Ofstedal, M. B., & Chang, M.-C. (1992). Types of supports for the aged and their providers in Taiwan. *Comparative study of the elderly in Asia* (Research Report 92–14). Ann Arbor: University of Michigan Population Studies Center.

International Labour Office (ILO). (1989). *From pyramid to pillar. Population change and social security in Europe.* Geneva: International Labour Organization.

Kamo, Y. (1988). A note on elderly living arrangements in Japan and the United States. *Research on Aging, 10,* 297–305.

Kasper, J. D. (1988). *Aging alone. Profiles and projections.* Baltimore: Commonwealth Fund Commission on Elderly People Living Alone.

Kendig, H. L. (1990). Comparative perspectives on housing, aging, and social structure. In R. H. Binstock & L. K. George (Eds.), *Handbook of aging and the social sciences* (3rd ed., pp. 288–306). San Diego: Academic Press.

Kinsella, K. (1988). *Aging in the third world* (U.S. Bureau of the Census International Population Report No. 79). Washington, DC: U.S. Government Printing Office.

Kinsella, K., & Taeuber, C. (1993). *An aging world II* (U.S. Bureau of the Census International Population Report No. P95/92–3). Washington, DC: U.S. Government Printing Office.

Korea Institute for Health and Social Affairs. (1991). *KIHASA Bulletin,* Number 21. Seoul: Institute for Health and Social Affairs.

Lee, Y.-J., & Palloni, A. (1992). Changes in the family status of elderly women in Korea. *Demography, 29,* 69–92.

Liu, K., & Manton, K. (1985, June). *Disability and long-term care.* Paper presented at the workshop on Methodologies of Forecasting Life and Active Life Expectancy, U.S. National Institute on Aging, Bethesda, MD.

Lopez, A. (1990, July). Mortality trends in the ECE region: Prospects and implications. *Seminar on Demographic and Economic Consequences and Implications of Changing Population Age Structures.* Ottawa: United Nations Economic Commission for Europe.

Mancini, J. A., & Blieszner, R. (1989). Aging parents and adult children: Research themes in intergenerational relations. *Journal of Marriage and the Family, 51,* 275–290.

Martin, L. (1988). *Living arrangements of the elderly in Fiji, Korea, Malaysia, and the Philippines.* Honolulu: East-West Population Institute.

Myers, G. (1990). Cross-national patterns and trends in marital status among the elderly. In *Populations agees et revolution grise* (pp. 469–481). Louvain-La-Neuve: Université Catholique de Louvain.

Myers, G. (1992). Demographic aging and family support for older persons. In H. L. Kendig, A. Hashimoto, & L. Coppard (Eds.), *Family support for the elderly: The international experience* (pp. 31–68). Oxford: Oxford University Press.

Myers, G., & Nathanson, C. (1982). Aging and the family. *World Health Statistics Quarterly, 35,* 225–238.

Omran, A. R. (1971). The epidemiologic transition: A theory of the epidemiology of population change. *Milbank Memorial Fund Quarterly, 49,* 509–538.

Pan American Health Organization (PAHO). (1990). *Health conditions in the Americas* (1990 ed., Vol. 1). Washington, DC: Pan American Health Organization.

Rice, D. (1984). Long-term care of the elderly and the disabled. *Long-term care and social security* (Studies and Research No. 21). Geneva: International Social Security Association.

Robine, J.-M. (1991, June). *Changes in health conditions over time.* Fourth workgroup meeting of the International Research Network on Healthy Life Expectancy, Noordwijkerhout, the Netherlands.

Ruggles, S. (1987). *Prolonged connections: The rise of the extended family in nineteenth-century England and America.* Madison: University of Wisconsin Press.

Siampos, G. (1990). Trends and future prospects of the female overlife by regions in Europe. *Statistical Journal of the United Nations Economic Commission for Europe, 7,* 13–25.

Sodei, T. (1991). *Elderly people living alone in Japan.* New York: International Leadership Center on Longevity and Society.

Stone, L., & Ng, E. (1992). *New approaches to decomposition of population aging.* Ottawa: Analytical Studies Branch, Statistics Canada.

Sundstrom, G. (1986). Family and state: Recent trends in the care of the aged in Sweden. *Ageing and Society, 6,* 169–196.

Taeuber, C. (1992). *Sixty-five plus in America* (U.S. Bureau of the Census Current Population Reports, Series P-23, No. 178). Washington, DC: U.S. Government Printing Office.

Tu, E. J. C., & Liang, J. (1988, April). *Demographic transition, kinship structure and population aging in Taiwan.* Paper presented at the Annual Meeting of the Population Association of America, New Orleans.

United Nations Department of International Economic and Social Affairs (UNDIESA). (1985). *The world aging situation: Strategies and policies.* New York: United Nations.

United Nations Department of International Economic and Social Affairs (UNDIESA). (1988, September). *Economic and social implications of population aging.* Proceedings of the International Symposium on Population Structure and Development, Tokyo. New York: United Nations. (ST/ESA/SER.R/85)

United Nations Department of International Economic and Social Affairs (UNDIESA). (1990, October). *Overview of recent research findings on population aging and the family.* United Nations International Conference on Aging Populations in the Context of the Family, Kitakyushu. New York: United Nations. (IESA/P/AC.33/6)

Wada, S. (1988). Daily life in later life in the changing Japanese context. In K. Altergott (Ed.), *Daily life in later life: Comparative perspectives* (pp. 205–228). Newbury Park, CA: Sage.

Way, P. (1984). *Issues and implications of the aging Japanese population* (Center for International Research Staff Paper No. 6). Washington, DC: U.S. Bureau of the Census.

Wolf, D. (1990). Household patterns of older women: Some international comparisons. *Research on Aging, 12,* 463–486.

Wolf, D. A. (1988). Kinship and family support in aging societies. *Economic and social implications of population aging* (pp. 305–330). New York: United Nations (ST/ESA/SER.R/85).

Yang, H. (1992). Population dynamics and kinship of the Chinese rural elderly: A microsimulation study. *Journal of Cross-Cultural Gerontology, 7,* 135–150.

Yi, Z. (1988). Changing demographic characteristics and the family status of Chinese women. *Population Studies, 42,* 183–203.

Programming for Family Care of Elderly Dependents: Mandates, Incentives, and Service Rationing

Amanda Smith Barusch

Over a decade ago, the World Assembly on Aging convened in Vienna to bring attention to an emerging demographic imperative. The world's population is aging at an unprecedented rate, with the most dramatic growth observed in developing nations. The International Institute on Aging estimated that between 1985 and 2025, the aging population would increase 77 percent in developed nations and 207 percent in developing nations (Kosburg, 1992). These estimates are consistent with those developed by the United Nations (1988) and demographer George C. Myers (1990). The final report of the World Assembly noted the central role that families play in support of dependent elderly people (United Nations, 1982).

Yet families in both developed and developing countries are experiencing changes that seriously limit the resources available for elder care. In developed countries women are entering the work force in unprecedented numbers. In developing countries economic growth draws young adults away from rural areas, leaving dependent elderly people to fend for themselves (Kosburg, 1992; Tracy, 1991). So although nations may vary in the degree to which cultural norms enforce the notion of filial piety, they have in common economic pressures that make fulfillment of this norm difficult for modern families.

Caught between the increased need for care of dependent elders and the diminished family capacity to provide such care, many have struggled to define the appropriate response of government....

This [reading] examines three approaches [to expanding social care for the elderly]: filial support legislation, incentives for family care, and service rationing provisions. These methods were chosen for their ubiquity—they have been either contemplated or applied in many nations. They share a common intent: to place the burden of care for dependent elderly people on family instead of government. Material used is drawn in part from a survey completed by ... specialists [on aging] in 33 countries throughout the world.

International Survey

The key informant approach was used to collect information about programs and policies throughout the world. Surveys were mailed to 137 organizations serving elderly people. The source of addresses was a directory prepared in 1988 by the American Association for International Aging, *An International Directory of Organizations in Aging*. The directory is divided into five sections: Africa, Asia and the Pacific, Europe, Latin America and the Caribbean, and North America. Respondents were selected to optimize geographic coverage and to ensure inclusion of representatives from both developed and developing nations.

TABLE 1 *Survey Response Rate by Region*

Region	Surveys Mailed	Responses Received	Response Rate (%)
Africa	22	5	23
Asia and the Pacific	47	7	15
Europe	41	15	37
Latin America and the Caribbean	20	5	25
North America	7	1	14
Total	137	33	24

Table 1 shows the number of surveys sent, the number returned, and the response rate in each region.

The overall response rate was 24 percent, with the highest rate in Europe and the lowest in North America and Asia....

Respondents were asked to describe and evaluate developments in their countries by responding to a survey consisting of four sections: (1) policies regarding family care of sick elderly relatives, (2) incentives for family care, (3) services for families who care for elderly relatives, and (4) innovations in their area. This article presents results from the first three sections.

Clearly, interpretation of the responses to this survey cannot imply that they are representative of the world, the region, or even the countries involved. Instead, they represent the viewpoints of professionals throughout the world who are committed to serving the elderly population. For cross-national comparisons, the key informant approach has limited value because of inconsistencies in the level of awareness and expertise of respondents. The advantage of this approach lies in its ability to tap the opinions and values of respondents. So, although not definitive, this material is intriguing in the diverse approaches and views represented.

Mandated Family Responsibility

In the United States, experience with mandated family responsibility has involved sporadic passage of "filial support statutes" at the state and local levels. With their roots in the Elizabethan Poor Laws, they embody the notion that responsibility for dependent and indigent people rests primarily with their families. A 1989 review found filial support laws on the books in 30 states (Bulcroft, Van Leynseele, & Borgatta, 1989). They were found under various statutory categories, primarily family law and poor law.

Under these laws, children's income is considered when determining an older person's eligibility for public relief or social services. Administration of the laws has proven costly in terms of both dollars and family relations (Newman, 1980). Dissatisfaction with the statutes led to the exclusion of family income from eligibility determination in the Medicaid program. Yet proposals to amend this aspect of the program continue to surface in response to escalating Medicaid expenses (Gibson, 1984).

The People's Republic of China uses a well-known approach to mandated filial responsibility. Under the 1982 constitution, adult children have the duty to support their parents. Furthermore, the 1980 Marriage Law establishes a legal right for older parents who lose the ability to earn a living to demand that children (or, if the children are dead, grandchildren) pay for their support. The 1981 Criminal Code provides for up to five years in prison for flagrant failure to comply with filial support statutes (Tout, 1989). Further examples of filial support statutes have been found in Morocco and Tunisia (Tout, 1989), as well as Japan and Singapore (Martin, 1988).

Survey Results Those responding to the international survey were asked about laws and regulations that require family members to provide care for sick and disabled elderly family members. Five respondents indicated that their countries (Romania, Hungary, Argentina, the Republic of Georgia, and Germany) had such laws or regulations.

As in the United States, these laws are commonly part of the code governing either family relations or social welfare programs. Filial responsibility may also be dictated by the nation's constitution. In Romania filial responsibility is established in the Family Code (Law No. 4/1953) and the Elderly Care Contribution Decrees (Nos. 253/1971 and 46/1982). In Argentina, it is part of the Code of Law. The German Law on Social

Welfare requires that adult children contribute to the costs of caring for their parents as much as they are able. Both the Republic of Georgia and Hungary have family responsibility embedded in their constitutions. In Romania the law establishes the "obligation of the family members to provide care to elderly who become sick or disabled, either directly or by financial contribution."

Notably, parental behavior can influence enforcement of filial responsibility. The Republic of Georgia makes filial responsibility contingent on parental performance. Under the Civil Code of Georgia, children may be exempted from their duties to support their parents if the court establishes that the parents have not fulfilled their parental duties. The parent who has been thus deprived of parental rights loses the right to "demand . . . alimony from children." In Argentina, children who have been disinherited are relieved of their statutory responsibility to care for their parents.

Respondents were also asked how the law was enforced. The respondent from the Republic of Georgia indicated that the law was rarely enforced. Romanian law stipulates that when a family member fails to meet his or her obligation to a parent, a certain amount of the family member's income will be deducted and given to the elderly person. In Hungary and Argentina, government agencies are charged with enforcement.

Respondents disagreed about the effect of these policies on families. When asked whether the law "helps families take care of the elderly," those from Romania and Argentina said no, whereas those from Hungary and the Republic of Georgia said yes. The Romanian respondent said, "The laws/regulations lay down only the obligation of the family members to provide care to the elderly, without any direct payments or tax incentives to these family members." The Argentinean respondent noted that "The elderly people generally do not use the law and the institutions try to avoid it because if they appeal to it the affective links would be harmed." The respondent from Georgia supported the law, saying it clarifies or regulates otherwise disputable questions.

Discussion The impact of filial support laws on family care of elderly members has not been systematically investigated. Their popularity and effectiveness depend on cultural norms and family resources. In the People's Republic of China, with an ancient Confucian tradition of filial responsibility, these laws are compatible with dominant norms. Furthermore, most couples live with the husband's parents for extended periods after marriage. Requiring that a son provide care to his mother should she become ill may be more feasible under these circumstances.

In the United States filial support legislation has not been popular. After surveying the enforcement of these laws in 30 states, Bulcroft and colleagues (1989) noted erratic patterns of enforcement and concluded that "laws that are enforced irregularly are thought of as 'bad' laws," (p. 389), capricious in their enforcement, and not likely to survive court tests, especially on appeal. On completion of her review of the history of filial support laws at the state level, Newman (1980) concluded that "neither intuition nor existing evidence suggests that family solidarity is developed or nurtured by mandated financial contributions" (p. 65).

Gibson (1984) noted proposals to make filial support part of the Medicaid program, adding that "these requirements [that hold families responsible for needy elderly relatives] have been abolished in most Western industrialized nations" (p. 104). She mentioned that in Yugoslavia these laws were rarely enforced. Findings of this study indicate this was also true of the Republic of Georgia.

Incentives for Family Care

Financial incentives for relatives to provide care include either direct payments or tax incentives for family caregivers. In the United States both are available. Several states use the Social Services Block Grant (SSBG) to provide payment to family members who provide home care. In some counties in California, 70 percent of home care services paid for under SSBG were provided by family members (Talley, 1980).

U.S. tax statutes allow medical deductions for some of the expenses associated with family care. Presently this applies to medical expenses in excess of 5 percent of the taxpayer's income and payment for medicine and drugs above 1 percent. (The graduated rates applied here effectively deny medical deductions to all but those experiencing major or catastrophic illness.) In addition, the dependent exemption and tax care credits may, in some circumstances, be used by family caregivers.

Two states, Idaho and Arizona, have tax incentive programs specifically designed to encourage family care (Sherwood & Morris, 1991). In addition to the standard dependent exemption, Idaho provides either a tax deduction of $1,000 or a refundable tax credit of $100 to family members other than a spouse who provide a home and at least 50 percent of the support for an elderly relative. Arizona permits family or friends to itemize and claim a deduction on state income tax for money spent on health-related services for someone who is 65 or older....

Survey Results Respondents to the survey were asked to describe tax or direct payment incentives for family care. The respondent from Great Britain reported that tax benefits were available. Those from six countries (Germany, Norway, Sweden, Australia, the Republic of Georgia, and Scotland) reported direct payments. And those from two (Argentina and Japan) reported both. Countries that provided direct payments typically gave limited support to families providing home care to compensate for lost wages.

When asked whether these payments or benefits helped families take care of elderly relatives, all respondents answered yes. Several noted that the payments were small, offering only limited support. Still, they compensated to some extent for lost income, and, as the respondent from the Republic of Georgia noted, they "increase attention of state and society to family problems where family members are providing care to elderly persons."

Discussion Although incentive policies were popular among respondents to this survey, both tax benefits and direct payments have disadvantages. Tax benefits are easy to administer and (at least in the United States) politically attractive. But they present serious equity problems. These benefits do not phase out as gross income rises, so they have more value to those with higher incomes. The largest benefits go to those who need them least, whereas those with the lowest incomes (who have no tax liability) receive no benefits.

Direct payments may compensate family members for wages lost to family care. Or they may transform the generous child caring for a beloved parent into an underpaid employee of the state. When caregiving becomes a job, family members may be deprived of its most salient reward, the sense of "doing right" or of "giving a gift" to a loved one. As Gilbert (1984) pointed out, "natural inducements may be reinforced by a dash of economic reward.... At some point, concrete economic incentives to provide social care will override the more ethereal inducements of sympathy, sentiment, and expectation" (p. 27).

Another critique of direct payments was provided by the respondent from Germany, who expressed her concern that "introducing a financial bonus for making family care attractive becomes hostile against all chances for the rehabilitation of elderly people" (Lehr, 1991, p. 12). This critique focused on proposals in her country that would increase the amount of the caregiver payment as the patient's dependency became greater.

Furthermore, these incentive policies may have limited value as incentives. The offer of small rewards may have little or no effect on a family member's decision to assume the overwhelming task of caring for an elderly relative. Incentives can operate in two ways: They may enhance the likelihood that a person will make a choice, or they may encourage a person to persist in a choice already made. It is unlikely that either tax benefits or direct payments contribute to a family member's initial choice to provide care. Other factors, such as emotional bonds (Cicirelli, 1981), filial obligation, social support, and personal resources must weigh more heavily.

The inducements offered through these policies might encourage those already providing care to continue or increase their efforts. Sherwood and Morris (1991) suggested that tax inducements can "influence the behavior of families in caring for their vulnerable elderly relatives" (p. 5). In their study, people who used tax benefits in Idaho and Arizona reported spending significantly more money for elder care than did those in a comparison group of those eligible who did not claim the benefit. Further, family members claiming the benefit in Idaho reported providing more hands-on supportive services. Without experimental design, it is impossible to determine whether tax inducements caused greater expenditures or whether greater expenses motivated people to claim tax benefits.

Rationing Services

When service allocation reflects the view that services should be provided only as a last resort, the rationing process often dictates withholding services from elderly people who have access to family care. In the United States, caseworkers often withhold supportive services when a family member might provide care (Gibson, 1992). In addition, access to meals-on-wheels and homemaker services for men has been inversely related to availability of a female spouse to provide care. In Great Britain, publicly funded home help is made available primarily to low-income elderly people with heavy disabilities who live alone (Doty, 1988). Sundstrum (1987) noted signs that this type of rationing was occurring in Sweden.

Restricting access to Medicaid nursing home beds is another means of enforcing family responsibility among those unable to afford private care in the United States (Gibson, 1984). Restriction of nursing home access to "economic elites" is also practiced in South Africa (Mamabolo, 1991).

Survey Results The survey presented a list of eight services—nutrition support, visiting nurse, home health aide, housekeeping assistance, counseling, transportation, respite care in the home, and respite care in another place—and asked respondents to describe eligibility requirements for services available in their areas. Respondents in two countries indicated that services were not provided when family members were present. In Japan, home help services are available only to elderly people who live alone (Hashimoto & Kendig, 1992). In Sweden, the notion of family responsibility is found in Section 6 of the Social Services Act, which specifies the responsibility of spouses. This provision states that the municipality has to help "if the need cannot be solved otherwise." But, my informant indicated, "living with someone has been interpreted by administrative courts as 'otherwise.'" Careful analysis of regulations governing social services in many countries would probably reveal similar provisions. Further, several respondents indicated that services were allocated by professionals according to an older person's need.

Discussion Gibson (1992) noted the common fear in industrialized nations that provision of community-based services will cause family members to reduce their involvement in caring for elderly relatives. This fear is reflected in rationing provisions that withhold services when potential family caregivers are present. Such provisions have been noted in the United States and Great Britain. Results of this study indicate they are also present in Japan and Sweden.

Yet there is considerable evidence that families do not reduce the amount of care they provide when services are offered. Gibson (1992) reviewed many studies in support of this conclusion. Studies conducted in Canada (Havens, 1987), Denmark (Friis, Thursz, & Vigilante, 1979), Australia (Kendig, 1986), and the United States support this finding (Barusch & Miller, 1986; Mathematica Policy Research, 1986). In fact, recent work in the United States examining what family members do when relieved of their caregiving responsibilities suggests that their primary activity is paid employment (Lund, Casserta, & Wright, 1991).

Another reason for withholding services when caregivers are present may be the desire to use scarce public resources to support those in greatest need. The assumption is that everyone without access to family care has greater need than anyone with a potential caregiver. This approach to needs assessment is appealing in its simplicity, yet it obscures important determinants of need. A family history of abuse, neglect, or indifference may leave the elderly person deprived of services and in desperate need. Similarly, strong bonds of friendship may insulate a senior citizen from the need usually associated with lack of family.

Conclusion

Any program or policy designed to support family care of elderly dependents is likely to find a receptive audience in policymakers throughout the world. Yet it is important to distinguish between methods designed simply to increase the amount of family care provided and those that address the quality and long-term consequences of that care.

The three approaches to increasing family care reviewed here have subtle, unanticipated costs. Mandating family care, although intuitively appealing, may bring high administrative costs and threaten intergenerational bonds. Incentives for family caregivers may be ineffective as inducements (when rewards are small), or they may jeopardize the affective benefits of caregiving and foster dependency among elderly people (when rewards are large). Service rationing reflects a common concern that families will withdraw their support if professional care is provided. Yet there is strong evidence that this is not so. Such rationing may result in denial of services to families in great need.

It is time for human services professionals to reconsider policies that require or encourage families to provide care. The promise of reduced public long-term-care expenditures is alluring. But the costs of these policies are typically borne by the disenfranchised—children whose mother cannot attend to their needs because of the intense care required by her dying parent or women who leave the work force, giving up pension rights and risking poverty in old age, to provide care for an elderly parent. Ultimately these private burdens have societal consequences, but these are far removed from the policy-making process.

Rather than attempt simply to increase family involvement in caring for elderly people, [we] must address the quality and long-term consequences of the caregiving experience. Our growing knowledge of the total cost of family care offers a compelling argument in favor of interventions that provide universal access to a continuum of care alternatives that facilitate rational health care decision making by families and that empower and sustain family members who choose to care for elderly relatives. Human services advocates should redirect their enthusiasm for family care to support initiatives in these areas.

REFERENCES

American Association for International Aging. (1988). *An international directory of organizations in aging.* Washington, DC: Author.

Barusch, A. S., & Miller, L. S. (1986). The effect of services on family assistance to the frail elderly. *Journal of Social Service Research, 9,* 31–45.

Bulcroft, K., Van Leynseele, J., & Borgatta, E. F. (1989). Filial responsibility laws: Issues and state statutes. *Research on Aging, 11,* 374–393.

Cicirelli, V. G. (1981). *Helping elderly parents: The role of adult children.* Boston: Auburn House.

Doty, P. (1988). Long-term care in international perspective. *Health Care Financing Review* (Suppl.), 145–155.

Friis, H., Thursz, D., & Vigilante, J. L. (1979). The aged in Denmark: Social programmes. In M. I. Teicher (Ed.), *Reaching the aged: Social services in 44 countries* (p. 206). Beverly Hills, CA: Sage Publications.

Gibson, M. J. (1984). Family support patterns, policies and programs. In C. Nusberg (Ed.), *Innovative aging programs abroad: Implications for the United States* (pp. 159–198). Westport, CT: Greenwood Press.

Gibson, M. J. (1992). Public health and social policy. In H. L. Kendig, A. Hashimoto, & L. C. Coppard (Eds.), *Family support for the elderly: The international experience* (pp. 88–114). Oxford, England: Oxford University Press.

Gilbert, N. (1984). *Capitalism and the welfare state: Dilemmas of social benevolence.* New Haven, CT: Yale University Press.

Hashimoto, A., & Kendig, H. L. (1992). Aging in international perspective. In H. L. Kendig, A. Hashimoto, & L. C. Coppard (Eds.), *Family support for the elderly: The international experience* (pp. 3–14). Oxford, England: Oxford University Press.

Havens, B. (1987). Issues in the delivery of long-term care services in the United States: A Canadian response. In M. J. Gibson (Ed.), *Proceedings of the U.S./Canadian expert group meeting in policies for midlife and older women.* Washington, DC: American Association of Retired Persons.

Kendig, H. L. (1986). Toward integrated community care for the disabled aged: The Australian case. *Australian Journal of Social Issues, 21*(2), 75–91.

Kosburg, J. (1992). *Family care of the elderly: Social and cultural changes.* Newbury Park, CA: Sage Publications.

Lehr, U. (1991, November). *Family care: Chances and problems.* Paper presented at the Second International Symposium on Aging, Tokyo.

Lund, D., Casserta, M., & Wright, S. (1991, November). *Caregivers' use of day care respite time.* Paper presented at 44th Annual Scientific Meeting of the Gerontological Society of America, San Francisco.

Mamabolo, N. (1991, December). *Innovative support programmes for caregivers.* Paper presented at the Third International Conference on Care of the Elderly, Hong Kong.

Martin, J. G. (1988). The aging of Asia. *Journal of Gerontology: Social Sciences, 43*(4), S99–S113.

Mathematica Policy Research. (1986). *National long-term care channeling demonstration: Analysis of the benefits and costs of channeling* (Tech. Rep. No. TR-86b-12). Washington, DC: U.S. Department of Health and Human Services.

Myers, G. C. (1990). Demography of aging. In R. H. Binstock & L. K. George (Eds.), *Handbook of aging and the social sciences* (3rd ed., pp. 19–44). New York: Academic Press.

Newman, S. J. (1980). Government policy and the living arrangements of the elderly. *Home Health Care Services Quarterly, 1*(4), 59–71.

Nusberg, C. (with M. J. Gibson & S. Peace). (1984). *Innovative aging programs abroad: Implications for the United States.* Westport, CT: Greenwood Press.

Sherwood, S., & Morris, S. (1991, December). *Care provision of elders and tax incentive programs.* Paper presented at the Third International Conference on Aging Policies and Services, Hong Kong.

Sundstrom, G. (1987). *Old age care in Sweden.* Lund, Sweden: Swedish Institute.

Talley, L. A. (1980). *Tax incentives for care of elderly family members.* Washington, DC: Library of Congress, Congressional Research Service.

Tout, K. (1989). *Aging in developing countries.* New York: Oxford University Press for HelpAge International.

Tracy, M. B. (1974). Constant-attendance allowances for non-work related disability. *Social Security Bulletin, 37,* 32–37.

Tracy, M. B. (1991). *Social policies for the elderly in the third world.* New York: Greenwood Press.

United Nations. (1982). *Report of the World Assembly on Aging,* Vienna, 26 July to 6 August 1982. New York: Author.

United Nations. (1988). *Economic and social implications of population aging* (ST/ESA/SER.R/85). New York: Author.

United Nations Economic and Social Council. (1989). *Second review and appraisal of the implementation of the International Plan of Action on Aging: Report of the Secretary-General* (E/1989/13). New York: Author.

CHAPTER 8

Questions for Writing, Reflection, and Debate

READING 41

Some Economic Complexities of Child Care Provided by Grandmothers • *Harriet B. Presser*

1. How do families benefit from having grandparents provide child care? What are the social, psychological, and economic costs associated with this arrangement?
2. In the article, how do the images of grandparents who provide care differ from the stereotype of the doting grandmother who cares for grandchildren while her son or daughter works?
3. This article clearly shows child care as an important role that binds the generations together. What social factors could increase or decrease the availability of the older generation to contribute in this way?

READING 42

Legacy, Aging, and Succession in Farm Families • *Norah C. Keating*

4. How might being the successor to a family farm or business alter one's relationship to and interactions with parents and siblings?
5. How has the transfer of such assets been complicated by women's entry into the labor market and by the increasing life span of individuals?
6. Which do you think has a harder or easier transition into retirement, the older generation on family farms or in other businesses?

READING 43

Social Security and Medicare Policy: A Personal, Intergenerational Story • *Mal Schechter*

7. How do programs such as Social Security and Medicare provide security for extended family as well as for recipients?
8. What key services for the elderly are not covered by Medicare? What impact can this omission have on the well-being of families?
9. Why do Americans view Social Security and Medicare differently from Food Stamps and Medicaid? How do eligibility requirements and funding sources contribute to these attitudes?
10. Why is the three-legged stool—Social Security, employment benefits, and savings—a shaky place to be sitting? How has the weakening of this stool affected interactions among the generations?

READING 44

Aging and the Family: Present and Future Demographic Issues • *Kevin Kinsella*

11. What factors contribute to the differences in age and sex composition of the population in developing and developed countries?
12. What social factors influence the proportion of the elderly who live on their own, with family, or in institutions?

13. Why is it difficult to predict kin availability from demographic data? How might this affect public policy?

READING 45

Programming for Family Care of Elderly Dependents: Mandates, Incentives, and Service Rationing • *Amanda Smith Barusch*

14. Three policy initiatives are explored in this piece: mandating that children care for elderly parents, providing incentives for them to do so, and withholding public services to the elderly that have access to family. Do you think any of these policies would improve the way the elderly are cared for in the United Sates? Explain.
15. What are some of the assumptions underlying each of the policies mentioned in the previous question?
16. The goal of all these policies is to increase familial care for the elderly. Do you think policy can motivate this kind of behavior? Why or why not?

9

Social Change and Family Policy

How are politics shaping family life?

The term *family values* has recently been added to a list of ill-defined terms Americans use to describe family life. Early in my teaching career, I had many interesting discussions with my students, trying to establish a definition of *love*. As times changed, I added to the debate, trying to define *family*. Now that the term *family values* jumps out at us from every type of media, I challenge students to attempt to understand the journalists' interpretation of this term and to see if it bears any resemblance to their definition.

The conclusion my classes inevitably reach is that the meanings of *love, family*, and *family values* are all open to interpretation. Thus we also recognize that proclamations declaring the "demise of the family" and a need to return to "family values" are based on the author's or speaker's definitions of those terms. The impassioned nature of such pronouncements conveys sincere concern. But this concern is grounded in a perception that the undeniable changes occurring in family life threaten the author's ideal. More important is the connection made between these changes in family and changes in the wider society. The "breakdown" of "the family" has been held responsible for a phenomenal number of social ills, including crime, drug abuse, suicide, and decreasing church attendance.

C. Wright Mills (1959) makes an interesting distinction between personal troubles and public issues. He views troubles as private matters, occurring when an individual experiences a threat to cherished values. *Troubles* stand as distinct from social issues, which are in the public arena. With *social issues*, historic or cultural trends threaten values cherished by the wider public. "Often there is debate about what that value really is and about what it is that threatens it" (p. 8). People "often sense that older ways of feeling and thinking have collapsed and that newer beginnings are ambiguous to the point of moral stasis" (p. 4) . Using this distinction to decide if there is really a crisis in family life, we must first decide what values are challenged and how widely they are held.

I have selected readings for this chapter that offer various perspectives on the current debate regarding "family decline" and "declining family values." As you read, try to identify the implicit definitions the authors use and the values that underpin their positions. This is a relatively easy task in Judith Stacey's "Virtual Social Science and the Politics of Family Values." If you are new to the study of family from a sociological perspective, this reading will bring you "up to speed" on the current controversies among those who study family life. Even if you never knew there is such a field as sociology of the family, you will undoubtedly recognize the debate and many of the names in this reading. It is hard to escape this debate among academics as it has been appropriated by politicians and the media.

Stacey concludes, in this first reading, that the debate on family decline is siphoning off resources and energy that could be devoted to understanding the realities of family life. After you have read this piece, think about your own family values and how they differ from others in your family, class, and community. How important is it to debate these differences, or do you agree that the debate distracts from understanding how change is shaping family life?

The second reading in the chapter, "Family Privacy: Issues and Concepts" by Felix Berardo, provides an example of considering the effects of social change on important aspects of family life. I like the way Berardo uses Goffman's concepts of front and back stage to illustrate the importance of privacy in family life. This piece takes the real-life issue of privacy and guides us through a very complex underlying issue. Each society must grapple with how to balance the family's need for privacy with the need for social control. Who should be allowed to marry, how much freedom should parents have in the way they treat their children, what are grounds for divorce—all these and many other aspects of family life are subject to a balance of personal privacy and the public need to know.

The final reading, by Carol Smart, "Wishful Thinking and Harmful Tinkering? Sociological Reflections on Family Policy," describes an example of British policy attempting to reform family life. As you will see, these policies are not that far removed from similar activities in the United States. Smart suggests that the futility of this approach keeps proponents of such policies on a treadmill, perpetually reforming policy in new attempts to shape family life.

This piece brings to light interesting underlying assumptions that greatly influence family policy. If we assume that changes in family life are caused by changes in individual values, then the way to "fix" families is to punish those individuals who don't comply with society's expectations. If, however, we assume that changes in family life are the result of changes in social conditions, then we must look to "fix" society, not individuals. Smart adds yet a third possible assumption. Maybe the individuals in families are actually the engine for change in society. As families take on new structures and roles, they create new norms, which eventually become the status quo, forcing social institutions to adapt to the changes initiated by families. This reading offers a thought-provoking ending to the readings although it by no means brings closure to the questions presented here.

The dynamic and personal nature of family life means that the questions at the end of this and other chapters have no easy answers. But that does not decrease the importance of asking them. As you struggle with these questions, don't put them in the "too hard" basket simply because there is no one best answer. If we agree with the possibility that through the way we live in families we may in fact be creating the society we live in, then we owe it to ourselves and others to consider how we live our family life.

REFERENCE

Mills, C. Wright. 1959. *The Sociological Imagination.* New York: Oxford University Press.

Virtual Social Science and the Politics of Family Values

Judith Stacey

> From the wild Irish slums of the 19th-century Eastern seaboard to the riot-torn suburbs of Los Angeles, there is one unmistakable lesson in American history: a community that allows a large number of young men to grow up in broken families, dominated by women, never acquiring any stable relationship to male authority, never acquiring any set of rational expectations about the future—that community asks for and gets chaos. Crime, violence, unrest, unrestrained lashing out at the whole social structure—that is not only to be expected; it is very near to inevitable.
>
> —*Daniel Patrick Moynihan, 1965*

> The way a male becomes a man is by supporting his children.... What (the Democrats) cannot accept is that government proposals have failed. It is the family that can rebuild America.... The dissolution of the family, and in particular, the absence of fathers in the lives of millions of America's children is the single most critical threat (to our future).
>
> —*Dan Quayle, September 8, 1994*

> That is a disaster. It is wrong. And someone has to say again, "It is simply not right. You shouldn't have a baby before you're ready, and you shouldn't have a baby when you're not married."
>
> —*President Bill Clinton, September 9, 1994*

No doubt many of the social scientists who have regrouped outside the academy to wage a secular campaign for family values sincerely worry about the deteriorating welfare of our children and society. Nonetheless, their extra-curricular efforts respond as well to major shifts in the politics of knowledge within the academy. Just when the postmodern family condition was supplanting the 1950s version of the modern family system whose conventions and values continue to haunt contemporary culture, so too were new approaches to knowledge challenging the ruling frameworks of modernist, 1950s social science to which the family-values campaigners adhere. The gap in the degree of prestige that modern family sociology now enjoys in popular, as compared with professional, domains is also considerable, and growing. Meanwhile, fin-de-millennium politicians of disparate ideological hues have come to perceive significant rewards in family-values discourse, as the two excerpts above from back-to-back, nearly interchangeable 1994 election season speeches by Dan Quayle and President Clinton indicate. A peculiar conjuncture of political and academic dislocations has opened the door for collaboration between mainstream social scientists and electoral politicians. Revisionist family-values scholars are supplying substantial ideology, rhetoric, and legitimacy to post–Cold War "New" Democrats and the vanishing species of "moderate" Republicans alike.

Playing hooky from our besieged and tarnished ivory towers, numerous social scientists, now including myself, are waging public cultural combat, for weighty political stakes, over the sources of and remedies to the fall of the modern family system. As the

United States approaches the third Christian millennium, the everyday practices and conditions of social and material decline are increasingly at odds with a mythic discourse of family values.

A Bedtime Fable for the American Century

Once upon a fabulized time, half a century ago, there was a lucky land where families with names like Truman and Eisenhower presided over a world of Nelsons, Cleavers, and Rileys. Men and women married, made love and (in that proper order) produced gurgling, Gerber babies. It was a land where, as God and Nature had ordained, men were men and women were ladies. Fathers worked outside the home to support their wives and children, and mothers worked inside the home without pay to support their husbands and to cultivate healthy, industrious, above-average children. Streets and neighborhoods were safe and tidy. This land was the strongest, wealthiest, freest, and fairest in the world. Its virtuous leaders, heroic soldiers, and dazzling technology defended all the freedom-loving people on the planet from an evil empire which had no respect for freedom or families. A source of envy and inspiration, the leaders and citizens of this blessed land had good reason to feel confident and proud.

And then, as so often happens in fairy tales, evil came to this magical land. Sometime during the mid-1960s, a toxic serpent wriggled its way close to the pretty picket fences guarding those Edenic gardens. One prescient, Jeremiah Daniel Patrick Moynihan,[1] detected the canny snake and tried to alert his placid country*men* to the dangers of family decline. Making a pilgrimage from Harvard to the White House, he chanted about ominous signs and consequences of "a tangle of pathology" festering in cities that suburban commuters and their ladies-in-waiting had abandoned for the crabgrass frontier. Promiscuity, unwed motherhood, and fatherless families, he warned, would undermine domestic tranquility and wreak social havoc. Keening only to the tune of black keys, however, this Pied Piper's song fell flat, inciting displeasure and rebuke.

It seemed that overnight, those spoiled Gerber babies had turned into rebellious, disrespectful youth who spurned authority, tradition, and conformity, and scorned the national wealth, power, and imperial status in which their elders exulted. Rejecting their parents' gray flannel suits and Miss American ideals, as well as their monogamous, nuclear families, they generated a counterculture and a sexual revolution, and they built unruly social movements demanding student rights, free speech, racial justice, peace, liberation for women, and for homosexuals. Long-haired, unisex-clad youth smoked dope and marched in demonstrations shouting slogans like "Question Authority," "Girls Say Yes to Boys Who Say No," "Smash Monogamy," "Black Is Beautiful," "Power to the People," "Make Love, Not War," "Sisterhood is Powerful," and "Liberation Now." Far from heeding Moynihan's warning, many young women drew inspiration from the "black matriarchs" he had condemned, condemning Moynihan instead for blaming the victims.

Disrupting families and campuses, the young people confused and divided their parents and teachers, even seducing some foolish elders into emulating their sexual and social experiments. But the thankless arrogance of these privileged youth, their unkempt appearance, provocative antics, and amorality also enraged many, inciting a right-wing, wishful, moral majority to form its own backlash social movement to restore family and moral order.

And so it happened that harmony, prosperity, security, and confidence disappeared from this once most fortunate land. After decimating Black communities, the serpent of family decline slithered under the picket fences, where it spewed its venom on white, middle-class families as well. Men no longer knew what it meant to be men, and women had neither time nor inclination to be ladies. Ozzie had trouble finding secure work. He was accused of neglecting, abusing, or oppressing his wife and children. Harriet no longer stayed home with the children. She too worked outside the home for pay, albeit less pay. Ozzie and Harriet sued for divorce. Harriet decided she could choose to have children with or without a marriage certificate, with or without an Ozzie, or perhaps even with a Rozzie. The clairvoyant Daniel Patrick Moynihan found himself vindicated at last, as political candidates from both ruling parties joined his hymns of praise to Ozzie and Harriet Nelson and rebuked the selfish family practices of their rebellious stepchild, Murphy Brown.

The era of the modern family system had come to an end, and few felt sanguine about the postmodern family condition that had succeeded it. Unaccustomed to a state of perpetual instability and definitional crisis, the populace split its behavior from its beliefs. Many who contributed actively to such postmodern family statistics as divorce, remarriage, blended families, single parenthood, joint custody, abortion, domestic partnership, two-career households, and the like still yearned nostalgically for the "Father Knows Best" world they had lost. "Today," in the United States, as historian John Gillis so aptly puts it, "the anticipation and memory of family means more to people than its immediate reality. It is through the families we live *by* that we achieve the transcendence that compensates for the tensions and frustrations of the families we live *with*."[2] Not only have the fabled modern families we live *by* become more compelling than the messy, improvisational, patchwork bonds of postmodern family life, but as my bedtime story hints, because they function as pivotal elements in our distinctive national consciousness, these symbolic families are also far more stable than any in which future generations ever dwelled.

Similar evidence of the decline of the modern family system appears throughout the postindustrialized world, and for similar reasons, but thus far, in no other society has the decline incited responses so volatile, ideological, divisive, or so politically mobilized and influential as in the U.S.[3] Only here, where the welfare state was always underdeveloped and is now devolving, where religious fervor and populist movements flourish and organized labor languishes, has the beloved bedtime fable begun to evoke so many nightmares. Now the crisis of family order incites acrimonious conflicts in every imaginable arena—from television sitcoms to Congress, from the Boy Scouts of America to the United States Marines, from local school boards to multinational corporations, from art museums to health insurance underwriters, from Peoria to Cairo, and from political conventions to social science conferences.

Social Science Marching On

> Contrary to expectations many of you may have of historians, I'm not here to tell you that we have seen it all before. The current obsession with family values seems to me, if not entirely new, then peculiar to the late twentieth century, and I will argue that what we are experiencing is yet another dimension of what David Harvey has called the "postmodern condition," an example of what Anthony Giddens has identified as the late twentieth century capacity for intimacy at a distance.
>
> —*John Gillis, "What's Behind the Family Values Debate?"*

> The essential integrity of at least a large proportion of American family social scientists is evidenced by the fact that as the evidence accumulated on the effects of family changes, the originally sanguine views of the changes began to change to concern.
>
> —*Norval Glenn, "The Re-evaluation of Family Change by American Social Scientists"*

Just as no new family system has yet succeeded in attaining the status which the modern, male-breadwinner, nuclear family order enjoyed at mid-century, likewise the ruling intellectual frameworks of 1950s sociology have been dethroned, but not supplanted. Probably no academic discipline felt the disruptive impact of the social movements of the 1960s and early 1970s more strongly than sociology. Seeking to understand and critique their own society and to explore alternatives, activist students with left-wing commitments, including draft resistors in search of student deferments, entered sociology in droves. Infatuated with romantic versions of Marxism, Leninism, Trotskyism, Maoism, and other radical theories, radical young sociologists rejected most mainstream social theories as apologias for racism and imperialistic ventures and for the conformity and false promises of the Cold War era.

A feminist onslaught on the discipline pursued the left-wing attacks on modernization theory and on the functionalist theories of Talcott Parsons that justified the gender order of the modern family as functional for a modern industrial society. During the late 1960s and early 1970s, the grass roots women's movement spurred a wave of feminists to enter academic careers, where the liberal cast and the diffuse intellectual boundaries of sociology attracted many, like myself, to the discipline. Feminists organized professional caucuses, conferences, journals, and research sections in the American Sociological Association, rapidly transforming the demographics, intellectual preoccupations, and the leadership of both the professional organization and the discipline. The explicitly feminist Sex and Gender section, formed in the early 1970s, rapidly outdistanced the older, far more mainstream section on family sociology in membership, intellectual dynamism, and appeal. By 1993, it had become the largest research section in the ASA.[4]

Left-wing and feminist interventions challenged the value-free, scientific pretensions of sociology throughout the discipline, but few subfields were quite so dislocated as family sociology. During the mid-1970s, sociology of the family experienced what Canadian sociologist David Cheal terms, "a Big Bang in which feminism played a conspicuous part."[5] I share Cheal's view that the "explosion blew the field apart, and the separate pieces have been flying off in different directions ever since." The modernist sociology that Talcott Parsons and his colleagues practiced had posited the universality of the nuclear family and theorized that the gender structure of its male breadwinner-female homemaker genus evinced an ideally evolved, "functional fit" with modern industrial society and political democracy.[6] In 1963, sociologist William J. Goode's *World Revolution and Family Patterns* predicted that modernization would accomplish the global diffusion of the superior Western variety of family life and, thereby, of the democratic society it was thought to nurture.[7]

Feminist scholars, however, rapidly subjected 1950s families and family sociology alike to trenchant critique. Influenced by demographic evidence of rapid family transformations in the United States, by countercultural communal experiments and by the anti-housewife ethos of the early women's liberation movement, feminist scholars challenged the implicitly ethnocentric and androcentric foundations of the prevailing theories about family life. Betty Friedan's scathing attack on "the functionalist freeze" in *The Feminine Mystique* had directly launched such a project by blaming this brand of social science for entrapping overeducated, suburban homemakers in the "problem that has no name."[8] It was but a short leap from there to sociologist Jessie Bernard's academic work on His and Her marriages,[9] or from incendiary movement classics like Pat Mainardi's "The Politics of Housework" and Ann Koedt's "The Myth of the Vaginal Orgasm," to scholarly treatments like Ann Oakley's *The Sociology of Housework* and Gayle Rubin's "The Traffic in Women."[10]

Meanwhile, outside the embattled groves of academe, a right-wing profamily movement rapidly polarized popular discourse on family change into feminist vs antifeminist, left vs right, and fundamentalist vs secular humanist camps. This forced the largely liberal ranks of mainstream family scholars, many of them predisposed to sympathize with Moynihan's earlier, ill-timed critique of "Negro family decay," to confront uncomfortable ideological choices.[11] Initially most accommodated their work to liberal feminist values, but the conditions under which they did so sowed seeds of resentment that would come full term in the backlash against political correctness of the 1990s. During the 1960s and 1970s most scholars in the subfield appeared to support liberal feminist critiques of the modern nuclear family. Many mainstream family scholars expressed a tolerant, relativistic stance toward various aspects of family experimentation and the sexual revolution of the time, and generally the findings of the studies they conducted on subjects like divorce, maternal employment, day care, single parenthood, and sexual experimentation gave comfort to the rising numbers of people involved in such practices.

Norval Glenn, a prominent senior family sociologist in the United States who is now a member of the

Council on Families in America, surveyed this disciplinary history as participant-observer during lectures he delivered in 1994 in conjunction with the UN International Year of The Family. "Social scientists in the United States generally took a sanguine view of the family changes that started or accelerated in the mid-1960s. Although the label of 'family decline' was often attached by social scientists to the family changes that occurred early in the century, the prevailing view in the 1960s and 1970s was that the family was only adapting to new circumstances, not declining."[12]

Sociologists in that period promoted a Pollyannish assessment of rising divorce rates, interpreting the trend both "as a sign that marriage had become more, not less, important," to adults, because they were no longer willing to settle for unhappy unions, and as beneficial in the long run for children released from the hostile environment of an unhappily married parental home.[13] Glenn claims that he and his colleagues felt "strongly inclined to express positive views of recent family changes," then, because the changes coincided with a feminist movement that "viewed family change and the trend toward gender equality as parts of the same bundle." "Being human," sociologists sought "the approval of those whose opinions matter to them, and those persons are largely liberals" who embrace "the ideal of male–female equality."[14] Andrew Cherlin, a prominent demographer and family sociologist whose work does not support the neo-family-values campaign, confirms this assessment of feminist influence on family sociologists: "It is above all the wish to avoid sounding like an antifeminist, I think, that causes liberals to downplay the costs of the recent trends."[15]

Since the late 1970s, however, the ideological force field within which scholars investigate the consequences of family change has become increasingly conservative. Speaking to prominent governmental, religious, and academic bodies in Australia, Glenn applauded social scientists for a voluminous "second thoughts" literature that recants earlier uncritical stances on family change, especially concerning the social effects of divorce, fatherlessness, and single motherhood. Psychologist Judith Wallerstein's widely popularized work,[16] which finds that divorce inflicts substantial, lasting, and harmful effects on children, has been particularly influential, as has demographer Sara McLanahan's revised assessment that single-parent families harm children.[17] Adding his voice to "most of the more prominent family social science researchers," who, he claims, now evaluate such changes "in distinctly negative terms," Glenn recounts his personal conversion to the now, "virtually unanimous," social scientific view that, "the best family situation for children and adolescents is one in which there is a successful, intact marriage of the biological (or adoptive) parents," and that single-parent and stepfamilies are "far less than ideal."[18]

Whereas Glenn perceptively recognized that feminists and liberal ideology influenced social scientists' earlier, less critical appraisals of family change, he shifts to an "objectivist," scientific narrative to account for their more pessimistic verdicts in the 1990s. His tribute to "the essential integrity" of family social scientists (included in an above epigraph), claims that the weight of cumulative data compelled this intellectual conversion. I believe, in contrast, that just as during the 1960s and 1970s, reconfigurations of power and knowledge provide more illuminating explanations. The metamorphosis in the opinions of family scholars occurred as a New World Order of global capitalism and economic crisis brought the glory days of sociology as a discipline to a halt, while the modernist theories that had sustained its liberal, humanitarian ethos began to falter in postmodernist, neoliberal, and postfeminist currents.

If feminism unleashed a "Big Bang" in family sociology, postmodern theories have yet to provoke much more than a whimper. Even though most family scholars are enmeshed in an ideological crisis over the meaning of family under postmodern conditions of patchwork intimacy, very few participate "in discussions about so-called postmodern conditions of society and thought."[19] Like most sociologists in the United States, they remain remote from these discourses, still wedded to a view of knowledge that considers the meaning of texts, images, and facts to transparently reflect an objective, external reality. Feminist sociologists, along with some European theorists like Michel Foucault, Anthony Giddens, and Pierre Bourdieu, and occasional visits from colleagues in other disciplines, like historian John Gillis, who spoke at a session on the family-values debate at the 1994 meetings of the American Sociological Association, provide mainstream sociologists with their principal, and often unwelcome, exposure to such questions.[20]

Fending off the intellectual challenge of Gillis's reading of postmodern family developments on the ASA panel, for example, Popenoe resorted to an anti-intellectual, populist mode that has become widespread in the family-values campaign. Amidst lingering echoes of appreciative applause for Gillis's paper, Popenoe mounted the podium of the Hotel Bonaventure's cavernous auditorium in Los Angeles and declared, "I guess I'm the right wing here, but that's not true outside of sociology."[21] Popenoe positioned himself against the elite world of academia and with "most Americans [who] agree with what I just said, except for the liberal intelligentsia I've been battling."[22]

An apparent lack of awareness about the changed historical conditions for knowledge production in which we now operate renders many family social scientists, particularly senior white males, responsive to the rewards of the revisionist campaign. Since the 1970s, so many women have entered the discipline of sociology that we have begun to outnumber men in the profession, but the massive feminist Sex and Gender research section of the American Sociological Association remains a female ghetto—alien, unfamiliar territory to most mainstream family scholars.[23] Academic feminism has been institutionalized, thereby achieving some respectability, but also, as any student familiar with Max Weber's classical analysis of the routinization of charisma might have predicted, becoming less threatening.[24] It is not surprising that displaced male scholars might now feel freer to expel feminist perspectives they once had been force-fed and never fully digested.

Moreover, the collapse of communism has provoked a crisis in Marxism and a loss of faith in materialist explanations for social change, with particularly strong effects in sociology. The nation's generalized right-wing political shift, the diffusion of postfeminist culture[25] and the organized movement against "political correctness" in higher education have offered family scholars compelling incentives to hopscotch over challenges to academic family theory posed by the postmodern family condition to land directly into the public political fray.

The centrist campaign for family values allows displaced, formerly liberal scholars a means to reclaim positions of intellectual authority without appearing to be antifeminist. Adopting the post-feminist rhetoric of "a new familism," the campaign distances itself from reactionaries intent on restoring Ozzie and Harriet to the frayed upholstery of their suburban throne. New familism, in Glenn's formulation, represents, "a return to the belief that stable marriages, two-parent families, and putting children's needs before those of adults are desirable and important. It differs from the older familism in its espousal of male-female equality and the rejection of economic dependence as a basis for marital stability."[26] Migrating ideologically from Moynihan to Friedan and then even farther back than Friedan herself traveled when she published *The Second Stage* in 1982,[27] revisionists applaud tenuous signs "at the mass level" of a "return toward more traditional family values (excluding the ideal of male dominance)."[28]

A few social scientists have found the centrist campaign a route to considerable public influence, media celebrity, and even academic attention. For example, speaking at a symposium, "Gender Equality, Children and the Family: Evolving Scandinavian and American Social Policy," at University of California, Berkeley in April of 1994, Popenoe acknowledged that his 1988 book, *Disturbing the Nest*, which criticizes the impact of social democratic policies on family change in Sweden, had received a chilly response from the Swedes: "My book did not start a dialogue in Sweden. I wasn't even invited back." As if to confirm Popenoe's self-report, prominent Scandinavian scholars and officials in attendance gave his critical analysis of Scandinavian family policy a dismissive response. Karin Stoltenberg, the Director General of the Norwegian Ministry of Children and Family Affairs, for example, termed "insane" Popenoe's belief that welfare-state policies were the source of rising divorce rates in the Nordic countries.[29] Likewise, an anti-intellectual polemic that Popenoe delivered at the August 1994 American Sociological Association Meetings in Los Angeles confronted nearly a solid wall of disapproval from co-panelists and the audience. Session organizer

Frank Furstenberg, a prominent demographer and family sociologist in the United States, chastised Popenoe's "unhelpful us/them approach" to sociologists and the family values debate. In response, Popenoe portrayed most sociologists as out of touch with popular concerns and invited listeners sympathetic with his views to accompany him to a nearby conference sponsored by the Communitarian Network.

Since Popenoe became a major organizer of the centrist family-values campaign, he has been invited again and again to deliver his lament for family decline in venues that range from the "MacNeil-Lehrer News Hour" to the U.S. Department of Transportation, from the *New York Times* to the *Chronicle of Higher Education*, in addition to academic conferences and meetings such as at U.C. Berkeley and the American Sociological Association. Blankenhorn, who often claims the mantle of social science despite his lack of an advanced degree in any of the social science disciplines, has achieved even greater celebrity with the National Fatherhood Tour he launched in conjunction with the release of his 1995 book *Fatherless America*. From "Oprah" to CNN's "Talkback," from the cover pages of *Time* magazine to syndicated feature stories in hundreds of local newspapers, Blankenhorn has blazed an extramural trail to academic podiums.[30] Likewise, the family-values campaign has offered Galston, Etzioni, and their colleagues access to extensive public and professional recognition. Etzioni's public campaign as founder and director of the communitarian movement, for example, coincided with his successful bid to become president of the ASA, and Galston took leave from his professorship at the University of Maryland to serve as Deputy Assistant to President Clinton until June of 1995.

The rhetoric of the few women scholars, like Jean Bethke Elshtain and Sylvia Hewlett, who are visibly active in the revisionist campaign, suggests that they harbor more personalized resentments against academic feminists. Hewlett has blamed the anti-maternalist ethos of early second-wave feminism for compounding the tribulations she suffered when she was a pregnant assistant professor of economics at Barnard College in the 1970s.[31] Elshtain complains, more plausibly, that she is "hooted out of the room" by feminists whenever she talks "about not ceding the issue of family values to the right."[32] In fact, few academic feminists do sympathize with Elshtain's support for heterosexual marital privilege or her disapproval of single motherhood. Feminists have personal and political stakes in these judgements as profound as Elshtain's, and so few respond to her public challenges with scholarly dispassion.[33]

It seems ironic, and in my view unlucky, that challenges to knowledge induced by radical family changes and by feminism, which decentered mainstream family social scientists within the academy, propelled quite a few of them onto center stage in the public sphere where they speak to the broad audiences feminists used to address. There, aloof from even the modest constraints of academic peer review, they deploy social scientific authority to influence political responses to postmodern family struggles by disseminating selective readings of the very kind of modernist research on family change that feminist and other critical sociologists imagined we had discredited. In an era of academic retrenchment when universities like the University of Rochester, Washington University, Yale University, and San Diego State University have abolished or slashed their sociology departments, it should not be surprising that podiums outside the ivy pastures beckon sociologists with more gratifying rewards. Modernist family social scientists (and pretenders, like Blankenhorn) can often enhance their academic status in the public domain, where they enjoy much more intellectual esteem and influence than do most postmodernist or feminist theorists.

They have developed an extramural social science apparatus with which they wage their cultural crusade for centrist family values. . . . [T]hrough interlocking networks of think tanks, organizations, periodicals, and policy institutes, these social scientists have been constructing a virtual scholarly and popular consensus in the media in support of the very narrative about universal family values that succumbed to feminist and other forms of scrutiny in academia. Saturating the media-beltway world they have come to inhabit with the ideology of new familism, they misleadingly maintain that social science has confirmed Moynihan's warning about the socially destructive effects of single motherhood, illegitimacy, and fatherless families.

In reality, scholars do not now, and likely will not ever, achieve consensus on the relative significance that family structure, material circumstances, the quality of parental relationships, and psychological factors play in shaping children's lives. While it is true that most family sociologists do express some uneasiness over mounting rates of single parenthood, the predominant scholarly view is that single-parent families are more often the consequence than the cause of poverty, unemployment, emotional distress, and other negative correlates.[34] Ironically, a book about the history of family policy in Sweden by right-wing historian and family-values champion Allan Carlson criticizes Gunnar and Alva Myrdal for their "use and abuse of social science" in promoting the kinds of cutting edge social democratic family policies in their nation of which Carlson disapproves. Yet Carlson's critique of this practice applies at least as well to the contemporary family-values campaign in the United States that he supports: "In short, it is difficult to see social science in this episode as little more than a new tool for rhetorical control and political advantage. Weak and inconsistent data, confusion over cause and effect, and avoidance of experimentation proved to be no obstacles to the construction and implementation of policy."[35]

The contemporary family-values campaign in the United States, which Carlson supports, mixes a flawed modernist framework with an unsophisticated notion of culture. It presumes that the truth about the relative merits and effects of diverse family structures—be they intact married-couple families, stepfamilies, single-parent families, extended families, adoptive families, not to mention gay families—is straightforward, knowable, and extricable from its social, economic, and political context. Although some revisionists, like Glenn, concede that at times (always past times), ideological and cultural convictions interfere with the capacity of social scientists to perceive this truth and temporarily distort social scientific knowledge, truth and virtue still triumph in the end. In this view, most social scientists are sufficiently scientific to listen when the data speak to them in robust and uniform tones.

A second dubious assumption of the family-values campaign is that truth is timeless as well as singular: a happy marriage in the 1990s is the same as one in the 1950s, and divorce has the identical, negative effects. Culture intervenes simply to affect the frequency of these structures by rendering each more or less attractive or despised, and by sustaining or subverting individual submission to a regime of duty, propriety, and self-sacrifice. Culture functions as a grab-bag category—a black hole ready to absorb all messy unexplained causes and consequences. Here it trumps material circumstances, collective struggles, and institutional constraints as the reason for the decline of stable and happy marriages and families documented by rising rates of divorce, unwed motherhood, and deadbeat dads. Culture becomes an unproblematic, remarkably flexible category from which individuals, like savvy shoppers, can select timeless garments like marital commitment, fidelity and responsibility and discard their unfashionable accessories—like male dominance.

Somewhat regretfully perhaps, most of the social scientists active in the centrist family-values campaign have accepted the demise of the 1950s-style, male breadwinner family and the likely permanence of some level of postmodern family instability, diversity, and change.[36] Recognizing that working mothers, at least, are "here to stay," they promote a new (post)familism that evades the power and justice conflicts embedded in family transformation.[37] Following the successful example set by right-wing intellectuals, centrist family social scientists have regrouped outside the academy to provide a middle course between ideologies of the religious right on one side and feminism and gay liberation on the other.[38] They proffer eager politicians a social science narrative to compete with naturalist and divine justifications for the contested modern gender and family regime.

Sitcom Sociology for a Disaffected Electorate

Post–Cold War politicians from both moribund parties have compelling cause to grasp at this outstretched academic hand. With a shrinking, increasingly cynical, electorate, one described in a study by the Times Mirror Center for the People and the Press[39] as, "angry, self-absorbed, and politically unanchored," the volatile balance of electoral power rests in the hands of those elusive, "neglected, middle-class" voters, who are disproportionately white.[40] The Democratic Party, weakened by the erosion of its traditional liberal and working-class base, has particularly urgent need to court this constituency. The neo-family-values campaign offers "New Democrats" a way to exploit the ideological stranglehold that religious, right-wing, profamily crusaders have secured on the Republican Party in their efforts to lure Reagan Democratic defectors back to the fold. As one political journalist quipped, "Democrats Find the Right's Stuff: Family Values."[41]

During the 1994 election season, Republicans, for their part, worked actively to shed the unpopular, intolerant, profamily image emblazoned on the national unconscious by their televised 1992 national convention. Dan Quayle, William Bennett, and even leaders of the right-wing Christian Coalition, retreated from the militant profamily rhetoric they had imposed on the 1992 party platform, such as hardline opposition to abortion and gay rights. Asked for his views on abortion and homosexuality just before he addressed the Christian Coalition in September 1994, Quayle told reporters, "That's their choice."[42] Bennett's speech to the Christian conference advised participants to constrain their homophobic passions: "I understand the aversion to homosexuality. But if you look in terms of damage to the children of America, you cannot compare the homosexual movement, the gay rights movement, what that has done in damage to what divorce has done to this society."[43] Quayle defended this political regression toward the mean in explicitly instrumental terms: "The political situation has changed in this country. There's not the political support to make it illegal, so we should focus on reducing the number of abortions, and we want to change attitudes."[44]

Mirror-image speeches delivered during the 1994 election season by Quayle and Clinton previewed campaign rhetoric for the last presidential election of the millennium as the electorate observes prime-time combat for the family-values crown. However, this is a riskier game show for the Democrats than for the Republicans. After all, President Clinton, son of Virginia Kelley's colorful postmodern marital history, competes with a credibility gap that his own dalliances exacerbate. Richard Sennett suggests that "the popular language of 'family values' and of 'values' per se is a barely disguised language of sexual prohibitions," which imagines "the breakdown of family values and community standards to be synonymous with sexual explicitness."[45] Voters overwhelmingly view Republicans rather than Democrats as defenders of this symbolic domain.[46]

The rhetoric of family values provides an infinitely malleable symbolic resource that is understandably irresistible to politicians from both major parties in the age of corporate-sponsored, mass-media politics. The vague, but resonant language of family values functions more like potent images than like verbal communications subject to rational debate. Little wonder, therefore, that Murphy Brown (but not Candice Bergen), enjoyed star billing in the 1992 presidential campaign when Quayle castigated the sitcom heroine for glamorizing unwed motherhood. In a moment of supreme irony, anchorwoman Murphy took to the sitcom airwaves to chastise the former vice president for being out of touch with the problems of *real* families. Millions of voters watched this well-hyped episode and the ensuing responses to Murphy's sermon provided by Quayle and the small group of single mothers he had selected to join him in viewing this electoral spectacle, on camera.[47] Most political commentators at the time echoed the sitcom heroine's scorn for the vice president's inability to distinguish virtual from actual families. Because the Quayle-Brown spectacle underscored the message of the Republicans' 1992 nominating convention that the grand old party was out of touch with ordinary families, it assisted Clinton's slim margin of victory.

Breathtakingly soon, however, Dan Quayle began to enjoy the last laugh, as even President Clinton joined the "Dan Quayle was Right" brigades. Quayle's campaign against single mothers scored such a dramatic comeback victory over Clinton, the reputed comeback king, because the former vice president's campaign scriptwriters were quick to grasp the "virtual" rather than factual character of contemporary family-values talk. They recognized that Murphy

Brown could function symbolically as a wayward stepdaughter of Ozzie and Harriet Nelson, the mythic couple who lodge, much larger than life, in collective nostalgia for the world of 1950s families.

The 1950s was the moment of origin of the fable of virtual family values. Those halcyon days of the modern nuclear family were also the years when television became a mass medium, indeed an obligatory new member of "the family." From its hallowed living room perch, the magic box broadcast the first generation of domestic sitcoms, emblazoning idealized portraits of middle-class family dynamics into the national unconscious. From "Ozzie and Harriet" to "Murphy Brown," from "Amos and Andy" to "The Cosby Show," from "The Life of Riley" to "Roseanne," to world of TV sitcoms saturates popular imagery of family life. Ozzie and Harriet and their kin serve as the Edenic families of our century's bedtime fable, because the apogee of the modern family system coincided with television's own origins and Golden Age.

Family sitcom programming was created in the post–World War II period to construct a mass viewing audience for the nascent television industry and its corporate sponsors. The programs did not simply reflect, or even just romanticize, the existing structure and values of the family audience they sought to entertain. Rather, as cultural historians have demonstrated, the introduction of television played an active role in constructing, and later in deconstructing, the boundaries of the isolated nuclear family it depicted in such sentimental tones.[48] Because the 1950s was also the first Cold War decade, the years when the United States emerged as the dominant global superpower, images of an invincible family and nation mingle inextricably in national imagery of the "good old days."[49] Clinton, Quayle, Newt Gingrich, and the primary constituencies of the electorate they address, as well as many of their academic counselors, all were reared in the first generation of families who learned to spend their evening hours huddled alone together in their families, watching family TV, in their newly conceived "family rooms," designed as small shrines for the magic box.

This semiotic history of family sitcom TV, which evolved while the modern family it celebrated devolved, renders the idiom of family values a potent, inescapably visual and emotional register. Addressing emotional, rather than rational, frequencies, family-values discourse offers politicians and populace a brilliant defense mechanism with which to displace anxieties over race, gender, sexual, and class antagonisms that were unleashed as the modern family regime collapsed. No wonder that as the century ends, "it's all in the family."

During the 1994 and 1996 political seasons, the most popular sitcom social science script furnishes simple, emotionally resonant motivations and resolutions for those spectacles of routine fin-de-siècle violence, crime, and social decay that the networks broadcast nightly. Serial killers, crack babies, gang rapists, carjackers, dope dealers, drive-by shooters, school dropouts, welfare queens, arsonists, wife-beaters, child abusers, sex offenders, kidnappers, runaways, pregnant teens, gang warriors, homeless vagrants, terrorists—all social pathologies begin in a broken home. Her parents divorced, or they never married. His mother was hooked on welfare and drugs, or she dumped him in daycare. No one taught them family values. We need to stop coddling these criminals and con artists. From the punitive, anti-crime fervor of "three strikes and you're out" to "two tykes and you're out" welfare caps, family-values ideology plays to the privatistic, anti-government sentiments and the moralistic and vindictive appetites of our dismal, late millennial political culture.

An improbable alliance of academic and political networks produce and sponsor this sitcom sociology which is increasingly discordant with the diverse images of family life that now characterize contemporary domestic sitcom programs. Fending off competing political networks on the channels to their right and their left, mainstream family scholars and electoral candidates hope to keep the public tuned to the center of the political dial. Unwilling or unable to analyze the social sources of postmodern family and civic disorder, or to address the manifold injustices these upheavals expose and intensify, they resort to reruns of old family favorites. Religious and naturalist treatments of virtuous family order continue to play to substantial numbers of viewers in their specialized market niches. However, aging scholars, allied with New Democrats and moderate Republicans alike, have hitched their hopes for robust Nielson and ballot-box ratings to narratives featuring a prodigal society returning to conjugal family virtue after suffering the painful consequences of self-indulgent rebellions. The production company has assembled a postmodern pastiche of social science, fable, advanced technology and (dis)simulation to script and enact the serial melodrama. The plot-line, imagery, and production values owe more to television archives, and to power and knowledge shifts in the academy, economy, and polity, than they do to ethnographic or analytical acumen. This is the season for sitcom sociology—an effort to distract a disaffected public from the dire familial and social realities that the United States confronts as an ignoble century expires.

NOTES

1. Moynihan, *The Negro Family.*
2. Gillis, "What's Behind the Debate on Family Values?"
3. However, recently politicians in Britain and elsewhere in Europe have initiated efforts to stir public concern over single motherhood, divorce and family instability. See Tuula Gordon, "Single Women and Familism," and Jane Millar, "State, Family and Personal Responsibility: The Changing Balance for Lone Mothers in the United Kingdom." For example, the headline story run by London's *Daily Express* on June 23, 1994, Paul Crosbie, "The Crumbling of Family life," seems a direct replica of U.S. rhetoric.
4. In 1994 the Sex and Gender section had 1271 members, which was more than 200 members greater than the section on Medical Sociology, which is currently the second largest specialty section of the ASA. (Data provided by American Sociological Association.)
5. Cheal, *Family and the State of Theory,* 8.
6. Parsons and Bales, *Family, Socialization and Interaction Process.*
7. Goode, *World Revolution and Family Patterns.*
8. Friedan, *The Feminine Mystique.*
9. Bernard, *The Future of Marriage.*
10. Mainardi, "The Politics of Housework"; Koedt, "The Myth of the Vaginal Orgasm"; Oakley, *The Sociology of Housework,* and Rubin, "The Traffic in Women: Notes on the Political Economy of Sex."
11. After all, Moynihan's analysis of Black family decline had built quite directly on the sociological work of E. Franklin Frazier, and even *Tally's Corner,* Elliot Liebow's decidedly liberal and sympathetic ethnography of urban Black families, supported the thesis of Black family pathology.
12. Glenn, "The Re-evaluation of Family Change by American Social Scientists," 2.
13. Ibid., 3.
14. Ibid., 4–5.
15. Cherlin, *Marriage, Divorce, Remarriage,* 138.
16. Wallerstein and Blakeslee, *Second Chances: Men, Women, and Children a Decade After Divorce.*
17. McLanahan, *The Consequences of Single Motherhood*; and McLanahan and Bumpass, "Intergenerational Consequences of Family Disruption."
18. Glenn, 10.
19. Marcus, "Power/Knowledge Shifts in America's Present Fin-De-Siècle: A Proposal for a School of American Research Advanced Seminar." The papers produced for the seminar, which was held at the S.A.R. in Santa Fe, NM, in November 1994, will appear in Marcus, ed., *New Locations.*
20. For example, as a speaker on an evening plenary session attended by more than one thousand sociologists at the American Sociological Association meetings in Los Angeles in August 1994, Patricia Hill Collins delivered a polemical assault on postmodern theory and was rewarded with rapturous applause.
21. Popenoe, "What's Behind the Family Values Debate?"
22. Ibid.
23. While the proportion of doctoral degrees awarded to women in all fields in the U.S. rose only from 30% in 1980 to 36% in 1990 (National Research Council 1991), in sociology, women's share of Ph.D. degrees rose much more substantially, from 33% in 1977 to 51% in 1989 (National Science Foundation 1991). The American Sociological Association reports that by 1991, 57% of students enrolled in doctoral programs in sociology in the U.S. were women as were 53% of the members of the A.S.A. The association does not report data on the proportion of women and men in each of the specialty sections, but all who have attended sessions and meetings of the Sex and Gender section and the Family section can confirm the contrast in the relative paucity of men in the former. Sociologist Barrie Thorne, a past president of the Sex and

Gender section and a past vice-president of the ASA reports that men have never comprised even 10% of the membership in the former (personal communication).

24. Weber, *The Sociology of Charismatic Authority.*
25. See Rosenfelt and Stacey, "Second Thoughts on the Second Wave."
26. Glenn, "The Reevaluation of Family Change," 12.
27. Friedan's *The Second Stage* celebrated the reappearance of familism among feminists. For an early critical discussion of the emergence of this perspective within feminism, see Stacey, "The New Conservative Feminism," and "Are Feminists Afraid to Leave Home?"
28. Glenn, "Reevaluation of Family Change," 12.
29. See Sanqvist and Andersson, "Thriving Families in the Swedish Welfare State."
30. For example, he co-convened an academic conference on family change at Stanford University and then co-edited the resulting conference volume (Blankenhorn, Elshtain and Bayme 1990). Blankenhorn also spoke at a Santa Clara University conference, "Ethics, Public Policy and the Future of the Family," on April 18, 1995.
31. Hewlett, *A Lesser Life.*
32. Quoted in Winkler, "Communitarians Move Their Ideas Outside Academic Arena."
33. However, Elshtain herself rarely forfeits an opportunity to raise the ante of animus. In a letter to the editor which *The Nation* published in abridged form, Elshtain "hooted" her response to my critical discussion of the centrist campaign, "The New Family-Values Crusaders," that had appeared in the magazine in quite personalized, and purple, prose: "I was pleased to see that Judith Stacey is on the job. Not having been attacked by Ms. Stacey in print for a few years, I wondered if she was still up to the ideological stalking so characteristic of her efforts a few years back. Then, if memory serves, Stacey expended a good bit of energy before finally landing on the label, "New Conservative Feminist" for a few of us and now, to my surprise, I get to be something brand new and exciting, a "New Family Values Crusader." Stacey seems to believe nothing can be discussed unless you first line people up in team jerseys. This labeling fetishism, alas, has long been characteristic of segments of the sectarian left and this same infantilizing urgency has infected a good bit of academic feminism."
34. For example Cherlin; Furstenburg and Cherlin, *Divided Families;* and even McLanahan.
35. Carlson, *The Swedish Experiment in Family Politics,* 194.
36. Glenn, for example, depicts himself as "among the relatively small number of American family social scientists who believe that some of the changes can be halted if not reversed. Reconstitution of the American family of the 1950s—a goal of some conservatives—is indeed unrealistic and, in my view, undesirable." "Revaluation of Family Change," 11–12.
37. *Here to Stay* is the title of Mary Jo Bane's early, misleadingly optimistic assessment of nuclear family stability in the U.S. that is a frequent foil for the family values campaign.
38. Messer-Davidow, "Manufacturing the Attack on Liberalized Higher Education."
39. Berke, "U.S. Voters Focus on Selves, Poll Says."
40. Yoachum, "Small Minority Voter Turnout a Product of Apathy and Anger." In California, whites constituted only 57% of the adult population in the state, but they cast 83% of the votes in the 1994 election.
41. Brownstein, "Democrats Find the Right's Stuff: Family Values."
42. Berke, "Two Top Republicans Soften Their Tone."
43. Ibid.
44. Ibid.
45. Sennett, "The New Censorship," 490.
46. Indeed, survey data indicate an alarming anomaly. On the one hand, since the 1980s "the electorate as a whole moved in a clearly liberal direction on three issues besides gay rights: abortion, the role of women, and, to a lesser extent, on the role of government in guaranteeing jobs and living standards." (Strand and Sherrill, "Electoral Bugaboos? The Impact of Attitudes Toward Gay Rights and Feminism on the 1992 Presidential Vote.") Nonetheless, in a November 1994 poll sponsored by the Democratic Leadership Council, more voters identified the Republicans than the Democrats as the party that "would do a better job strengthening families." Greenberg Research.
47. Fiske, *Media Matters.*
48. Spigel, *Make Room for TV,* and "From the dark ages to the golden age;" also Taylor, *Primetime Families.*
49. See May, *Homeward Bound.*

EPIGRAPH SOURCES

Epigraphs on page 450

1. Senator Moynihan quoting his 1965 work in a fund-raising letter mailed in October 1994.
2. Quayle quoted in Susan Yoachum, "Quayle Talks Tough on Fatherhood," *San Francisco Chronicle*, September 9, 1994: A1, 17.
3. President Clinton quoted in, "In Baptist Talk Clinton Stresses Moral Themes," *New York Times*, September 10, 1994: A1, 6.

Epigraphs on page 452

1. John Gillis, "What's Behind the Debate on Family Values?" Paper delivered at American Sociological Association Meetings, Los Angeles, August 6, 1994.
2. Norval Glenn, "The Re-evaluation of Family Change by American Social Scientists," unpublished paper delivered in Australia, 1994, available from author.

REFERENCES

Bane, Mary Jo. *Here to Stay: American Families in the Twentieth Century.* New York: Basic Books, 1976

Berke, Richard L. "Two Top Republicans Soften Their Tone." *New York Times*, September 17, 1994a. A8.

———. "U.S. Voter Focus on Selves, Poll Says." *New York Times*, September 21, 1994b. A12.

Bernard, Jessie. *The Future of Marriage.* New York: World Publishing, 1972.

Blankenhorn, David, Jean Bethke Elshtain and Steven Bayme, eds. *Rebuilding the Nest: A New Commitment to the American Family.* Milwaukee: Family Service America, 1990.

Brownstein, Ronald. 1994. "Democrats Find the Right's Stuff: Family Values." *Los Angeles Times*, August 1, 1994. A1, A23.

Carlson, Allan. *The Swedish Experiment in Family Politics: The Myrdals and the Interwar Population Crisis.* New Brunswick: Transaction, 1990.

Cheal, David. *Family and the State of Theory.* New York: Harvester Wheatsheaf, 1991.

Chira, Susan. *Marriage, Divorce, Remarriage.* Revised edition, 1992. Cambridge: Harvard University Press, 1981.

Fiske, John. *Media Matters: Everyday Culture and Political Change.* Minneapolis: University of Minnesota Press, 1994.

Frazier, E. Franklin. *The Negro Family in the United States.* Chicago: University of Chicago Press, 1939.

Friedan, Betty. *The Feminine Mystique.* New York: Norton, 1963.

———. *The Second Stage.* New York: Summit Books, 1981.

Furstenberg, Jr., Frank, and Andrew J. Cherlin. *Divided Families: What Happens to Children When Parents Part.* Cambridge: Harvard University Press, 1991.

Gillis, John. "What's Beyond the Debate on Family Values?" Paper delivered at the annual meetings of the American Sociological Association. Los Angeles, August 6, 1994.

Glenn, Norval D. "The Re-evaluation of Family Change by American Social Scientists." Paper presented to Committee for the International Year of the Family of the Catholic Archdiocese of Melbourne, 1994.

Goode, William J. *World Revolution and Family Patterns.* New York: Free Press, 1963.

Gordon, Tuula. 1994. "Single Women and Familism: Challenge from the Margins." *European Journal of Women's Studies.* 1(2): 165–82.

Hewlet, Sylvia Ann. *A Lesser Life: The Myth of Women's Liberation in America.* New York: William Morrow, 1986.

Koedt, Anne. "The Myth of the Vaginal Orgasm," in *Liberation Now! Writings From the Women's Liberation Movement.* New York: Dell, 1970.

Liebow, Elliot. *Tally's Corner.* Boston: Little, Brown and Co., 1967.

Mainardi, Pat. "The Politics of Housework," in *Sisterhood Is Powerful: An Anthology of Writings from the Women's Liberation Movement,* ed. Robin Morgan. New York: Vintage, 1970.

Marcus, George. 1993. *Power/Knowledge Shifts in America's Present Fin-De-Siècle: A Proposal for a School of American Research Advanced Seminar.*

May, Elaine. *Homeward Bound: American Families in the Cold War Era.* New York: Basic Books, 1988.

McLanahan, Sara. 1994. "The Consequences of Single Motherhood." *The American Prospect.* n. 18.

Messer-Davidow, Ellen. 1993. "Manufacturing the Attack on Liberalized Higher Education." *Social Text* 36: 40–80.

Millar, Jane. 1994. "State, Family and Personal Responsibility: The Changing Balance for Lone Mothers in the United Kingdom." *Feminist Review* 48: 24–39.

Moynihan, Daniel Patrick. *The Negro Family: The Case for National Action.* Washington, D.C.: U.S. Department of Labor, 1965.

Oakley, Ann. *The Sociology of Housework.* New York: Pantheon, 1974.

Parsons, Talcott, and Robert Bales. *Family, Socialization and Interaction Process.* Glencoe, IL: Free Press, 1955.

Popenoe, David. 1994. "What's Behind the Family Values Debate?" Paper presented at the annual meetings of the American Sociological Association. Los Angeles, August 7, 1994.

Rosenfelt, Deborah, and Judith Stacey. 1987. "Second Thoughts on the Second Wave." *Feminist Studies* 13(2): 341–61.

Rubin, Gayle. "The Traffic in Women: Notes on the 'Political Economy' of Sex." In *Toward an Anthropology of Women,* ed. Rayna Reiter. New York: Monthly Review, 1975.

Sandqvist, Karin, and Bengt-Erik Andersson. 1992. "Thriving Families in the Swedish Welfare State." *Public Interest* n.109: 114–17.

Sennet, Richard. 1994. "The New Censorship." *Contemporary Sociology* 23(4): 487–91.

Spigel, Lynn. *Make Room for TV: Television and the Family Ideal in Postwar America.* Chicago: University of Chicago Press, 1992.

Stacey, Judith. 1983. "The New Conservative Feminism." *Feminist Studies* 9(3): 559–83.

———. "Are Feminists Afraid to Leave Home? The Challenge of Profamily Feminism." In *What Is Feminism?* Juliet Mitchell and Ann Oakley, ed. London: Basil Blackwell, 1986.

Strand, Douglas Alan, and Kenneth Sherrill. 1993. "Electoral Bugaboos? The Impact of Attitudes Toward Gay Rights and Feminism on the 1992 Presidential Vote." Paper delivered at the annual meetings of the American Political Science Association, Washington, D.C.

Wallerstein, Judith S., and Sandra Blakeslee. *Second Chances: Men, Women, and Children a Decade After Divorce.* New York: Ticknor & Fields, 1989.

Weber, Max. "The Sociology of Charismatic Authority." In *From Max Weber: Essays in Sociology.* Hans Gerth and C. Wright Mills, ed. 1958 edition, New York: Oxford University Press, 1946.

Winkler, Karen. "Communitarians Move Their Ideas Outside Academic Arena." *Chronicle of Higher Education,* April 21, 1993, A7.

Yoachum, Susan. 1994. "Small Minority Voter Turnout a Product of Apathy and Anger." *San Francisco Chronicle,* 22 September, A4.

Family Privacy: Issues and Concepts

Felix M. Berardo

For some decades now, growing numbers of Americans have been voicing concerns over widespread invasions of their personal lives. Frequently, such intrusions occur without the permission of and contrary to the desires of those under surveillance. Personal information about others is purchased and exchanged in a far-reaching information economy in which data "collected in one context can be reused in entirely unanticipated and even hostile ways without the knowledge or consent of the individuals involved (Bernstein, 1997, p. 3). The mass media and popular literature have long documented extreme examples of privacy infringement. There is a spreading awareness of the undesirable consequences of such transgressions for the individual, the family, and society (Alderman & Kennedy, 1995).

Heated debates over issues of privacy are regularly renewed in newspapers and magazines, on television, and across the vast Internet (Quittner, 1997). The intensity of the rhetoric is exacerbated by interventions now possible through modern telecommunication systems, including a pervasive computer technology as well as a rapidly expanding and sophisticated biomedical technology such as genetic testing.[1] This escalating argument suggests that this issue strikes at the core of the American value system, with its strong emphasis on democracy, individualism, freedom, and personal autonomy.[2]

Past Scientific Research on Privacy

Despite the importance attached to privacy, until recently, social scientists devoted relatively little attention to the subject. In the early 1960s, for example, sociologists were urged to undertake empirical explorations of public and personal definitions of this concept for the analysis of social problems (Bates, 1964). However, they and other social scientists were slow in undertaking this definitional task. With few exceptions, this area was still largely unexplored a decade later (Berardo, 1974). In the late 1970s, a special issue of the *Journal of Social Issues* was devoted to this topic, with the editor's hope that it would stimulate interest and debate and also legitimate and direct the study of privacy as a behavioral phenomenon (Margolis, 1977, p. 2). Twenty years later, again with some notable exceptions, that hope has yet to be fully realized (Cheal, 1991; Fahey, 1995; Nock, 1993).

Even within the specialized area of family sociology, it was observed early on that whereas socioeconomic and ethnic variation in family structure and behavior had undergone considerable research, analyses of variations in privacy by social class or ethnicity remained to be done (Laslett, 1973, p. 480). In the past, the major efforts of the family specialists focused primarily on the effects of spatial dimensions of the home and family size and density on member interaction and privacy. Thus, knowledge of the social structural conditions that maximize or minimize familial autonomy and the impact such conditions have on enactment of domestic roles was and remains rather limited.

This article concentrates on the role of privacy within the context of marriage and family. It attempts to elaborate on both the functional and dysfunctional aspects of family autonomy. The presentation begins with a general discussion of *invisibility* and its relation to marital and familial role enactments. Questions are raised concerning the growing capacity of society to penetrate that important but intangible circle surrounding family life and to make public the activities of its members. In light of the limited research on family privacy, an expansion of conceptual and empirical efforts in this important area is called for (Fahey, 1995).

The Invisibility of Marriage

An interesting aspect of marriage and family life in the United States is its invisibility, or the general unobservability of the psychosocial dimensions of this intimate unit of social organization. Although individuals appear to know a great deal about *their* particular families, they may know very little about the structural-interactional modes of other people's families (Cuber & Haroff, 1965, p. 36). Such invisibility often is reflected by a "pluralistic ignorance" about marriage and family life norms and role expectations. This concept describes a situation in which there are often only vague, indirect, and erroneous perceptions of one another's family attitudes, values, and behavior.

Married couples tend to actively conceal various aspects of their relationship from public view. This exclusive nature of the marriage relationship and the concealment of family transactions in general has been depicted as being analogous to a closed corporation by "presenting a common front of solidarity to the world, handling internal differences in private, protecting the reputation of members by keeping family secrets and standing together under attack" (Hill, 1958, p. 139). External agencies are permitted only selective accessibility to the internal transactions of the family unit. Consequently, relatively little is known about how other families function and the degree to which they meet normative expectations. Therefore, it is difficult to ascertain what are acceptable or successful patterns of family life. Under such conditions, diverse marital and familial norms become elusive and difficult to characterize with precision.

Of course, this does not mean that there is no interest in other people's marital and familial relationships. The vast amount of time spent in gossip and the general attentiveness given to events in other people's families is highly indicative of this pervasive interest. The popularity of past and current television shows that revolve around the lives of particular families, soap operas that exaggerate complex issues of domestic life, and the ubiquitous talk shows, is not accidental. Such programs often mirror problems and issues that remain most closely guarded in real life. A common characteristic of such shows is that they give viewers the impression that they are privy to the most intimate details of family life.

Invisibility and the Backstage of Family Life

Much of the intimacy of family life remains hidden. Most of what is seen in this guarded domain are the "front stage" performances. These consist of behavior characterized by public conformity to the role expectations and norms that society imposes on its members (Goffman, 1959). This front-stage activity is considered functional for the individual, the family, and society. Not only do front-stage performances serve to maintain societal organization by allowing the influence of others to guide individual behavior but they also provide an important psychological function, namely, making the individual feel she or he is significant to others and they are to her or him, thereby creating a concomitant tendency to contribute to the common good.[3]

If one is attempting to learn about marital roles, however, observations of front-stage performances will provide insufficient and very often distorted images and information. To complete the picture, one needs to know what goes on behind the scenes in the "backstage" where the family life drama is played out. The importance of the backstage as a prism of family life reflects the historical shift—both in a physical and psychological sense—from the visible public sphere to what has become known as the private family domain. For example, in an earlier period in our history, most middle-class homes had front porches, and much of the activity there was open to public view.

> The neighbors, who spent a good deal of time on *their* front porches, could see and hear much of what went on.... The relatively public front porch gave way to the more private screened back

> porch, sun room, or deck. In most suburban, middle-class communities today, the front lawn . . . is mainly for ceremonial display. Not much of family life is visible from the street. Outdoor activities take place in the backyard, which is far more private than the front porch was in our grandparent's generation. (Bird & Melville, 1994, p. 278)

Private spaces were created within the modern household as well, through the construction of special-purpose rooms, like the family room or separate bedrooms. Thus, today's homes "reflect and encourage the belief that maximum privacy, not only between the family and the surrounding community but within the house as well, is something that is highly desirable" (Bird & Melville, 1994, p. 278).

The backstage is that place—usually the home—where members no longer have to worry about presenting a common front of solidarity but can give vent to their real feelings and selves. It is within this inner sanctum that the true nature of the group's interactions is revealed. The backstage is closely guarded by its constituents, and only a few selected outsiders are allowed access to the activities that take place there. The indiscernibility of marriage and family life refers primarily to what transpires in this region. The privacy that it affords serves a major function of sustaining marital and family stability.

Functionality of the Backstage: Privacy

Sociologists have long observed that some mechanism for monitoring the activities of individuals and groups is essential in order that a society can function adequately. For example, Merton (1968) pointed out that privacy is not merely an individual propensity but a necessary prerequisite for the effective functioning of a social structure.

> *Some* measure of leeway in conforming to role expectation is presupposed in all groups. To have to meet the strict requirements of a role at all times, without some degree of deviation, is to experience insufficient allowances for individual differences in capacity and training and for situational exigencies which make strict conformity extremely difficult. (p. 397)

More particularly, full surveillance of activities in a group would become psychologically overwhelming and, as a consequence, dysfunctional for the maintenance and stability of the group as a whole. What is being stressed here is that if all behavior were observable, then to enforce infractions against the norms would be impossible. To live under the fear that every action was being detected would be intolerable for the members of society. Thus, although group or societal norms may serve as guidelines for behavior, they are not and cannot always be rigidly adhered to.

Privacy allows for a certain degree of flexibility in fulfilling the demands of the society. For instance, it permits the parent a level of freedom in the way he or she learns and enacts that role, even though there are general prescriptions of performance and expectations laid down by society. At the same time, it is recognized that situations in which social transactions remain completely immune from observability, and therefore accountability, have the potential for allowing deviant behavior to accumulate and to depart widely from prevailing norms. The reality of this potential is dramatically illustrated by the thousands of cases of child abuse or neglect that occur yearly in the United States, some large proportion of which go undetected or are never reported. Similarly hidden are other forms of domestic violence, including the large number of abused wives subjected to repeated but hidden brutality within the home (Cheal, 1991, p. 90). At some point, therefore, the shield of invisibility surrounding family life becomes problematic in that it can become a source of socially patterned, or even institutionalized, evasions of rules.

Merton (1968) posits that there is an optimum of observability that is essential for the successful functioning of groups—including the family—and society. Indeed, when this optimum is exceeded and there is too much surveillance, "the autonomy of a person is experienced as threatened by having no private—that is to say, wholly separate and secret—life, immune to observations by others" (p. 397).

The Need for Marital and Familial Privacy

These concepts have direct applications to contemporary family life in the United States. One can argue that invisibility and the privacy it affords are important functional requirements for the effective operation of marital and familial systems. They allow for fluctuations of behavior around the norms that are established by society. This particular functionality of privacy was noted decades ago by Waller (1951). He observed that the newly married couple is expected to succeed. Therefore, society provides a looking-glass self that supports the front of solidarity that the couple erects.

> The true nature of marital interaction is hidden even from the closest of friends. In the presence of others, the pair presents a common point of view, an almost taken-for-granted position of marital harmony.... And it is very important, likewise, that outsiders assume the marriage relation to be one of amity and accord; this too, is taken for granted and not commented upon.... This tacit assumption of an intimacy too great to need any proof and too common to be commented upon is ... one of the roots of solidarity. (pp. 325–326)

As Waller (1951) and others (Brown-Smith, 1998 ...) correctly perceived, the secrets of married life are among its most crucial assets and as such are closely guarded.

The impulse to maintain a public facade of solidarity persists even in those marriages that are deteriorating. Even when outsiders become aware that the marriage is on the decline, they rarely feel it appropriate to intrude. They recognize implicitly that once the marital difficulty is made public, the process of deterioration and alienation may be accelerated (Mayer, 1966).

Functions of Privacy for the Family

The ability to separate family life from others serves several important functions (Bates, 1964). One function is the *self-protective* one. For example, a couple having marital problems is protected from public scrutiny and allowed to conceal their difficulties while they work out a resolution, or until such time as the couple sees fit to expose them. A second function is the provision of a certain degree of *latitude* in adhering to the societal norms that govern family behavior. Thus, parents are charged with the responsibility of socializing children to enact adult roles, but the manner in which husbands and wives meet this obligation varies considerably. Parents decide how—within broad limits—to fulfill this and other societal expectations on the basis of their own personalities, capabilities, and value orientations. Third, privacy functions as a *buffer* between social pressures on the couple and their responses to them. This allows the marital pair to experiment, to make mistakes, to reveal motives, to express feelings, and to engage in actions that, if disclosed, might prove humiliating or provoke the application of sanctions by others. In addition, privacy affords protection for self-imposed punishment. For example, one partner can say "Well, if they knew all the circumstances they might understand, but it's too risky to explain. I at least can make some case for myself." Thus, a family or its individual members can use privacy as insulation from both internal and external sanctions. Fourth, separation from others also performs an *ameliorative* function for family members. "After bruising contact with the world, privacy may be required within which self-esteem can be restored" (Bates, 1964, p. 433).

In sum, privacy allows a family sufficient autonomy from disruptive extrafamilial scrutiny to foster a feeling of group cohesiveness, thereby enhancing solidarity and impelling them to act as a unified group in resisting outside interference in their affairs. Privacy protects the family's general welfare; when it becomes threatened, the welfare of the family becomes endangered.

Disadvantages of Family Privacy and Secrecy

Certain limitations or disadvantages may accrue from the nonobservability of marital and familial life. For example, if couples know very little about the marital interaction of their contemporaries, they will tend to use other points of reference—such as the images of family life projected through the mass media or, as is more often the case, their recollection of their parents' marriages—as models on which to pattern their own behavior. Whether or not these models represent distorted or constricted pictures of reality, once assumed as points of reference, they can have both direct and indirect effects on marital and familial interaction.

Ignorance of the behavioral repertoires of other couples can severely restrict the range of reference groups available to spouses who are seeking to achieve adequate role performance. Marital invisibility prevents them from making comparisons or judgments as to the appropriateness of their own role enactments. Thus, some have suggested that increasing the visibility of other people's marriages and family life would make a positive contribution to learning and stability by extending the range of reference groups against which reality testing could take place.

In sum, the argument is that increased accessibility to the family lives of others would influence the manner in which people perceive and perform similar roles within their own families. However, there is no guarantee that people will imitate other people's patterns of behavior; in fact, they may choose to reject them. Because American society is composed of a number of subcultural elements, and because subcultures by definition have their own sets of norms and values, what might be conceived as proper marital and familial role performance in one subculture may be rejected as inappropriate by another. Unfortunately, our empirical knowledge of the ways in which privacy is structured among different subcultures is embarrassingly meager.

The Erosion of Privacy: Impact on the Family

A general erosion of privacy appears to be taking place in the United States. Current articles in newspapers and national magazines abound with concern over what is viewed as a widespread trend of infringement into personal spheres of existence. One only has to "surf" via the various "search engines" on the Internet to realize the continuous and extraordinary information stream now being devoted to controversial issues concerning unwanted intrusions and ways designed to prevent them (Quittner, 1997). Successive polls conducted by Louis Harris and associates in 1994 and 1995 found that four out of five Americans are concerned about threats to their personal privacy (as cited in Bernstein, 1997, p. 6). This general encroachment has a serious potential for invading the backstage of contemporary family life. Virtually every information-seeking agency in the country attempts to cross the boundaries that separate personal and familial data from the public and is doing so with varying degrees of success.

The view has been advanced that privacy is increasingly devalued in our information-driven economy, which exhibits an almost insatiable hunger for data about our personal lives, so much so that "individual privacy is looking more and more like an endangered natural resource" (Bernstein, 1997, p. 2). The phenomenal growth of data collection or storage enterprises has accelerated the ongoing and far-reaching national debate over what many consider to be highly invasive and inappropriate threats to personal autonomy. Alderman and Kennedy (1995), who were interested in the legal principles underlying the right to privacy, conclude that "With so much information available at a keystroke, it is now inescapable that there will be times when what is whispered in the closet will indeed be shouted from the housetops" (p. 332).

Incidents of external excursions into the marital and familial spheres raise a serious question concerning the boundary that separates parental prerogatives from societal interests. This question is frequently evident in cases involving the propriety or impropriety of governmental interference with the autonomy of the parents to control the upbringing of their children. The issue is whether the right of privacy embraces the parent–child relationship. Under the well-established *parens patriae* doctrine, both the state and the parents have a vested interest in the welfare of children. However, although parents are generally afforded maximum latitude in child rearing, there is no immunity from state interference. The parens patriae doctrine has encountered widespread challenge, and courts are increasingly asked to rule on the conflicting claims of the state, the parents, and the children as to who shall be the ultimate arbiter of the latter's fate.[4]

At what point do external agencies expose parental behavior? Where is the demarcation between a married couple's right to privacy and compulsory public discourse of beliefs and practices? Is there some discernible point at which family invisibility becomes dysfunctional with respect to its members as well as to societal values and goals? In this connection, note Slater's (1970) thesis that as we secure greater degrees of privacy, our lives become increasingly problematic.

> We seek more and more privacy, and feel more and more alienated and lonely when we get it. What accidental contacts we do have, furthermore, seem more intrusive, not only because they are unsought but because they are unconnected with any familiar pattern of interdependence. (p. 7)

In his view, technological change, mobility, and our peculiarly individualistic ethos have combined to seriously rupture the bonds that hold us together as a family, a community, and a nation. Indeed, some have long ago argued that as a result of such factors, America was well on its way to becoming a nation of strangers (Packard, 1972). In more recent times, several scholars have argued that the contemporary emphasis on self-development, individualism, and the need for privacy was in fact detrimental to family life and impeded interest in and pursuit of neighborhood and community ties (Bellah, Madsen, Sullivan, Swidler, & Tipton, 1985; C. Lasch, 1977; D. Lasch, 1979).

It can be and has been argued that invasions of personal and familial privacy are justified. Too much privacy can allow deviant behavior to accumulate behind a shield that hides antisocial conduct and thereby reduces adequate social control. The question is, of course, how much is too much privacy? It has been suggested that both too little and too much privacy can result in eccentric, inadequate, and inappropriate role enactments. If this is true, how does one determine what is the optimum level of privacy? And, perhaps more important, who makes this determination?

Needed Research on Privacy and the Family

This presentation began with the general observation that systematic social scientific research on privacy and on its relationships to family structure and interaction has been slow to materialize, and that much needed research in this area remains to be undertaken. A major problem has been the difficulties associated with conceptualizing a sensitive and multidimensional phenomenon and arriving at operational definitions that are scientifically viable and connotatively valid. Family privacy is perceived and experienced differently by individual members. Fahey's (1995) concept of multiple and sometimes overlapping "zones of privacy" and the diversity in form and function such zones display suggests a much needed research agenda for understanding the public or private transactions of families. Moreover, as Fahey (1995) and Cheal (1991) have cautioned, the contradictory literature surrounding the public/private dichotomy itself is problematic, with shifting and ambiguous meanings that must be resolved if we are to better understand the role of privacy in structuring the family.

As we noted at the outset, with few exceptions, family analysts have not been engaged in a systematic effort to specify the dimensions of family privacy. Some years ago, a beginning was made by Hess and Handel (1959; Handel & Whitchurch, 1994). They developed several concepts including patterns of separateness and connectedness, congruence of images, testimony exchanges, family themes, and boundary setting and maintenance. Some of these concepts denote mechanisms for regulating the transactions and experiences of the family world internally and for determining what aspects of the external world are admissible and under what conditions (Handel, 1967, pp. 10–13). Systematic efforts using this analytic framework are still sorely needed to specify how families meet these conditions within the context of privacy.

Kantor and Lehr (1975) developed a sophisticated "cybernetic-like model" that includes several useful concepts for analyzing the ways that families negotiate social space. For example, all families and each subsystem within families attempt to establish "zones of safety" to demarcate their interior and exterior spaces. The "purpose of these safety zones is usually the protection of property, privacy, and the relationships among family members, rather than the guarding of physical safety" (p. 42). Their theoretical and empirical explorations of the dimensions of a social space offer a complex framework that can be employed in research designed to investigate how families develop, maintain, and defend their territories.

Work on privacy as a multidimensional communication construct has implications for family research. Burgoon et al. (1989) have focused attention on "how people close themselves off to others and regulate their degree of accessibility, either individually or as members of dyads and other social units" (p. 132). They analyze communicative acts that are privacy invading, communication strategies that are used to restore privacy when it has been violated, and how relationship type affects communication of privacy. Their delineation of the distinguishing properties of four major dimensions of privacy—physical, social or interactional, psychological, and informational—have heuristic potential for issues related to the modern family's attempt to maintain a balance between accessibility and inaccessibility.

Similarly, research needs to be undertaken on the consequences of disclosure for marital and familial stability. The issue needs to be specified not only in terms of whether disclosures to outsiders strengthen or weaken spousal ties but also with respect to "the *conditions under which* these divergent effects take place" (Mayer, 1967, p. 34). His initial research on lower- and middle-class wives uses a model that elaborates *accommodative* and *corrective* approaches to marital problems. This original work related these two orientations to the functions and consequences of disclosure and was a step in the right direction, but much more certainly remains to be done.

Analyses of the significance of privacy for nuclear family, as opposed to alternative family structures is another area of empirical investigation that needs to be developed. It has often been observed, for example, that in preliterate societies where an extended family structure predominates, there is little that occurs in the family that is not observable to others.

> For the primitive, society is largely an extension of family relationships.... These relationships are not shut out by high walls of brick and mortar.... In most primitive societies each home spreads into another and the households intermingle in a communal life and without privacy, or the desire for it. (Evans-Pritchard, 1965, p. 49)

The implications of structural differences in household composition over generations and within the context of privacy has only recently come under empirical scrutiny by family scientists.

Laslett's (1973) analysis of the family as a public and private institution is illustrative of this connection. She cites data to support the hypothesis that the emergence of the private family, characterized by relatively limited access and greater social control over the observability of member behavior, is a relatively recent development in the United States. The evidence suggests that in an earlier period of our history, the enactment of family roles within the nuclear household was visible to a larger audience than at present. This greater visibility, which allowed more social control and wider support for traditional family role behavior, was a result of such factors as the combination of economic activities in or near the home with family life within the household. Moreover, there was a constricted structural arrangement of dwelling space—for example, few rooms, with limited sources of heating. In addition, there were nonkin who had access to the household as resident members—servants, boarders and lodgers, apprentices, and children from other families who had been "put out" to be reared in a different household—whose presence diminished the degree of privacy available.

Over time, each of these factors has been altered by structural changes that occurred in the United States as a result of industrialization, urbanization, and other forces. Among these changes were the separation of economic activities from the home, changes in architectural styles and practices that allowed expanded and segmented structural arrangements of living space, and a significant reduction in the number of nonkin resident members in the home. The result of these changes has been an increase in family privacy and, at the same time, a reduction in social control over and social support of the traditional definitions of family role performances.

Laslett (1973) posited that if these historical trends of increased privacy and decreased social control continue, they will support a continuation of alternative family forms and lifestyles—provided that external forces, such as changes enacted through the legal system, do not move in the direction of imposing greater conformity to societal expectations. Her analysis is particularly provocative, not only because of its treatment of the relationship between social change and privacy within different historical contexts but also because of the various hypotheses it suggests for empirical inquiry.

However, the efforts of family researchers to date have centered on privacy at the intrafamilial level. The role of the family dwelling in encouraging or inhibiting priorities granted to privacy and member interaction has received particular attention.[5] An assumption underlying such investigations is that the physical opportunities available for privacy have significant implications for determining the quantity and the quality of interpersonal relations that arise within the family group. The ways in which spatial limitations imposed by the family dwelling structure patterns of family privacy need considerably more analysis that has been undertaken thus far.

We need more studies of the effects of privacy deprivation and crowding on individual and family functioning as well as on personality development. Such research will need to take into account several influential variables, such as family composition with respect to the age and sex of children, arrangement of space, stage of the family life cycle, and differential privacy orientations of individual members. In the latter connection, it is important to remember that some members are more privacy oriented than others and, indeed, some may seek or require more solitude or detachment than others. These and other variables have yet to be incorporated into a sophisticated model that would order family interactional constellations in some predictable fashion.

Conclusion

It has been argued that the right of the family to control the amount and type of information it is willing to reveal to others, and when and to whom it is disposed to make such disclosure, is being challenged. The point at which the delicate balance between an optimum of invisibility and an optimum of observability is threatened is being approached. When this point is reached, the family may find itself struggling fiercely to protect its right to be left alone in order to survive. Without adequate insulation, this basic unit will lose much of the special intimacy that holds it together; it would not be able to effectively carry out its functions.

There are no ready-made solutions to prevent the rising encroachment on family boundaries. As we have noted, some degree of surveillance is necessary to enforce societal rules and taboos. At the same time, domestic privacy and the functions it performs are equally essential to the maintenance of the larger community. Thus, like most problems of significance, the issue of how to protect the interior of family life from unnecessary external intrusions does not yield to simple resolutions. Whatever solutions ultimately emerge, they will have to take into account the minimum needs of married couples and their children to have private space—a "haven in a heartless world" to use C. Lasch's (1977) well-known phrase—where they can regroup and recoup from their daily and sometimes harsh encounters with the outside world.

NOTES

1. In 1997, the chair of the House Task Force on Medical Records and Genetic Privacy suggested that genetic testing could become the civil rights issue of the 21st century. In his words, "Genetic information is personal, powerful, permanent and sensitive. It not only affects the individual, but it also has an impact on offspring and other blood relatives. Genetic privacy must be protected" (Stearns, 1997, p. 9A).
2. A related and perhaps equally important issue, which cannot be dealt with in this article, is the growing restrictions on the freedom of choice in our everyday lives. New attempts to regulate personal lifestyles may

reflect paternalism and a declining tolerance for the choices of others. The result may be a diminished civility and common sense (McGovern, 1997).

3. In this connection, Motz (1965) has noted,

> People who faithfully play their parts exhibit personal and civic responsibility. The rules make life predictable and safe, confine ad libs within acceptable limits, control violence and emotional tangents, and allow the show to go on and the day's work to be done. Thus, the challenging game of maintaining front relates unique personalities to one another and unites them in activity and into a nation." (p. 29)

4. The courts have rather consistently upheld the *parens patriae* doctrine. Consequently, there is a widespread pattern in the United States of statutory regulation of family life—for example, laws governing compulsory education for the child, child labor, emergency medical care, delinquency and dependency, parental neglect or child abuse, divorce and adoption, mandatory testing of newborns for the HIV virus, and so on. A current example of such regulation is the lengthy debate by state and federal legislators over the proposed Family Privacy Protection Act of 1995. With certain exemptions, it declares that in conducting a program or activity funded in whole or in part by the federal government, a person may not—without prior written parental or guardian consent (or, if the minor is emancipated, without the minor's own prior consent)—require or otherwise seek the response of the minor to a survey or questionnaire intended to elicit, or having the effect of eliciting, information concerning parental political affiliations or beliefs; mental or psychological problems; sexual behavior or attitudes; illegal, antisocial, or self-incriminating behavior; appraisals of other individuals with whom the minor has a familial relationship, relationships legally recognized as privileged, such as those with lawyers, physicians, and clergy; or religious affiliations or beliefs. The Privacy Act of 1974, viewed by many as inadequate, was similarly enacted to deal with the steady encroachment on privacy in this country. It is still not against the law for an employer to collect information on the private and family life of employees (see Chartbrand, 1996).

5. It needs to be emphasized that most studies on this topic concentrate on intrafamilial privacy. Much more research is needed that focuses on the interfamilial dimensions of privacy. Such analyses could operate out of an institutional frame of reference and examine the interplay between the family as a total unit and other societal structures. Hopefully, such an approach will ultimately lead to cross-cultural comparisons, which are almost totally absent from the current literature on privacy and the family.

REFERENCES

Alderman, E., & Kennedy, C. (1995). *The right to privacy.* New York: Knopf.

Bates, A. P. (1964). Privacy: A useful concept? *Social Forces, 42,* 429–434.

Bellah, R., Madsen, R., Sullivan, W., Swidler, A., & Tipton, S. (1985). *Habits of the heart: Individualism and commitment in American society.* Berkeley: University of California Press.

Berardo, F. M. (1974). Marital invisibility and family privacy. In Daniel H. Carson (Ed.). *Man–environment interactions: Evaluations and applications* (pp. 5–11). New York: Environmental Design Research Association.

Bernstein, N. (1997, June 12). Lives on file: Privacy devalued in information economy. *The New York Times,* pp. 1–14.

Bird, G., & Melville, K. (1994). *Families and intimate relationships.* New York: McGraw-Hill.

Brown-Smith, N. (1998). Family secrets. *Journal of Family Issues, 19,* 23–24.

Burgoon, J. K., Parrott, R., LePoire, B. A., Kelleyu, D. L., Walther, J. B., & Perry, D. (1989). Maintaining and restoring privacy through communication in different types of relationships. *Journal of Social and Personal Relationships, 6,* 131–158.

Chartbrand, S. (1996, May 19). What your employer knows about you. *The New York Times,* pp. 1–4.

Cheal, D. (1991). *Family and the state of theory.* Toronto, Canada: University of Toronto Press.

Cuber, J. F., & Haroff, P. B. (1965). *The significant Americans.* New York: Appleton-Century.

Evans-Pritchard, E. E. (1965). *The position of women in primitive societies and other essays in social anthropology.* New York: Free Press.

Fahey, T. (1995). Privacy and the family: Conceptual and empirical reflections. *Sociology, 29,* 687–703.

Goffman, E. (1959). *The presentation of self in everyday life.* Garden City, NY: Doubleday.

Handel, G. (Ed.). (1967). *The psychological interior of the family.* Chicago: Aldine.

Handel, G., & Whitchurch, G. G. (Eds.). (1994). *The psychosocial interior of the family* (4th ed.). New York: Aldine.

Hess, R. D., & Handel, G. (1959). *Family works.* Chicago: University of Chicago Press.

Hill, R. (1958). Social stresses on the family. *Social Casework, 39,* 139–150.

Kantor, D., & Lehr, W. (1975). *Inside the family: Toward a theory of family process.* San Francisco: Jossey-Bass.

Lasch, C. (1977). *Haven in a heartless world.* New York: Basic Books.

Lasch, D. (1979). *The culture of narcissism.* New York: Norton.

Laslett, B. (1973). The family as a public and private institution: An historical perspective. *Journal of Marriage and the Family, 35,* 480–492.

Margolis, S. T. (1977). Introduction. *Journal of Social Issues, 33,* 1–4.

Mayer, J. E. (1966). *The discourse of marital problems: The "knowledgeability" of lower and middle class wives.* New York: Community Service Society.

Mayer, J. E. (1967). People's imagery of other families. *Family Process, 8,* 27–36.

McGovern, G. (1997, August 14). Whose life is it? *The New York Times,* pp. 1–2.

Merton, R. K. (1968). *Social Theory and Social Structure.* Glencoe, IL: Free Press.

Motz, A. B. (1965). The family as a company of players. *Transaction, 2,* 27–30.

Nock, S. L. (1993). *The costs of privacy: Surveillance and reputation in America.* New York: Aldine.

Packard, V. (1972). *A nation of strangers.* New York: David McKay.

Quittner, J. (1997, August 15). Invasion of privacy. *Time,* pp. 28–35.

Slater, P. (1970). *The pursuit of loneliness.* Boston: Beacon.

Stearns, C. (1997, August 16). Genetic blueprint and your privacy. *The Gainesville Sun,* p. 9A.

Waller, W. (1951). *The family: A dynamic interpretation* (revised by R. Hill). New York: Dryden.

Wishful Thinking and Harmful Tinkering? Sociological Reflections on Family Policy

Carol Smart

Introduction: Wishful Thinking[1]

It is a noted feature of what we might call late modernity that change itself is no longer a defining characteristic so much as the rapidity of change itself. We have, I think, a sense of our social world being transformed at a pace which would have been inconceivable in 1955 let alone 1905. I want to suggest that this sense of rapid change has given the family a renewed status in our modern (Western) concerns about the disintegration of traditional life and values. Such concerns form a core element of the so-called underclass debate and have become manifest in calls for a reversal of policies which are seen as undermining (supposedly) traditional family forms (Morgan, P., 1995; Dennis and Erdos, 1993). It is, of course, not new for the idea of a traditional family to be symbolically depicted as the calm in the middle of a storm, or the haven in a heartless world (Barrett and McIntosh, 1982). In this schema it is the proper traditional family that forms the bulwark against anarchy and crime: an idealized vision of the family emerges in which it is seen as the source of decency and morality. The Victorians idealized the middle-class family in this way, and sought to transform the working-class family into a replica of this specific household organization. But in the 1990s there is not simply a desire to continue the incomplete work of transforming working-class (and now also black and lone-parent) families into replicas of middle-class families, but also a wish to turn the clock back on *all* family life in order to recapture an era when the family was the solution to social ills rather than source of social problems.

We are witnessing a renewed emphasis on the family as the one institution which *should not* change, even though it is taken for granted that everything else around the family is changing radically. The Thatcherite "revolution," followed by the more intensive privatization policies of John Major, have changed key elements of industry, the welfare state and service provision. These changes have brought about important transformations to everyday lives which emphasize individualism, private sources of security and health, the dominance of the client/provider relationship in all things and an insistence on private, rather than public, provision. The dominant political response to those who have objected to these developments and who have attempted to revitalize ideas about public means of meeting "needs," socialized childcare and education, free higher education, etc., has been to suggest that such practices merely restrict competitiveness and prevent the individual from becoming self-sufficient. Ultimately such policies are said to prevent Britain from becoming more competitive in world markets. I need not dwell on this particular debate, which will be familiar.[2] But I do wish to point to an irony in this rush to embrace the restructuring of society according to market forces, and that is the way in which, *at the level of political rhetoric,* it is always assumed that the family should stand aside from this turbulence and should not become a location of client/provider exchanges or internal markets.

More than this, it is assumed that in order for these economic and social changes to work well, the family must stand still, preferably in some idealized and frozen moment which approximates to the 1950s (prior to mass immigration, crime, illegitimacy, no fault divorce, women's rights and so on). Thus, for example, the policy on community care rests entirely on a model of the family where there is a wife or mother at home with the spare capacity to take on extra care work. Although I fully appreciate that government policies on the family do not necessarily work as intended and that they often have unforeseen consequences, it is still the case that at the *level of rhetoric* the family has been constructed as the one site where changes should not occur and where change is seen as positively undesirable unless it is in a backwards direction. Qvortrup (1995) has made precisely this point in relation to the relationship between parents and children in families. He argues:

> We appear to stick to an unrevised family ideology—i.e., unchanged relations between generations—as if nothing had happened in the world around us. We appear to insist on the omnipotence of the family as if the revolution in societal developments had nothing to do with children and their lives. (Qvortrup, 1995: 194)

It is Qvortrup's argument that this myopic view has actually given rise to a decline in children's material circumstances because major structural changes have actually occurred which make it more difficult for parents to care for children and for children to explore the world and to have direct experiences of matters outside the family or the school. In a similar vein, I want to argue that whilst there is a dominance of wishful thinking on the unchanging nature of family life and whilst policy often refuses directly to facilitate change in the private sphere, change is nonetheless necessarily occurring. But this change is now constantly construed as illegitimate and undesirable and is popularly depicted as arising from "unbridled" individualism (Dench, 1996) or from a lack of moral restraint. Changes that are occurring because of social, historical or cultural changes are constantly reduced, in popular discourse, to symptoms of individual moral decline. In other words, families are changing but the public debate divorces these changes from other social transformations and then seeks to admonish family members for their failure to stand still whilst the conditions that supported the family in the past are demolished.[3]

It has now also become commonplace for policies affecting the family to be introduced in terms of how they will *re*stabilize the family, or how they will *re*inforce parental responsibility. There is a dominant inferential framework into which policy is inscribed, and this is the idea of *getting back* to a specific family form. Whilst everything else is rushing forward into a bright and thrusting universe involving total quality control, flexible workforces, mission statements and competitive tendering, the family is meant to embrace old values of altruism, unpaid labor, implicit contracts and co-operation. Quite specifically, changes to the law on divorce in England and Wales throughout this century have almost always been introduced as measures to stabilize marriage and the family. In the 1950s the Royal Commission on Marriage and Divorce (1955) argued that if they could find the evidence that easier divorce destabilized marriage, they would recommend that all divorce should be banned (Smart, 1984). In the 1960s the crucial Divorce Reform Act (1969) was steered through Parliament on the back of the argument that abolishing matrimonial fault, and making the grounds for divorce less arduous, would stabilize the institution of marriage. It was said that allowing people to divorce was the only way to allow them to *re-marry* and thus have legitimate children by their new relationships. Meanwhile, in the 1990s, the Lord Chancellor introduced the government's White Paper, *Looking to the Future* (1995), which proposed various changes to the law of divorce, as a measure that "supports the institution of marriage" (p. 73).

There is clearly an unresolved tension here. Governments have altered the law on divorce in substantial ways, most particularly since the end of the second world war. Each time government has acted, it has promised to stabilize the family. Then, some years later, it has introduced a new measure to re-stabilize the family because it has discovered that the previous measures either failed or actually seemed to accelerate the destabilization of the family. So the recent history of family law reform is marked by a repetitive incantation of the ideal of the unchanging, stable family and its preservation, alongside the development of new legal policies on divorce and marital breakdown which increasingly normalize the process. Only recently has this pattern been broken. In its progress through the Commons, the Family Law Bill was challenged by Tory backbenchers for being another measure which promised more family stability whilst in reality offering "easier" divorce. In consequence the Bill was strengthened in order to further delay divorce proceedings and to try to encourage reconciliation. Thus the Act when it comes into force is meant to be yet another measure which will recreate the ideal family of the 1950s.

There is therefore a kind of collective handwringing over the state of the family and marriage at present. The family, which should be a supplier of model citizens, is seen as, all too often, producing the underclass. Broken homes are said to produce social misfits who, in turn, produce broken homes. The debate has been modelled by two dominant discourses. One arises from a post-war psychology which seeks to find the cause of social problems in the psyches of individuals who do not conform to a moral or statistical norm.[4] The other arises from the politico-moralism of the New Right which recognizes only one kind of family organization as legitimate. What I therefore wish to do in this paper is briefly to look at how certain new contributions to the sociology of family life are attempting to explain changes in family forms and personal relationships. These new contributions have attempted to link changes to family life with wider social changes and therefore offer an important way out of the dominant frameworks that now inform family policy. There is a sense in which family policy has been hijacked by a specific political agenda which is quite impervious to broad sociological interpretations of family life. So I want to reintroduce some of these ideas and will then go on to map family law, as a key instance of family policy, onto these sociological models of family life. This will reveal a surprising congruence between sociological concepts of change and the law's shift towards the "private ordering" of the exit from marriage. However, I will then suggest that major flaws in this new sociological approach produce an inability to account for recent dramatic changes to family policy *in practice* whilst nonetheless offering some account of why these changes may be meeting with resistance to new policy directions. I cannot hope to answer the question of whether new family policies are "preserving" or stabilizing the family, but I do hope to reintroduce a sociological perspective into a field which is increasingly weighted towards moral invective[5] and a reliance on individualistic explanations (e.g. people have become selfish and uncaring, or people now need to be trained in parenthood).

Sociological Accounts of Family Change

In the past the dominant sociological explanations of family and household change have used structuralist models. That is to say, the family has been conceptualized as a social institution which is part of a system of inter-related institutions which change and shift in relation to one another. Typically, although not inevitably, within this model it is the economic sphere which has been treated as the engine of change. Thus, for example, as economic changes demanded child and women's labor in intensified units of production, it has been argued that the working-class family disintegrated. Or, it has been argued that in the more recent past, the demand for women's labor has led to changing patterns of gender relationships within the family. In this sort of model, family change was always seen as the outcome of something else. Feminist explanations of family formation and change have similarly tended to understand the family within this structuralist framework, although substituting the concept of patriarchy for capitalism. Of course such accounts have become increasingly sophisticated and nuanced, and we need not assume that all rely on some kind of simple economic determinism. Davidoff and Hall's (1987) work is a good example of a sociological history of the modern middle-class family which incorporates the economic, the ideological, the religious, gender power, child development and so on. But, in general, within sociology we have seen the family as something which is affected by other social changes rather than as a source of change itself. I think that this in part accounts for why the family has never been of much real interest to sociology[6] (although it has of course been of great interest to feminist work). The epiphenomenal nature of the family always meant that for most sociologists it was not where the "action" was to be found (see Morgan, D., 1996 for a fine exception to this).

Interestingly, some social historians have tended to give greater primacy to the family, particularly within the sentiments school. This approach has tended to argue that relationships between men and women, or between parents and children, have shifted because of changing sentiments between family members and that this in turn has given rise to wider social formations. Thus, historians such as Shorter (1976) have looked at the rise of affection and love between husbands and wives, whilst others like Pollock (1983) argue that love and affection were always an important element of parent/child relations, rather than a twentieth-century development. These historians have placed great emphasis on the significance of "mentalities," whereas sociology, having been dominated by a materialist approach, has tended to eschew such considerations.

However, recently Giddens (1992) has contributed to the theorization of family change. He is concerned to discuss the rise of the "separating and divorcing" society and seeks to explain this change in terms of mentalities. This is, of course, congruent with his project of reintroducing the concept of agency into sociological accounts and it marks an important shift in mainstream sociological thinking because it allows us to take account of the social effects of intimate behavior whilst treating this behavior as socially and historically located. For Giddens, like the social historians of the sentiments school, the key factor in changing familial relationships in the past has been the rise of intimacy. Of course, the idea of the rise of the companionate marriage cannot be attributed solely to Giddens, and historians of the family have long debated precisely when love, or romantic attachment, began to be a feature of marriage and the family. Precise timings are hotly debated, as are the issues of whether such histories can only ever speak of the ruling classes.[7] But, nonetheless, there is a consensus that in modern societies, marriage ceased to be a matter of economics or status for the large part of the population. Alongside this change, there has also been the change in the status of children, or, what is often called the gradual invention of childhood.[8] Thus, it has gradually come to be assumed that children are to be loved, nurtured and protected (although some like Pollock argue that this is not new).

The importance of Giddens' work for my discussion here, however, is not to do with the reiteration of this debate on the *rise* of intimacy, but the way in which he extends it into the present in order to explain high rates of divorce and changes in sexual morality. He also seeks to explain the internal qualities of modern marriage/relationships and argues that the changes we are witnessing are an effect of changing mentalities rather than simply changing material circumstances. His focus therefore shifts towards questions of agency rather than structures. In his formulation family members become active agents rather than resembling the furniture in a household which is moved around by external forces.

Giddens introduces two new concepts to encapsulate the key changes he identifies. The first is the idea of the *pure relationship*. By this he means:

> a social relationship is entered into for its own sake, for what can be derived by each person from a sustained association with another; and which is continued only in so far as it is thought by both parties to deliver enough satisfactions for each individual to stay within it. (Giddens, 1992: 58)

The second concept is that of *confluent love* which Giddens contrasts with the recent idea of romantic love. He states, "Confluent love is active, contingent love, and therefore jars with the 'for-ever,' 'one-and-only' qualities of the romantic love complex" (Giddens, 1992: 61). He argues that there is currently an ongoing clash between the romantic love complex and the pure relationship/confluent love. Under the romantic love complex he argues that individuals sought out the perfect partner, the Mr Right of the Barbara Cartland novel (if you were a heterosexual woman). With confluent love, however, one seeks out the perfect relationship and if one person does not provide it, one moves on until one finds it.

Importantly, Giddens stresses the idea of negotiations between couples; he also stresses that confluent love does not require a couple to be heterosexual. Thus he might argue that a woman does not necessarily feel she must put up with being treated like a doormat because she is a wife (i.e., because of her status). Rather she can decide that as an individual she will not put up with it and will find another partner. He also argues that couples can create for themselves the normative order of their relationship. Thus if a couple agree to a certain set of boundaries, the important element is sticking to what is agreed rather than following general or traditional rules which are presumed to accompany one's status as husband or wife.

There are strong similarities here with the work of Beck (1992). Beck argues that the advancement of modernization and the shift away from traditional life-styles has led to a need for bonding. He argues, "The need for a shared inner life, as expressed in the ideal of marriage and bonding, is not a primeval need. It *grows* with the losses that individualization brings as the obverse of its opportunities" (1992: 105, emphasis in original). He goes on to argue that conflict between husband and wife is a reflection of the conflicts inherent in modern society. Structural change has led to intensive individualization in all spheres, including marriage. Yet the effect of this on the individual is to create a yearning for bonding, but this is stressful precisely because of the pressures toward individualization. Giddens expresses this in a slightly different way. He puts much emphasis on the idea of the "project of the self" in which modern individuals require constantly to improve or remake themselves. Under such a regime, he argues that relationships need to be constantly reappraised, remade or remaindered.

To support this argument Giddens refers to the empirical work of Janet Finch on family obligations. Finch (1989) and later Finch and Mason (1993) have pointed out that modern kinship relationships are constantly negotiated and that family obligations cannot be assumed to follow some set of rules which ordain who should do what for whom. Thus, for example, they discovered that individuals felt obligations to kin in relation to how they felt about the kin concerned and not in relation to an abstract set of rules. They thus suggest a loosening of a taken-for-grantedness of kinship obligations and an emphasis on earned trust or regard. Beck refers to this phenomenon as the *reflexive* nature of social ties. He suggests people now make choices about their social ties rather than simply presuming familial ties and relationships. There is, he suggests, the opportunity for people to plan their own biographies. Significantly though, Beck stresses that the opportunities to do so, and the ability to be reflexive about one's relationships, are not evenly distributed across the population. He therefore avoids the trap of forgetting about social inequalities[9] and, unlike Giddens, specifies the significance of social class to his analysis.

Both Giddens and Beck place a lot of emphasis on the way in which the family/marriage has not really transformed itself sufficiently to accommodate changes which have particularly affected or have been brought about by women. Again, quoting Beck:

> The contradictions between female expectation of equality and the reality of inequality, and between male slogans of mutual responsibility and the retention of the old role assignments, are sharpening and will determine the future development in the thoroughly contradictory variety of their expressions in politics and in private. Thus we are situated at the very *beginning* of a liberation from the opportunities and contradictions. Consciousness has rushed ahead of conditions. (1992: 104, emphasis in original)

Thus, it is argued that, precisely because the family has failed to change enough, women are increasingly discontented and feel most sharply the lack of congruence between expectations and reality. What is also significant in the quotation above is the argument that "consciousness has rushed ahead of conditions." This would, of course, be unthinkable in orthodox Marxist accounts of the family, but for Giddens and Beck it is this consciousness which makes women in particular, agents of change.

In fact, Giddens posits the idea that the change from romantic love to confluent love has been driven largely by women. He suggests that, "Men are the lag-

TABLE 1 *Mapping family law onto the Giddens model of family change*

	Sociological Model	Family Law/Policy Development
Phase 1 pre C18th	Arranged marriage and courtly love	No divorce (only ecclesiastical separation). Father right.
Phase 2 C19th/C20th	Romantic love	Fault based/limited divorce. Ideology of motherhood.
Phase 3 1960–90	Confluent love (Individualism, reflexive social relationships)	Clean break divorce. Presumption of gender equality.

gards in the transitions now occurring — and in a certain sense have been so ever since the late eighteenth century" (1992: 59). Men, he suggests, have largely remained stuck within a nineteenth-century framework of masculinity,[10] whilst women have not only broken out of old fashioned femininity, but have excelled in the modern realm of intimacy. Women wish to negotiate not only because they are better able to express emotion and affection than men, but because the old rules of marriage and kinship were, of course, so loaded against them.

These accounts of changing familial and kinship dynamics are not without their problems as I shall argue below. However, they are useful in the way that they link the family and households to wider developments, whilst incorporating ideas of change emanating from within the private sphere as well as from without. Both Beck and Giddens avoid a position which assumes that intimate or kin relationships simply respond to economic or other changes elsewhere whilst retaining a sociological account of agency rather than slipping into ahistoricism or psychologism. As far as family policy is concerned, such accounts are significant because of the way in which they avoid an increasing tendency to explain divorce in terms of individual inadequacies or cycles of deprivation. Importantly, Giddens and Beck do not isolate the private sphere, indeed they really transcend the old notions of private and public sphere at the analytical level. Thus the significance of sociology's new found interest in families is that it can provide a counter balance to the epistemological basis of much politically motivated modern family policy which precisely sees the family as influenced only by an internal moral decline and a kind of self-generated individualism.

Gidden's account is, however, extremely general, sketching out broad impressions rather than providing the detail of specific changes. He operates as the grand theoretician rather than journeyman who works amongst the conflicting and ambiguous details. He therefore benefits from a clarity of vision which often becomes a bit blurred at lower levels of abstraction. So what I propose to do in the next section is to map Giddens' schema onto the development of family law in England and Wales. What I hope to show is that although there is much that is useful in his analysis, there are some very important details that he overlooks which ultimately show the limits of his reconceptualization of intimacy.

I shall say little about the first phase except that under the system of arranged marriage, the institution was essentially an economic contract which was indissoluble. Although this regime affected primarily the upper classes, elements of the family law established under this regime have lasted into the twentieth century and have therefore had an impact upon a very wide populace. Although family law gradually became secular, it lent very heavily upon the ecclesiastical presumption of the irrevocable nature of the marriage contract. The wife was, in law, *femme covert de baron*, which is to say that she was completely under the formal, legal control of her husband. She had absolutely no legal rights to her children, nor to property.

Under the second phase we see the rigid restrictions on divorce and separation gradually soften. As marriage starts to be based more universally on the concept of love and companionship, marriage itself becomes dissolvable if the conditions conducive to such love are missing. Thus domestic violence becomes grounds for separation, divorce on the grounds of adultery alone is extended to women too, and desertion and cruelty become additional grounds for divorce. However, divorce is only possible on the basis of matrimonial fault and hence the idea of the guilty and the innocent party. Father right remains a dominant presumption when the custody of children was considered, but this began to weaken in the face of the ideology of mother love. Slowly, in the twentieth century, it became a presumption that mothers would retain the care of children on divorce although the doctrine of fault was also a determining factor in decisions about children. Thus the guilty wife might also be deemed to be a guilty or inadequate mother and not infrequently could lose her children. In this second phase we can see that the meaning of marriage gradually changed. It became more of a personal institution, albeit heavily guarded by the state and legal machinery. The position of women changed as the ideology of motherhood developed, but economic dependence remained a key feature of marriage for most women until the 1970s.

The third phase coincides with Giddens' ideas of the growth of confluent love. This is the period when divorce rates increase dramatically and where cohabitation becomes more of a normal pattern. Divorce becomes relatively easy, technically speaking at least, and demographers start to speak of serial monogamy with patterns of marriage, divorce and remarriage becoming common. Significantly in the 1980s there grew up the idea of the clean break on divorce. Although this was primarily a concept which arose from the Men's Rights Movement, it also coincided with the views of many women who wished to cease all contact with, and to end their economic dependency on, their former spouses. The dominant family law policy at this stage was to facilitate divorce, to absolve men of their financial obligations to their first families, and to encourage remarriage where ever possible (especially if a lone mother was dependent upon state benefits) (Smart, 1984). This fits extremely well with Giddens' idea of confluent love because divorce law allowed couples to put their past mistakes behind them and to turn over a fresh sheet to start again without unpleasant, lingering financial and emotional ties. In particular, the law became increasingly forward looking rather than encouraging a backward looking rehearsal of past wrongs and abuses. The idea that one could start again was paramount and thus there was legal policy support for the idea of the project of the self in which one could remake oneself and start another life.

Family law also came to presume that there should be equality between spouses and sought to sweep away presumptions that women could not/should not go out to work, that husbands should support wives and that only mothers could rear children. The material reality facing most mothers was largely disregarded in the rhetorical celebration of equality.

On the face of it, therefore, there is a very close congruence between these sociological accounts and the way in which family law, at least until the mid-1980s, managed serial monogamy. Many of the changes that have occurred in the policy field have come about precisely to meet the changes that were occurring in the family or as a consequence of pressure from groups disadvantaged by the outmoded nature of family law at various points over the last century. Thus, as Giddens would recognize, it has been women campaigners who have pressed for many of the changes to bring family law out of the nineteenth century. But only a decade after the high point of the principle of the clean break and no fault divorce, family law/policy is starting to look very different and hence completely out of step with ideas of confluent love and reflexive social ties. There are two key elements that we need to consider in explaining this rapid onset of incongruity between the sociological account which has identified profound changes and the new direction in policy which suddenly seems determined to stem these changes. The first is whether Giddens has overlooked some vital element of family life and intimate relations or has actually misunderstood some of the mechanisms of change. The second is whether family law has started to become more actively engaged in purposive social engineering in order to redirect change away from the current tendency. Let us first consider whether there are problems with the way in which Giddens has theorized intimate relationships.

Some Problems with the Giddens Thesis

Whilst I find Giddens' insights compelling I find I have a sense of dis-ease as his arguments on the new basis of intimate relationships develop. Giddens depicts a scenario in which couples agree to part because they find their relationships unfulfilling. But what does this concept of lack of fulfillment actually mean? It suggests an existential state in which some interior void is not filled, or perhaps, more prosaically, it suggests that a partner fails to meet implicit or explicit expectations on domestic labor, in relation to sexuality, or in relation to other forms of compatibility. No doubt these things may be very important elements in a failed relationship, but the way in which Giddens emphasizes these "causes" obscures other, less socially acceptable, reasons for women, in particular, to leave their partners. I am referring to violence, bullying, economic deprivation and various forms of cruelty. Although Giddens' account does not exclude these (especially in his reference to men as emotional Luddites), these grimmer elements of married life and intimate relationships do not occupy much of his analysis. One therefore gets the impression of a rather harmonious field of intimacy in which people may cool in their affections and then withdraw from each other on terms of equality.

He also gives the impression that concluding relationships is relatively straightforward and that people do indeed simply move on to the next relationship. But in this depiction he misses a vital element of many of these relationships, namely that they may have given rise to children. Giddens talks only of couples and seems to think that, in their negotiations, they have only themselves to think about. Children do not feature in his consideration of intimate relationships. Whilst we know that on divorce many fathers lose contact with their children and start "new" families, we also know that many do not (and I shall say more on this below). We also know that when faced with a lack of fulfillment, or even with routine violence, many women actually stay in their marriages/relationships, because of their children and not because they have not comprehended the full implications of confluent love.

Children alter things considerably, moreover they too are agents in the family, not simply family assets to be moved around. And what is more, children are not just any kin such as cousins or grandparents. They cannot, at least at first, enter into the play of reflexive social ties, they simply do not have such choices to make. Moreover, their dependency on their parents means that adults cannot treat them as they might other kith and kin. The care that they need and the love that they give and receive does not, I would suggest, fall into (quite) the same moral category as the contingent love of which Giddens speaks. Nor can dependent children fit easily into the schema of negotiated obligations discussed by Finch and Mason (1993). (Indeed Finch and Mason do not focus on dependent children and thus do not presume to include them in their argument.) It therefore becomes problematic to build theories of contingent love or reflexive social ties as if they capture the nature of intimacy under modernity, unless of course we presume that children are unimportant or insignificant to social and intimate life.

The question is, therefore, whether we reject the core of this thesis on intimacy. I think there are reasons to retain these ideas, even though they must be modified by the inclusion of children into the schema. Although I am not at all certain that we know why more people divorce and separate now, we do know that on divorce or separation many people would like a "clean break." Former spouses prefer not to have contact with each other, fathers lose contact with their children and many seem highly resistant to any financial commitment to their first families. It is also clear that prior to the introduction of the Child Support Agency, courts were perfectly content to allow impoverished fathers to abandon paying child support expressly in order to enable them to start fresh relationships and families (Maclean, 1994). The preservation of second marriages was the focus of much practice at the grass-roots level, even if it was never government policy. It was thought sensible to allow men to move on relatively unencumbered in order that they might properly support their second families. Drawing on Finch and Giddens we could say that fathers in this sort of situation were negotiating their obligations at a personal level and deciding against an obligation to their children, in favor of a new relationship. We can see that the promised satisfactions of a new relationship outweighed the obligations of fatherhood—at least in its economic aspects.

There are therefore good reasons for recognizing that the idea of confluent love has some explanatory power. But, just because many people wanted a "clean break" and even though many may have achieved it, it is quite inappropriate to assume that this kind of emotional and practical "clean break" ever became a dominant pattern of post-divorce relationships. Again, it is the presence of children which makes the "clean break" almost impossible. Research on divorce has shown that there are instances where wives would dearly like to see no more of their former spouses, but continue to do so for the sake of the children (Smart, 1990; Kier and Lewis, 1995). There is also evidence of divorced parents who refrain from forming new relationships precisely because of their children, or because they have negotiated a shared caring arrangement with a former partner that they do not want to disturb.[11] Children stop confluent love in its tracks. Equally we might say that the Child Support Agency is also attempting to do this in its efforts to reattach men to their children, at least financially.

Giddens' notion of confluent love may therefore have captured a wish that many express but which in practice few actually achieve, at least once they become parents. It may even have reflected a possibility for some divorced parents in the 1970s and 1980s. But this, it seems to me, was a short-lived moment that we cannot safely project into the future as Giddens seems to do. If we look more closely at what is happening "on the ground" we can see that there are two linked reasons why this trajectory is likely to be short lived. The first is that some men no longer seem willing to leave behind the children of their families. It is too early to fully comprehend this development or to predict what it may come to mean, but certainly Giddens is wrong to assume that fatherhood may be unchanging when he refers to men as Luddites. The second reason is that, whether or not men and women want to change, family law has taken very definite steps towards redefining the normative expectations of post-divorce parenting in the 1990s. It is to this that I shall now turn.

New Departures in Family Law: Harmful Tinkering?

I have suggested above that we can see developments in family law which reflect the broad changes that we can detect in family forms. There is a sense in which policy has both followed and facilitated changes, even if at every stage it has uttered its hope to preserve the family and marriage. But it is now possible that we are witnessing a moment when policy is trying to impose a new direction on the trajectory that Giddens has identified. There have been two major pieces of legislation, the Child Support Act 1992 and the Children Act 1989, and a further reform of the law of divorce is currently under way (Lord Chancellor's Department, 1995). The Family Law Act when it is implemented will make divorce more difficult and will take every opportunity to encourage couples to stay together. The Child Support Act[12] has already been mentioned as a measure which attempts to defeat the transfer of economic resources from a first family to a second family, and thus to defeat the presumption that one has a right to move on to a new relationship regardless of one's financial circumstances.

The Children Act has attracted less attention but is nonetheless part of trend towards changing the fundamental nature of divorce. Contained within the Act are three newly articulated principles. The first is the principle of "non-intervention." This means that the courts are reluctant to become involved in matters over children and actively encourage parents to negotiate outcomes without the need for a court order. The second is the principle of joint parenting. This means that the old idea of one parent having "custody" and the other having "access" is abandoned in favor of a system where parents simply go on being parents with the same legal duties and obligations as existed during the marriage. The terms custody and access are replaced by residence and contact. The third principle is the welfare of the child. This is not a new principle,[13] but in this Act it has come to be synonymous with the idea of the right of a child to have two parents. This means that joint parenting is meant to be an active sharing of the upbringing of the child in which both parents are as much involved as they were before the divorce. I want to suggest that these new principles have in fact introduced a new marriage contract by another name. This new marriage contract ends the possibility of confluent love for mothers (although not necessarily for fathers)—by which I mean that it ends the possibility of divorce finishing a relationship with a person one no longer loves or cares for. To appreciate this point it is necessary to understand what is entailed in joint parenting.

Research on how parents negotiate over the children on divorce or separation has indicated that, under the old divorce law, access required a lot of work on the part of the custodial parent (almost always the mother) (Smart, 1990; 1991). It required negotiations with the father, with the children, with schools. It required careful planning of holidays and all sorts of sport or other leisure activities. It required the packing of toys and equipment and the washing of clothes. It also required a lot of emotional work with children who might be reluctant to go or who may be difficult to settle on their return. Access did not just happen. In many ways access was, for some mothers, far more work than co-residence had been and also required more negotiating with the father than before. It is clear therefore that under the new legislation joint parenting will require an ongoing relationship—precisely the continuation of a relationship that divorce is meant to end. This is what I mean by the introduction of a new marriage contract. More correctly it is really a joint parenting contract and it is becoming indelible. As one solicitor in our current research put it,[14]

> I think it's important to get across to people that, okay, there's been a marital breakdown or a relationship breakdown and that will inevitably change things substantially but they will still continue as a family. I would present it as a way of working towards evolving a family which is a continuing family, but a family in a different form without talking about responsibilities and rights.

The problem that spouses face is that this is not what they expect from divorce. This is not to say that many do not wish to jointly parent after divorce. But for some this causes problems. In these instances we are witnessing very direct intervention by the courts and by solicitors to try to ensure that divorcing and separating couples do conform to the new orthodoxy.[15] It is interesting to see how rapidly this changing normative order is being established.

Below I outline two almost identical legal cases over contact/access issues which occurred only three years apart. In the first case the mother is seen to be a good mother, working in the interests of her child. In the second case the mother, who is behaving in exactly the same way, is seen as working to damage the welfare of her child. In the first case she was allowed to form a fresh relationship and to have a clean break, in the second she was not and the weight of state intervention was brought to bear on her such that her second relationship became unstable.

Re SM (A Minor) (Natural Father: Access) [1991] 2 FLR 333

In this case a woman had an illegitimate child by a man from whom she parted before the birth. He saw the child regularly for nearly two years. She then remarried her former husband and contact between the child and the "natural" father ceased. When the "natural" father applied for access (old Act) the magistrates granted it. The mother appealed. The case was heard by the President of the Family Division who ruled in favor of the mother who had argued that the child was now in a stable family unit and that to introduce the "natural" father would destabilize the unit and disturb the child who knew him as an uncle.

Re R (A Minor) (Contact) [1993] 2 FLR 762

This involved a married couple but otherwise the facts are similar. That is to say, a married woman had a child by her husband, but she left him soon after the birth and went to live with another man who raised the child as his own. The child in this case had been brought up to believe that her mother's cohabitee was her father and she did not know her "natural" father who had not seen her for four years. The mother argued the same case as the mother in Re SM (namely that the child was now in a stable unit and should not be disrupted) but was unsuccessful. Butler-Sloss LJ argued that it was the right of a child to know the truth as to the identity of its "natural" father and that it was the right of a child to have a relationship with both "natural" parents. Thus, against the wishes of the mother, she ordered a child psychiatrist to work with the child to overcome the trauma of discovering the truth of her parentage and that a guardian ad litem should be appointed to advise the court on when contact could be resumed with the "natural" father.

In the first instance the mother is held to be doing the best for the child in providing a stable family and she is allowed her clean break, as is the child. In the second, a mere three years later, the mother is labelled implacably hostile and is seen as failing to recognize the true interests of her child.[16] This mother is obliged to restart a relationship which will endure for as long as the child's natural father wishes. These cases reflect an important shift in the normative order governing divorce.[17] In a remarkably short space of time the idea that one can turn over a new leaf and start again has been redefined as a form of selfish individualism generated by a combination of moral decline and feminist inspired self-interest. The disdain with which such an aspiration is now met obliterates completely the social history that gave rise to what was once a perfectly legitimate desire associated with divorce, namely the desire to separate from a spouse or former sexual partner.

It is also important to recognize that the new reaction to this aspiration is only mobilized when it is voiced by mothers. The newly invoked implacably hostile parent is nearly always the mother. The father who simply chooses not to see his children is not defined as implacably irresponsible and as deserving compulsion. Moreover, it would appear that this label of implacable hostility is being applied even to mothers who have been the victims of severe physical violence from their husbands. Thus, in precisely those cases where a woman might hope most strongly and most reasonably that divorce would actually mean the severing of a relationship with a violent man, she is likely to find herself labelled as hostile, as unable to recognize the welfare of her children, and forced to maintain contact, even if indirectly (Hester and Radford, 1996). She cannot avail herself of the clean break even if she has extremely good reasons for needing to do so.

Conclusion

If we accept the sociological thesis that changes in family life cannot be separated from other social changes and that they are also related to struggles within the sphere of intimacy, then we should perhaps be concerned when policy on the family seems to ignore such developments in order to impose a new direction on intimate relationships. Developments like the private law provisions of The Children Act and the forthcoming Family Law Act seek to, in the first instance, change the end of marriage into an ongoing parenting arrangement and, in the second, to delay and hamper the severing of the marital relationship in the first place. In political terms, divorce has been redefined as a social problem whilst for the many it is the only solution to a problem marriage. Problem marriages must, in turn, be understood in terms of wider social processes such as changing employment patterns, changes to welfare provision, changes in expectations and so on. Thus both marriage and divorce are forms of social action not simply "unbridled individual choice" (Dench, 1996). The problem with the increasingly commonplace formulation offered by authors like Dench is that they seek to explain changes within personal relationships solely in terms of one factor, namely changing individual values. These normative shifts however are depicted as free floating (hence the idea of unbridled choice) rather than situated in cultural time and place. Such ideas are seen to spring from the individual's lack of connection to the moral community, not from an ethical engagement with changing social conditions. In this sense such ideas are profoundly unsociological even if they are now politically popular. But ultimately they are immensely unhelpful unless all one is interested in is rhetoric.

As Giddens has argued, confluent love (with all its faults as a concept) is one outcome of a number of social processes and changes in which women and men have been agents. But it is not reducible to personal inadequacies or a lack of morality. As Finch and Mason have shown, people in fact think very hard about ethical matters and the fact that they do not make decisions which follow old rules of kinship obligation does not mean they have become unethical. Rather, ethics are negotiated. So it may be that social engineering which seeks to blame individuals for wider social changes simply misses the point and ends up becoming little more than a harmful tinkering on the edges of large-scale social change which needs to be addressed in a more fundamental way.

We are witnessing an important clash of historical forces and not just a few selfish fathers and hostile mothers. Yet the push towards mediation (see Lord Chancellor's Department, 1995) consistently defines these issues as individual ones rather than sociological ones. Divorce continues to be seen as a personal event and a symptom of individual failure. Few people are asking wider questions such as how can joint parenting be achieved under current social conditions? Or how can we achieve joint parenting on divorce when there are no structural supports for it even during marriage? We need also to ask whether policy can now insist that women go on co-operating with men they wish to divorce when, for at least a century, women have been struggling to leave violent and oppressive relationships. Many of the changes we have seen to the legal policy on marriage and divorce this century have come about because of a ground swell of public opinion or as a result of a social movement such as the Women's Movement. Women fought for the right to divorce,[18] the right to the custody of children, the right to be protected from domestic violence, the right to a share in the matrimonial home and so on. Many of these changes were related in complex ways to other social changes. But neither The Children Act nor The Family Law Act has arisen from the demands of a social movement or swings in public opinion.[19] Rather they are forms of legislation which appear to derive from the concerns of professional child welfare specialists and, in that sense, are imposed from above, or from narrow political concerns to reduce the cost of legal aid and to appear to make divorce more difficult for the many. They seem to be out of step with more fundamental social changes and are particularly incongruent with established assumptions about the availability of divorce and the presumption that divorce does in fact mean a real separation. In the attempt to impose a different normative order on married life under the mistaken assumption that patterns of divorce arise from a modern immorality, it may be that marriage itself will continue to decline in "popularity" and that rather than achieving a renewed compliance, this combination of wishful thinking and harmful tinkering will only produce the outcome that is ironically the most unwelcome to those who long for the mythical golden age of family life.

NOTES

1. An earlier version of this paper was originally presented as the Jacqueline Burgoyne Memorial Lecture at the Sheffield Hallam University, 24 February 1995.
2. It is part of this debate to argue that the political goal of rolling back the state (especially the welfare state) has ironically produced even greater state involvement and support through the rise of unemployment etc. Although this is the case, we can also witness significant changes brought about by privatization and by the introduction of internal markets in areas such as the provision of social and health care.
3. I cannot now remember an instance in the recent past where any documented shifts in family life or family demography have been greeted positively or with enthusiasm.
4. I am thinking here of the persistent search by psychologists to find differences between the children of divorced parents and children of married parents. It is my view that in the future we will regard this obsessive search as being as useful and as meaningful as the obsessive search for differences in intelligence between peoples of different ethnic backgrounds. The attempt by science to establish these differences may eventually be reinterpreted as a normative attempt to persuade us of the inevitability of harm arising from social change. It may also be seen as a method by

which science supports a specific family form as the only desirable and natural household arrangement in much the same way that science once provided evidence that it was unnatural for women to be educated or to join the professions.

5. Qvortrup (1995) has argued that "what is needed is not moralistic criticism, but rather a critical morality." At present there seems to be a surfeit of the former in the field of divorce.
6. This is ironically mirrored in the policy sphere where, for example, there have been demands for a Minister for the Family or for Family Impact Statements. Such demands assume that the primary motivation for change is economic and that we need to know what impact fiscal changes or Health Service changes will have on the family. We rarely ask what impact changes in the family will have on the provision of Health Services or on the viability of economic plans.
7. See Laslett (1977), Shorter (1976) and Flandrin (1979) amongst others.
8. See Pollock (1983), Ariès (1962) and James and Prout (1990) for conflicting views on this debate.
9. Indeed, much of Beck's work is concerned with understanding how inequalities are now organized given his argument that the standard Marxist/sociological concepts of social class are no longer adequate to grasp the new forms of social division and differentiation.
10. I think that Beck would differ here and I too would argue that masculinity has changed considerably. In particular I think we can see important shifts in the meaning of fatherhood which Giddens seems to be unaware of.
11. See Smart (1990). We are also finding evidence of this in an ongoing ESRC funded research project entitled The Legal and Moral Ordering of Households in Transition.
12. I am fully aware of the criticisms of the CSA and the idea that it is really a cynical measure to recoup income for the Treasury. However, this does not deflect from the argument I wish to make here.
13. I am of course aware that this principle has been stated for over a century in family law. However, it is my belief that it is only since the Children Act of 1989 that this has really had any *real* effect on the operation of the divorce law and child support.
14. The current research (see note 11) involves interviewing sixty parents on two separate occasions and thirty solicitors, in order to explore the workings of the Children Act 1989, private law provisions.
15. Let me state at this stage, I am not necessarily concerned with whether the basic principles of the new orthodoxy are correct or not. Rather I am concerned that they change dramatically the nature of divorce and many people's expectations. They are also a very direct intervention under the rubric of an Act which proclaims the benefit of non-intervention!
16. I am aware that in the first case the child was illegitimate and that the mother went on to marry the social father and that these elements would have carried weight. It was the converse in the second case because the child had not been illegitimate and the mother had not married the child's social father. But there is evidence that the issue of marital status of the father and mother is no longer the crucial issue, rather what seems to matter is the question of biological parenthood.
17. This is not the only evidence available on this issue. See, for example, Neale and Smart (forthcoming) and Hester and Radford (1996).
18. I am of course aware that not all women's organizations agreed that divorce *per se* would be helpful to women. Some preferred the route of reforming marriage. However, few felt that the situation of marriage was at all satisfactory for women.
19. The Men's Rights Movement was active to some extent prior to the introduction of this Act, but this movement did not want the reforms the Act eventually introduced and has since criticized them. The Men's Rights Movement was more successful in the 1980s when it brought about the clean break principle for financial matters on divorce.

REFERENCES

P. Ariès (1962), *Centuries of Childhood*, Jonathan Cape, London.

M. Barrett and M. McIntosh (1982), *The Anti-Social Family*, Verso Books, London.

U. Beck (1992), *The Risk Society*, Sage, London.

L. Davidoff and C. Hall (1987), *Family Fortunes*, Hutchinson, London.

G. Dench (1996), "Men without a mission," *The Times Higher*, 28 June.

N. Dennis and G. Erdos (1993), *Families Without Fatherhood*, Institute of Economic Affairs, London.

J. Finch (1989), *Family Obligations and Social Change*, Polity Press, Cambridge.

J. Finch and J. Mason (1993), *Negotiating Family Responsibilities*, Routledge, London.

J.-L. Flandrin (1979), *Families in Former Times*, Cambridge University Press, Cambridge.

A. Giddens (1992), *The Transformation of Intimacy*, Polity Press, Cambridge.

M. Hester and L. Radford (1996), *Domestic Violence and Child Contact Arrangements in England and Denmark*, Report to the Joseph Rowntree Foundation and the Nuffield Foundation, School for Policy Studies, University of Bristol.

A. James and A. Prout (1990), *Constructing and Reconstructing Childhood*, The Falmer Press, Brighton.

C. Kier and C. Lewis (1995), "Family dissolution: mothers' accounts" in J. Brannen and M. O'Brien (eds.), *Childhood and Parenthood*, Institute of Education, London.

P. Laslett (1977), *Family Life and Illicit Love in Earlier Generations*, Cambridge University Press, Cambridge.

Lord Chancellor's Department (1995), *Looking to the Future*, Cm 2799, HMSO, London.

M. Maclean (1994), "The making of the Child Support Act of 1991: policy making at the intersection of law and social policy," *Journal of Law and Society*, 21:4, 505–19.

D. Morgan (1996), *Family Connections*, Polity Press, Cambridge.

P. Morgan (1995), *Farewell to the Family?* Institute of Economic Affairs, London.

B. Neale and C. Smart (forthcoming 1997), "Experiments with parenthood?" *Sociology*.

L. Pollock (1983), *Forgotten Children*, Cambridge University Press, Cambridge.

J. Qvortrup (1995), "Childhood and modern society: a paradoxical relationship?" J. Brannen and M. O'Brien (eds.), *Childhood and Parenthood*, Institute of Education, London.

Royal Commission on Marriage and Divorce, 1951–1955 Report (1956), Cmnd 9678, London.

E. Shorter (1976), *The Making of the Modern Family*, Collins, London.

C. Smart (1984), *The Ties That Bind*, Routledge and Kegan Paul, London.

C. Smart (1990), *The Legal and Moral Ordering of Child Custody*, Report to the Nuffield Foundation, University of Warwick.

C. Smart (1991), "The legal and moral ordering of child custody," *Journal of Law and Society*, 18:4, 485–500.

CHAPTER 9

Questions for Writing, Reflection, and Debate

READING 46

Virtual Social Science and the Politics of Family Values • *Judith Stacey*

1. Do you think family researchers arrive at their value position as a result of the findings from their studies, or do they conduct studies to support the value positions they bring to their work?
2. Many people attribute the prosperity of the 1950s and the social ills of today to the family structure prevalent at the time. What is the flaw in assuming a causal relationship between family structure and the country's level of well-being?
3. What are the potential benefits and dangers of mixing academics and politicians?
4. What does Stacey mean by the term "sitcom sociology"? What does she see as its dangers?

READING 47

Family Privacy: Issues and Concepts • *Felix M. Berardo*

5. How can family privacy be both functional and dysfunctional?
6. How is family privacy affected by structural conditions? What elements of contemporary culture are invading the backstage of family life? With what effect? How was the family backstage invaded by elements of culture at past points in history? In other cultures?
7. What are the methodological implications for research of placing a high value on family privacy?
8. How can society balance family privacy with the need for social control?

READING 48

Wishful Thinking and Harmful Tinkering? Sociological Reflections on Family Policy • *Carol Smart*

9. Smart suggests that we accept change as an inevitable corollary to progress in most aspects of social life, yet we expect family life to remain stable. What could motivate people to hold such a position?
10. What are the implications of attributing family change to individualism and lack of moral restraint rather than changes in society?
11. What is the difference between viewing changes in family as due to social change as opposed to a source of social change? How do social norms that guide family life change?
12. Do you think society can use policy to reform family life by legislating rights and responsibilities? Is it possible for policy to facilitate rather than follow change?
13. The author suggests that changes in marriage, divorce, and parenting might be viewed as a form of social action resulting in a change in social values, rather than a result of changing individual values. Do you agree?

INDEX

G

H

M

S

Y

Z

LaVergne, TN USA
12 December 2009
166784LV00003B/61/A
9 780761 986102